Applied Hydrometeorology

Applied Hydrometeorology

Editor: Peter Spencer

STATES
ACADEMIC PRESS
www.statesacademicpress.com

States Academic Press,
109 South 5th Street,
Brooklyn, NY 11249, USA

Visit us on the World Wide Web at:
www.statesacademicpress.com

ISBN: 978-1-63989-053-8 (Hardback)

Cataloging-in-Publication Data

Applied hydrometeorology / edited by Peter Spencer.
p. cm.
Includes bibliographical references and index.
ISBN 978-1-63989-053-8
1. Hydrometeorology. 2. Hydrology. 3. Meteorology. I. Spencer, Peter.
GB2805 .A56 2022
551.57--dc23

Table of Contents

Preface

This book was inspired by the evolution of our times; to answer the curiosity of inquisitive minds. Many developments have occurred across the globe in the recent past which has transformed the progress in the field.

The sub-field of hydrology and meteorology that focuses on the water and energy transfer between the surface of land and the lower atmosphere is known as hydrometeorology. It also deals with the water cycle and water budget related problems. It is concerned with the determination of rainfall possibilities, region and time distribution of rainfall, and evaporation. Hydrometeorology emphasizes on determining the connection between meteorological variables and the maximum precipitation that reaches the land. It helps in flood control and designing water-usage structures such as dams and reservoirs. This book elucidates the concepts and innovative models around prospective developments with respect to hydrometeorology. Coherent flow of topics, student-friendly language and extensive use of examples make this book an invaluable source of knowledge. The various studies that are constantly contributing towards advancing technologies and evolution of this field are examined in detail.

This book was developed from a mere concept to drafts to chapters and finally compiled together as a complete text to benefit the readers across all nations. To ensure the quality of the content we instilled two significant steps in our procedure. The first was to appoint an editorial team that would verify the data and statistics provided in the book and also select the most appropriate and valuable contributions from the plentiful contributions we received from authors worldwide. The next step was to appoint an expert of the topic as the Editor-in-Chief, who would head the project and finally make the necessary amendments and modifications to make the text reader-friendly. I was then commissioned to examine all the material to present the topics in the most comprehensible and productive format.

I would like to take this opportunity to thank all the contributing authors who were supportive enough to contribute their time and knowledge to this project. I also wish to convey my regards to my family who have been extremely supportive during the entire project.

Editor

1

Climate change and stream temperature projections in the Columbia River basin: habitat implications of spatial variation in hydrologic drivers

D. L. Ficklin[1,7], B. L. Barnhart[2], J. H. Knouft[3,4], I. T. Stewart[5], E. P. Maurer[6], S. L. Letsinger[7], and G. W. Whittaker[2]

[1]Department of Geography, Indiana University, 701 E. Kirkwood Ave., Bloomington, IN 47405, USA
[2]Agricultural Research Service, United States Department of Agriculture, 3450 SW Campus Way, Corvallis, OR 97333, USA
[3]Department of Biology, Saint Louis University, 3507 Laclede Ave., St. Louis, MO 63103, USA
[4]Center for Environmental Sciences, Saint Louis University, 3507 Laclede Ave., St. Louis, MO 63103, USA
[5]Department of Environmental Studies and Sciences, Santa Clara University, 500 El Camino Real, Santa Clara, CA 95053, USA
[6]Civil Engineering Department, Santa Clara University, 500 El Camino Real, Santa Clara, CA 95053, USA
[7]Center for Geospatial Data Analysis, Indiana Geological Survey, 611 N. Walnut Grove, Bloomington, IN 47405, USA

Correspondence to: D. L. Ficklin (dficklin@indiana.edu)

Abstract. Water temperature is a primary physical factor regulating the persistence and distribution of aquatic taxa. Considering projected increases in air temperature and changes in precipitation in the coming century, accurate assessment of suitable thermal habitats in freshwater systems is critical for predicting aquatic species' responses to changes in climate and for guiding adaptation strategies. We use a hydrologic model coupled with a stream temperature model and downscaled general circulation model outputs to explore the spatially and temporally varying changes in stream temperature for the late 21st century at the subbasin and ecological province scale for the Columbia River basin (CRB). On average, stream temperatures are projected to increase 3.5 °C for the spring, 5.2 °C for the summer, 2.7 °C for the fall, and 1.6 °C for the winter. While results indicate changes in stream temperature are correlated with changes in air temperature, our results also capture the important, and often ignored, influence of hydrological processes on changes in stream temperature. Decreases in future snowcover will result in increased thermal sensitivity within regions that were previously buffered by the cooling effect of flow originating as snowmelt. Other hydrological components, such as precipitation, surface runoff, lateral soil water flow, and groundwater inflow, are negatively correlated to increases in stream temperature depending on the ecological province and season. At the ecological province scale, the largest increase in annual stream temperature was within the Mountain Snake ecological province, which is characterized by migratory coldwater fish species. Stream temperature changes varied seasonally with the largest projected stream temperature increases occurring during the spring and summer for all ecological provinces. Our results indicate that stream temperatures are driven by local processes and ultimately require a physically explicit modeling approach to accurately characterize the habitat regulating the distribution and diversity of aquatic taxa.

1 Introduction

The temporal and spatial variability of stream temperature is a primary regulator of the life history, behavior, ecological interactions, and distribution of most aquatic species (Peterson and Kwak, 1999). For example, metabolic processes in ectothermic freshwater organisms (e.g., fishes, amphibians, invertebrates) are directly regulated by water temperature (Angilletta, 2009), and thus the persistence of populations and the rate of energy flow through aquatic ecosystems is dependent on the thermal characteristics of a local habitat (Woodward et al., 2010). Moreover, much like terrestrial species, the timing of important life history traits such as reproduction and migration is heavily dependent on seasonal thermal regimes (Johnson et al., 2009; Woodward et al., 2010). Additionally, stream temperature plays a large role in chemical kinetic rates and is important for governing stream management for recreation as well as urban and industrial water supplies. Therefore, to better understand hydrologic systems and to better manage water resources in a changing environment, it is critical to predict the potential effects of climate variability and change on stream temperature, and to characterize how these changes affect the distribution and diversity of freshwater taxa.

Potential impacts of climate change on stream temperatures have been widely estimated using field investigations and modeling studies (Webb and Nobilis, 1994; Mohseni et al., 2003; Caissie, 2006; Hari et al., 2006; Nelson and Palmer, 2007; Webb et al., 2008; Isaak et al., 2010; van Vliet et al., 2011; Null et al., 2013; Ficklin et al., 2013). At larger spatial scales, regional regression models have been used to predict the impacts of climate change on stream temperatures (Mohseni et al., 1998, 1999; Mohseni and Stefan, 1999; Erickson and Stefan, 2000; Bogan et al., 2003; Webb et al., 2003; Stefan and Preud'homme, 1993). However, regression methods are not sufficient predictors of stream temperature because they do not account for hydrologic component inputs to the stream such as snowmelt, groundwater, and surface runoff (Constantz et al., 1994; Constantz, 1998; Pekarova et al., 2008; Ficklin et al., 2012; MacDonald et al., 2014). Neglecting these components severely limits the ability of regression-based models to accurately predict spatial variability in stream temperature changes since the contributions of different sources to streamflow will be modified in a changing climate. Ignoring the distinct characteristics of different sources to streamflow therefore negatively impacts the assessment of the effects of climate change on aquatic biodiversity at landscape (and larger) scales.

To adequately capture the role of changing hydrology from a changing climate on stream temperature, numerical (Isaak et al., 2010; Kim and Chapra, 1997; Sinokrot and Stefan, 1994) and analytical (Null et al., 2013; Tang and Keen, 2009; Edinger et al., 1974) stream temperature models, in conjunction with hydrologic models, have been applied with success. These models allow stream temperature assessments at the local or regional level. For example, our previous work in the Sierra Nevada mountain range in California found subbasin-scale stream temperature differences from region-to-region largely from localized changes in hydrology from changes in climate. Additionally, Null et al. (2013) found increasing stream temperatures with increasing elevation due to the transition from snow- to rain-dominated, an effect opposite what would be predicted by a model based solely on air temperature.

The primary objectives of this work are to (1) predict changes in stream temperature over the coming century across the Columbia River basin (CRB) at the ecological province level, (2) identify the contribution of specific hydrological components (such as snowmelt, surface water runoff, etc.) to the overall heat and water budget across the watershed, and (3) add to the literature regarding the role of changing hydrology on changes in stream temperature. Specifically, we aim to demonstrate the extent to which future changes in hydrology – streamflow, surface runoff, snowmelt, groundwater inflow, and lateral soil flow as simulated using global climate projections at the subbasin scale – could critically affect changes in localized stream temperatures, which are of high importance for aquatic species. The Columbia River basin is a snowmelt-dominated region, where projected increases in global air temperatures are expected to result in early snowmelt runoff. These changes lead to reduced late spring and summer water discharges that change the thermal content of stream flow. Moreover, previous stream temperature assessments indicate that the Columbia River basin is sensitive to changes in climate (Mantua et al., 2010; Chang and Psaris, 2013; Luce et al., 2014); these sensitivities vary spatially and are governed in part by the land use, hydroclimate and topographic variables of the local region (Chang and Psaris, 2013).

We use a landscape-scale hydrological model – the Soil and Water Assessment Tool (SWAT; Arnold et al., 1998) – combined with a stream temperature model that simulates stream temperature based on the effects of subbasin air temperature and hydrology (Ficklin et al., 2012). The SWAT model efficiently represents snowmelt and runoff processes, and also incorporates a full range of water quality processes (Gassman et al., 2007). SWAT has been found to accurately simulate streamflow in regions where snowmelt dominates the hydrology (Wang and Melesse, 2005; Watson and Putz, 2012; Zang et al., 2012). Downscaled output from seven general circulation model (or global climate models, GCMs) using one representative concentration pathway (RCP) associated with a trajectory of future greenhouse gas accumulation in the atmosphere for the late 21st century was used to drive the calibrated SWAT model at the subbasin-scale. For all Columbia River basin ecological provinces, we spatially and temporally explore the changes in stream temperature, and interpret these changes with respect to changes in the hydrologic system.

2 Materials and methods

2.1 Study area

The CRB encompasses portions of seven states in the western United States and the Canadian province of British Columbia. The CRB for this study is defined as the area that flows into the The Dalles, Oregon (Fig. 1) and has a surface area of 613 634 km^2. The water resources in the CRB have been extensively developed in the past 70 years for hydroelectric power, agricultural irrigation, and urban use. The CRB study area has been extensively discussed in Hatcher and Jones (2013), Mantua et al. (2010), and Payne et al. (2004).

Subbasins were aggregated into ecological provinces according to designations of the Northwest Habitat Institute (N.H.I., 2008). Ecological provinces are delineated based on species composition within the region and environmental conditions. Because the ecological provinces do not expand into Canada, we extrapolated the boundaries based on watershed delineations. The ecoprovince areas (Fig. 1) for this study average 68 000 km^2 and range from 300 km^2 (Columbia Gorge) to 145 000 km^2 (Mountain Columbia). For descriptive purposes, we further characterize ecological provinces as either "warmwater" (Centrarchidae – bass, bluegill, crappie; Percidae – perch, walleye), "coldwater migratory" (Salmonidae – salmon, steelhead, trout], and "coldwater non-migratory" (Salmonidae – trout, whitefish) (Table 2), based on predominant focal fish species (N.H.I., 2008).

2.2 Modeling stream flow and water quality using SWAT

We used the SWAT model coupled with a stream temperature model to predict streamflow and stream temperature throughout the Columbia River basin at an average spatial resolution of 250 km^2. SWAT is an integrative, mechanistic model that utilizes inputs of daily weather, topography, land use, and soil type to simulate the spatial and temporal dynamics of climate, hydrology, plant growth, and erosion (Arnold et al., 1998). Within SWAT, surface runoff and soil water infiltration were simulated using the modified Curve Number method (Neitsch et al., 2005). The Penman–Monteith method was used to estimate potential evapotranspiration. Stream temperature was simulated using the Ficklin et al. (2012) SWAT stream temperature model that uses local air temperature and hydrology for stream temperature estimation:

$$T_{\text{w,local}} = \frac{(0.1 \cdot (\text{sub\_snow}) + \left(T_{\text{gw}} \cdot \text{sub\_gw}\right) + \lambda \left(T_{\text{air,lag}} \cdot (\text{sub\_surq} + \text{sub\_latq})\right)}{\text{sub\_wyld}}, \quad (1)$$

where sub_snow is the snowmelt contribution to streamflow within the subbasin (m^3), sub_gw is the groundwater contribution to streamflow within the subbasin (m^3), sub_surq is the surface water runoff contribution to streamflow within the subbasin (m^3), sub_latq is the soil water lateral flow contribution to streamflow within the subbasin (m^3), sub_wyld is the total water yield (all contributing hydrologic components) contribution to streamflow within the subbasin (m^3), T_{gw} is the groundwater temperature (°C; annual average input by user), $T_{\text{air,lag}}$ is the average daily air temperature with

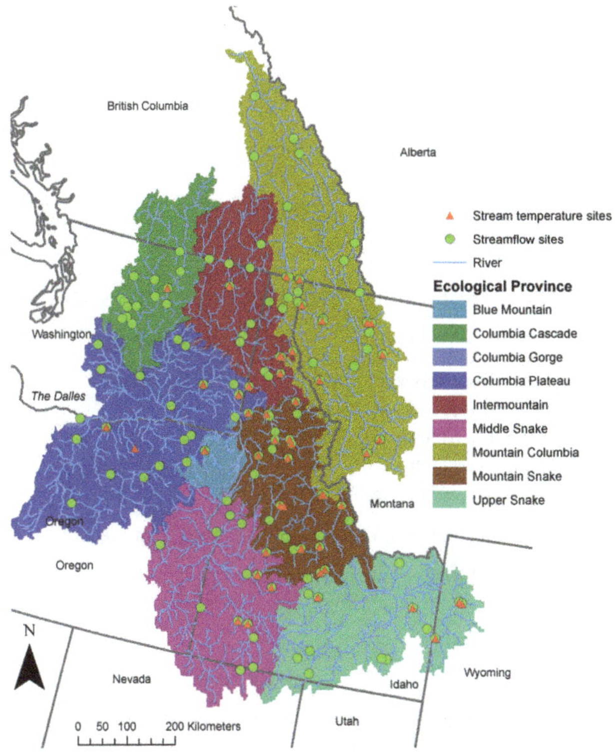

Figure 1. Columbia River basin study area ecological provinces with streamflow and stream temperature gauges for calibration.

a lag (°C), and λ is a calibration coefficient relating to the relative contribution of the surface water runoff and lateral soil water flow to the local water temperature and is included to aid in calibration in case of improper hydrologic model calibration. The lag (days) is incorporated to allow the effects of delayed surface runoff and soil water flow into the stream. The 0.1 in Eq. (1) represents the assumed temperature of snowmelt (0.1 °C).

After the stream temperature of the local contributing water is determined, the stream temperature before the effects of air temperature is determined by

$$T_{\text{water}_{\text{intial}}} = \frac{T_{\text{w,upstream}} \cdot (Q_{\text{outlet}} - \text{sub\_wyld}) + (T_{\text{w,local}} \cdot \text{sub\_wyld})}{Q_{\text{outlet}}} \quad (2)$$

where $T_{\text{w,upstream}}$ is the temperature of the streamflow entering the subbasin (°C) and Q_{outlet} is the streamflow discharge at the outlet of the subbasin.

The final stream temperature is calculated by adding a change to the initial stream temperature in the subbasin from differences between stream and air temperature and travel time of water through the subbasin. Depending on T_{air}, the final stream temperature is estimated as

$$T_{\text{water}} = T_{\text{water}_{\text{intial}}} + \left(T_{\text{air}} - T_{\text{water}_{\text{intial}}}\right) \cdot K \cdot (\text{TT}) \text{ if } T_{\text{air}} > 0, \quad (3)$$

$$T_{\text{water}} = T_{\text{water}_{\text{intial}}} + \left((T_{\text{air}} + \varepsilon) - T_{\text{water}_{\text{intial}}}\right) \cdot K \cdot (\text{TT}) \text{ if } T_{\text{air}} < 0, \quad (4)$$

where T_{air} is the average daily air temperature (°C), K is a calibration conductivity parameter, TT is the travel time of

Table 1. Coupled Model Intercomparison Project – phase 5 general circulation models used in this study.

Modeling group	CMIP5 model
Canadian Centre for Climate Modeling & Analysis	canesm2
Météo-France/Centre National de Recherches Météorologiques, France	cnrm-cm5
Geophysical Fluid Dynamics Laboratory, USA	gfdl-cm3
Institut Pierre Simon Laplace, France	ipsl-cm5a-mr
Center for Climate System Research (The University of Tokyo), National Institute for Environmental Studies, and Frontier Research Center for Global Change (JAMSTEC), Japan	miroc5
Max Planck Institute for Meteorology, Germany	mpi-esm-lr
Meteorological Research Institute, Japan	mri-cgcm3

water through the subbasin (hours) and is calculated from the SWAT simulations, and ε is an air temperature addition coefficient (°C), which was included to account for water temperature pulses when T_{air} is below 0 °C. For the case when the effects of T_{air} and the hydrologic contributions are such that the final is $T_{water} < 0$ °C, the stream temperature model sets T_{water} to 0.1 °C. T_{water} is also assumed to be the temperature of water discharge to the downstream subbasin, and is further routed along the stream network. The calibration parameter, K, acts as a proxy for reach-specific adjustment of the radiative forcing, such as shading due to a vegetation canopy or geomorphic changes resulting in differing geometry. Additional details regarding the stream temperature model can be found in Ficklin et al. (2012).

Based on our previous work throughout the western United States (Ficklin et al., 2012), the stream temperature model is highly sensitive to changes in λ (the calibration coefficient for the surface runoff and lateral soil water flow contributions to streamflow) and K (calibration conductivity parameter between air and stream temperature). Previous work also indicates that simulated stream temperatures are sensitive to changes in hydrologic components from increases in air temperature. For example, shifting snowmelt earlier into the winter buffered the effects of increasing air temperature, resulting in only a minor increase in stream temperature. Stream temperature in the late spring/early summer, however, decreased from increases in snowmelt. Increasing groundwater streamflow inputs decreased stream temperatures from the increase in cool water from groundwater. These results are contingent on the volume and timing of the various hydrologic components. For example, the larger the increase in groundwater flow volume to streamflow, the larger the decrease in stream temperature. Further discussion on the stream temperature model sensitivity can be found in Ficklin et al. (2012).

Table 2. Summary of streamflow calibration statistics.

	Calibration		Validation	
	Average	Std. dev.	Average	Std. dev.
NS[a]	0.69	0.13	0.64	0.13
$R^{2,b}$	0.75	0.10	0.75	0.08
Φ^c	0.62	0.15	0.65	0.13

[a] NS: Nash–Sutcliffe coefficient; [b] R^2: coefficient of determination; [c] Φ: coefficient of determination multiplied by slope of regression line, b.

2.3 Input data

SWAT input parameter values for topography, land cover, and soils data were compiled from freely available federal and state databases. A 30 m digital elevation model (USGS) formed the basis for watershed and sub-basin delineation. Soil properties were obtained from the STATSGO soil data set. The 2001 National Land Cover Database was used for land cover/land use. Meteorological data (air temperature, precipitation, and wind speed) were extracted from Maurer et al. (2002) and relative humidity and solar radiation were generated within SWAT (Neitsch et al., 2005).The Columbia River basin natural flow data that were used for streamflow calibration were obtained from output from a calibrated Variable Infiltration Capacity Model (VIC) model (from http://cses.washington.edu/) and the United States Geological Survey Hydro-Climatic Data Network (HCDN; Slack et al., 1993). These data represent streamflow that would occur if no reservoirs or streamflow diversions were present within the basin. The HCDN is a hydrologic data set developed to study surface water conditions throughout the United States that only fluctuate with changes in local climatic conditions and is therefore apt for use in climate change studies (Maurer et al., 2014). SWAT was run at a monthly time step.

Climatic projections from seven GCMs (Table 1) and one RCP (8.5) were input into the calibrated SWAT model. Daily downscaled output from the seven GCMs (RCP 8.5) were obtained from the Downscaled CMIP3 and CMIP5 Climate and Hydrology Projections archive (Maurer et al., 2014). RCP 8.5 represents the highest increase in radiative forcing of the Coupled Model Intercomparison Project – phase 5 (CMIP5; Taylor et al., 2011) projections, and is based on an increased radiative forcing of 8.5 Wm^{-2} (relative to pre-industrial values) at the end of the 21st century. Downscaling was achieved using the daily bias-corrected and constructed analogs (BCCA) method (Maurer et al., 2010). In summary, the BCCA procedure consists of two steps. The first step is a bias correction using a quantile mapping technique which is applied to raw GCM output. Quantile mapping bias correction has been widely and successfully used in climate model downscaling (Wood et al., 2004). The bias correction step is followed by spatial downscaling using a constructed analogues approach for each day using a linear combination of days drawn from the historic record (Hidalgo et al., 2008). Maurer et al. (2010) found that the BCCA method consistently outperformed the bias-correction/spatial-downscaling method (BCSD) and the constructed analogues (CA) approach in capturing the daily large-scale skill and translating it to simulated streamflows that accurately reproduced historical streamflows.

2.4 SWAT streamflow calibration

The program Sequential Uncertainty Fitting Version 2 (SUFI-2; Abbaspour et al., 2007) was used to automatically calibrate SWAT streamflow at 104 sites in the Columbia River basin (Fig. 1). Initial and default SWAT model parameters were varied simultaneously until an optimal solution was met. Three statistics were used to evaluate model efficiency: (1) the Nash–Sutcliffe coefficient (Nash and Sutcliffe, 1970), (2) the coefficient of determination (R^2), and (3) a modified efficiency criterion (Φ). Φ is the result of the coefficient of determination, R^2, multiplied by the regression line slope, m (Krause et al., 2005). This statistic captures the discrepancy in the magnitude of the observed and simulated streamflow (captured by m) as well as the dynamics (captured by R^2). For all previously mentioned statistics, a perfect simulation is represented by a value of 1. A split-sample approach was used for calibration and validation, and the calibration and validation periods differed at each streamflow gauge depending on streamflow data availability.

2.5 SWAT stream temperature calibration

Monthly stream temperatures were predicted using the SWAT stream temperature model of Ficklin et al. (2012). This model includes the effects of hydrologic component inputs (e.g., snowmelt, groundwater, and surface runoff) on stream temperature. Previous studies have demonstrated that this stream temperature model performs better than linear regressions that use air temperature alone (Ficklin et al., 2013; Barnhart et al., 2014). The model requires four calibration parameters for each subbasin in the SWAT setup. Since the model is not incorporated into the previously mentioned SWAT-CUP software, we utilized the steady-state S-metric evolutionary multi-objective optimization algorithm (SMS-EMOA) to calibrate the stream temperature parameters after hydrologic calibration was performed (Emmerich et al., 2005; Beume et al., 2007). SMS-EMOA is an efficient and effective Pareto optimization evolutionary algorithm for finding solutions to multi-objective optimization problems. The algorithm seeks optimal solutions that maximize the hypervolume (S-metric) – which can be thought of as the volume of dominated space – and has been theoretically proven to converge to the Pareto set (Fleischer, 2003; Emmerich et al., 2005; Beume et al., 2007). For a recent application, see Stagge and Moglen (2014).

For this study, SMS-EMOA was used to seek the optimal set of calibration parameters to reduce the differences between simulated stream temperatures from SWAT and observed values. Observed stream temperatures were obtained from 50 sites within the Columbia River basin between 1970 and 1992. Four calibration parameters for each subbasin were adjusted using the algorithm, and three objectives were specified including the RMSE values for the January–April, May–August, and September–December time periods to match the stream temperature rising limb, peak, and falling limb. Further objective functions were intentionally omitted to simplify the analysis. This decision is justified by the limited range of stream temperatures matched by the algorithm. Conversely, hydrological calibration attempts to match flows that vary over orders of magnitude and therefore require additional objectives to match all portions of the hydrograph. Convergence of the stream temperature calibration algorithm was assumed to be met when the S-metric did not vary more than 1 % between three generations. The final set of solutions exhibited trade-offs between the three objective functions; therefore, a single solution – more specifically, a single set of calibration parameters – was then chosen from this set to be used in the calibrated SWAT simulation.

2.6 Statistical analyses

The impacts of potential climate change on streamflow and hydrologic components were evaluated by comparing historical time period (1961–1990) simulations to those using the GCMs in Table 1 for the late 21st century (2080s; 2081–2099). When describing the ensemble average (or standard deviation) of a time period (i.e., late 21st century), this value is the average (or standard deviation) of the seven CMIP5 GCMs for this time period. Months are lumped into seasons for temporal analysis and are defined as spring (April–June), summer (July–September), fall (October and November), and winter (December–March). These seasons are defined to capture the snowmelt and dry/low flow seasons. Pearson correlations using a bootstrap method were used to measure the relationship between annual and seasonal changes in stream temperature and individual hydroclimatological components. A total of 10 000 bootstrap correlation iterations were run. Statistical significance was determined at the $\alpha = 0.05$ level. For statistical significance, the 5th and 95th percentiles of the bootstrap correlation iterations must agree on the correlation sign (+ or −). If the lower (higher) end of our confidence interval is above (below) zero, we can conclude that the correlation between stream temperature and hydroclimatological component change is significant at the $\alpha = 0.05$ level (two-tailed). Additionally, with changes in climate, it can be expected that the drying of streams will occur. In this study, streams that have no flow for an extended time period of the year (and thus have no stream temperature) are removed from

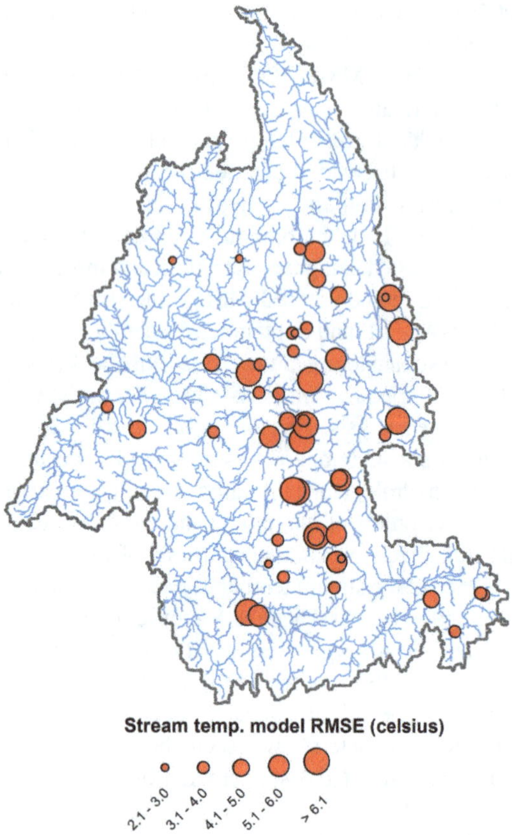

Figure 2. Root mean square errors of the simulated and observed stream temperatures.

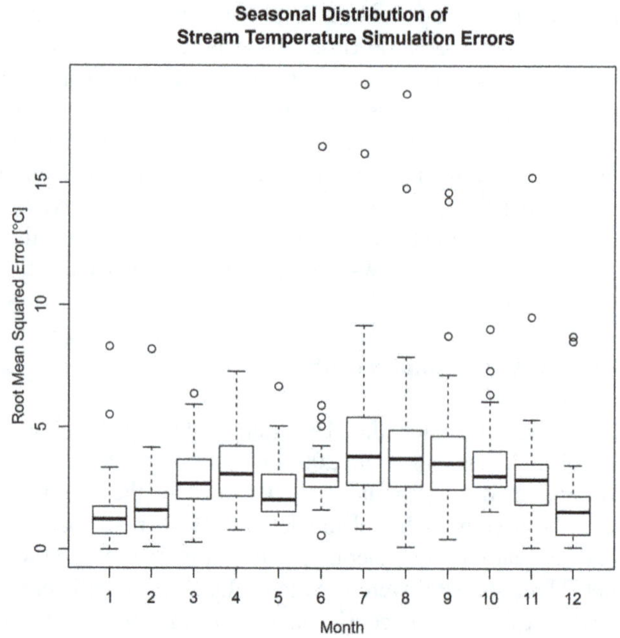

Figure 3. Monthly stream temperature error distributions for all stream temperature gauges. This modified "box and whisker" plot shows 1st and 3rd quartiles (Q_1 and Q_3) as the lower and upper box edges, respectively, and medians as bold lines. The lower whisker is defined by $Q_1 - 1.5 \cdot (Q_3 - Q_1)$, and the upper whisker is defined by $Q_3 + 1.5 \cdot (Q_3 - Q_1)$. Outliers are designated as open circles.

the stream temperature analyses, but since drying streams are an important barrier for aquatic species' migration, they will be discussed.

3 Results

3.1 Hydrologic model calibration

The Nash–Sutcliffe coefficient (NS), R^2, and Φ average and standard deviation values for the calibration and validation time periods are shown in Table 2. Overall, the model efficiency statistics show that the SWAT model adequately simulated streamflow compared to observations. The average NS coefficient for the calibration and validation period was 0.69 and 0.64, respectively, with a standard deviation of 0.13 for the calibration period and 0.13 for the validation period. This indicates that a large portion of the NS values for both time periods varied only 0.13 around their respective means, which is still within acceptable NS limits (Moriasi et al., 2007). The other model efficiency statistics, R^2 and Φ, indicate similar model performance.

3.2 Stream temperature model calibration

After SWAT was calibrated for discharge, the model was used within the SMS-EMOA algorithm to calibrate the stream temperature model. RMSE values between observed and simulated daily stream temperatures range from 2 to 5 °C for the majority of observation sites. The resulting monthly RMSE values for each site are shown in Fig. 2. No distinct spatial distributions of the magnitude of errors are present. Errors distinguished by month of year were also quantified (Fig. 3). Errors are largest during the summer months of July through September. Lowest RMSE values were present between December and February. Also, the model gives highly unrealistic (RMSE > 15 °C) results for a moderate number of points, especially during summer months. This is due to low values of discharge within reaches during the summer months. Stream temperature is strongly inversely dependent on streamflow, and very small values of discharge cause the model to produce uncharacteristically high stream temperature simulation values. The calibrated stream temperature model parameters can be found in the supplemental information.

3.3 Temperature and precipitation projections

Ensemble average projections of maximum and minimum air temperature and precipitation, as compared to the his-

Table 3. Stream temperature changes and focal fish species groups for the Columbia River basin ecological provinces during the 2080s.

Ecological province	Spring (°C)	Summer (°C)	Fall (°C)	Winter (°C)	Annual (°C)	Focal fish species
Blue Mountain	3.7	5.3	3.2	2.1	3.5	coldwater migratory
Columbia Cascades	2.6	4.1	2.0	1.2	2.4	coldwater migratory
Columbia Plateau	2.0	3.8	2.0	1.5	2.2	warmwater
Intermountain	3.3	5.0	2.7	1.5	3.0	warmwater
Middle Snake	2.4	3.7	2.3	1.4	2.2	coldwater migratory
Mountain Columbia	3.6	5.0	2.4	1.5	3.1	coldwater non-migratory
Mountain Snake	5.0	7.0	4.0	2.1	4.3	coldwater migratory
Upper Snake	4.3	6.0	3.3	1.6	3.6	coldwater non-migratory

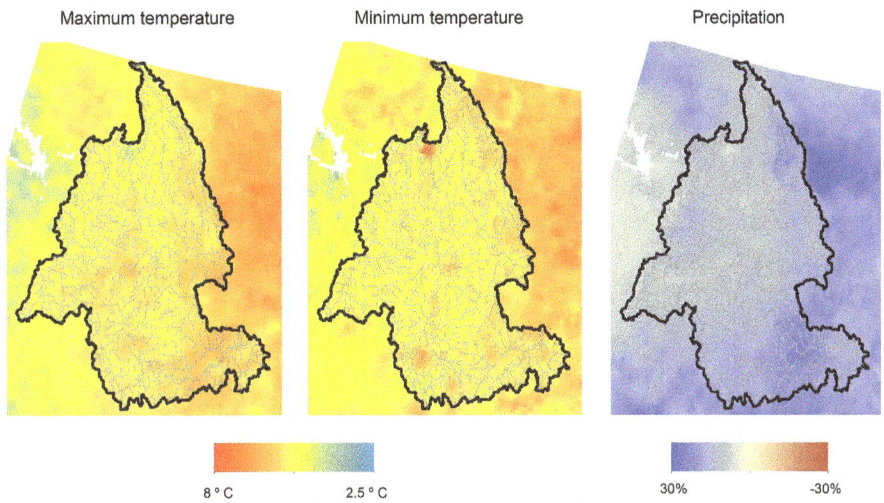

Figure 4. Changes in average precipitation and air temperature (maximum and minimum) for the end of the 21st century as compared to the historical time period.

torical time period, are shown in Fig. 4. Overall, the maximum and minimum air temperatures vary spatially throughout the CRB, with an average ensemble increase of 5.5 °C for maximum air temperature and 5.4 °C for minimum air temperature. All GCMs agreed that air temperature is expected to increase by the end of the 21st century. Precipitation projections, on the other hand, varied between downscaled GCM projections, with an overall average of a 14.4 % increase compared to the historical time period.

3.4 Stream temperature projections

Figures 5 and 6 display the spring/summer and fall/winter historical and projected stream temperatures for the CRB. Simulated stream temperatures are projected to increase throughout the CRB, with the largest increases occurring in the east-central portion of the CRB. On average, stream temperatures are projected to increase 3.5 °C for the spring, 5.2 °C for the summer, 2.7 °C for the fall, and 1.6 °C for the winter. It is important to note that a large number of subbasins were removed from this analysis due to no-flow conditions (i.e., running completely dry or icing-up) from changes in climate (hatched areas in Figs. 5 and 6). Of these, winter had the largest number of subbasins removed from the analysis (31 %), followed by fall (18 %), summer (16 %), and spring (15 %). The average period of subbasins with no-flow conditions is projected to be 34 %, or 81 months out of the 240 months for the 2080s time period. We consider these subbasins to not be reliable refugia for aquatic species.

Simulated stream temperature changes also vary at the ecological province scale (Table 3). At the annual timescale, the largest stream temperature increases (4.3 °C) occurred within the Mountain Snake ecological province, which is characterized by cold-water migratory fish species. The largest interannual variation around the mean occurred in the Upper Snake ecological province, which is characterized by non-migratory coldwater species, with a ±3.8 °C standard deviation. Important differences between ecological provinces occurred at the seasonal time scale. Overall, the largest spring increases in stream temperature occurred in the Mountain Snake (5.0 °C) and Upper Snake (4.3 °C), both containing coldwater species. The largest summer tem-

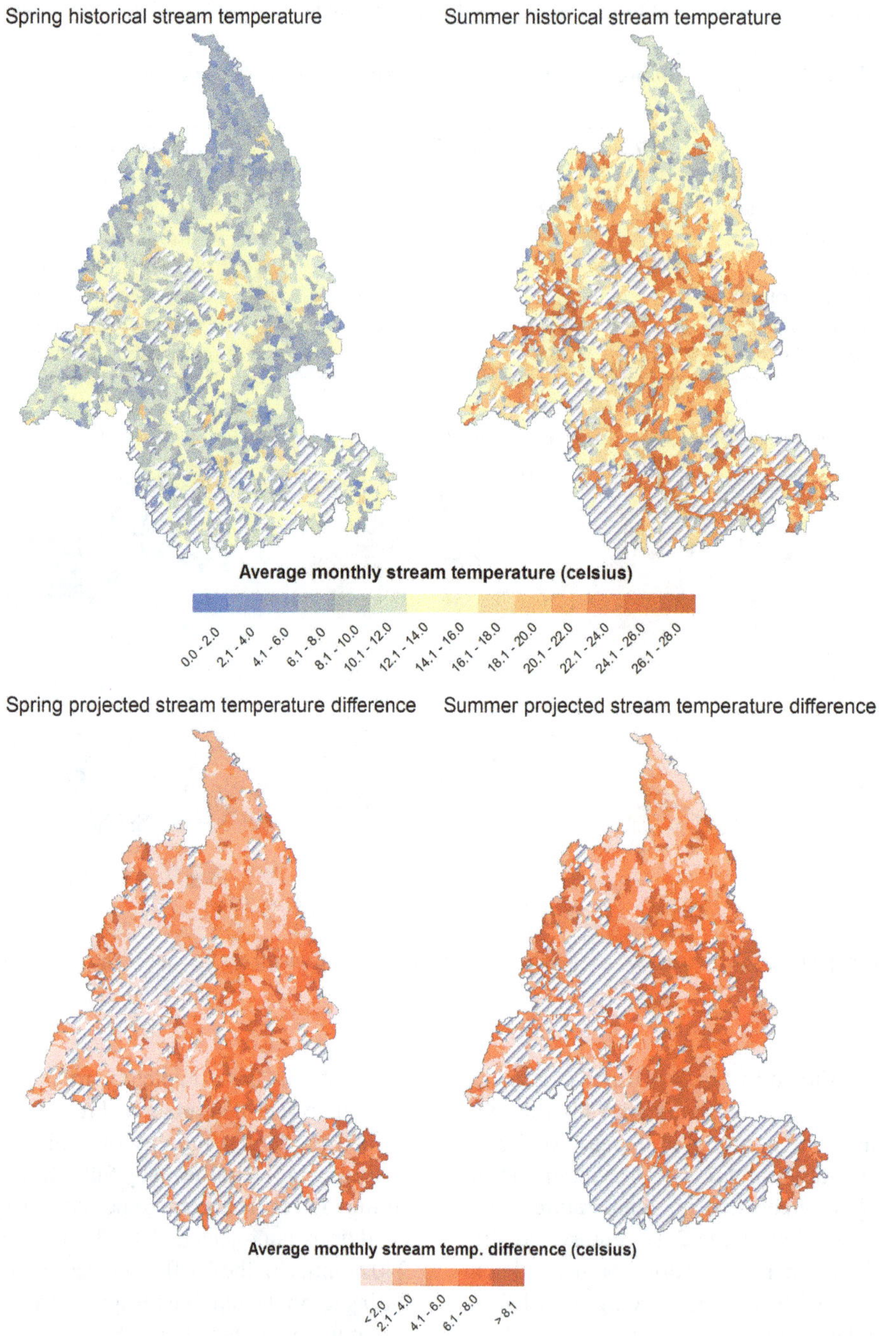

Figure 5. Spring and summer historical and projected stream temperatures at the subbasin level. Hatched subbasins indicate that drying occurred under climate projections and were removed from analyses.

perature increase compared to the historical time period was for the Mountain Snake ecological province with a 7 °C increase in average monthly stream temperature, followed by Upper Snake (6 °C), Blue Mountain (5.3 °C), Intermountain (5.0 °C), and Mountain Columbia (5.0 °C), indicating that ecological provinces with coldwater species will experience some of the largest increases in stream temperature in the basin. These large increases are expected during the summer because air temperature is at its highest and streamflow is at its lowest.

Fall and winter had the smallest increases in stream temperature including a CRB average of 2.9 °C for fall and 1.6 °C for winter. This was expected because this is when air temperatures are the lowest, and cold precipitation recharge and streamflow are highest, resisting stream temperature increases. The basins with the highest stream temperature in-

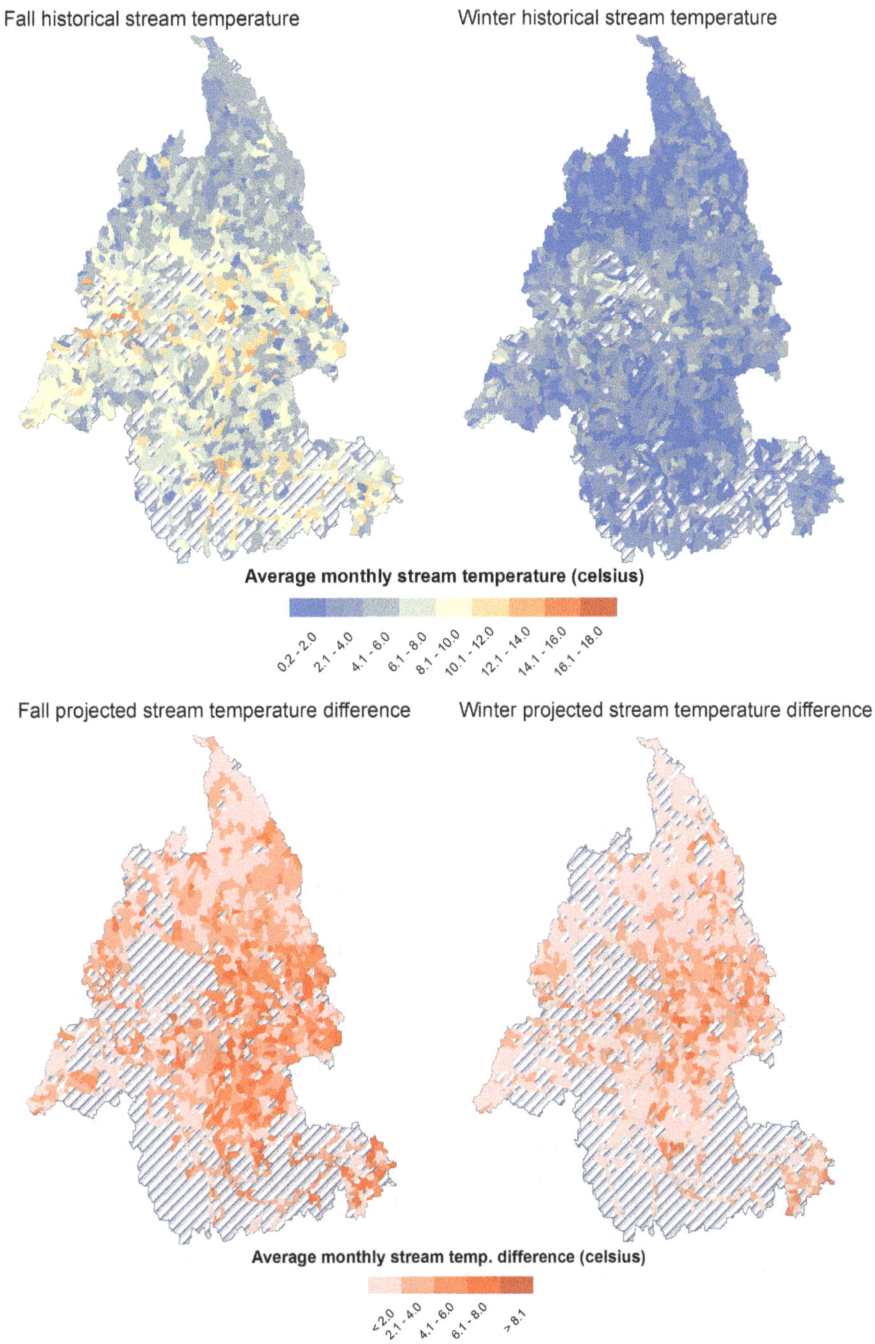

Figure 6. Fall and winter historical and projected stream temperatures at the subbasin level. Hatched subbasins indicate that drying occurred under climate projections and were removed from analyses.

creases for the fall and winter time period were the Mountain Snake and Blue Mountain (4.0 and 2.1 °C).

3.5 Sensitivities of stream temperature changes to air temperature

We define TS_{max} and TS_{min} as the thermal sensitivity or stream temperature change per 1 °C of maximum or minimum air temperature change. For the entire CRB and the water year annual timescale, the value for the average TS_{max} is 0.6 and that for TS_{min} is 0.86, demonstrating that, on average, the increases in stream temperature seen by the 2080s are to a larger degree tied to future changes in minimum air temperatures (Table 4). On the seasonal timescale, stream temperature changes during the summer were the most sensitive to changes in maximum air temperature, with TS_{max}

Table 4. Sensitivities of stream temperature changes to changes in maximum and minimum air temperatures for the Columbia River basin during the 2080s.

Ecological province	Spring ($°C°C^{-1}$)	Summer ($°C°C^{-1}$)	Fall ($°C°C^{-1}$)	Winter ($°C°C^{-1}$)	Annual ($°C°C^{-1}$)
Maximum air temperature					
Blue Mountain	0.7	0.5	0.8	0.4	0.6
Columbia Cascades	0.5	0.7	0.7	0.3	0.6
Columbia Plateau	0.5	0.4	0.7	0.0	0.4
Intermountain	0.7	0.8	1.1	0.6	0.8
Middle Snake	0.5	0.5	0.8	0.9	0.7
Mountain Columbia	0.4	0.7	0.7	0.3	0.5
Mountain Snake	0.7	1.0	1.0	0.0	0.7
Upper Snake	0.6	0.7	0.8	0.3	0.6
Minimum air temperature					
Blue Mountain	0.7	0.7	0.9	0.0	0.6
Columbia Cascades	0.2	0.7	0.8	1.4	0.7
Columbia Plateau	0.2	0.6	0.8	0.4	0.5
Intermountain	0.7	0.9	0.8	0.0	0.6
Middle Snake	0.8	0.9	1.0	0.5	0.6
Mountain Columbia	0.3	0.9	0.6	0.2	0.5
Mountain Snake	0.7	1.1	1.0	0.5	0.8
Upper Snake	0.8	1.2	0.9	0.5	0.9

equal to 0.8, followed by spring (0.7), fall (0.5), and winter (0.3). For minimum air temperature sensitivities, however, spring values of TS_{min} were the highest of all seasons, equal to 0.9, followed by summer (0.8), fall (0.5), and winter (0.3). Air temperature sensitivities varied by ecological province as well as by season. At the annual and seasonal timescales the Intermountain, Middle Snake, and Mountain Snake ecological provinces exhibited the highest values of TS_{max}.

For minimum air temperatures, the ecological provinces that were the most sensitive were Columbia Cascade, Mountain Snake, and Upper Snake. Summer once again had the highest overall TS_{min} values. However, the largest individual TS_{min} values were found in the winter and spring seasons, with the Columbia Cascades exhibiting a value of 1.4 in the winter and the Mountain Snake and Upper Snake exhibiting TS_{min} values of 1.1 and 1.2 in the spring. Overall, it can be seen that spring has higher TS_{min} values than TS_{max}, a possible artifact of snowmelt (see Sect. 4).

3.6 Sensitivities of stream temperature to changes in hydroclimatological components

3.6.1 Correlations at the Columbia River basin scale

At the CRB scale, all stream temperature changes were significantly correlated to all hydroclimatic components during the spring and fall seasons for the 2080s (Table 5), suggesting that during these seasons stream temperatures are highly sensitive to changing environments. For summer, groundwater inflow change was the only variable not significantly correlated to stream temperature changes. For winter, streamflow and groundwater inflow changes were the only variables not significantly correlated to stream temperature changes (see Sect. 4).

Table 5. Pearson correlations between stream temperature and individual hydroclimatological changes for the entire Columbia River basin during the 2080s.

Hydroclimatological component	Spring	Summer	Fall	Winter
Maximum air temperature	0.67	0.61	0.49	0.36
Minimum air temperature	0.65	0.61	0.47	0.34
Precipitation	−0.51	−0.50	−0.36	−0.20
Streamflow	0.08	0.07	−0.10	−0.02*
Snowmelt	−0.36	0.10	−0.31	−0.26
Surface runoff	−0.39	−0.08	−0.30	−0.28
Groundwater inflow	−0.24	−0.04*	−0.12	0.00*
Lateral soil flow	−0.42	−0.32	−0.36	−0.07

* indicates there was *no* significant correlation at $p = 0.05$.

3.6.2 Correlations at the ecological province scale

Correlations between stream temperature and hydroclimatological components at the seasonal timescale and ecological province spatial scale for the 2080s suggest that multiple hydroclimatological components affect stream temperatures (Fig. 7). As expected, maximum and minimum air temper-

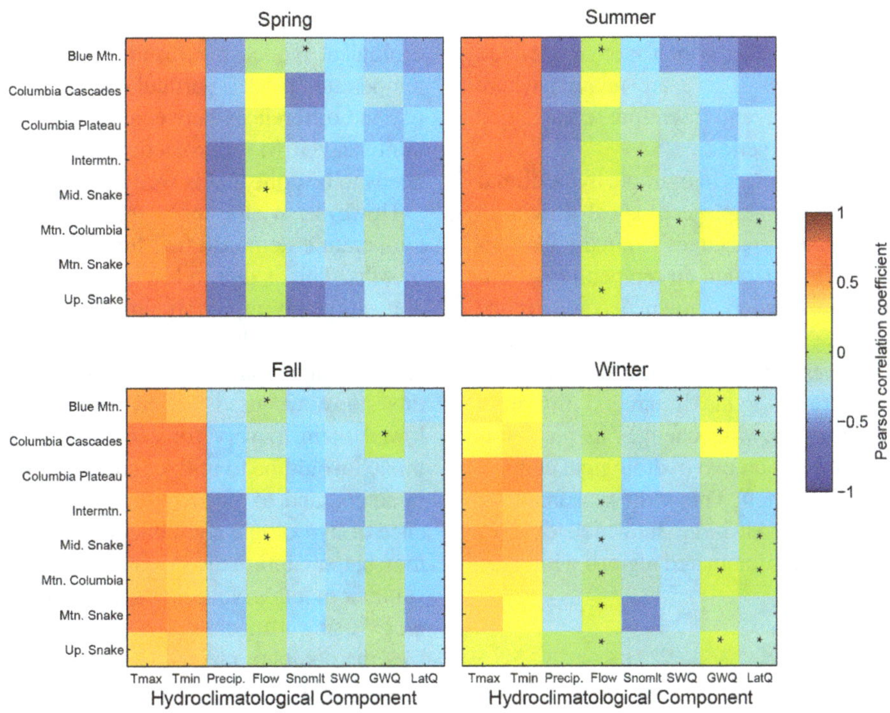

Figure 7. Pearson correlations between changes in stream temperature and hydroclimatological components for the Columbia River basin ecological provinces. T_{max} = maximum air temperature; T_{min} = minimum air temperature; Precip. = precipitation; Flow = streamflow; Snomlt = snowmelt; SW_Q = surface water runoff; GW_Q = groundwater inflow; Lat_Q = lateral soil flow. Asterisks represent *no* significant correlation at $p = 0.05$.

atures were significantly positively correlated to changes in stream temperatures for all seasons and nearly all ecological provinces. The only two ecological provinces where no significant correlations were found between air and stream temperature were the Blue Mountain and Upper Snake provinces (see Discussion and conclusions), which are characterized by migratory salmonids and non-migratory salmonids, respectively. Additionally, precipitation changes were negatively correlated to stream temperature changes for all seasons and nearly all ecological provinces.

For spring, nearly all hydroclimatological components were significantly correlated to stream temperature changes for each ecological province. Streamflow changes were not correlated to stream temperature changes within the Blue Mountain, Intermountain, and Upper Snake ecological provinces, which are characterized by warmwater species, migratory coldwater salmonids, and non-migratory coldwater salmonids, respectively. We also found that snowmelt changes within the Blue Mountain ecological province were not correlated to stream temperature changes. However, within the Blue Mountain ecological province we found that snowmelt is not a large portion of the hydrological cycle during this season.

For the summer season, no relationships were found for streamflow, snowmelt, surface runoff, and groundwater inflows within multiple ecological provinces. Overall, streamflow was found to be significantly correlated with stream temperature within the Columbia Cascades and Middle Snake, which are characterized by coldwater migratory salmonids, and Mountain Columbia, which is characterized by non-migratory coldwater salmonids, ecological provinces. Within the Columbia Plateau, Intermountain, and Mountain Columbia ecological provinces, we found snowmelt to still be a large portion of the hydrological cycle, thus any reductions of snowmelt do not significantly affect stream temperature. Lastly, surface runoff and groundwater inflows were not significantly correlated to the stream temperature changes in the Mountain Columbia and Upper Snake ecological provinces and the Mountain Snake ecological province, respectively. Within these regions we did not find large changes in surface runoff or groundwater inflows.

For the fall season, we found that changes in stream temperature within the Blue Mountain ecological province, which is characterized by migratory coldwater salmonids, are only positively correlated to changes in maximum and minimum air temperature, and thus lose their ties to the other hydrology-related components. Note also that during the fall season, groundwater inflow changes become a non-significant factor in stream temperature changes for five out of the eight ecological provinces. The only ecological provinces where groundwater inflow changes were significantly correlated to stream temperature changes were the

Columbia Plateau, Intermountain, characterized by warmwater species, and the Middle Snake, which is characterized by coldwater migratory species. These are regions where groundwater inflows increased and therefore contributed cooling effects during this time period.

During the winter season, changes in multiple hydroclimatological components within multiple ecological provinces are not significantly correlated to changes in stream temperature. Generally, changes in maximum air temperature, minimum air temperature, precipitation, snowmelt, and surface runoff are still significantly correlated to changes in stream temperature. These relationships make sense because during the winter season, increases in maximum and minimum air temperatures in conjunction with changes in precipitation will have the largest effects on two hydrological components: snowmelt and surface runoff. This is the season where snowmelt-dominated regions with large snowmelt components may perhaps become rain-dominated regions with large surface runoff components.

4 Discussion and conclusions

The importance of stream temperature to aquatic species distributions, interactions, behavior, and persistence is well documented (Matthews, 1998), particularly for coldwater-adapted taxa such as trout and salmon (Milner et al., 2003; McCullough, 1999). Considering predicted increases in air temperature in the coming century, accurate assessment of suitable thermal habitats is critical for predicting species' responses to changes in climate. Accordingly, recent research has investigated the potential impacts of climate change on aquatic taxa by explicitly incorporating regression-based stream temperature predictions into ecological models (Britton et al., 2010; Al-Chokhachy et al., 2013). While simplified regression studies may boast low RMSE values between simulated and observed stream temperatures, the relatively broad spatial scale of many of these studies (Mohseni et al., 2003), neglects the variety of local hydrological systems that are driven differentially by the array of inputs to each system (e.g., snowmelt, groundwater, runoff). The resulting stream temperature model inaccuracies from this approach, clustered in particular regions, can be particularly problematic when investigating local population responses and range shifts at the edge of species distributions. Our results highlight this issue by characterizing the varied relative contributions of different hydrological component inputs among ecological provinces and suggest the complex system-level regulation of stream temperature.

As with any modeling study, modeling errors originate from multiple sources. Wilby and Harris (2006) discuss these aforementioned uncertainties in detail and ranked their importance in decreasing order as follows: differences in GCM output, downscaling methods, hydrological model structure, hydrological model parameters, and then greenhouse gas emission scenario. While their work was performed for a hydrological model, the results still hold true for our stream temperature model. Particular to this study, in order to quantify the differences between errors due to parameter uncertainty and GCM (or projection) uncertainty, much more work needs to be done and is well beyond the scope of this work.

However, we do note that our simulations for stream temperature demonstrated higher errors during the summer months. This is due to low and fluctuating discharge values that ultimately affect stream temperature. Also, it is likely due to the fact that hydrologic components may influence stream temperature differently during different seasons. For this study, we used annual calibration parameters and allowed them to vary for each subbasin. An alternative approach would be to utilize seasonally varying calibration parameters, and to analyze the dynamic (i.e., seasonal) influence of hydrologic components on stream temperature. This may better capture the stream temperature fluctuations in the summer months. Nonetheless, our spatially resolved methodology using a mechanistic model, SWAT, better characterizes the complex processes of stream temperature throughout the CRB by accounting for the hydrologic components contributing to stream temperature and its variation.

Within the CRB, Wenger et al. (2013) used air temperature as a surrogate for stream temperature to predict the response of Bull trout (Salmonidae: *Salvelinus confluentus*) to predicted changes in climate, while Beer and Anderson (2013) used air–stream temperature relationships to predict the impacts of climate change on salmonid life histories. These approaches are common (Britton et al., 2010; Tisseuil et al., 2012; Al-Chokhachy et al., 2013), yet overlook important differences in the inputs influencing stream temperature across the basin. For example, our results suggest that hydrologic contributions from snowmelt are relatively important drivers of stream temperature within ecological provinces with primarily non-migratory coldwater focal fish species. The influence of snowmelt tends to buffer stream temperatures against increases in air temperature during the year relative to other areas in the watershed. In this case, a regression-based approach to estimating stream temperature or the use of air temperature as a surrogate for stream temperature will tend to overestimate stream temperature, and thus underestimate the amount of suitable thermal habitat for coldwater species. In addition, decreases in snowcover (and snowmelt) in the future will result in increased thermal sensitivity within these formerly buffered regions. For example, current stream temperatures in the Mountain Snake ecological province are buffered by relatively high levels of snowmelt, yet decreases in future snowcover are predicted to result in this province experiencing the greatest seasonal and annual increases in stream temperature in the coming century.

Some of the relationships between stream temperature and hydroclimatic changes at the CRB scale were expected, such as increases in maximum air temperature and minimum air temperature resulting in increases in stream temperature, which were significant for all seasons for the entire CRB. This relationship is well established and many models have been developed solely based on air–stream temperature relationships (Stefan and Preud'homme, 1993; Mohseni and Stefan, 1999). Also, a decrease in precipitation led to an increase in stream temperature, largely because greater runoff and infiltration leads to larger volumes of water in the stream channel, and thus increases the amount of energy needed to heat the water. Precipitation changes had the largest negative correlations during the spring and summer seasons, followed by fall and winter. Both surface runoff and lateral soil flow changes follow the same correlation patterns as precipitation, as both are inherently tied to the amount of incoming precipitation. Additionally, streamflow is tied to all hydrological components within the subbasin and the incoming streamflow that is entering the streamflow reach. Since streamflow is a mix of incoming hydrologic components, it is difficult to determine correlations. However, much research has assumed that streamflow and stream temperature changes are inversely correlated (van Vliet et al., 2011). The correlations within this study were significant and positively correlated for the spring, summer, and fall seasons; however, all correlations were below 0.10, which suggests the correlations were relatively minor, especially compared to other components.

Snowmelt changes were negatively correlated during the spring, fall, and winter seasons, and positively correlated during the summer season. A decrease in snowmelt will lead to an increase in stream temperature because the cooling effect that snowmelt has on stream temperature is no longer present. In summer, snowmelt and stream temperature were positively correlated (albeit not significantly), suggesting the counterintuitive notion that an increase in snowmelt led to an increase in stream temperature. This can be explained largely because snowmelt changes did not occur at all in 975 (60 % of the subbasins with streamflow) of the CRB subbasins, while for spring, fall, and winter, these values were 89 (5 %), 50 (3 %) and 48 (3 %), respectively. These observations suggest that snowmelt is still a component of the hydrologic cycle during the summer season.

Lastly, groundwater inflow changes to the stream channel were negatively correlated to stream temperature change at the CRB scale for the spring and fall seasons. This also makes sense, as groundwater temperature is generally cooler than the stream temperature of the water already within the channel. Quite often, stream temperature variations of cool water are used for tracer studies to determine where surface and groundwater flows are exchanging water (Anderson, 2005; Constantz et al., 2003). However, no significant correlation was found during the summer, when groundwater is a large source of stream flow. This is likely because groundwater is the main source of water for this season so any climate-induced changes in groundwater will not have a major effect on stream temperature. For example, if 85 % of the streamflow comes from groundwater, and is then decreased to 75 %, the change in stream temperature is not likely to significantly change. Additionally, no groundwater inflow change correlations were found for the winter season.

Species' responses to stream temperature occur within populations and are based on local environmental conditions. Consequently, accurate assessment of local variation in stream temperature is critical and only possible when local system drivers are accurately represented in stream temperature models. While stream temperature is primarily influenced by air temperature, this study emphasized the important effects of other contributors (e.g., runoff, groundwater, snowmelt) that are represented differently across the CRB. Also, we have characterized the ecological provinces by warmwater and coldwater focal fish species, which was done for qualitative biological assessments and not as a predictive approach. However, these groupings have provided important information regarding factors driving differential variation in stream temperatures across seasons in the context of the biological groups experiencing particular stream temperature changes. River basins encompass a spatially heterogeneous array of biological communities, and these communities are regulated by a spatially heterogeneous array of environmental conditions. These environmental conditions are driven by local processes and require a systems-based approach to accurately characterize the habitat regulating the distribution and diversity of aquatic taxa.

Acknowledgements. The authors gratefully acknowledge financial support for this work from the US Environmental Protection Agency through EPA STAR grant no. RD-83419101-0, the Environmental Protection Agency's Science to Achieve Results (STARs) Consequences of Global Change for Water Quality program (EPA-G2008-STAR-D2), and from the National Science Foundation (DEB-0844644). We acknowledge the World Climate Research Programme's Working Group on Coupled Modelling, which is responsible for CMIP, and we thank the climate modeling groups (listed in Table 1 of this paper) for producing and making available their model output. For CMIP, the US Department of Energy's Program for Climate Model Diagnosis and Intercomparison provides coordinating support and led development of software infrastructure in partnership with the Global Organization for Earth System Science Portals. Additionally, this material is based upon work supported by the National Science Foundation under grant no. CNS-0723054.

Edited by: J. Liu

References

Abbaspour, K. C., Yang, J., Maximov, I., Siber, R., Bogner, K., Mieleitner, J., Zobrist, J., and Srinivasan, R.: Modelling hydrology and water quality in the pre-alpine/alpine Thur watershed using SWAT, J. Hydrol., 333, 413–430, 2007.

Al-Chokhachy, R., Alder, J., Hostetler, S., Gresswell, R., and Shepard, B.: Thermal controls of yellowstone cutthroat trout and invasive fishes under climate change, Global Change Biol., 19, 3069–3081, 2013.

Anderson, M. P.: Heat as a ground water tracer, Ground Water, 43, 951–968, 2005.

Angilletta, M. J.: Thermal adaptation: a theoretical and empirical synthesis, Oxford University Press, Oxford, 2009.

Arnold, J. G., Srinivasan, R., Muttiah, R. S., and Williams, J. R.: Large Area Hydrologic Modeling and Assessment Part I: Model Development, J. Am. Water Resour. Assoc., 34, 73–89, 1998.

Barnhart, B. L., Whittaker, G. W., and Ficklin, D. L.: Improved Stream Temperautre Simulations in SWAT Using NSGA-II For Automatic Multi-Site Calibration, T. ASABE, 57, 517–530, doi:10.13031/trans.57.10472, 2014.

Beer, W. N. and Anderson, J. J.: Sensitivity of salmonid freshwater life history in western US streams to future climate conditions, Global Change Biol., 19, 2547–2556, 2013.

Beume, N., Naujoks, B., and Emmerich, M.: SMS-EMOA: Multiobjective selection based on dominated hypervolume, Eur. J. Operat. Res., 181, 1653–1669, 2007.

Bogan, T., Mohseni, O., and Stefan, H. G.: Stream temperature-equilibrium temperature relationship, Water Resour. Res., 39, 1245, doi:10.1029/2003WR002034, 2003.

Britton, J., Cucherousset, J., Davies, G., Godard, M., and Copp, G.: Non-native fishes and climate change: predicting species responses to warming temperatures in a temperate region, Freshwater Biol., 55, 1130–1141, 2010.

Caissie, D.: The thermal regime of rivers: a review, Freshwater Biol., 51, 1389–1406, 2006.

Chang, H. and Psaris, M.: Local landscape predictors of maximum stream temperature and thermal sensitivity in the Columbia River Basin, USA, Sci. Total Environ., 461, 587–600, 2013.

Constantz, J.: Interaction between stream temperature, streamflow, and groundwater exchanges in alpine streams, Water Resour. Res., 34, 1609–1615, 1998.

Constantz, J., Thomas, C. L., and Zellweger, G.: Influence of diurnal variations in stream temperature on streamflow loss and groundwater recharge, Water Resour. Res., 30, 3253–3264, 1994.

Constantz, J., Cox, M. H., and Su, G. W.: Comparison of heat and bromide as ground water tracers near streams, Ground Water, 41, 647–656, 2003.

Edinger, J. E., Brady, D. K., and Geyer, J. C.: Heat exchange and transport in the environment, in: Heat exchange and transport in the environment, Johns Hopkins University, Baltimore, Maryland, USA, 1974.

Emmerich, M., Beume, N., and Naujoks, B.: An EMO algorithm using the hypervolume measure as selection criterion, Evolutionary Multi-Criterion Optimization, Springer-Verlag, Berlin, Heidelberg, 62–76, 2005.

Erickson, T. R. and Stefan, H. G.: Linear Air/Water Temperature Correlations for Streams during Open Water Periods, J. Hydrol. Eng., 5, 317–321, 2000.

Ficklin, D. L., Luo, Y., Stewart, I. T., and Maurer, E. P.: Development and application of a hydroclimatological stream temperature model within the Soil and Water Assessment Tool, Water Resour. Res., 48, W01511, doi:10.1029/2011WR011256, 2012.

Ficklin, D. L., Stewart, I. T., and Maurer, E. P.: Effects of climate change on stream temperature, dissolved oxygen, and sediment concentration in the Sierra Nevada in California, Water Resour. Res., 49, 2765–2782, 2013.

Fleischer, M.: The measure of Pareto optima applications to multiobjective metaheuristics, Evolutionary multi-criterion optimization, Springer-Verlag, Berlin, Heidelberg, 519–533, 2003.

Gassman, P. W., Reyes, M. R., Green, C. H., and Arnold, J. G.: The Soil and Water Assessment Tool: Historical Development, Applications, and Future Research Directions, T. ASABE, 50, 1211–1250, 2007.

Hari, R. E., Livingstone, D. M., Siber, R., Burkhardt-Holm, P., and Guettinger, H.: Consequences of climatic change for water temperature and brown trout populations in Alpine rivers and streams, Global Change Biol., 12, 10–26, 2006.

Hatcher, K. L. and Jones, J. A.: Climate and Streamflow Trends in the Columbia River Basin: Evidence for Ecological and Engineering Resilience to Climate Change, Atmosphere-Ocean, 51, 436–455, 2013.

Hidalgo, H. G., Dettinger, M. D., and Cayan, D. R.: Downscaling with constructed analogues: daily precipitation and temperature fields over the United States., California Energy Commission, Public Interest Energy Research Program, Sacramento, CA, 2008.

Isaak, D. J., Luce, C. H., Rieman, B. E., Nagel, D. E., Peterson, E. E., Horan, D. L., Parkes, S., and Chandler, G. L.: Effects of climate change and wildfire on stream temperatures and salmonid thermal habitat in a mountain river network, Ecol. Appl., 20, 1350–1371, 2010.

Johnson, A. C.,Acreman, M. C., Dunbar, M. J., Feist, S. W., Giacomello, A. M., Gozlan, R. E., Hinsley, S. A., Ibbotson, A. T., Jarvie, H. P., Jones, J. I., Longshawb, M., Maberly, S. C., Marsh, T. J., Neal, C., Newman, J. R., Nunn, M. A., Pickup, R. W., Reynard, N. S., Sullivan, C. A., Sumpter, J. P., and Williams, R. J.: The British river of the future: how climate change and human activity might affect two contrasting river ecosystems in England, Sci. Total Environ., 407 4787–4798, 2009.

Kim, K. S. and Chapra, S. C.: Temperature model for highly transient shallow streams, J. Hydraul. Eng., 123, 30–40, 1997.

Krause, P., Boyle, D. P., and Bäse, F.: Comparison of different efficiency criteria for hydrological model assessment, Adv. Geosci., 5, 89–97, doi:10.5194/adgeo-5-89-2005, 2005.

Luce, C., Staab, B., Kramer, M., Wenger, S., Isaak, D., and McConnell.: Sensitivity of summer stream temperatures to climate variability in the Pacific Northwest, Water Resour. Res., 50, 3428–3443, 2014.

MacDonald, R. J., Boon, S., Byrne, J. M., and Silins, U.: A comparison of surface and subsurface controls on summer temperature in a headwater stream, Hydrol. Process., 28, 2338–2347, 2014.

Mantua, N., Tohver, I., and Hamlet, A.: Climate change impacts on streamflow extremes and summertime stream temperature and their possible consequences for freshwater salmon habitat in Washington State, Climatic Change, 102, 187–223, 2010.

Matthews, W. J.: Patterns in freshwater fish ecology, Springer, London, UK, 1998.

Maurer, E. P., Wood, A. W., Adam, J. C., Lettenmaier, D. P., and Nijssen, B.: A long-term hydrologically-based data set of land surface fluxes and states for the conterminous United States, J. Climate, 15, 3237–3251, 2002.

Maurer, E. P., Hidalgo, H. G., Das, T., Dettinger, M. D., and Cayan, D. R.: The utility of daily large-scale climate data in the assessment of climate change impacts on daily streamflow in California, Hydrol. Earth Syst. Sci., 14, 1125–1138, doi:10.5194/hess-14-1125-2010, 2010.

nia, Hydrol. Earth Syst. Sci., 14, 1125–1138, doi:10.5194/hess-14-1125-2010, 2010.

Maurer, E. P., Brekke, L., Pruitt, T., Thrasher, B., Long, J., Duffy, P., Dettinger, M., Cayan, D., and Arnold, J.: An enhanced archive facilitating climate impacts and adaptation analysis, B. Am. Meteorol. Soc., doi:10.1175/BAMS-D-13-00126.1, in press, 2014.

McCullough, D.: A Review and Synthesis of Effects of Alterations to the Water Temperature Regime on Freshwater Life Stages of Salmonids, with Special Reference to Chinook Salmon, prepared for the US Environmental Protection Agency Region 10, published as EPA 910-R-99-010, Columbia Intertribal Fisheries Commission, Portland, OR,1999.

Milner, N., Elliott, J., Armstrong, J., Gardiner, R., Welton, J., and Ladle, M.: The natural control of salmon and trout populations in streams, Fishe. Res., 62, 111–125, 2003.

Mohseni, O. and Stefan, H. G.: Stream temperature/air temperature relationship: a physical interpretation, J. Hydrol., 218, 128–141, 1999.

Mohseni, O., Stefan, H. G., and Erickson, T. R.: A nonlinear regression model for weekly stream temperatures, Water Resour. Res., 34, 2685–2692, 1998.

Mohseni, O., Erickson, T. R., and Stefan, H. G.: Sensitivity of stream temperatures in the United States to air temperatures projected under a global warming scenario, Water Resour. Res., 35, 3723–3733, 1999.

Mohseni, O., Stefan, H. G., and Eaton, J. G.: Global Warming and Potential Changes in Fish Habitat in U.S. Streams, Climatic Change, 59, 389–409, 2003.

Moriasi, D. N., Arnold, J. G., Liew, M. W. V., Bingner, R. L., Harmel, R. D., and Veith, T. L.: Model Evaluation Guidelines for Systematic Quantification of Accuracy in Watershed Simulations, T. ASABE, 50, 885–900, 2007.

Nash, J. E. and Sutcliffe, J. V.: River flow forecasting through conceptual models part I – A discussion of principles, J. Hydrol., 10, 282–290, 1970.

Neitsch, S. L., Arnold, J. G., Kiniry, J. R., Williams, J. R., and King, K. W.: Soil and Water Assessment Tool Theoretical Documentation: Version 2005, Texas Water Resources Institute, College Station, TX, 2005.

Nelson, K. C. and Palmer, M. A.: Stream Temperature Surges Under Urbanization and Climate Change: Data, Models, and Responses1, J. Am. Water Resour. Assoc., 43, 440–452, 2007.

N.H.I.: Northwest Habitat Institute web site: http://www.nwhi.org, last access: December 2014, Northwest Habitat Institute, Corvallis, OR, 2008.

Null, S. E., Viers, J. H., Deas, M. L., Tanaka, S. K., and Mount, J. F.: Stream temperature sensitivity to climate warming in California's Sierra Nevada: impacts to coldwater habitat, Climatic Change, 116, 149–170, 2013.

Payne, J. T., Wood, A. W., Hamlet, A. F., Palmer, R. N., and Lettenmaier, D. P.: Mitigating the effects of climate change on the water resources of the Columbia River Basin, Climatic Change, 62, 233–256, 2004.

Pekarova, P., Halmova, D., Miklanek, P., Onderka, M., Pekar, J., and Skoda, P.: Is the Water Temperature of the Danube River at Bratislava, Slovakia, Rising?, J. Hydrometeorol., 9, 1115–1122, 2008.

Peterson, J. T. and Kwak, T. J.: Modeling the effects of land use and climate change on riverine smallmouth bass, Ecol. Appl., 9, 1391–1404, 1999.

Sinokrot, B. A. and Stefan, H. G.: Stream water-temperature sensitivity to weather and bed parameters, J. Hydraul. Eng., 120, 722–736, 1994.

Slack, J. R., Lumb, A. M., and Landwehr, J. M.: Hydroclimatic data network (HCDN): A US Geological Survey streamflow data set for the United States for the study of climate variation, 1874–1988, US Geol. Surv. Open File Rep., US Geological Survey, 92–129, 1993.

Stagge, J. H. and Moglen, G. E.: Evolutionary Algorithm Optimization of a Multi-Reservoir System with Long Lag Times, J. Hydrol. Eng., 95, 1011–1019, doi:10.1175/BAMS-D-13-00126.1, 2014.

Stefan, H. G. and Preud'homme, E. B.: Stream Temperature Estimation from Air Temperature, J. Am. Water Resour. Assoc., 29, 27–45, 1993.

Tang, H. and Keen, T. R.: Analytical solutions for open-channel temperature response to unsteady thermal discharge and boundary heating, J. Hydraul. Eng., 135, 327–332, 2009.

Taylor, K. E., Stouffer, R. J., and Meehl, G. A.: An Overview of CMIP5 and the Experiment Design, B. Am. Meteorol. Soc., 93, 485–498, 2011.

Tisseuil, C., Leprieur, F., Grenouillet, G., Vrac, M., and Lek, S.: Projected impacts of climate change on spatio-temporal patterns of freshwater fish beta diversity: a deconstructing approach, Global Ecol. Biogeogr., 21, 1213–1222, 2012.

van Vliet, M. T. H., Ludwig, F., Zwolsman, J. J. G., Weedon, G. P., and Kabat, P.: Global river temperatures and sensitivity to atmospheric warming and changes in river flow, Water Resour. Res., 47, W02544, doi:10.1029/2010WR009198, 2011.

Wang, X. and Melesse, A. M.: Evaluation of the SWAT model's snowmelt hydrology in a northwestern Minnesota watershed, T. ASABE, 48, 1359–1376, 2005.

Watson, B. M. and Putz, G.: Comparison of temperature-index snowmelt models for use within an operational water quality model, J. Environ. Qual., 43, 199–207, 2012.

Webb, B. W. and Nobilis, F.: Water temperature behaviour in the River Danube during the twentieth century, Hydrobiologia, 291, 105–113, 1994.

Webb, B. W., Clack, P. D., and Walling, D. E.: Water–air temperature relationships in a Devon river system and the role of flow, Hydrol. Process., 17, 3069–3084, 2003.

Webb, B. W., Hannah, D. M., Moore, R. D., Brown, L. E., and Nobilis, F.: Recent advances in stream and river temperature research, Hydrol. Process., 22, 902–918, doi:10.1002/hyp.6994, 2008.

Wenger, S. J., Som, N. A., Dauwalter, D. C., Isaak, D. J., Neville, H. M., Luce, C. H., Dunham, J. B., Young, M. K., Fausch, K. D., and Rieman, B. E.: Probabilistic accounting of uncertainty in forecasts of species distributions under climate change, Global Change Biol., 19, 3343–3354, doi:10.1111/gcb.12294, 2013.

Wilby, R. L. and Harris, I.: A framework for assessing uncertainties in climate change impacts: low-flow scenarios for the River Thames, UK, Water Resour. Res., 42, W02419, doi:10.1029/2005WR004065, 2006.

Wood, A. W., Leung, L. R., Sridhar, V., and Lettenmaier, D. P.: Hydrologic implications of dynamical and statistical approaches to downscaling climate model outputs, Climatic Change, 62, 189–216, 2004.

Woodward, G., Perkins, D. M., and Brown, L. E.: Climate change and freshwater ecosystems: impacts across multiple levels of organization, Philos. T. Biol. Sci., 365, 2093–2106, 2010.

Zang, C. F., Liu, J., van der Velde, M., and Kraxner, F.: Assessment of spatial and temporal patterns of green and blue water flows under natural conditions in inland river basins in Northwest China, Hydrol. Earth Syst. Sci., 16, 2859–2870, doi:10.5194/hess-16-2859-2012, 2012.

2

Combined impacts of ENSO and MJO on the 2015 growing season drought on the Canadian Prairies

Zhenhua Li[1,2], Yanping Li[1], Barrie Bonsal[3], Alan H. Manson[2], and Lucia Scaff[1]

[1]Global Institute for Water Security, University of Saskatchewan, Saskatoon, Saskatchewan, Canada
[2]Institute of Space and Atmospheric Studies, University of Saskatchewan, Saskatoon, Saskatchewan, Canada
[3]National Hydrology Research Centre, Environment and Climate Change Canada, Saskatoon, Saskatchewan, Canada

Correspondence: Yanping Li (yanping.li@usask.ca) and Zhenhua Li (zhenhua.li@usask.ca)

Abstract. Warm-season precipitation on the Canadian Prairies plays a crucial role in agricultural production. This research investigates how the early summer 2015 drought across the Canadian Prairies is related to the tropical Pacific forcing. The significant deficit of precipitation in May and June 2015 coincided with a warm phase of the El Niño–Southern Oscillation (ENSO) and a negative phase of Madden–Julian Oscillation (MJO)-4 index, which favour a positive geopotential height (GPH) anomaly in western Canada. Our further investigation during the instrumental record (1979–2016) shows that warm-season precipitation in the Canadian Prairies and the corresponding atmospheric circulation anomalies over western Canada teleconnected with the lower boundary conditions in the tropical western Pacific. Our results indicate that MJO can play a crucial role in determining the summer precipitation anomaly in the western Canadian Prairies when the equatorial central Pacific is warmer than normal (NINO4 > 0) and MJO is more active. This teleconnection is due to the propagation of a stationary Rossby wave that is generated in the MJO-4 index region. When the tropical convection around MJO-4 index region (western tropical Pacific, centred over 140° E) is more active than normal (NINO4 > 0), Rossby wave trains originate from the western Pacific with wavenumbers determined by the background mean wind and meridional absolute vorticity gradient. Under warm NINO4 conditions waves are generated with smaller wavenumbers compared to cold NINO4 conditions. These waves under warm NINO4 can propagate into the mid-latitudes over North America, causing a persistent anomalous ridge in the upper level over western Canada, which favours dry conditions over the region.

1 Introduction

The Canadian Prairies depend on summer precipitation especially during the early to mid-growing season (May through August) when the majority of annual precipitation normally occurs (e.g., Bonsal et al., 1993). High natural variability in growing season precipitation causes periodic occurrences of extreme precipitation (Li et al., 2017; Liu et al., 2016) and droughts that are often associated with reduced agriculture yields, low streamflow, and increased occurrence of forest fires (Wheaton et al., 2005; Bonsal and Regier, 2007). Drought events with great environmental and economic impacts on the Canadian Prairies have occurred in 1961, 1988, 2001–2002, and as recent as 2015 (Dey, 1982; Liu et al., 2004; Bonsal et al., 1999; Wheaton et al., 2005; Shabbar et al., 2011; Bonsal et al., 2013; Szeto et al., 2016). The subseasonal forecast of precipitation for the growing season is crucial for the agriculture, water resource management, and the economy of the region. Therefore, an investigation into the causes of inter-annual variability in the growing season precipitation of the Canadian Prairies is needed.

Low precipitation and extended dry periods on the Canadian Prairies are often associated with an upper-level ridge and a persistent high pressure centred over the region (Dey, 1982; Liu et al., 2004). These prolonged atmospheric anomalies often concurred with abnormal boundary layer conditions such as a large-scale sea surface temperature (SST) anomalies in the Pacific Ocean (Shabbar and Skinner, 2004). Large-scale oscillation in the SST anomalies in the Pacific Ocean, namely El Niño and the Pacific Decadal Oscillation (PDO), can affect the hydroclimatic pattern in summer over North America, although the strongest impacts of these boundary conditions occur during the boreal winter. Inter-annual variability such as El Niño–Southern Oscillation (ENSO) is linked with extended droughts in the Prairies (Bonsal et al., 1999; Shabbar and Skinner, 2004). Interdecadal oscillations such as the PDO and the Atlantic Multidecadal Oscillation (AMO) also affect the seasonal temperature and precipitation in the Canadian Prairies (Shabbar et al., 2011).

ENSO's relationship with the Canadian Prairies' precipitation has been studied extensively. Previous investigations (e.g., Shabbar et al., 2011) have found that El Niño events are associated with a summer moisture deficit in western Canada while La Niña events cause an abundance of moisture in far western Canada (British Columbia and Yukon). However, they also noted that, although tropical SST variability accounted for some aspects of the large-scale circulation anomalies that influence the Canadian Prairies meteorological drought, a consistent and clear-cut relationship was not found. The warm phase of ENSO often favours drought in this region, especially during the growing season after the mature phase of El Niño (Bonsal and Lawford, 1999; Shabbar and Skinner, 2004). The positive North Pacific Mode (NPM, Hartmann et al., 2015) like the North Pacific SST anomaly pattern often follows a matured El Niño, and the accompanying atmospheric ridging leads to extended dry spells over the Prairies during the growing season (Bonsal and Lawford, 1999). Furthermore, in association with the recent North Pacific SST anomaly from 2013 to 2014, researchers have attributed the precipitation deficit in California during 2013 to the anomalous upper-level ridge over western North America (Wang et al., 2014; Szeto et al., 2016).

The aforementioned SST variations mostly vary on inter-annual and decadal scales. Another important factor that affects the weather patterns in North America is the Madden–Julian Oscillation (MJO), an intra-seasonal (40–90 days) oscillation in convection and precipitation pattern over the tropics (Madden and Julian, 1971; Zhang, 2005; Riddle et al., 2013; Carbone and Li, 2015). MJO is a coupled atmosphere–ocean oscillation involving convection and large-scale equatorial waves, which produces an eastward propagation of the tropical convection anomaly (Madden and Julian, 1971). The MJO affects the winter temperature and precipitation in North America and Europe through its impact on moisture transport associated with the "Pineapple Express" and its effects on the North Atlantic Oscillation and stratospheric polar vortex (Cassou, 2008; Garfinkel et al., 2012; Rodney et al., 2013). MJO is also connected to the summer precipitation anomalies in the southwestern United States (Lorenz and Hartmann, 2006). During the warm season, MJO's impact on the Canadian Prairies' precipitation has not been thoroughly investigated as MJO's amplitude is weak during spring and early summer. The amplitude of MJO in spring and early summer is related to the inter-annual variation of the tropical SST, especially the SST in the central Pacific (Hendon et al., 2007; Marshall et al., 2016). MJO in terms of the Real-time Multivariate MJO index (RMM, Wheeler and Hendon, 2004) was extremely strong in the early spring of 2015 with a positive PDO-like SST anomaly in the central Pacific and, at the same time, El Niño started to strengthen.

MJO activities in the western Pacific under the modulation of inter-annual SST variability have the potential to act together with ENSO and impact mid-tropospheric circulation over western Canada and, thus, warm-season precipitation over the Canadian Prairies. The goal of this study is to demonstrate that MJO has contributed to the 2015 growing season drought in the Canadian Prairies through the propagation of stationary Rossby waves. Subsequently, further investigations are carried out to determine if similar relationships exist in association with other summer extreme precipitation events during the instrumental record (1979–2016). Section 2 provides the datasets and methodology used in this paper while Sect. 3 presents the analysis of the upper-level circulation anomaly and SST pattern associated with the 2015 drought. This is followed by the examination of the effects of central Pacific SST anomalies and MJO on the summer precipitation in the Canadian Prairies. The mechanism by which MJO affects summer precipitation when equatorial central Pacific SST is warmer than normal is discussed in Sect. 4, followed by the summary and concluding remarks in Sect. 5.

2 Data and methodology

Multiple observation and reanalysis datasets are used to investigate the circulation anomalies associated with the Canadian Prairies' growing season (May–August) precipitation. The observed precipitation is taken from the Climate Prediction Center (CPC) Merged Analysis of Precipitation (CMAP) dataset (Xie and Arkin, 1997; NOAA CPC, 2017a). Geopotential height (GPH) fields from the National Centers for Environmental Predictions (NCEP) Reanalysis (Kalnay et al., 1996; NOAA NCEP, 2017) and the European Centre for Medium-Range Weather Forecast's (ECMWF) ERA-Interim reanalysis (Dee et al., 2011; ECMWF, 2017) are used to analyze the mid- and upper-level (500 and 200 hPa) atmospheric circulation patterns.

To represent the central Pacific SST anomaly, the NINO4 SST index (Rayner et al., 2003; NOAA, 2017b) from the CPC of the National Oceanic and Atmospheric Administration (NOAA) is used since the NINO4 region is near the central Pacific and spans over the dateline (5° S–5° N, 160° E–150° W). Multivariate ENSO Index (NOAA, 2017c) data are retrieved from NOAA's Climate Data Center (CDC) website and are used to determine the ENSO phase (Wolter, 1987; Wolter and Timlin, 1993). In particular, the El Niño condition is defined when the monthly mean index of MEI is larger than 0.5 (Andrews et al., 2004).

The Real-time Multivariate MJO series (RMM1 and RMM2; Bureau of Meteorology of Australia, 2017) developed by Wheeler and Hendon (2004) are used to identify periods of strong MJO activity as the MJO amplitudes are directly calculated by the square root of RMM1 + RMM2. For MJO intensities over the investigated regions, we used the monthly averaged pentad MJO indices from the NOAA CPC's MJO index (Xue et al., 2002; NOAA CPC, 2017b), which have 10 indices representing locations around the globe. The CPC's MJO index is based on extended empirical orthogonal function (EEOF) analysis on pentad velocity potential at 200 hPa. A total of 10 MJO indices on a daily scale are constructed by projecting the daily (00:00 UTC) velocity potential anomalies at 200 hPa (CHI200) onto the 10 time-lagged patterns of the first EEOF of pentad CHI200 anomalies (Xue et al., 2002). Negative values of 10 MJO indices correspond to enhanced convection in the 10 regions centred on 20° E, 70° E, 80° E, 100° E, 120° E, 140° E, 160° E, 120° W, 40° W, and 10° W in the tropics. MJO indices usually vary between −2 and 2, with negative values indicating above-average convective activities in the corresponding region. Because boreal summer usually corresponds to a period of a weaker amplitude of MJO than the winter, we chose the monthly mean value of −0.3 as the criterion of strong convection which is connected to MJO as the indices generally vary between −1 and 1. An MJO-4 index (centred on 140° E) of less than −0.3 was considered a relatively strong convection in the western Pacific, which has been found to be a source region of stationary Rossby waves (Simmons and Hoskins, 1980). SST observations include the Extended Reconstructed Sea Surface Temperature (ERSST) v4 (NOAA, 2017a; Huang et al., 2015). Outgoing Longwave Radiation (NOAA, 2017d) data from NOAA Interpolated Outgoing Longwave Radiation are used to derive the composite of anomalies of OLR for a certain phase of MJO.

Our study focuses on the growing season precipitation in the provinces of Alberta and Saskatchewan in the Canadian Prairies, where the largest deficits were observed in 2015. Specifically, the regional mean precipitation over 115–102.5° W, 50–57.5° N is used (boxed area in Fig. 1, top panel) to represent the Canadian Prairies east of the Rocky Mountains and south of the boreal forest. The chosen region also covers most of the arable land in the Canadian Prairies. Considering the unique MJO-4 and NINO4 indices for 2015, the relationship between the Prairies' warm-season (May–August) precipitation with MJO-4 and ENSO during the instrumental records is investigated using correlation and regression. Though the dry months of the 2015 growing season are May and June when MJO-4 was in negative phase, we want to study the statistical relationship between MJO-4 and the Prairies' precipitation in the whole period of the growing season (May–August). The possible mechanism behind the correlation between MJO-4 and the Prairies' warm-season precipitation under El Niño condition is further investigated by analyzing the upper-level circulation associated with convection in the tropical Pacific and stationary Rossby waves in mid-latitudes.

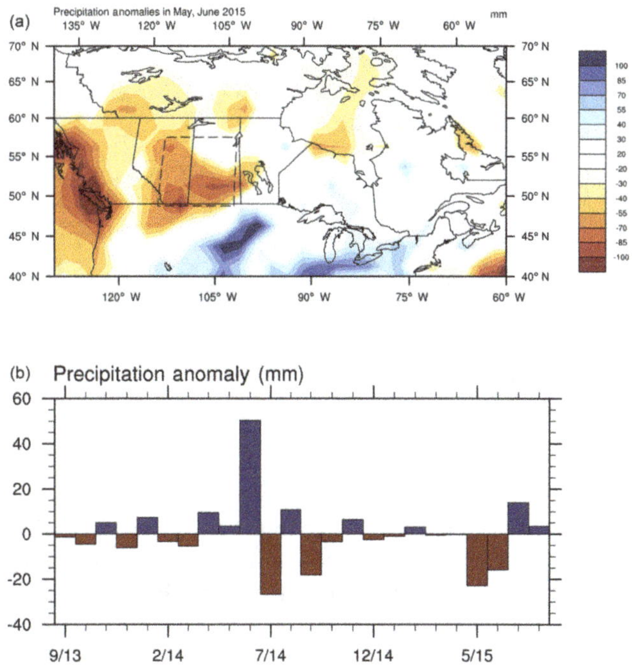

Figure 1. (a) Precipitation anomalies (mm) from CMAP over the region (115–102.5° W, 50–57.5° N) during May and June 2015. **(b)** Time series of monthly precipitation anomaly over the boxed region (shown in **a**) between September 2013 and August 2015.

3 Results

3.1 The 2015 summer drought

Almost all of western Canada including British Columbia, the southern Northwest Territories, Alberta, and Saskatchewan had negative precipitation anomalies during May and June 2015. The top plot in Fig. 1 shows the precipitation anomaly in percentage relative to the climatology (1981–2010 long-term mean) in Canada during May and June 2015. The bottom plot in Fig. 1 presents the monthly precipitation anomaly averaged over the region encompassed by the dashed lines (top panel in Fig. 1). The average annual cycle of the regional precipitation has a dry period between February and May and June has the largest precipitation in all months. The May and June 2015 precipitation deficit was also accompanied by a relatively dry period from February to April (Fig. 1 and Szeto et al., 2016), which added to the drought conditions.

The 500 hPa geopotential height anomaly averaged in May and June is examined together with the SST anomaly and ENSO, as well as MJO-4 indices for 2014 and 2015. The 500 hPa GPH anomaly for May and June 2015 shows strong positive anomalies near Alaska and the British Columbia coast (Fig. 2), which is consistent with the findings for other episodes of growing season droughts (e.g., Dey, 1982; Bonsal and Wheaton, 2005). Accompanying this anomalous ridge are above-normal SSTs in the northeast Pacific off

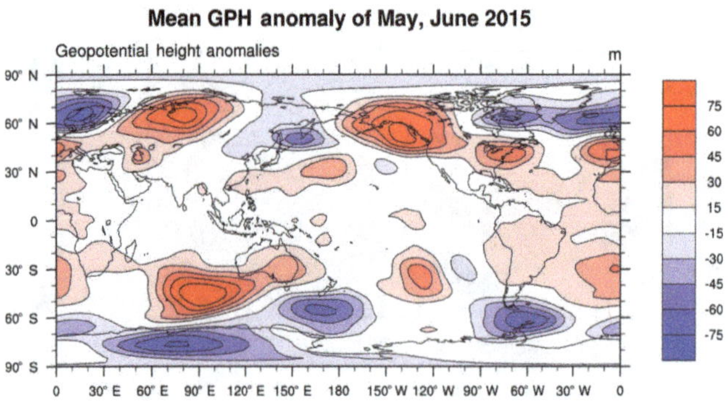

Figure 2. NCEP GPH anomaly at 500 hPa during May and June 2015 when the precipitation deficit was the largest.

the coast of North America and the central-eastern Pacific (Fig. 3). Both ENSO and the NPM are in positive phases that correspond to a warmer SST near the Pacific coast of North America, consistent with the positive GPH anomalies in western Canada and Alaska. The ridge in Alaska/Bering Strait and the one near the British Columbia coast have been previously associated with El Niño and the North Pacific SST anomaly such as NPM (Shabbar et al., 2011). The monthly mean anomalous ridge prevents storms from reaching the British Columbia coast and the Canadian Prairies, causing extended dry spells. Therefore, the GPH anomaly in the early growing season in 2015 is consistent with the precipitation anomaly in these regions. The anomalous upper-level ridge in the western United States and Canada in 2014 and 2015 has also been associated with the developing El Niño and the other main components of Pacific SST variation such as NPM by several recent studies (Hartmann et al., 2015; Lee et al., 2015; Li et al., 2017).

The SST anomaly and the associated oscillations and modes, especially ENSO, show consistent agreement with the observed GPH anomaly pattern. The average SST anomaly during the growing season (May–June, July–August) of 2015 shows a persistent strong positive anomaly in the northeast and eastern equatorial Pacific (Fig. 3), which corresponds to the warm phase of NPM and ENSO. SSTs in the eastern tropical Pacific warmed increasingly since the end of 2014 and qualified as an El Niño in early 2015. The NPM became positive in fall 2013, turned exceptionally strong in 2014, and persisted to 2015 (Hartmann, 2015). The anomalous ridge is concurrent with strong SST anomalies in the tropical Pacific and extratropical North Pacific. NPM, as the third EOF of Pacific SST (30° S–65° N), has also a strong connection to the anomalous ridge in western North America and trough in the eastern US and Canada in the 2013–2014 winter (Hartmann, 2015; Lee et al., 2015). During the ENSO-neutral condition in 2013 and 2014, the precursor of ENSO, the so-called "footprinting" mechanism is considered to cause this anomalous ridge in western North America (Wang et al., 2014).

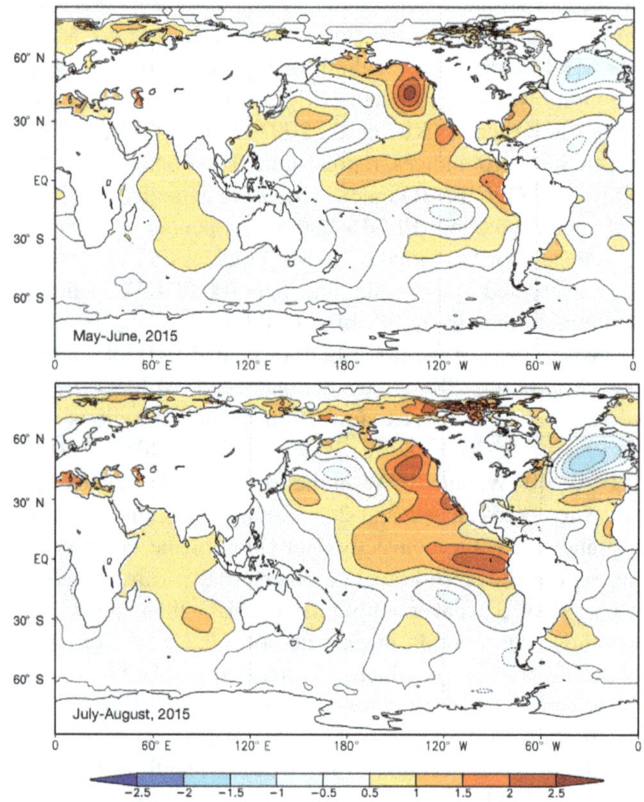

Figure 3. The mean SST anomaly (°C) from ERSST v4 for May–June and July–August 2015.

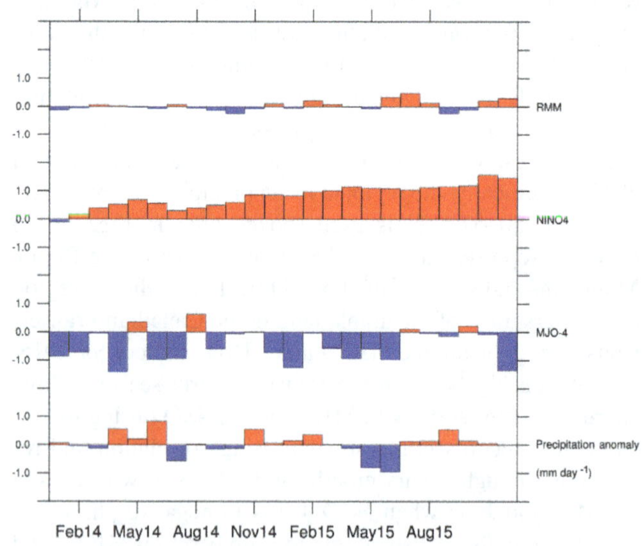

Figure 4. RMM amplitude anomaly, NINO4, MJO-4 indices, and precipitation anomaly of the Canadian Prairies from January 2014 to December 2015.

The variation of the Canadian Prairies' precipitation and its relationship with NINO4 and MJOs are shown in Fig. 4. The time series of the monthly RMM amplitude, NINO4 in-

Table 1. Correlation between mean precipitation anomaly in the Prairies from CMAP and MEI, as well as MJO-4 indices. MJO indices and CMAP are from 1979 to 2016.

	Correlation	*p* value	No. of sample
MEI	−0.096	0.24	156
MJO-4	0.18	0.023	156
MJO-4(NINO4 > 0)	0.33	0.0015	90
MJO-4(NINO4 < 0)	−0.01	0.94	66

dex, MJO-4 indices, and the Canadian Prairies' precipitation anomaly from January 2014 to December 2015 shows the atmospheric–oceanic circulation indices for the drought in 2015. In May and June 2015, the western Pacific witnessed a strong MJO-4 negative index, whereas in July the MJO-4 index became positive. This corresponds well with the precipitation anomaly in Fig. 1. As shown in Fig. 3, El Niño continued to strengthen in July and August 2015, while at the same time the MJO-4 index increased. The increase in the MJO-4 index indicated that the active convection associated with MJO moved away from the tropical western Pacific region and propagated eastward into the central Pacific. Coincident with this change in MJO, the precipitation in the Canadian Prairies then returned to slightly above normal in July.

The good correspondence of MJO-4 and the negative precipitation anomaly suggests a link between MJO and Prairies precipitation during the growing season. Although El Niño and the associated northeast Pacific SST warm anomaly (i.e., NPM) in summer 2015 can be a contributing factor for the persistent upper-level ridge over the west coast of Canada (Shabbar et al., 2011), it cannot fully explain the drought condition in western Canada, as these SSTs do not guarantee a prolonged dry spell as shown by correlation analysis (Table 1). The negative MJO-4 index concurred with the negative anomaly of the Prairies' growing season precipitation in 2015, which prompts the investigation of their relationship with the instrumental records.

3.2 Instrumental record

El Niño and its associated North Pacific SST anomaly may contribute to extended dry spells in Canadian Prairies after the mature phase of El Niño (Bonsal et al., 1993) on an inter-annual timescale. ENSO, however, is not a strong intra-seasonal-to-seasonal predictor of Canadian Prairies summer precipitation. The lack of a strong correlation between the Prairies' precipitation and ENSO index can be caused by the fact that many factors can affect the Prairies' precipitation on a seasonal and sub-seasonal scale. Shabbar and Skinner (2004) showed the connection between the warm phase of ENSO and western Canadian drought through singular value decomposition analysis. However, they also found

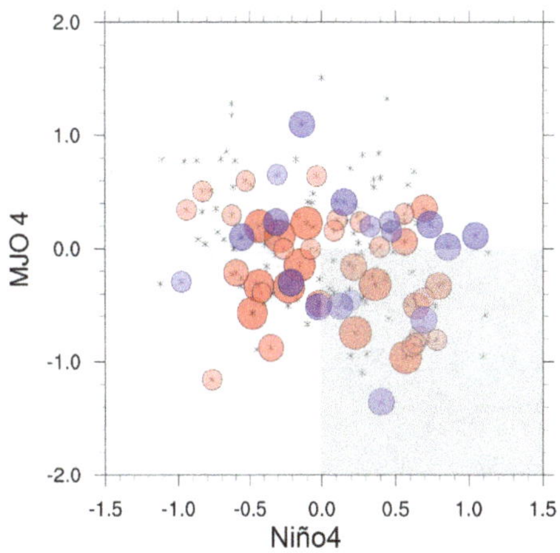

Figure 5. The scatter plot of the monthly precipitation anomaly (mm month^{-1}) as a function of MJO-4 and NINO4. Each asterisk represents a month from May to August 1979–2016. Circled asterisks denote a month with precipitation anomaly larger than 18 mm month^{-1}. The blue circles are months with positive precipitation anomaly and the red circles are months with negative precipitation anomaly. The sizes of circles denote the magnitudes of the anomalies (large > 30 mm month^{-1}, medium > 24 mm month^{-1}, small > 18 mm month^{-1}). The shaded area denotes NINO4 > 0 and MJO-4 index < 0.

other modes of SST variation (e.g., the positive phase of PDO) can produce a wet condition in the Prairies. Here we present a new result showing that, under warm central Pacific SST conditions (NINO4 > 0), a certain phase of MJO, which connected to the active convection in the tropical western Pacific (Li and Carbone, 2012), plays an important role in modulating the growing season precipitation in the Canadian Prairies.

The correlation coefficients between the mean regional precipitation anomaly over Canadian Prairies and MJO-4 indices and MEI from May to August are shown in Table 1. The correlation between MEI alone and the precipitation anomalies is not significant. The correlation between MJO-4 and precipitation in the Prairies is 0.18 with a p value of 0.023, which indicates that stronger tropical convection in the equatorial region centred around 140° E favours less precipitation in the Canadian Prairies from May to August. When NINO4 is larger than 0, the correlation between MJO-4 and the growing season precipitation is 0.33 with a p value of 0.0015. Conversely, the correlation between MJO-4 and Canadian Prairies precipitation is −0.01 when NINO4 < 0.

The scatter plot in Fig. 5 shows the distribution of monthly precipitation anomaly versus the MJO-4 index and NINO4 index. Circled asterisks denote a month with a precipitation anomaly larger than 18 mm month^{-1} and the red (blue) circles denote a negative (positive) precipitation anomaly. The criterion for precipitation anomaly to be emphasized by the circles is roughly one-third of the mean monthly precipitation in the growing season. The size of the circle repre-

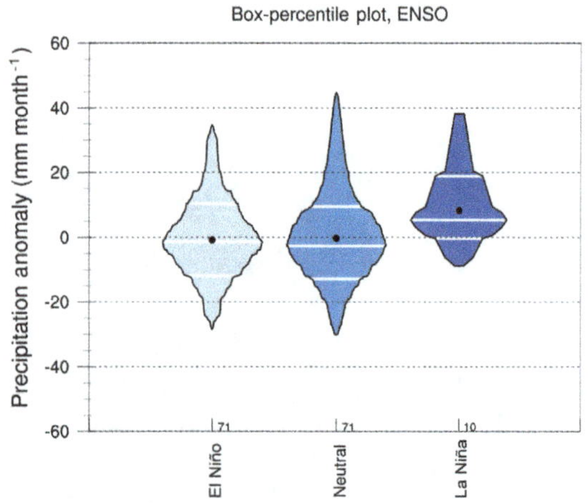

Figure 6. The box-percentile plot of the Canadian Prairies precipitation anomaly during growing season under different ENSO conditions.

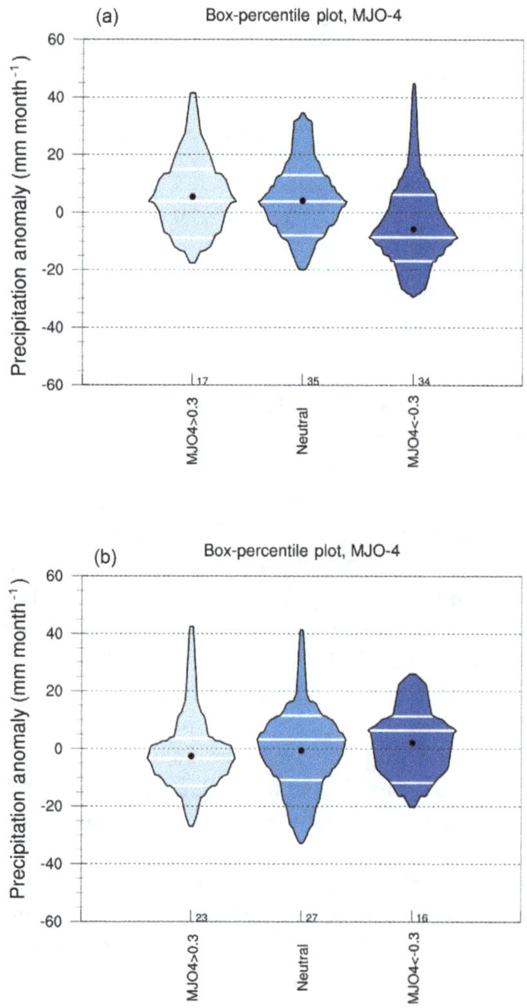

Figure 7. Box-percentile plots of the Canadian Prairies' precipitation anomaly during growing season versus MJO-4 under warm NINO4 (NINO4 > 0, **a**) and cold NINO4 (NINO4 < 0, **b**) SST condition.

sents the magnitude of the monthly precipitation anomalies with 6 mm month^{-1} interval. The bottom-right quadrant, indicated by shading, shows that negative MJO-4 corresponds to many more dry months than wet months under NINO4 > 0 conditions. We noticed that some significant dry months are not in the shaded area, which corresponds to the dry months occurring during La Niña or in the period after the mature phase of El Niño (Bonsal et al., 1999). Summer drought in the Prairies can occur in both phases of ENSO or any other teleconnection indices. For example, for the summer drought that happened in the Prairies from 1999 to 2005, the large-scale anomalous patterns of SST first showed La Niña conditions and then became a weak El Niño in the latter half of the period (Hanesiak et al., 2011). Bonsal and Wheaton (2005) showed that the tropospheric atmospheric circulation patterns in 2001 and 2002 lacked the typical meridional flow in the North Pacific and North America during the drought in western Canada. Their results show that the drought in 1999–2005 was related to the expansion of the continuous drought that happened in the US to the north.

The impact of ENSO on the growing season precipitation over the Canadian Prairies is investigated through Fig. 6. The box-percentile plot in Fig. 6 shows the distribution of monthly Canadian Prairies' precipitation anomalies from May to August along with different ENSO conditions. In general, under El Niño and neutral ENSO conditions, the precipitation anomalies are centred around 0, and there is no bias toward either end. Under La Niña condition, the mean precipitation has a positive bias. There are only 10 summer months under La Niña condition, whereas there are 71 months under El Niño and neutral conditions.

The distributions of precipitation anomalies versus the MJO-4 index under different ENSO conditions are shown in Fig. 7. For NINO4 > 0, the precipitation anomaly has a negative tendency when MJO-4 < −0.3. With NINO4 < 0, there is no negative tendency for MJO-4 < −0.3. Therefore, Figs. 6 and 7 agree with the significant correlation between precipitation and MJO-4 under NINO4 > 0, relative to ENSO in Table 1.

The correlation between MJO-4 and the Prairies' precipitation during the growing season leads us to investigate the underlying circulation anomalies. Figure 8 presents the regressed stream function and wind field at 200 hPa in the mid-latitudes (north of 30° N) on the negative MJO-4 index from May to August under warm NINO4 SST condition (NINO4 > 0.5). In the tropics (10° S–20° N), during the Northern Hemisphere summer, the OLR, velocity potential, and divergent wind vector are presented. Only regression patterns having p values lower than 0.05 are plotted for OLR and velocity potential. The negative MJO-4 index corresponds to a negative anomaly in OLR, stronger convection, and larger-than-average divergence at 200 hPa in the region centre around 150° E. The strong convection anomaly centres around 150° E, 5° N with divergent wind extending well into the subtropics in the Northern Hemisphere. The positive GPH or stream function anomaly extended from Japan to the central Pacific is associated with the enhanced convection and divergence in the upper troposphere over the western tropical–subtropical Pacific. A Rossby wave train linked to the OLR anomaly and strong divergence in the western Pacific propagate eastward into North America in the extratropics. To better demonstrate the propagation of the wave train, we conducted a ray-tracing experiment of stationary Rossby wave following the non-divergent barotropic Rossby wave theory of Hoskins and Karoly (1981) and Hoskins and Ambrizzi (1993). Equation (1) describes the group velocity, which represents the propagation of wave activity. C_{gx} and C_{gy} are the group velocity components on zonal and meridional directions; $\overline{U}$ and $\overline{V}$ are the mean zonal and meridional winds; q is the mean absolute vorticity; K, k, and l are the total wave number, zonal wavenumber, and meridional wavenumber, respectively. The ray path is integrated using a fourth-order Runge–Kutta method.

$$C_{gx} = \overline{U} + \frac{(k^2 - l^2)q_y - 2klq_x}{K^4}$$
$$C_{gy} = \overline{V} + \frac{(k^2 - l^2)q_x + 2klq_y}{K^4} \tag{1}$$

Under the average conditions in May–August derived from ERA-Interim at 200 hPa with NINO4 > 0.5 or NINO4 < −0.5, we released rays with a total wavenumber matching with the mean flow at the extratropical location of the OLR anomaly (140–150° E, 25–30° N). For quasi-stationary waves, the wavenumber is determined by the basic zonal flow and background absolute vorticity gradient through the Rossby wave dispersion relation. For the NINO4 > 0.5 May–August condition, K = 4.14. With this total wavenumber and launching angle from 0 to 60° relative to the zonal direction, Rossby wave rays (coloured by red, orange to blue according to their angle from 0 to 60°) released at 140° W, 20° N can propagate successfully to western Canada for those with smaller launching angles (< 30 °) as shown in the bottom plot in Fig. 9. With NINO4 < −0.5, the zonal wind in the source region is weaker, and the meridional gradient of absolute vorticity is stronger due to its southern position relative to the subtropical jet. The total wavenumber for stationary Rossby waves is 6.2, determined by the mean May–August condition for NINO4 < −0.5. The waves with shorter wavelength tend to be evanescent near the source region as shown in the top plot in Fig. 9. However, there is no significant difference in ray path under the NINO4 < −0.5 condition compared to NINO4 > 0.5 if the source wavenumbers are set to the same value (results not shown). The changes in the mean conditions in the midlatitudes away from the source region from El Niño to La Niña are not sufficient to alter the propagation condition for quasi-stationary Rossby waves.

4 Discussion

The summer of 2015 is the first summer after the development of El Niño during the 2014–2015 winter. Though the upper-level GPH pattern, seen in summer 2015, can be attributed to the SST modes in the Pacific, namely ENSO and NPM, the precipitation in the western Canadian Prairies is not strongly correlated with either. Bonsal and Lawford (1999) found that more extended dry spells tend to occur in the Canadian Prairies during the second summer following the mature stage of the El Niño events. The winter precipitation in Canada has a strong connection to ENSO (Shabbar et al., 1997), whereas summer precipitation, in most regions of western Canada (except the coast of British Columbia and Southern Alberta), does not have a significant correlation with ENSO. This is consistent with our investigation using instrumental records from 1948 to 2016.

Growing season precipitation in the Canadian Prairies is affected by many factors. Precipitation deficits can occur under various circulation and lower boundary conditions. Thus, it is not expected that a universal condition for all the significant droughts in the region can be identified. In fact, extreme drought events have been found in both El Niño and La Niña years. A previous study by Bonsal and Lawford (1999) indicates the meteorological drought often occurs after the mature phase of El Niño, which is not the case for 2015. The associated anomaly in the North Pacific represented by the NPM positive phase is consistent with their results. The direct linkage between ENSO and the summer precipitation in the Canadian Prairies is not clear. In fact, the correlation between MEI and the precipitation in the investigated region is −0.096 ($p = 0.239$, sample size = 152). The investigated region's growing season precipitation does not possess a significant correlation with ENSO, which is consistent with other researchers' findings (Dai and Wigley, 2000).

The regression pattern is consistent with stationary Rossby wave theory as shown in a hierarchy of theoretical and modelling studies (Karoly et al., 1989; Simmons et al., 1983; Hoskins and Ambrizzi, 1993; Ambrizzi and Hoskins, 1997; Held et al., 2002). A similar wave train extends from the western Pacific toward extratropical South America but at lower latitudes compared to its counterpart in the Northern Hemisphere (not shown). The node of the wave train in western Canada and the Pacific Northwest of the US corresponds to an anomalous ridge, which is in phase of El Niño forcing. When the convection in the region associated with MJO-4 is weaker than normal (MJO-4 > 0), a wave train with the opposite sign will reach western Canada, which then counteracts the El Niño forcing. Thus, the weak correlation between Canadian Prairies precipitation and ENSO is understandable as MJO plays an additional role that enhances or cancels out the GPH anomaly caused by El Niño.

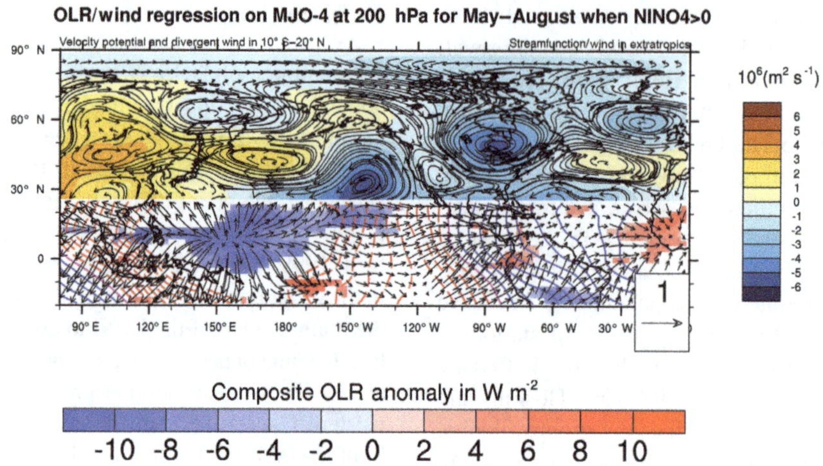

Figure 8. The regression of stream function and wind field in the extratropics on negative MJO-4 for May–August with NINO4 > 0.5 condition. In the tropics, the regression of OLR, velocity potential, and divergent wind on negative MJO-4 indices for May–August with NINO4 > 0.5 condition. The shaded region for the tropical OLR has p value < 0.05. Blue shading indicates active convection region. Red dashed contour and solid blue contour corresponds to negative and positive velocity potential, respectively.

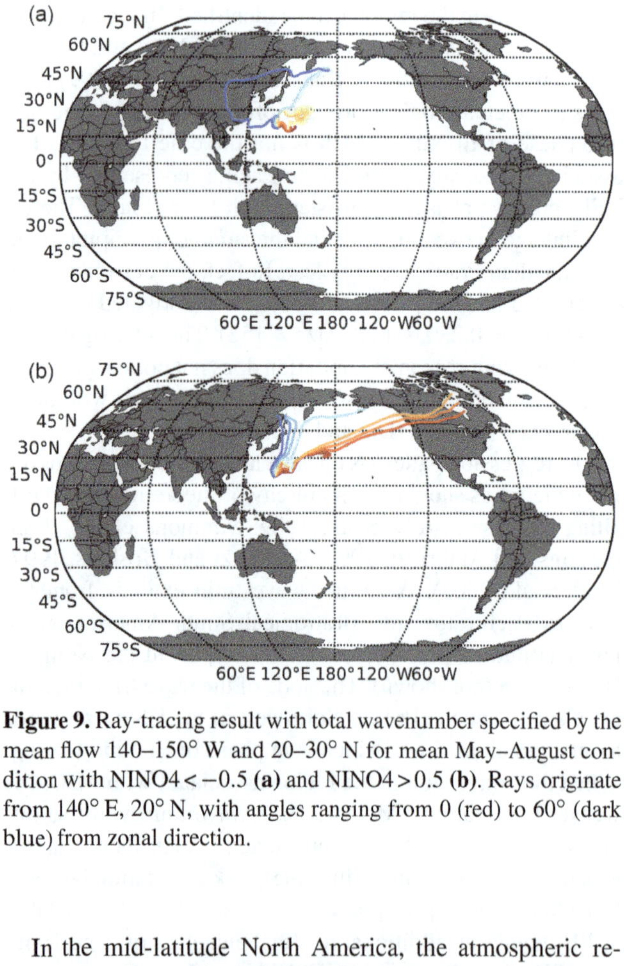

Figure 9. Ray-tracing result with total wavenumber specified by the mean flow 140–150° W and 20–30° N for mean May–August condition with NINO4 < −0.5 **(a)** and NINO4 > 0.5 **(b)**. Rays originate from 140° E, 20° N, with angles ranging from 0 (red) to 60° (dark blue) from zonal direction.

In the mid-latitude North America, the atmospheric response to the tropical forcing in the western Pacific depends on the mean circulation condition associated with tropical SST. Intraseasonal tropical convection oscillation in the western Pacific associated with the MJO-4 index cannot determine the sign of the precipitation anomaly in the Prairies alone. Both warm SST in the central Pacific and strong tropical convection in the western Pacific and Maritime Continent are essential to cause a significant precipitation deficit in the western Canadian Prairies. Warm SST in the central Pacific causes an eastward expansion of Pacific warm pool that favours enhanced MJO activity in the western-central Pacific (Hendon et al., 1999; Marshall et al., 2016). As shown by the ray-tracing result, the NINO4 also affects the wavenumber of the quasi-stationary Rossby waves over the source region in the western Pacific. Under warm NINO4, the wavenumbers tend to be smaller due to stronger westerly winds in the source region and these waves can propagate northeastward into western Canada. Conversely, from May to August under cold NINO4, the westerly flow is weaker, and the meridional vorticity gradient is stronger in the subtropics near the source region. These mean flow conditions correspond to waves with lager wavenumbers that cannot propagate across the dateline.

In the year 2015, the SST anomaly in the Pacific (e.g., ENSO, NPM) coincided with the anomalous ridge on the west coast of Canada. This positive GPH anomaly was associated with the strong negative MJO4 indices, it then caused a blocking pattern, and suppressed precipitation in the Canadian Prairies in the early summer through the mechanism discussed above. Although the El Niño continued to strengthen in July and August 2015, the active convection associated with MJO in the western Pacific propagated eastward into the central Pacific. As the convection in the western Pacific/Maritime Continent waned, the precipitation in the Canadian Prairies returned to slightly above normal in July.

5 Conclusions

The cause of the 2015 summer precipitation deficit in the western Canadian Prairies is investigated in relation to atmospheric circulation anomalies, SST, and the intraseasonal tropical convection oscillation, MJO. The drought in western Canada is immediately related to an anomalous upper-level ridge that persisted over the west coast of Canada and Alaska since fall 2014. This ridge was likely associated with a developing El Niño that was enhanced by the MJO.

In general, MJO-4 indices demonstrated significant correlation with the meteorological drought over the Canadian Prairies from May to August when the SST in the central Pacific is warm (NINO4 > 0), which also corresponds to a stronger MJO amplitude in boreal summer. Our study discovered that MJO phase/strength is connected to the anomalous ridge over western Canada through the propagation of stationary Rossby waves from the western Pacific when NINO4 is positive. Though seasonally MJO is weaker in summer, the spring and early summer MJO amplitudes are larger than normal when the central Pacific SST is warmer than normal (NINO4 > 0). The teleconnection between the Canadian Prairies precipitation deficit and MJO is stronger when NINO4 is positive. The underlying cause of this significant correlation between MJO-4 indices and the prairie precipitation in May–August is a stationary Rossby wave train originating from the Maritime Continent and western Pacific which propagates into Canada. The ray-tracing experiments show the main difference between a warm phase of NINO4 and a cold phase is the changes in stationary Rossby wave wavenumber over the source region. Under NINO4 > 0.5 May–August conditions, the total wavenumber is about 4 and can propagate into western Canada if the waves in the experiments are oriented relatively zonally. Compared to NINO4 > 0.5, NINO4 < −0.5 corresponds to a weaker zonal wind and stronger meridional gradient of absolute vorticity in the subtropics of the source region (140–150° E). Hence, the wavenumbers of stationary Rossby waves from the source region under NINO4 < −0.5 are larger (about 6), and these waves fail to reach the western hemisphere. The intra-seasonal predictability of the growing season precipitation in the Canadian Prairies can be potentially improved by including the MJO amplitude and phase factors for medium-range and intra-seasonal projection in addition to the ENSO effect especially when the central-Pacific SST is warm.

Data availability. All the data used in this study are publicly available. For the atmospheric circulation field, both the ERA-Interim dataset (ECMWF, 2017) and the NCEP Reanalysis (NOAA NCEP, 2017) are publicly available through their corresponding institutions. The SST dataset ERSST v4 (NOAA, 2017a) and the SST related NINO4 index (NOAA, 2017b) are publicly available through the NOAA website. The precipitation dataset CMAP (NOAA CPC, 2017a) can be accessed publicly through the NOAA website. The MJO index from the NOAA CPC of the USA (NOAA CPC, 2017b) is available through the NOAA CPC's MJO website. The MJO index (Bureau of Meteorology of Australia, 2017) and archives on MJO are hosted by the Bureau of Meteorology of Australia. MEI (NOAA, 2017c) is publicly available online through the NOAA website. The Interpolated Outgoing Longwave Radiation (OLR) dataset (NOAA, 2017d) is also available online.

Author contributions. ZL and YL conceived of the presented idea. ZL developed the theory and performed the computations and analyses. BB and AM verified the analytical methods. YL supervised the findings of this work. LS improved the texts and figures. All the authors discussed the results and contributed to the final manuscript.

Competing interests. The authors declare that they have no conflict of interest.

Special issue statement. This article is part of the special issue "Understanding and predicting Earth system and hydrological change in cold regions". It is not associated with a conference.

Acknowledgements. We gratefully acknowledge the Natural Sciences and Engineering Research Council of Canada (NSERC) for funding the Changing Cold Regions Network (CCRN) through their Climate Change and Atmospheric Research (CCAR) initiative. Zhenhua Li is supported by the Probing the Atmosphere of the High Arctic project sponsored by the NSERC. Yanping Li gratefully acknowledges the support from the Global Institute of Water Security at the University of Saskatchewan. This research is also supported by Environment and Climate Change Canada (ECCC). We gratefully acknowledge the support from the Changing Cold Region Network (CCRN), Global Water Future (GWF) and Global Institute of Water Security (GIWS) at the University of Saskatchewan.

Edited by: John Hanesiak

References

Ambrizzi, T. and Hoskins, B. J.: Stationary Rossby-Wave Propagation in a Baroclinic Atmosphere, Q. J. Roy. Meteor. Soc., 123, 919–28, 1997.

Andrews, E. D., Antweiler, R. C., Neiman, P. J., and Ralph, F. M.: Influence of ENSO on Flood Frequency along the California Coast, J. Climate, 17, 337–348, https://doi.org/10.1175/1520-0442(2004)017<0337:IOEOFF>2.0.CO;2, 2004.

Bonsal, B. R., Chakravarti, A. K., and Lawford, R. G.: Teleconnections between North Pacific SST Anomalies and Growing Season Extended Dry Spells on the Canadian Prairies, Int. J. Climatol., 13, 865–878, 1993.

Bonsal, B. R., Zhang, X., and Hogg, W. D.: Canadian Prairie growing season precipitation variability and associated atmospheric circulation, Climate Res., 11, 191–208, 1999.

Bonsal, B. and Lawford, R.: Teleconnections between El Niño and La Niña Events and Summer Extended Dry Spells on the Canadian Prairies, Int. J. Climatol., 19, 1445–1458, 1999.

Bonsal, B. R. and Wheaton, E.: Atmospheric circulation comparisons between the 2001 and 2002 and the 1961 and 1988 Canadian Prairie droughts, Atmos. Ocean, 43, 163–172, 2005.

Bonsal, B. R. and Regier, M.: Historical Comparison of the 2001/2002 Drought in the Canadian Prairies, Clim. Res., 33, 229–242, 2007.

Bonsal, B. R., Aider, R., Gachon, P., and Lapp, S.: An Assessment of Canadian Prairie Drought: Past, Present, and Future, Clim. Dynam., 41, 501–516, 2013.

Bureau of Meteorology of Australia: MJO index, available at: http://www.bom.gov.au/climate/mjo/, last access: 10 April 2017.

Carbone, R. E. and Li, Y.: Tropical Oceanic Rainfall and Sea Surface Temperature Structure: Parsing Causation from Correlation in the MJO, J. Atmos. Sci., 72, 2703–2718. 2015.

Cassou, C.: Intraseasonal Interaction Between the Madden-Julian Oscillation and the North Atlantic Oscillation, Nature, 455, 523–527, 2008.

Dai, A. and Wigley, T. M. L.: Global Patterns of ENSO-Induced Precipitation, Geophys. Res. Lett., 27, 1283–1286, 2000.

Dee, D. P., Uppala, S. M., Simmons, A. J., Berrisford, P., Poli, P., Kobayashi, S., Andrae, U., Balmaseda, M. A., Balsamo, G., Bauer, P., Bechtold, P., Beljaars, A. C. M., van de Berg, L., Bidlot, J., Bormann, N., Delsol, C., Dragani, R., Fuentes, M., Geer, A. J., Haimberger, L., Healy, S. B., Hersbach, H., Hólm, E. V., Isaksen, L., Kållberg, P., Köhler, M., Matricardi, M., McNally, A. P., Monge-Sanz, B. M., Morcrette, J.-J., Park, B.-K., Peubey, C., de Rosnay, P., Tavolato, C., Thépaut, J.-N., and Vitart, F.: The ERA-Interim Reanalysis: Configuration and Performance of the Data Assimilation System, Q. J. Roy. Meteor. Soc., 137, 553–597, 2011.

Dey, B.: Nature and Possible Causes of Droughts on the Canadian Prairies-Case Studies, J. Climatol., 2, 233–249, 1982.

ECMWF: ERA-interim Dataset, available at: https://www.ecmwf.int/en/forecasts/datasets/archive-datasets/reanalysis-datasets/era-interim, last access: 15 August 2017.

Garfinkel, C. I., Feldstein, S. B., Waugh, D. W., Yoo, C., and Lee, S.: Observed Connection Between Stratospheric Sudden Warmings and the Madden-Julian Oscillation, Geophys. Res. Lett., 39, L18807, https://doi.org/10.1029/2012GL053144, 2012.

Hanesiak, J. M., Stewart, R. E., Bonsal, B. R., Harder, P., Lawford, R., Aider, R., Amiro, B. D., Atallah, E., Barr, A. G., Black, T. A., Bullock, P., Brimelow, J. C., Brown, R., Carmichael, H., Derksen, C., Flanagan, L. B., Gachon, P., Greene, H., Gyakum, J., Henson, W., Hogg, E. H., Kochtubajda, B., Leighton, H., Lin, C., Luo, Y., McCaughey, J. H., Meinert, A., Shabbar, A., Snelgrove, K., Szeto, K., Trishchenko, A., van der Kamp, G., Wang, S., Wen, L., Wheaton, E., Wielki, C., Yang, Y., Yirdaw, S., and Zha, T.: Characterization and summary of the 1999–2005 canadian prairie drought, Atmos. Ocean, 49, 421–452, https://doi.org/10.1080/07055900.2011.626757, 2011.

Hartmann, D. L.: Pacific Sea Surface Temperature and the Winter of 2014, Geophys. Res. Lett., 42, 1894–902, 2015.

Held, I. M., Ting, M., and Wang, H.: Northern Winter Stationary Waves: Theory and Modeling, J. Climate, 15, 2125–2144, 2002.

Hendon, H. H., Zhang, C., and Glick, J. D.: Interannual variation of the Madden-Julian Oscillation during Austral summer, J. Climate, 12, 2538–2550, 1999.

Hendon, H. H., Wheeler, M. C., and Zhang, C.: Seasonal dependence of the MJO-ENSO relationship, J. Climate, 20, 531–543, https://doi.org/10.1175/JCLI4003.1, 2007

Hoskins, B. J. and Ambrizzi, T.: Rossby Wave Propagation on a Realistic Longitudinally Varying Flow, J. Atmos. Sci., 50, 1661–1671, 1993.

Hoskins, B. J. and Karoly, D. J.: The Steady Linear Response of a Spherical Atmosphere to Thermal and Orographic Forcing, J. Atmos. Sci., 38, 1179–1196, 1981.

Hoskins, B. J. and Ambrizzi, T.: Rossby Wave Propagation on a Realistic Longitudinally Varying Flow, J. Atmos. Sci., 50, 1661–1671, 1993.

Huang, B., Banzon, V. F., Freeman, E., Lawrimore, J., Liu, W., Peterson, T. C., Smith, T. M., Thorne, P. W., Woodruff, S. D., and Zhang, H.-M.: Extended Reconstructed Sea Surface Temperature Version 4 (ERSST. v4). Part I: Upgrades and Intercomparisons, J. Climate, 28, 911–930, 2015.

Kalnay, E., Kanamitsu, M., Kistler, R., Collins, W., Deaven, D., Gandin, L., Iredell, M., Saha, S., White, G., Woollen, J., Zhu, Y., Chelliah, M., Ebisuzaki, W., Higgins, W., Janowiak, J., Mo, K., Ropelewski, C., Wang, J., Leetmaa, A., Reynolds, R., Jenne, R., and Joseph, D.: The NCEP/NCAR 40-Year Reanalysis Project, B. Am. Meteorol. Soc., 77, 437–471, 1996.

Karoly, D. J., Plumb, R. A., and Ting, M.: Examples of the Horizontal Propagation of Quasi-Stationary Waves, J. Atmos. Sci., 46, 2802–2811, 1989.

Lee, M. Y., Hong, C. C., and Hsu, H. H.: Compounding Effects of Warm Sea Surface Temperature and Reduced Sea Ice on the Extreme Circulation Over the Extratropical North Pacific and North America During the 2013–2014 Boreal winter, Geophys. Res. Lett., 42, 1612–1618, 2015.

Li, Y. and Carbone, R. E.: Excitation of rainfall over the tropical western Pacific, J. Atmos. Sci., 69, 2983–2994, 2012.

Li, Y., Szeto, K., Stewart, R., Theriault, J., Chen, L., Kochtubajda, B., Liu, A., Boodoo, S., Goodson, R., Mooney, C., and Kurkute, S.: The June 2013 Alberta Catastrophic Flooding: Water vapor transport analysis by WRF simulation, J. Hydrometeorol., 18, 2057–2078, 2017.

Li, Z., Manson, A., Li, Y., and Meek, C.: Circulation Characteristics of Persistent Cold Spells in Central-Eastern North America, J. Meteorol. Res., 31, 250–260, 2017.

Liu, J., Stewart, R. E., and Szeto, K. K.: Moisture Transport and Other Hydrometeorological Features Associated With the Severe 2000/01 Drought Over the Western and Central Canadian Prairies, J. Climate, 17, 305–319, 2004.

Liu, A., Mooney, C., Szeto, K., Thériault, J. M., Kochtubajda, B., Stewart, R. E., Boodoo, S., Goodson, R., Li, Y., and Pomeroy, J.: The June 2013 Alberta Catastrophic Flooding Event: Part 1 – Large scale features, Hydrol. Process., 2016, 4899–4916, 2016.

Lorenz, D. J. and Hartmann, D. L.: The Effect of the MJO on the North American Monsoon, J. Climate, 19, 333–343, https://doi.org/10.1175/JCLI3684.1, 2006.

Madden, R. A. and Julian, P. R.: Detection of a 40–50 Day Oscillation in the Zonal Wind in the Tropical Pacific, J. Atmos. Sci., 28, 702–708, 1971.

Marshall, A. G., Hendon, H. H., and Wang, G.: On the role of anomalous ocean surface temperatures for promoting the record Madden-Julian Oscillation in March 2015, Geophys. Res. Lett., 43, 472–481, 2016.

NOAA CPC: CMAP, available at: http://www.cpc.ncep.noaa.gov/products/global_precip/html/wpage.cmap.html, last access: 15 August 2017a.

NOAA CPC: MJO CPC index, available at: http://www.cpc.ncep.noaa.gov/products/precip/CWlink/MJO/mjo.shtml, last access: 15 August 2017b.

NOAA NCEP: NCEP Reanalysis, available at: https://www.esrl.noaa.gov/psd/data/gridded/data.ncep.reanalysis.html, last access: 15 August 2017.

NOAA: ERSST-V4, available at: https://www.ncdc.noaa.gov/data-access/marineocean-data/extended-reconstructed-sea-surface-temperature-ersst-v4, last access: 15 August 2017a.

NOAA: NINO4 index, available at: https://www.esrl.noaa.gov/psd/gcos_wgsp/Timeseries/Nino4/, last access: 15 August 2017b.

NOAA: MEI, available at: https://www.esrl.noaa.gov/psd/enso/mei/, last access: 15 August 2017c.

NOAA: Interpolated OLR, available at: https://www.esrl.noaa.gov/psd/data/gridded/data.interp_OLR.html, last access: 15 August 2017d.

Rayner, N. A., Parker, D. E., Horton, E. B., Folland, C. K., Alexander, L. V., Rowell, D. P., Kent, E. C., and Kaplan, A.: Global analyses of sea surface temperature, sea ice, and night marine air temperature since the late nineteenth century. J. Geophys. Res., 108, 4407, https://doi.org/10.1029/2002JD002670, 2003.

Riddle, E. E., Stoner, M. B., Johnson, N. C., L'Heureux, M. L., Collins, D. C., and Feldstein, S. B.: The Impact of the MJO on Clusters of Wintertime Circulation Anomalies Over the North American region, Clim. Dynam., 40, 1749–1766, 2013.

Rodney, M., Lin, H., and Derome, J.: Subseasonal Prediction of Wintertime North American Surface Air Temperature during Strong MJO Events, Mon. Weather Rev., 141, 2897–2909, https://doi.org/10.1175/MWR-D-12-00221.1, 2013.

Shabbar, A., Bonsal, B., and Khandekar, M.: Canadian precipitation patterns associated with the Southern Oscillation, J. Climate, 10, 3016–3027, 1997.

Shabbar, A. and Skinner, W.: Summer Drought Patterns in Canada and the Relationship to Global Sea Surface Temperatures, J. Climate, 17, 2866–2880, 2004.

Shabbar, A., Bonsal, B. R., and Szeto, K.: Atmospheric and Oceanic Variability Associated with Growing Season Droughts and Pluvials on the Canadian Prairies, Atmos. Ocean, 49, 339–355, 2011.

Simmons, A. J. and Hoskins, B. J.: Barotropic influences on the growth and decay of nonlinear baroclinic waves, J. Atmos. Sci., 37, 1679–1684, https://doi.org/10.1175/1520-0469(1980)037<1679:BIOTGA>2.0.CO;2, 1980.

Simmons, A. J., Wallace, J. M., and Branstator, G. W.: Barotropic Wave Propagation and Instability, and Atmospheric Teleconnection Patterns, J. Atmos. Sci., 40, 1363–1392, 1983.

Szeto, K., Zhang, X., White, R. E., and Brimelow, J.: The 2015 Extreme Drought in Western Canada, B. Am. Meteorol. Soc., 97, S42–S46, https://doi.org/10.1175/BAMS-D-16-0147.1, 2016.

Wang, S. Y., Hipps, L., Gillies, R. R., and Yoon, J.-H.: Probable Causes of the Abnormal Ridge Accompanying the 2013–2014, California Drought: ENSO Precursor and Anthropogenic Warming footprint, Geophys. Res. Lett., 41, 3220–3226, 2014.

Wheaton, E., Wittrock, V., Kulshreshtha, S., Koshida, G., Grant, C., Chipanshi, A., and Bonsal, B. R.: Lessons Learned from the Drought Years of 2001 and 2002: Synthesis Report, Agriculture and Agri-Food Canada, Saskatchewan Research Council Publ. No. 11602–46E03, Saskatoon, 2005.

Wheeler, M. C. and Hendon, H. H.: An all-season real-time multivariate MJO index: Development of an index for monitoring and prediction, Mon. Weather Rev., 132, 1917–1932, 2004.

Wolter, K.: The Southern Oscillation in surface circulation and climate over the tropical Atlantic, Eastern Pacific, and Indian Oceans as captured by cluster analysis, J. Clim. Appl. Meteorol., 26, 540–558, 1987.

Wolter, K. and Timlin, M. S.: Monitoring ENSO in COADS with a seasonally adjusted principal component index. Proc. of the 17th Climate Diagnostics Workshop, Norman, OK, NOAA/NMC/CAC, NSSL, Oklahoma Clim. Survey, CIMMS and the School of Meteor., Univ. of Oklahoma, 52–57, 1993.

Xie, P. and Arkin, P. A.: Global Precipitation: A 17-year Monthly Analysis Based on Gauge Observations, Satellite Estimates, and Numerical Model Outputs, B. Am. Meteorol. Soc., 78, 2539–2558, 1997.

Xue, Y., Higgins, W., and Kousky, V.: Influences of the Madden-Julian Oscillations on Temperature and Precipitation in North America during ENSO-neutral and Weak ENSO Winters, Proc. workshop on prospects for improved forecasts of weather and short-term climate variability on subseasonal (2 week to 2 month) time scales, 2002.

Zhang, C.: Madden-Julian Oscillation, Rev. Geophys., 43, RG2003, https://doi.org/10.1029/2004RG000158, 2005.

3

A simple temperature-based method to estimate heterogeneous frozen ground within a distributed watershed model

Michael L. Follum[1,2], **Jeffrey D. Niemann**[2], **Julie T. Parno**[3], **and Charles W. Downer**[1]

[1]Coastal and Hydraulics Laboratory, Vicksburg, MS 39180, USA
[2]Department of Civil Engineering, Colorado State University, Fort Collins, CO 80523, USA
[3]Cold Regions Research and Engineering Laboratory, Hanover, NH 03755, USA

Correspondence: Michael L. Follum (michael.l.follum@usace.army.mil)

Abstract. Frozen ground can be important to flood production and is often heterogeneous within a watershed due to spatial variations in the available energy, insulation by snowpack and ground cover, and the thermal and moisture properties of the soil. The widely used continuous frozen ground index (CFGI) model is a degree-day approach and identifies frozen ground using a simple frost index, which varies mainly with elevation through an elevation–temperature relationship. Similarly, snow depth and its insulating effect are also estimated based on elevation. The objective of this paper is to develop a model for frozen ground that (1) captures the spatial variations of frozen ground within a watershed, (2) allows the frozen ground model to be incorporated into a variety of watershed models, and (3) allows application in data sparse environments. To do this, we modify the existing CFGI method within the gridded surface subsurface hydrologic analysis watershed model. Among the modifications, the snowpack and frost indices are simulated by replacing air temperature (a surrogate for the available energy) with a radiation-derived temperature that aims to better represent spatial variations in available energy. Ground cover is also included as an additional insulator of the soil. Furthermore, the modified Berggren equation, which accounts for soil thermal conductivity and soil moisture, is used to convert the frost index into frost depth. The modified CFGI model is tested by application at six test sites within the Sleepers River experimental watershed in Vermont. Compared to the CFGI model, the modified CFGI model more accurately captures the variations in frozen ground between the sites, inter-annual variations in frozen ground depths at a given site, and the occurrence of frozen ground.

1 Introduction

Frozen ground (also known as frozen soil or soil frost) is important to predicting stormflows produced by certain watersheds (Shanley and Chalmers, 1999; McNamara et al., 1997; Prèvost et al., 1990; Woo, 1986). Several plot-scale studies have shown that frozen ground can impede infiltration and thus enhance runoff (Bayard et al., 2005; Dunne and Black, 1971; Stähli et al., 1999). Several of these studies have also shown that frozen ground is highly variable temporally and spatially (Campbell et al., 2010; Shanley and Chalmers, 1999; Stähli, 2017), which affects the amount and type of runoff (Wilcox et al., 1997). The presence, spatial pattern, and depth of frozen ground are driven by mass (water) and energy balances. The energy available from the atmosphere to thaw the soil is subject to the insulation of the snowpack (Pearson, 1920; Willis et al., 1961) and ground cover, including any vegetation, woody debris, and leaf litter (Brown, 1966; Diebold, 1938; Fahey and Lang, 1975; Sartz, 1973; Stähli, 2017). MacKinney (1929) found that ground cover reduced the depth of frost penetration by 40 % at a test site in Connecticut. Additionally, the presence and depth of frozen ground is affected by soil moisture (Fox, 1992; Willis et al., 1961) and the thermal conductivity of the soil (Farouki, 1981; Johansen, 1977).

Frozen ground has proven difficult to simulate within hydrologic models due to complex interactions of energy and water between the atmosphere, snowpack, and soil (Dun et al., 2010; Kennedy and Sharratt, 1998; Lin and McCool, 2006). Physically based models of frozen ground, such as the simultaneous heat and water (SHAW) model (Flerchinger and Saxton, 1989), the coupled heat and mass transfer model for soil–plant–atmosphere systems (COUP) (Jansson 2001; Jansson and Karlburg, 2010), and the distributed water–heat coupled (DWHC) model (Chen et al., 2007) have large parameter and forcing data requirements – such as wind speed, relative humidity, and short- and long-wave radiation – which restricts their applicability in many watershed. Additionally, these types of models either include, or are tightly coupled to soil moisture models, which can limit their applicability in models that do not explicitly simulate soil moisture content. To reduce data and parameter requirements and increase applicability, simple temperature-index or degree-day methods (Molnau and Bissell, 1983; Rekolainen and Posch, 1993) remain widely used within watershed models, including LISFLOOD (De Roo et al., 2001; Van Der Knijff et al., 2010), CREAMS (Rekolainen and Posch, 1993), and the gridded surface subsurface hydrologic analysis (GSSHA) model (Downer and Ogden, 2004). Degree-day approaches typically accumulate the daily average temperature as a frost index (°C-days). When the frost index exceeds a threshold, the soil is considered frozen and impermeable to infiltration. The sudden restriction on infiltration can be an incorrect assumption, especially in forested environments where frozen soils often still experience infiltration (Lindstrom et al., 2002; Nyberg et al., 2001; Shanley and Chalmers, 1999). A limitation of degree-day approaches is that they are often untested against observed frost data because the frost index is not a physical property that can be compared to measurements. However, degree-day methods have been successful in capturing increased runoff from frozen ground events (Molnau and Bissell, 1983), and higher frost index values have been shown to correlate to deeper frost depths (Vermette and Christopher, 2008; Vermette and Kanack, 2012). Spatial variations of frozen ground within degree-day methods are typically based on variations in temperature (which are estimated from an elevation–temperature relationship) and variations in snowpack insulation (which are also typically inferred from an elevation–temperature relationship). Such reliance on elevation may lead to errors because Stähli (2017) found no clear connection between elevation and presence of frozen ground at test sites in the Swiss pre-Alpine zone.

The objective of this paper is to develop a model for frozen ground that (1) captures the spatial variations of frozen ground within a watershed, (2) allows the frozen ground model to be incorporated into a variety of watershed models, and (3) allows application in data sparse environments where limited forcing data may prohibit use of energy balance methods. In this paper, we use the GSSHA watershed model and develop the frozen ground model by modifying the commonly used conceptual frozen ground index (CFGI) (Molnau and Bissell, 1983) method in four ways. First, the CFGI method is coupled to an improved snowpack model that more accurately captures the spatial heterogeneity of the snowpack. In past applications of GSSHA, the CFGI method has been coupled with a temperature-index (TI) snowpack model based on SNOW-17 (Anderson, 1973, 2006). However, Follum et al. (2015) proposed a radiation-derived temperature index (RTI) snow model that uses a proxy temperature instead of air temperature to represent the energy available to the snowpack. Compared to the TI model, the RTI model more directly includes the effects of shortwave radiation and canopy cover and was shown to better represent the spatial variations of snow cover and snow water equivalent (SWE) in the Senator Beck Basin in Colorado. The RTI model is adopted to simulate the snowpack in the present study. Second, the effects of shortwave radiation and canopy cover are included in the CFGI model when calculating the energy available at the snow or ground surface. These effects are included by using a similar radiation-derived proxy temperature when calculating the frost index. Third, the insulation effects of ground cover are included by modifying the frost index equation. Fourth, an option is included to compute frost depth as a function of the frost index value. The modified Berggren equation and similar Stefan equation have been previously used to estimate frost depth from degree-days (Carey and Woo, 2005; DeWalle and Rango, 2008; Fox, 1992; Woo et al., 2004); a similar approach is used here to convert the frost index to frost depth.

The following sections first describe the existing TI and CFGI models within GSSHA. The combination of these two models serves as the baseline or control case for the experiments. Then, the RTI snow model and the modified CFGI frozen ground model (referred to as modCFGI) are described. Finally, the results of the TI/CFGI model and RTI/modCFGI models are compared to each other and to observations of snow depth, SWE, and frost depth at the Sleepers River experimental watershed (SREW) in Vermont.

2 Methodology

2.1 TI snowpack model

The TI snow model was implemented into GSSHA by Follum et al. (2014), who provides additional information about the model. Although GSSHA allows a variable time step for multiple processes, it always uses an hourly time step (Δt) for snow calculations. GSSHA utilizes a structured grid in which each cell can have a different air temperature T_a (°C) and precipitation P (m h^{-1}). Air temperature is the primary driver of snowpack dynamics in the TI model and is estimated as follows:

$$T_a = T_g + \varnothing \left(E_g - E_c\right), \tag{1}$$

where T_g (°C) is the air temperature at a gage, $\varnothing$ is a linear lapse rate (°C km^{-1}), and E_g and E_c (m) are the elevations of the temperature gage and the grid cell where T_a is being calculated, respectively. Precipitation accumulates as SWE (m) when $T_a \leq T_{px}$, where T_{px} is the freezing point (0°C by default). The precipitation P is multiplied by a uniform multiplication factor (S_{cf}), which crudely represents snowpack sublimation and redistribution of snow due to wind (Anderson, 2006). The resultant effective precipitation (P_{eff}) is added to the SWE.

Before the snowpack begins to melt, its heat deficit (or cold content) must be overcome. The change in heat deficit ΔD_{t} (mm of SWE), due to a temperature difference between the snow surface and air, is calculated using the following equation:

$$\Delta D_{\text{t}} = N_{\text{mf, max}}\,(\Delta t/6)\left(M_{\text{f}}/M_{\text{f, max}}\right)(A_{\text{TI}} - T_{\text{sur}}), \tag{2}$$

where T_{sur} is the snow surface temperature, and A_{TI} is the antecedent temperature index (°C), which is calculated using T_{a} and the antecedent snow temperature index parameter A_{TIPM} (see Anderson, 2006, for details regarding T_{sur} and A_{TI}). $N_{\text{mf,max}}$ is the maximum negative melt factor (mm °C^{-1} (6 h)$^{-1}$), which is a parameter. M_{f} is the melt factor (mm °C$^{-1}\Delta t^{-1}$), which is calculated as follows:

$$M_{\text{f}} = (\Delta t/6)\left[S_{\text{v}}A_{\text{v}}\left(M_{\text{f, max}} - M_{\text{f, min}}\right) + M_{\text{f, min}}\right], \tag{3}$$

where S_{v} and A_{v} are seasonal melt adjustments that change by Julian day, and $M_{\text{f, max}}$ and $M_{\text{f, min}}$ are the maximum and minimum melt factors (mm °C^{-1} (6 h)$^{-1}$), which are parameters.

Once the heat deficit is overcome, SWE decreases as melt occurs. During normal conditions, the melt M (mm of SWE) is

$$M = \left[M_{\text{f}}(T_{\text{a}} - T_{\text{mbase}}) + 0.0125 P_{\text{eff}} f_{\text{r}} T_{\text{r}}\right]\Delta t, \tag{4}$$

where T_{mbase} is the temperature at which melt begins (0 °C by default), f_{r} is the fraction of any precipitation that is rain (assumed equal to 1 when $T_{\text{a}} > 0$°C, otherwise set to 0), and T_{r} is the precipitation temperature (assumed equal to T_{a} or 0°C, whichever is greater). During rain-on-snow events (more than 1.5 mm of rainfall in the previous 6 h), M is calculated from a simple energy balance:

$$M = \sigma\left[(T_{\text{a}} + 273)^4 - 273^4\right]\Delta t + 0.0125 P_{\text{eff}} f_{\text{r}} T_{\text{r}} + 8.5 f_u (\Delta t/6)\left[(r_{\text{h}} e_{\text{sat}} - 6.11) + 0.00057 P_{\text{a}} T_{\text{a}}\right], \tag{5}$$

where σ is the Stefan–Boltzmann constant, f_u is the average wind function (mm mb^{-1} (6 h)$^{-1}$) (see Anderson, 2006, for details), r_{h} is the relative humidity (assumed to be 0.9 during rain-on-snow events) (Anderson, 1973, 2006), P_{a} is atmospheric pressure (mb) (either measured or calculated from elevation) (Anderson, 2006), and e_{sat} is the saturation vapor pressure (mb) (calculated based on Smith, 1993). The ripeness of the snowpack affects the amount of melt that is released and is controlled by the liquid holding capacity L_{hc}, which is a specified percentage of the ice in the snowpack (Anderson, 2006).

For frozen ground calculations, the snow depth is needed from the snow model. The snow depth D_{s} (cm) is found from the SWE and the snowpack density. GSSHA uses the single-layer snow density functions from SNOW-17 (Anderson, 1976, 2006). The density of newly fallen snow ρ_{n} (gm cm^{-3}) varies between 0.05 ($T_{\text{a}} \leq -15$°C) and 0.15 ($T_{\text{a}} = 0$°C) according to the following equation:

$$\rho_{\text{n}} = 0.05 + 0.0017\,(T_{\text{a}} + 15)^{1.5}. \tag{6}$$

Increases in snowpack density ρ_x from compaction, destructive metamorphism, and melt metamorphism due to the presence of liquid water are calculated as (Koren et al., 1999):

$$\rho_{x,t} = \rho_{x,t-1}\left(\frac{e^{B_2}}{B_2}\right)e^{B_1} \tag{7}$$

where,

$$B_1 = c_3\, c_5\, \mathrm{d}t\, e^{c_4 T_{\text{s}} - c_x \beta\,(\rho_{x,t-1} - \rho_{\text{d}})}, \quad \text{and} \tag{8}$$

$$B_2 = W_{t-1}\, c_1\, \mathrm{d}t\, e^{0.08\, T_{\text{s}} - c_2 \rho_{x,\,t-1}}. \tag{9}$$

The variable t is an index for time, W is the ice portion of the snow pack (cm, $W = 100\, S_{\text{swe},\,t-1}$) where S_{swe} is the snow water equivalent on the ground in m, T_{s} is the average snow pack temperature (°C, calculated based on Anderson, 2006), and ρ_{d} is the threshold density above which destructive metamorphism decreases (ρ_{d} is set to 0.15 gm cm^{-1} based on Anderson, 2006). Finally, $\beta = 0$ if $\rho_{x,t-1} \leq \rho_{\text{d}}$, and $\beta = 1$ if $\rho_{x,t-1} > \rho_{\text{d}}$, $c_1 = 0.026\,\text{cm}^{-1}\,\text{h}^{-1}$, $c_2 = 21\,\text{cm}^3\,\text{gm}^{-1}$, $c_3 = 0.005\,\text{h}^{-1}$, $c_4 = 0.10\,°\text{C}^{-1}$, and $c_5 = 2.0$ if there is liquid water in the snowpack and $c_5 = 1.0$ if there is not (see Anderson, 1976, 2006, for details).

2.2 CFGI frozen ground model

The CFGI model was originally developed as a lumped model for flood forecasting in the Pacific Northwest, but it has been used in distributed models as well (De Roo et al., 2001; Van Der Knijff et al., 2010). The rationale of the CFGI method is that air temperature ultimately controls the ground temperature, but its impact is moderated by the insulating effects of any snowpack. The presence of frozen ground is determined by the frozen ground index F (°C-days), which is calculated by

$$F_{\text{t}} = F_{t-1}A - T_{\text{a,d}}\, e^{-0.4 K_{\text{s}} D_{\text{s}}}, \tag{10}$$

where $T_{\text{a,d}}$ is the average daily air temperature (°C), A is a daily decay coefficient, and K_{s} is the snow reduction coefficient (cm^{-1}). The daily decay coefficient (A) controls the persistence of the F values, and K_{s} controls the insulation from the snowpack. Molnau and Bissell (1983) recommended changing K_{s} depending on whether $T_{\text{a,d}}$ is above or below freezing (denoted as $K_{\text{s, T}_{\text{a}} > 0\,°\text{C}}$ and $K_{\text{s,T}_{\text{a}} < 0\,°\text{C}}$, respectively).

Higher values of F indicate a higher likelihood that the ground is frozen. Once F exceeds a specified threshold

($F_{threshold}$), the ground is considered frozen and infiltration is restricted. Molnau and Bissell (1983) found the ground to be frozen when $F > 83$ °C-days and thawed when $F < 56$ °C-days. When F is between these values, the ground could be either frozen or thawed. It is worth noting that F does not depend on soil moisture, which is known to affect the initialization and depth of frozen ground (Kurganova et al., 2007; Willis et al., 1961).

2.3 RTI snowpack model

The RTI model makes two modifications to the TI model: (1) it uses a radiation-derived temperature T_{rad} (°C) to better describe the available energy, and (2) it estimates spatially varying snowpack sublimation based on solar radiation approximations.

The RTI model replaces T_a in Eqs. (4) and (5) with a radiation-derived proxy temperature T_{rad} (°C). In those equations, T_a is used to conceptually represent the energy available to the snowpack. T_{rad} has a similar purpose but is intended to improve the estimation of available energy. T_{rad} is calculated by assuming that the radiation terms dominate the energy balance at the snow surface so that outgoing longwave radiation balances the net incoming shortwave and longwave radiation (Follum et al., 2015). Thus

$$R_{LW\uparrow} = R_{SW,\,net} + R_{LW\downarrow}, \tag{11}$$

where $R_{LW\uparrow}$ is outgoing longwave radiation, $R_{SW,net}$ is the net incoming shortwave radiation, and $R_{LW\downarrow}$ is the downwelling longwave radiation. The right side of Eq. (11) represents the energy that is supplied to the snowpack via the atmosphere. $R_{LW\uparrow}$ (W m^{-2}) is the radiative response of the snowpack to that energy. Using the Stefan–Boltzmann law, $R_{LW\uparrow}$ can be written in terms of temperature T_{rad}:

$$T_{rad} = \left(\frac{R_{SW,net} + R_{LW\downarrow}}{\varepsilon_{snow}\,\sigma}\right)^{1/4} - 273.15, \tag{12}$$

where ε_{snow} is the emissivity of snow (assumed to be 0.97) and σ is the Stefan–Boltzmann constant.

$R_{SW,net}$ is calculated as follows:

$$R_{SW,net} = (1 - \alpha_s) R_{SW\downarrow}, \tag{13}$$

where α_s is the albedo of the snowpack, which is calculated based on the time elapsed since the most recent snowfall and whether melt is occurring (Henneman and Stefan, 1999). $R_{SW\downarrow}$ is the incident shortwave radiation, which is calculated using the following:

$$R_{SW\downarrow} = R_{SW,0}\varphi_r\varphi_{atm}\varphi_c\varphi_v\varphi_s\varphi_t, \tag{14}$$

where $R_{SW,0}$ is the solar constant (Liou, 2002), φ_r accounts for distance from the Earth to the sun (based on Julian day, TVA, 1972), φ_{atm} accounts for atmospheric scattering (based on elevation, Allen et al., 2005), φ_c accounts for absorption by clouds (based on fractional cloud cover, TVA, 1972), φ_v accounts for vegetation (set equal to the vegetation transmission coefficient K_v (Bras, 1990), a vegetation-specific parameter ranging from 1.0 for no canopy coverage to 0.0 for complete canopy coverage), φ_s accounts for the slope/aspect of the terrain (based on latitude, slope, and azimuth angle, Duffie and Beckman, 1980), and φ_t accounts for topographic shading (based on elevation, azimuth angle, and solar elevation angle).

$R_{LW\downarrow}$ is calculated from the contributions of the atmosphere (including clouds) and the canopy:

$$R_{LW\downarrow} = \sigma\varepsilon_a(T_a + 273.15)^4\left(1.0 + 0.17\,N^2\right)(1 - F_c) + F_c\sigma\varepsilon_c\left(T_{canopy} + 273.15\right)^4, \tag{15}$$

where ε_a is the air emissivity (0.757 when snow is present based on Bras, 1990), N is the fractional cloud cover, F_c is the fractional canopy cover (estimated from leaf area index L_{AI} following, Liston and Elder, 2006; Pomeroy et al., 2002), ε_c is the canopy emissivity (assumed equal to 1 following Sicart et al., 2004), and T_{canopy} is the canopy temperature (°C) which is assumed equal to T_a following DeWalle and Rango (2008).

Because the TI model uses T_a to drive snowpack dynamics, those dynamics are only directly associated with the downwelling longwave radiation from the air, which is a component of $R_{LW\downarrow}$. Furthermore, the spatial variations in the available energy only depend on the variations of T_a, which are inferred from elevation. T_{rad} in the RTI model considers both $R_{SW,net}$ and $R_{LW\downarrow}$ and thus accounts for heterogeneity in topographic orientation and shading as well as canopy cover. The TI model partially accounts for seasonal variation in solar radiation and snow albedo by empirically adjusting M_f as shown in Eq. (3). In the RTI model, seasonal variations in solar radiation and snow albedo are included in T_{rad}, so a constant melt factor M_f is used (Follum et al., 2015).

The TI model uses a uniform multiplication factor (S_{sf}) that is applied to the precipitation to account for sublimation, but sublimation is known to vary spatially (Musselman et al., 2008; Rinehart et al., 2008; Veatch et al., 2009). Most sublimation methods depend on relative humidity and wind speed (e.g., Pomeroy, 1988; Liston and Elder, 2006), which are often unavailable in data sparse environments. However, Gustafson et al. (2010) linked differences in sublimation rates to the amount of solar radiation a location receives. In the RTI model a simple approach is used to estimate hourly sublimation rates S_{sub} (cm h^{-1}) as follows:

$$S_{sub} = S_{sub,\,d}\left(\frac{R_{SW\downarrow}}{R_{SW\downarrow,\,flat}}\right), \tag{16}$$

where $S_{sub,\,d}$ (cm d^{-1}) is the watershed-average daily maximum sublimation amount (a parameter), and $R_{SW\downarrow,\,flat}$ is the daily shortwave radiation for a flat cell within the watershed on a cloud-free day. Thus, locations with higher $R_{SW\downarrow}$ (e.g., open areas and south-facing slopes in the Northern Hemisphere) will have higher values of S_{sub}. The method neglects wind speed and relative humidity, but does vary sublimation rates based on spatial patterns of solar radiation.

2.4 modCFGI frozen ground model

The CFGI model is modified in three ways to create the modCFGI model. First, the average daily proxy temperature $T_{rad,d}$ is used in place of $T_{a,d}$ to represent available energy. Second, ground cover (leaf litter, woody debris, etc.) is included as an insulator in the frozen ground index. And third, an option is included to estimate frost depth based on the frozen ground index. The frost depth calculation is optional because it requires soil moisture estimates and may not be needed in many hydrologic models that only require the occurrence (not depth) of frozen ground.

The CFGI uses $T_{a,d}$ in Eq. (10) to represent the energy that is available to heat the ground surface. In the modCFGI model, $T_{a,d}$ is replaced with $T_{rad,d}$. $T_{rad,d}$ is calculated using α_s (see Eqs. 12 and 13) when snow is present, and the albedo of the land cover when snow is not present. By using $T_{rad,d}$, the modCFGI model is expected to better represent the spatial heterogeneity of energy supply due to variations in the topography and canopy cover within a watershed.

The insulation by the ground cover is included by modifying Eq. (10) which becomes

$$F_t = F_{t-1}A - T_{rad}\, e^{-0.4(K_s D_s + K_{gc} D_{gc})}, \quad (17)$$

where K_{gc} is the ground cover reduction coefficient (cm^{-1}) and D_{gc} is the depth of ground cover (cm). This formulation retains the original form of the CFGI model but includes insulation from both snowpack and ground cover. F can still be used to identify the occurrence of frozen ground, which may be sufficient for many hydrologic models. However, because F is not a measurable quantity, an option to extend modCFGI to calculate frost depth is also needed.

Frost depth is calculated using F and the modified Berggren equation. As originally proposed (and described by DeWalle and Rango, 2008), the Berggren equation relates the number of degree days in the freezing/thawing period U (°C-days) to the maximum frost depth Z_{max} (m) as follows:

$$Z_{max} = \lambda\left(48\, U\, \delta^{-1} \Omega_m\right)^{1/2}, \quad (18)$$

where λ is a dimensionless coefficient that accounts for changes in sensible heat of the soil, δ ($J\,m^{-3}$) is the latent heat of fusion of the soil, and Ω_m ($J\,m^{-1}\,h^{-1}\,°C^{-1}$) is the mean thermal conductivity of the frozen and unfrozen soil layers. The derivation and corresponding assumptions (i.e., linear soil temperature gradients, Aldrich Jr., 1956) do not reveal any major impediments to adapting this equation for a shorter time step. In addition, Fox (1992), Woo et al. (2004), and Carey and Woo (2005) have used a layered version of the Stefan equation, which is similar to Eq. (18) to simulate daily frost depths with daily input data. Thus, the modified Berggren equation is applied at a daily timescale and revised to become:

$$Z_d = \lambda[48\,(F - F_{threshold})\, \delta^{-1} \Omega_m]^{1/2}, \quad (19)$$

where Z_d is the depth of frozen ground (m). By using the difference between F and $F_{threshold}$, the degree-days of the current freezing/thawing period is utilized, which is similar to the use of U in the original equation. Z_d is only calculated once the ground begins to freeze (when $F > F_{threshold}$). Z_d deepens as F becomes increasingly larger than $F_{threshold}$. When F decreases (due to increasing T_{rad}), so does the thickness of frost depth. No frost remains when F falls below $F_{threshold}$.

For the original modified Berggren equation, λ can be estimated annually from Aldrich Jr. (1956) using U, the mean annual air temperature, and the soil water content ω (% of dry weight). Here, λ is calculated using daily differences between F and $F_{threshold}$, the mean annual air temperature, and daily ω values. Thus, soil moisture is included in the calculation of Z_d even though it is not included in the calculation of F. Furthermore, δ is estimated daily as

$$\delta = \delta_f \rho\, \omega / 100, \quad (20)$$

where δ_f is the latent heat of fusion of water ($0.334\,MJ\,kg^{-1}$ at 0°C) and ρ is the dry soil density. Ω_m is estimated as follows (Farouki, 1981; Johansen, 1977):

$$\Omega_m = \left(\Omega_{sat} - \Omega_{dry}\right)\omega + \Omega_{dry}, \quad (21)$$

where Ω_{dry} and Ω_{sat} are the thermal conductivity of dry and saturated soil, respectively. Ω_{sat} is calculated as the geometric mean of the conductivities of the materials within the soil profile (Farouki, 1981; Johansen, 1977):

$$\Omega_{sat} = \Omega_s^{(1-n_{total})}\, \Omega_{ice}^{(n_{ice})}\, \Omega_{water}^{(n_{total}-n_{ice})}, \quad (22)$$

where Ω_s, Ω_{ice}, and Ω_{water} are the thermal conductivity of solids, ice, and water, respectively (Farouki, 1981). n_{total} is the porosity, and n_{ice} is

$$n_{ice} = n_{total}\, Z_d / H, \quad (23)$$

where H (m) is the soil thickness.

3 Model application

3.1 Study area

The TI/CFGI and RTI/modCFGI models are tested at the W-3 sub-basin (Fig. 1) of the SREW. The study period is 1 October 2005 through 30 September 2010, which is water year (WY) 2006 through 2010. The SREW was

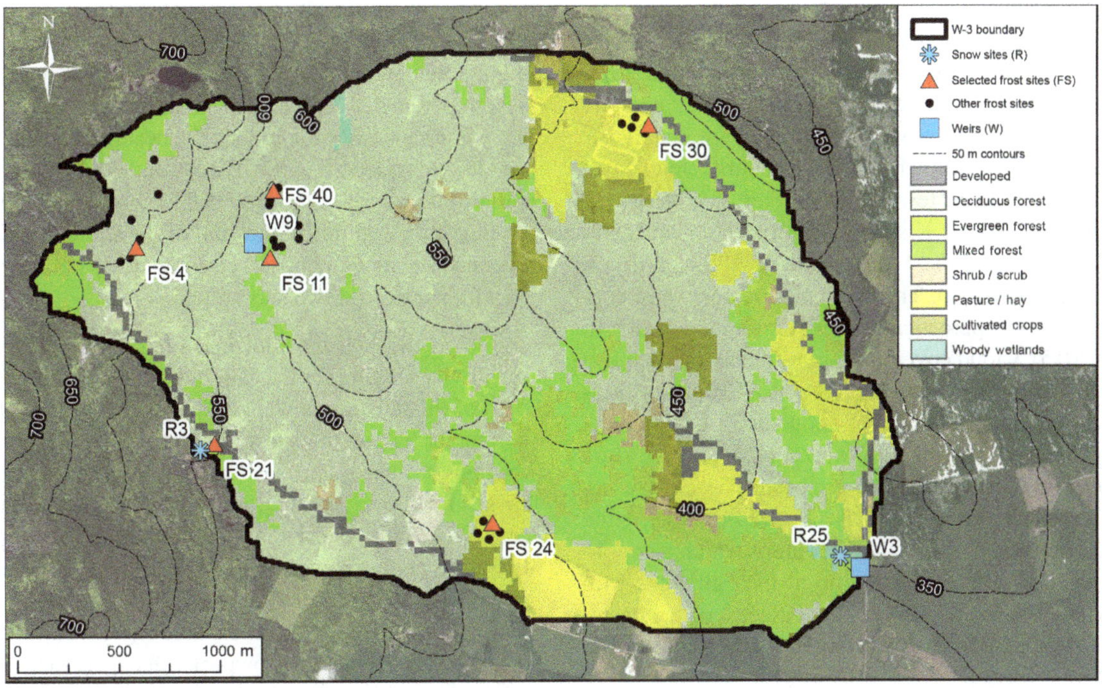

Figure 1. W-3 sub-basin in the Sleepers River experimental watershed. Sites used in this study are identified with red triangles and blue snowflakes. Basin delineation and elevation contours (m) are based on the 1/3 arcsec National Elevation Dataset, land cover classification is based on the 2006 National Land Cover Database, and sources of the background imagery include ESRI, DigitalGlobe, Earthstar Geographics, CNES/Airbus DS, GeoEye, USDA FSA, USGS, Getmapping, Aerogrid, IGN, IGP, and the GIS User Community.

founded in 1958 primarily for studies of snow accumulation, melt, and runoff (Anderson, 1973, 1976; Dunne and Black, 1970a, b; Dunne and Black, 1971; Shanley, 2000; Shanley and Chalmers, 1999). The W-3 sub-basin is located at 44°29′ N and 72°09′ W. Elevations range between 348 and 697 m, and the area is approximately 8.5 km^2 (based on the National Elevation Dataset, Gesch et al., 2002). The basin is primarily forested with deciduous (57.7 %), evergreen (7.8 %), and mixed (15.3 %) trees (based on the 2006 National Land Cover Database (NLCD), Fry et al., 2011). Approximately 14.6 % of the land cover is pasture/hay and cultivated crops. These open areas are typically below an elevation of 525 m, which is the approximate limit for viable agriculture (Shanley and Chalmers, 1999). The W-3 sub-basin is extensively gaged for both hydrometeorology and hydrology by the US Geological Survey (USGS) and collaborators from federal agencies and universities. Additional basin information and data are provided by Shanley et al. (1995), Shanley and Chalmers (1999), and the USGS website (https://nh.water.usgs.gov/project/sleepers/index.htm, last access: 7 November 2016).

Two snow sites and 35 frost sites within W-3 were monitored by the Vermont Field Office of the USGS. At the snow sites, SWE and snow depth were measured approximately weekly, and both sites are used in the present study. At the frost sites, snow depth and frost depth were measured periodically (between 0 and 14 measurements in a given winter). Frost depth was measured using CRREL-Gandahl frost tubes (Ricard et al., 1976), which are filled with a methylene blue solution. The frost depth is identified by a change in color within the tube (blue indicates thawed, clear indicates frozen). Vermette and Kanack (2012) provide images and descriptions of similar frost tubes, and Shanley and Chalmers (1999) provides detailed descriptions of the frost tubes at SREW. The frost sites (labeled FS in Fig. 1) are clustered in six parts of the watershed. For this paper, one site from each cluster (FS4, FS11, FS21, FS24, FS30, and FS40) was selected for analysis. The selected sites are far enough apart to be relatively independent but still capture the variations in elevation and land cover classification within the watershed.

3.2 Model inputs

The TI and CFGI models require hourly precipitation and temperature data, which were obtained from the USGS. Precipitation was measured at the W9 weir and R3 snow site (Fig. 1). The USGS then creates a single spatially averaged precipitation time series by weighting the measurements using the distribution of elevation (based on personal communication with James Shanley of the Vermont Field Office of the USGS on 14 November 2016). The W9 gage receives more weight because the watershed includes elevations both above and below this site. Hourly temperature was measured at the W9 site, which has an elevation of 520 m.

The RTI and modCFGI models also require cloud cover data, which were obtained from the National Centers for Environmental Information (NCEI, https://www.ncdc.noaa.gov/, last access: 7 November 2016). The hourly cloud cover classification data (clear, few clouds, broken sky, etc.) were collected at the Edward F. Knapp State Airport (44 km southwest of the basin) and the Morrisville-Stowe State Airport (36 km west of the basin). The classification data were converted to cloud cover percentages using the method from Follum et al. (2015). Cloud cover data are routinely measured at most airports in the US (data archived at NCEI) as well as many meteorological stations. For simulation of frost depth (and comparison to frost depth observations), soil moisture and evapotranspiration were also simulated. These two components additionally require hourly relative humidity, wind speed, and atmospheric pressure data, all of which were obtained from a meteorological station at the Fairbanks Museum in Saint Johnsbury, VT (11 km southeast of the basin) with missing values replaced with hourly data from the two airports.

All the models require elevation data to determine the spatial patterns of snow and frozen ground. W-3 was delineated using the 1/3 arcsec ($\sim$9 m) National Elevation Dataset (Gesch et al., 2002). The RTI and modCFGI models additionally require land cover classifications, which were obtained from the 2006 NLCD (Fry et al., 2011) and have a 30 m resolution. The classifications of some grid cells were changed to match the land covers observed in the field. In particular, the grid cell containing R3 was changed from deciduous forest to pasture/hay, FS11 was changed from mixed forest to evergreen forest, and FS21 was changed from developed to mixed forest. Both FS24 and FS30 are classified as pasture/hay, where FS24 is a managed pasture and FS30 is an unmanaged pasture (Ann Chalmers, Vermont Field Office of the USGS, personal communication, 15 November 2016). For example, during field observations in November 2016, FS24 had manure spread throughout the field, while FS30 was not fertilized.

Soil classification data are also required for calculating frost depth, and were obtained from the Digital General Soil Map of the United States (Soil Survey Staff, Natural Resources Conservation Service, United States Department of Agriculture, Web Soil Survey, available online at http://websoilsurvey.nrcs.usda.gov/, last access: 10 August 2016). Almost the entire W-3 basin is classified as fine sandy loam. The Watershed Modeling System (Aquaveo, 2013) was used to develop the GSSHA model with a 30 m structured grid. This resolution is adequate to capture the spatial heterogeneity of the basin while remaining computationally efficient.

3.3 Parameter estimation and calibration

The model-independent parameter estimation and uncertainly analysis (PEST) method (Doherty et al., 1994) was used to calibrate seven parameters in the TI model and eight parameters in the RTI model. PEST is a nonlinear local search parameter estimator that calibrates numerous parameters simultaneously to produce the best fit between simulated results and observations. WY 2006 and 2007 were used as the calibration period. The TI and RTI snow models were calibrated first to minimize the sum of the squared residuals between simulated and observed snow depths at the eight sites (six frost sites and two snow sites).

Table 1 displays the allowable range, calibrated value, and sensitivity ranking for the calibrated snow parameters. Goodness of fit statistics as well as description of the affects each parameter has on the snow simulations are described in the results and discussion section of this paper. The allowable ranges for A_{TIPM}, f_u, L_{hc}, $N_{\mathrm{mf,\,max}}$, M_{f}, $M_{\mathrm{f,\,max}}$, and $M_{\mathrm{f,\,min}}$ are based on physical limitations and typical ranges in the literature (Follum et al., 2015). L_{AI} can be estimated from seasonal and annual relationships to remotely sensed normalized difference vegetation index (NDVI) values (Wang et al., 2005). However, snowpack affects the measurement of greenness in high latitude regions (Beck et al., 2006). Thus, L_{AI} and K_{v} values were calibrated based on land cover classifications with forested land covers being categorized as deciduous forest (including deciduous forest, woody wetlands, and mixed forest) or evergreen forest. L_{AI} and K_{v} values for non-forested land cover classifications were set to 0.0 and 1.0, respectively. T_{px} and T_{mbase} were not calibrated (both set to 0 °C) because the temperature data were post-processed by the Vermont USGS and are expected to be accurate. By comparing the temperature measurements at W9 and the Fairbanks Museum (elevation of $\sim$212.4 m), $\varnothing$ was estimated at 6.6 °C km^{-1}. All snow density parameters are set based on Anderson (1973, 2006).

The PEST results indicate that the TI model's snow depths are most sensitive to S_{cf}, $M_{\mathrm{f,\,max}}$, A_{TIPM}, and $M_{\mathrm{f,min}}$. For the RTI model, snow depths are most sensitive to K_{v} (deciduous), A_{TIPM}, L_{AI}(evergreen), and M_{f} (Table 1). The calibrated deciduous K_{v} is near the top of the allowable range (1.0) and L_{AI} is near the bottom (0.103), indicating that the snow in the deciduous forest behaves similarly to the open pasture areas where $K_{\mathrm{v}} = 1$ and $L_{\mathrm{AI}} = 0$.

The CFGI and modCFGI frozen ground models were calibrated to minimize the sum of squared residuals between the simulated and observed frost depths at the six frost sites. For purposes of comparison the modified Berggren equation was also added to the CFGI model to calculate frost depth. Table 2 displays the allowable range, calibrated value, and sensitivity ranking of each calibrated frozen ground parameter. Goodness of fit statistics as well as description of affects each parameter has on the frost depth simulations are described in the results and discussion section. $F_{\mathrm{threshold}}$ was calibrated

Table 1. Allowable ranges and calibrated values for the TI and RTI model parameters using a model-independent parameter estimation and uncertainly analysis (PEST). Dashes indicate parameters that are not required in the associated model. The sensitivity ranking for each parameter is shown in parentheses.

Parameter	Units	Allowable range	Calibrated values TI	RTI
$M_{f,max}$	$mm\,°C^{-1}\,(6\,h)^{-1}$	0.001–2.400	1.017 (2)	–
$M_{f,min}$	$mm\,°C^{-1}\,(6\,h)^{-1}$	0.001–1.600	0.001 (4)	–
M_f	$mm\,°C^{-1}\,(6\,h)^{-1}$	0.001–2.400	–	0.391 (4)
S_{cf}	fraction	0.800–1.000	0.869 (1)	–
$S_{sub,\,d}$	$cm\,d^{-1}$	0.001–0.100	–	0.068 (6)
$N_{mf,max}$	$mm\,°C^{-1}\,(6\,h)^{-1}$	0.001–2.400	0.002 (6)	0.256 (8)
fu	$mm\,mb^{-1}\,(6\,h)^{-1}$	0.001–1.000	0.500 (7)	0.500 (10)
A_{TIPM}	fraction	0.001–1.000	1.000 (3)	0.992 (2)
L_{hc}	fraction	0.001–0.100	0.001 (5)	0.001 (9)
$K_{v,\,deciduous}$	fraction	0.200–1.000	–	0.969 (1)
$K_{v,\,evergreen}$	fraction	0.200–0.800	–	0.308 (7)
$L_{AI,\,deciduous}$	$m^2\,m^{-2}$	0.100–1.000	–	0.103 (5)
$L_{AI,\,evergreen}$	$m^2\,m^{-2}$	1.000–4.000	–	1.000 (3)

Table 2. Allowable ranges and calibrated values for the CFGI and modCFGI model parameters using PEST. Dashes indicate parameters that are not required in the associated model. The sensitivity ranking for the modCFGI parameters are shown in parentheses.

Parameter	Units	Allowable range	Calibrated values CFGI	modCFGI
$F_{threshold}$	°C - days	5.00–83.00	52.55	5.00 (3)
K_{gc}	cm	0.001–1.000	–	1.033 (1)
$K_{gc,FS24}$	cm	0.001–1.000	–	1.887 (2)
$K_{gc,FS30}$	cm	0.001–1.000	–	0.001 (4)

Table 3. Values of soil parameters used to calculate soil moisture in the single-layer Green and Ampt infiltration model.

Parameter	Units	Value
Saturated hydraulic conductivity	$cm\,h^{-1}$	2.040
Effective porosity	$cm^3\,cm^{-3}$	0.407
Residual water content	$cm^3\,cm^{-3}$	0.038
Field capacity	$cm^3\,cm^{-3}$	0.166
Wilting point	$cm^3\,cm^{-3}$	0.075
Capillary head	cm	8.570
Pore distribution arithmetic mean	$cm\,cm^{-1}$	0.466

for both the CFGI and modCFGI models with the upper range based on Molnau and Bissell (1983). Three K_{gc} values were calibrated for the modCFGI frozen ground model: one for the managed pasture site FS24 ($K_{gc,FS24}$), one for the unmanaged pasture site FS30 ($K_{gc,FS30}$), and one for all other frozen ground sites (K_{gc}).

Following Molnau and Bissell (1983), multiple combinations of A (0.8 and 0.97), and $K_{s,\,T_a<0\,°C}$ and $K_{s,\,T_a>0\,°C}$ (0.08, 0.2, and 0.5) values were tested with $A=0.97$, $K_{s,\,T_a<0\,°C}=0.08$, and $K_{s,\,T_a>0\,°C}=0.5$ producing frost indices that best replicate the rise and fall of the frost depth as well as the timing of the peak frost depth. Depth of ground cover for each land cover type was obtained from field observations in November 2016. Specifically, $D_{gc}=6\,cm$ for deciduous forest (fallen leaves), $D_{gc}=2\,cm$ for evergreen forest (fallen leaves), $D_{gc}=4\,cm$ for pasture (grass), and $D_{gc}=0\,cm$ for all other land cover types.

The modified Berggren equation requires soil moisture, which can be simulated using several methods in GSSHA (Downer and Ogden, 2006). To facilitate extension of these results to other hydrologic models, the commonly used single-layer Green and Ampt infiltration model (Green and Ampt, 1911) with soil moisture redistribution between rainfall events (Ogden and Saghafian, 1997) is utilized to calculate infiltration. Soil moisture is tracked using a simple bucket approach, accounting for infiltration, evapotranspiration, and groundwater recharge as described in Downer (2007). The soil layer thickness (H) is set to 0.5 m for both the soil moisture calculations and frost depth equations. Soil infiltration parameters are set based on published values for the W-3 soil type (Downer and Ogden, 2006; Rawls et al., 1982, 1983; Rawls and Brakensiek, 1985) and are shown in Table 3. Evapotranspiration, which can reduce the soil moisture, is simulated using a Penman–Monteith approach (Monteith, 1965, 1981) with parameters estimated based on land cover (Downer and Ogden, 2006). The dry soil density ($\rho=1137\,kg\,m^{-3}$) and dry soil thermal conductivity ($\Omega_{dry}=792\,J\,m^{-1}\,h^{-1}\,°C^{-1}$) are set based on measurements of fine sandy loam by Nikolaev et al. (2013).

For the CFGI model, the calibrated $F_{threshold}$ value (Table 2) is relatively close to the lower bound value of 56 °C-days found in Molnau and Bissell (1983). For the modCFGI model, the calibrated $F_{threshold}$ value is at the lower bound. The $F_{threshold}$ value is expected to be lower for the modCFGI model than the CFGI model. The modCFGI model incorporates the insulation by ground cover directly using K_{gc} and D_{gc}, whereas the CFGI model can only account for those effects by adjusting the $F_{threshold}$ value. It is also worth noting that $K_{gc,FS30}$ has a very low value (minimum of allowable range), which suggests that insulation from grass in an unmanaged pasture is very small. This could be the result of snow falling within the grass of the unmanaged pasture, thus making any insulating contribution from the grass very small.

4 Results and discussion

4.1 Snow depth and SWE (TI vs. RTI)

Figure 2 shows maps of simulated snow depth on 23 February 2007 from the TI and RTI snow models. The spatial variability in the TI snowpack is entirely based on elevation (due to the inference of local air temperature from elevation). Higher elevations have deeper snowpack due to lower air temperatures. The RTI snowpack also varies with elevation but shows variation due to land cover as well. In particular, pasture areas have slightly shallower snowpack than surrounding areas due to higher sublimation rates and higher absorbed shortwave radiation. North-facing slopes also have more snow than south-facing slopes due to lower absorbed shortwave radiation. Although no maps of observed snow depth are available for comparison, large-scale distributions of snowpack are known to be controlled by elevation, land cover, and slope/aspect (Fassnacht et al., 2017; Jost et al., 2007), which is more consistent with the RTI model.

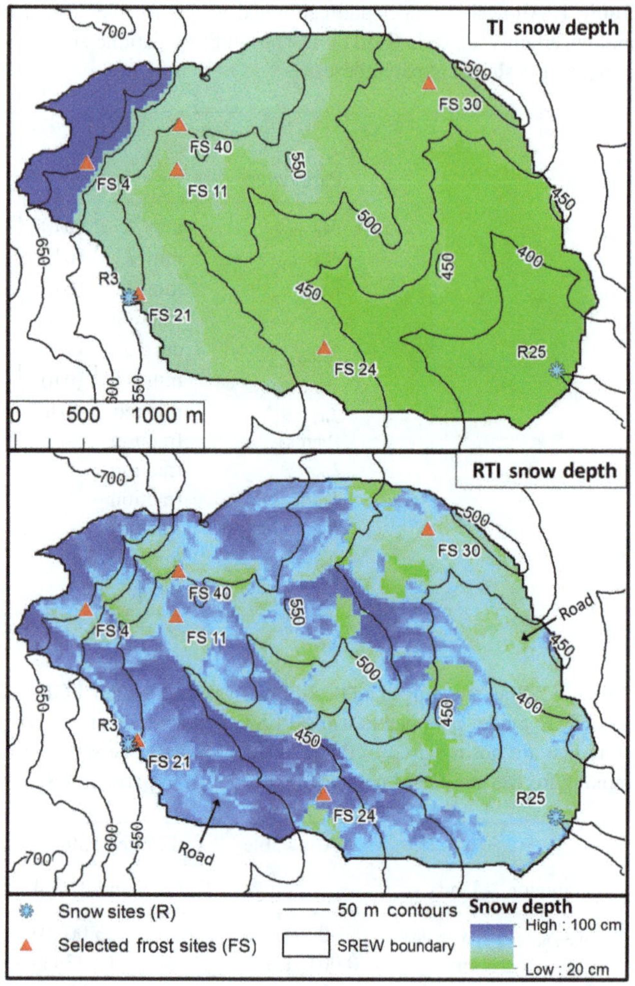

Figure 2. Simulated maps of snow depth (TI and RTI models) within the W-3 watershed for 23 February 2007. No observed maps of snow depth are available, but the map shows the differences between the temperature-based (TI) model and the modified (RTI) model.

Figure 3 shows the snow depths from the TI and RTI models at all eight test locations and compares them to the observations. Root mean squared error (RMSE) and Nash–Sutcliffe efficiency (NSE) are shown in Table 4 for the calibration period (WY 2006–2007), validation period (WY 2008–2010), and complete period (WY 2006–2010). The TI and RTI models track closely together at the eight test locations despite differences in the snow depth shown in Fig. 2. Differences between the TI and RTI snowpack at the test sites are small (Fig. 3 and Table 4). The RTI model performs slightly better than the TI model in overall average RMSE (15.69 vs. 15.71 cm), while the TI model performs slightly better in overall average NSE (0.58 vs. 0.53). The observed snow depth is relatively low in WY2008 and 2009 at two of the pasture sites (FS24 and FS30) compared to the other sites. Specifically in WY2008 the small snow depth observations are not captured within either model. The R3 site is also classified as pasture yet has a higher snowpack in WY2008 and 2009. The higher snowpack at this pasture site may be explained by the proximity of R3 to forested areas, which may reduce the wind and help preserve the snowpack. Neither model considers wind effects.

The snow depths from the two models are similar at each location (Fig. 3) because, on average, the available energy to melt snow (T_a in the TI model and T_{rad} in the RTI model) is similar (Fig. 4). However, the diurnal variation of T_{rad} is typically greater than that of T_a. T_{rad} is derived from a simple radiation balance (i.e., neglecting other terms in the thermal energy balance). Thus, T_{rad} is higher than T_a during the day due to high $R_{SW\downarrow}$ values, and it is typically lower than T_a at night because $R_{SW\downarrow}$ reduces to 0 and ε_a (set to 0.757) in Eq. (15) limits the affect T_a has on $R_{LW\downarrow}$ and therefore T_{rad}. As shown in Fig. 4, the available energy is also similar between these locations. The elevation difference between the highest and lowest elevation site is approximately 300 m, corresponding to a maximum temperature difference of approximately 2 °C

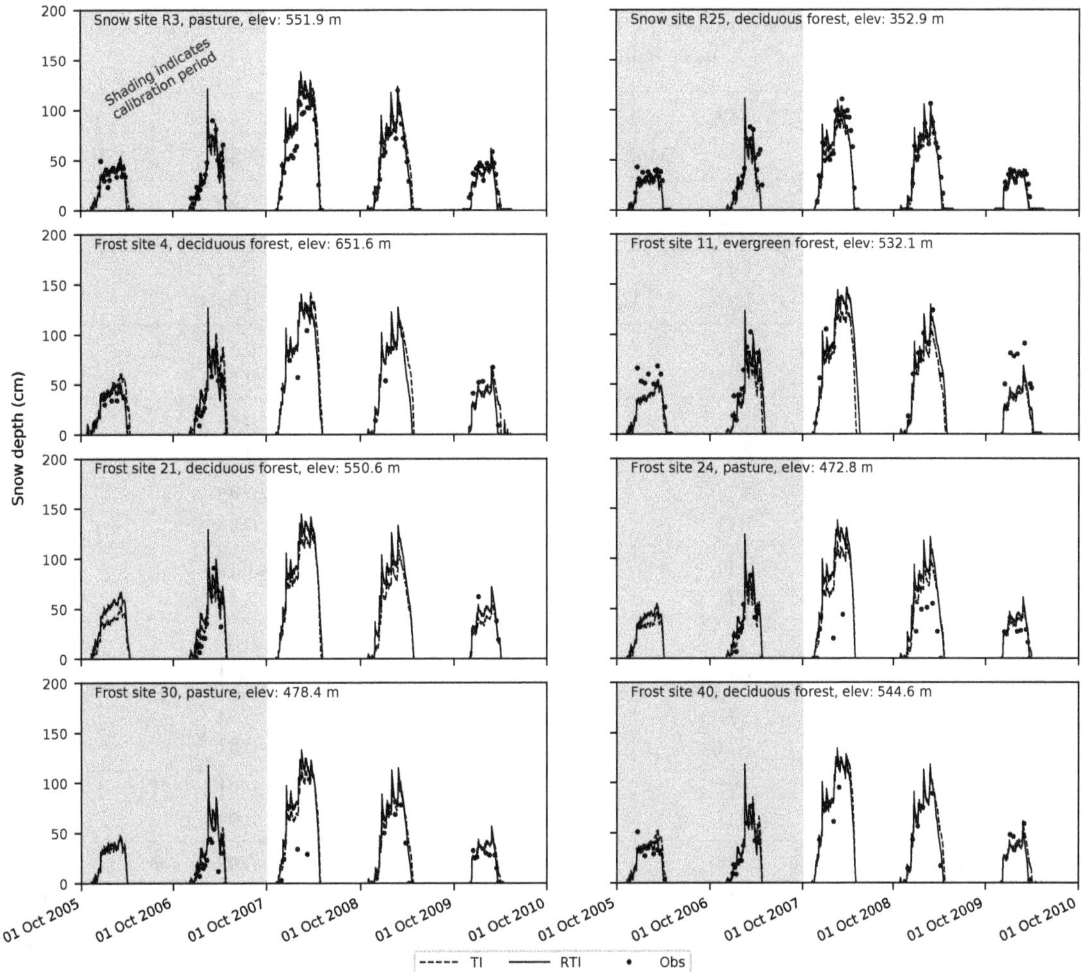

Figure 3. TI and RTI simulated snow depth at all eight test sites within the W-3 watershed.

between the sites. Also, the test sites are typically located on shallow slopes so topographic aspect has little influence on the energy available to melt the snowpack (i.e., T_{rad}). All land cover classifications except evergreen forest (FS11) have K_v values at or near 1 and L_{AI} values at or near 0, which reduces any variations due to land cover. T_{rad} at FS11 (evergreen forest) is different from the other seven sites because its low K_v value (0.308) reduces $R_{SW,net}$ during the day, and a high L_{AI} value (1.0) increases $R_{LW\downarrow}$ during day and night.

Figure 5 shows the simulated (both TI and RTI models) and observed SWE values, and Table 4 shows the associated performance metrics at the R3 and R25 snow sites. The TI and RTI models are only calibrated to snow depth, but SWE is calculated first and then combined with snow density to determine snow depth. Both models use the same method to calculate snow density. Both models also exhibit similar behavior and performance at the two sites, which is consistent with their similar snow depths discussed earlier (Fig. 3 and Table 4). Overall, this suggests that the snow density equations used within GSSHA are relatively accurate at the W-3 watershed. Thus, accurate estimates of snow depth typically correspond to accurate estimates of SWE as well.

4.2 Frost depth (CFGI vs. modCFGI)

Figure 6 shows simulated frost depth maps for 23 February 2007 using the CFGI and modCFGI models (no maps of observed frost depths are available for comparison). In the CFGI model, the frost depths mainly depend on elevation. Colder temperatures at higher elevations generally result in greater snowpack, which insulates the ground and produces smaller frost depths. However, at the beginning of the snow season when the snowpack is shallow, low temperatures at high elevations create deep frost in the higher elevations of the watershed. Later, deeper snowpack at high elevations insulate the ground, while the frost depth increases at lower elevations. This reversal in the elevation dependence can produce an inversion (localized minima in frost depth), as seen between the 500 and 650 m contour lines in Fig. 6. The modCFGI frost depth also has some elevation dependence, but

Table 4. Statistics for TI and RTI snow depth values at all eight test sites, and statistics for TI and RTI SWE values at the R3 and R25 snow test sites. Values are shown for calibration period (WY 2006–2007), validation period (WY 2008–2010), and overall (WY 2006–2010). RMSE values closer to zero and NSE values closer to one indicate better fit.

	Site	Land cover	Snow model	Calibration RMSE (cm)	Calibration NSE	Validation RMSE (cm)	Validation NSE	Overall RMSE (cm)	Overall NSE
Snow depth	R3	Pasture	TI	6.6	0.91	12.1	0.89	10.6	0.89
			RTI	8.4	0.86	11.9	0.89	10.9	0.89
	R25	Deciduous forest	TI	13	0.61	9.4	0.93	10.7	0.88
			RTI	12.2	0.65	9.2	0.93	10.3	0.89
	FS4	Deciduous forest	TI	18.1	0.3	17	0.57	17.7	0.51
			RTI	9.4	0.81	15.4	0.64	12.1	0.77
	FS11	Evergreen forest	TI	18.4	0.54	24.5	0.65	21.6	0.62
			RTI	15.8	0.66	19.3	0.78	17.6	0.75
	FS21	Deciduous forest	TI	12.1	0.82	13	0.45	12.3	0.79
			RTI	18.1	0.59	6.1	0.88	16.4	0.62
	FS24	Pasture	TI	10.1	0.85	26.2	−1.18	21.3	0.08
			RTI	10.7	0.83	33.2	−2.51	26.7	−0.44
	FS30	Pasture	TI	16.1	−0.12	22.6	0.09	20.3	0.06
			RTI	17	−0.26	24.9	−0.1	22.2	−0.12
	FS40	Deciduous forest	TI	9.3	0.74	13.1	0.74	11.2	0.78
			RTI	7.6	0.83	11.1	0.81	9.3	0.85
SWE	R3	Pasture	TI	2.1	0.9	3.3	0.91	3	0.91
			RTI	3.6	0.71	3.1	0.92	3.3	0.89
	R25	Deciduous forest	TI	5.2	0.29	2.7	0.92	3.7	0.83
			RTI	5	0.35	3.1	0.91	3.8	0.82

the spatial variation mainly follows land cover classification, which is similar to observations of frozen ground in the Swiss pre-Alpine zone (Stähli, 2017). This variation is partly due to the use of T_{rad} and the increased heterogeneity in the snow depth. The effect of snowpack can be seen by comparing hillslopes with the same land cover but different orientations, such as along the 500 m contour south of FS11. Lower T_{rad} values on northeast-facing slopes result in deeper snowpack than the southwest-facing slopes (Fig. 2). This deeper snowpack produces shallower frost depths on the northeast-facing slopes due to insulation by the snow. However, the spatial pattern of frost depth is more heavily affected by the land cover. Land cover's impact largely occurs through the associated ground cover. This effect can be seen by comparing the deep frost at the unmanaged pasture (near FS30) with the shallower frost depth at the deciduous forest areas near FS4, FS21, and FS40. The low ground cover reduction coefficient at the unmanaged pasture ($K_{gc,FS30}$) reduces the insulation from the ground cover, creating deeper frost compared to the deciduous forest areas. The larger than expected role of ground cover in the modCFGI model may occur because ground cover is present during the initiation, deepening, and decrease of frost depth, while the snowpack is much more variable throughout the season.

Figure 7 shows the frost depths from the CFGI and modCFGI models along with the frost depth observations. The RMSE and NSE values during the calibration, validation, and overall periods are shown in Table 5. The simulated frost depth remains more constant amongst the sites when using the CFGI model, which produces similar maximum frost depths for a given year independent of the land cover. The modCFGI results deviate considerably from the CFGI results, producing greater frost depths at the unmanaged pasture (FS30) and evergreen (FS11) sites and smaller frost depths at the deciduous (FS4, FS21, and FS40) and managed pasture (FS24) sites. These simulated differences between the sites are consistent with the observations. The decreased frost depth in the deciduous forest and managed pasture result from their high measured litter depth ($D_{gc} = 6\,\text{cm}$) and high reduction coefficient ($K_{gc,FS24} = 1.887\,\text{cm}^{-1}$), respectively. The two pasture sites (FS24 and FS30) differ considerably in the observed frost depth with FS30 consistently having deeper frost. This difference likely occurs because FS24 is managed and FS30 is not. With the exception of the validation period at FS30, the modCFGI model performs better

Table 5. Statistics for CFGI and modCFGI frost depth at all six frost sites. Values are shown for calibration period (WY 2006–2007), validation period (WY 2008–2010), and overall (WY 2006–2010). RMSE values closer to zero and NSE values closer to one indicate better fit. No frost was present at FS4 and FS21 during the validation period, resulting in an inability to calculate NSE. Statistics for a recalibrated modCFGI model without ground cover (labeled as "modCFGI no gc") are also shown.

Site	Land cover	Frost depth model	Calibration		Validation		Overall	
			RMSE (cm)	NSE	RMSE (cm)	NSE	RMSE (cm)	NSE
FS4	Deciduous forest	CFGI	8.2	−5.0	5.7	NA	7.2	−3.8
		modCFGI	2.5	0.4	0.2	NA	1.9	0.7
		modCFGI no gc	26.5	−62.9	16.0	NA	22.6	−45.9
FS11	Evergreen forest	CFGI	15.5	−1.2	9.9	−1.6	13.1	−0.6
		modCFGI	12.6	−0.5	8.2	−0.8	10.7	−0.1
		modCFGI no gc	17.5	−1.8	10.9	−2.1	14.6	−1.0
FS21	Deciduous forest	CFGI	12.5	−24.2	8.4	NA	11.8	−10.9
		modCFGI	3.9	−1.5	0.0	NA	3.5	−0.1
		modCFGI no gc	26.2	−109.8	14.3	NA	24.3	−49.4
FS24	Pasture	CFGI	17.5	−188.5	7.3	−12.3	12.4	−49.1
		modCFGI	1.4	−0.3	2.3	−0.3	2.0	−0.3
		modCFGI no gc	28.2	−490.0	13.5	−44.3	20.6	−137.4
FS30	Pasture	CFGI	27.5	−5.8	6.2	−2.4	17.7	0.2
		modCFGI	11.7	−0.2	24.9	−55.0	20.9	−0.1
		modCFGI no gc	18.1	−2.0	11.4	−10.8	14.4	0.5
FS40	Deciduous forest	CFGI	14.2	−22.6	10.1	−1642	12.6	−20.9
		modCFGI	3.4	−0.3	1.8	−52.1	2.8	−0.1
		modCFGI no gc	36.9	−157.9	21.8	−7631	31.2	−133.1

Table 6. Number of true positive (both simulated and observed data show frost depth), true negative (both simulated and observed data show no frost depth), false positive (simulated data shows frost depth but observed data does not), and false negative (simulated data shows no frost depth but observed data shows frost depth) occurrences during the entire test period. The Accuracy is the sum of the true positive and true negative divided by the total number of observations.

Site	Land cover	Elevation (m)	Model	True positive	True negative	False positive	False negative	Accuracy (%)
FS4	Deciduous forest	651.6	CFGI	9	12	4	3	75.0 %
			modCFGI	9	15	1	3	85.7 %
FS11	Evergreen forest	532.1	CFGI	24	2	0	21	55.3 %
			modCFGI	39	1	1	6	85.1 %
FS21	Deciduous forest	550.6	CFGI	10	3	1	1	86.7 %
			modCFGI	8	4	0	3	80.0 %
FS24	Pasture	472.8	CFGI	7	12	9	5	57.6 %
			modCFGI	6	20	1	6	78.8 %
FS30	Pasture	478.4	CFGI	16	8	0	10	70.6 %
			modCFGI	26	1	7	0	79.4 %
FS40	Deciduous forest	544.6	CFGI	13	9	11	1	64.7 %
			modCFGI	13	12	8	1	73.5 %
Total			CFGI	79	46	25	41	65.4 %
			modCFGI	101	53	18	19	80.6 %

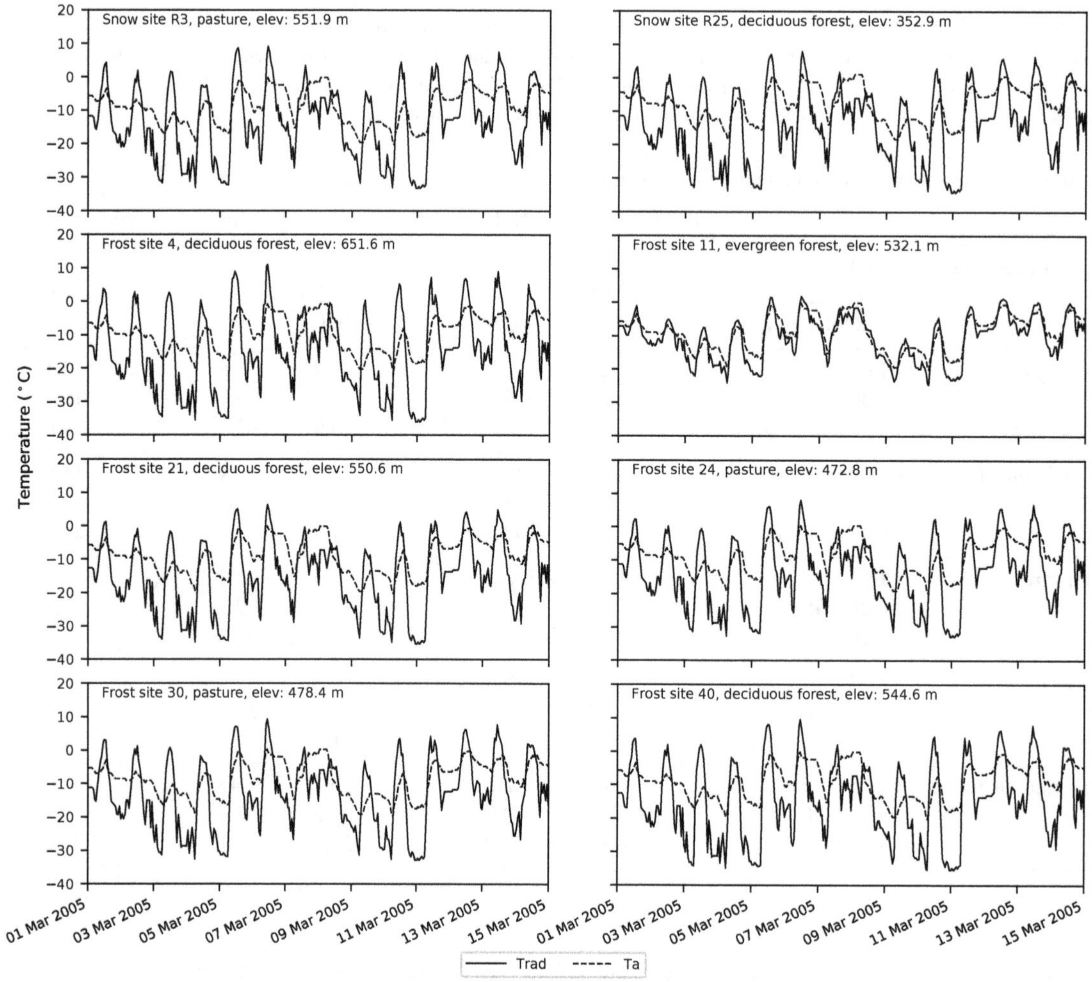

Figure 4. T_a and T_{rad} values at all eight test sites within the W-3 watershed between 1 and 15 March 2005.

(lower RMSE and higher NSE values) than the CFGI model. The difference in performance is most pronounced at the deciduous sites (FS4, FS21, and FS40) where the average overall NSE value is −11.9 for the CFGI model and 0.20 for the modCFGI model.

In hydrologic models, capturing the presence of frozen ground is important because even shallow frost with high moisture content (concrete frost) has the potential to impede infiltration (Dunne and Black, 1971). Therefore, the ability of the CFGI and modCFGI models to accurately capture the presence of frozen ground is evaluated. Whenever frost observations are available, the simulated frost depths are categorized as follows: true positive (both simulated and observed data show frost), true negative (both simulated and observed data show no frost), false positive (simulated data shows frost but observed data shows no frost), or false negative (simulated data shows no frost but observed data shows frost). Table 6 shows the number of observations in each category for each test site. The table also shows the model accuracy, which is calculated as the percent of the observations that are correctly classified (true positive or true negative). The CFGI and modCFGI models perform similarly in capturing true positives at FS4, FS21, FS24, and FS40, while modCFGI has more true positives at FS11 and FS30. The lower true positives and higher false negatives indicate that the CFGI model tends to underestimate the presence of frozen ground at FS11 and FS30. Overall, both the CFGI and modCFGI models capture most of the frozen ground events, with the modCFGI model performing better than the CFGI model at five sites and worse at one site (FS21). The average accuracy of the modCFGI model is 15.2 % higher than the CFGI model, with the largest increase in accuracy at FS11 (29.8 %).

A simple test is employed to explore the modification that contributes most to the increased accuracy of the modCFGI model. This test removes ground cover from the modCFGI model, recalibrates, and then compares the results to observations. When ground cover is removed, the calibrated $F_{threshold}$ value is 83 °C-days, which is at the top of the calibration range. This change indicates that ground cover has a

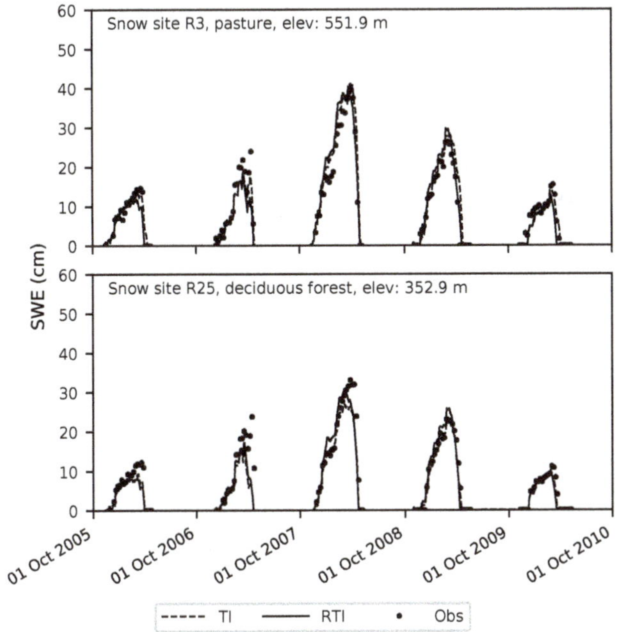

Figure 5. TI and RTI simulated SWE at R3 and R25 snow sites within the W-3 watershed.

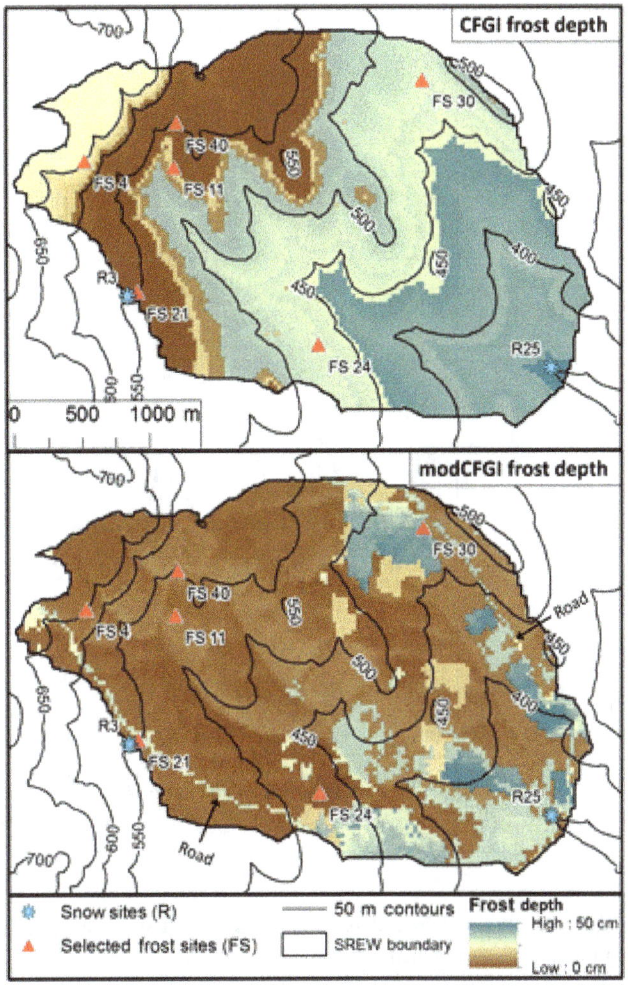

Figure 6. Simulated maps of frost depth (CFGI and modCFGI models) within the W-3 watershed for 23 February 2007. No observed maps of frost depth are available, but the map shows the differences between the temperature-based (CFGI) model and the modified (modCFGI) model.

large impact on the appropriate value of this threshold. Figure 8 shows the simulated frost depths using the modCFGI model with and without ground cover for each test site. Performance metrics for the modCFGI model with and without ground cover are shown in Table 5. Variability in frost depth between the sites is diminished when ground cover is removed, leading to large errors between simulated and observed frost depth. When ground cover is removed, the frost depth results decrease in accuracy (higher RMSE values and lower NSE values) compared to the complete modCFGI model. The only exception is the overall period at FS30, which is also the only site where the CFGI model outperforms the full modCFGI model. These results suggest that inclusion of ground cover is an important reason why the modCFGI model outperforms the CFGI model.

The sensitivity of the modCFGI results to soil moisture is also examined. Soil moisture does not affect the calculation of F, but it is included within the modified Berggren equation (Eqs. 18 and 19) in the calculation of δ (Eq. 20) and Ω_m (Eq. 21). Soil moisture was simulated using a single-layer Green and Ampt approach. However, no soil moisture measurements are available at any of the test sites to evaluate the accuracy of the simulated values. Sensitivity of the modCFGI model to volumetric soil moisture is tested by artificially setting the soil moisture to either the residual water content (θ_{low}) or the effective porosity (θ_{high}), which are the lower and upper bounds for soil moisture values within the model. Figure 9 shows the modeled frost depths from the modCFGI model using θ_{low}, θ_{high}, and the soil moisture from the Green and Ampt approach (θ_{sim}, which is identical to modCFGI in Figs. 7 and 8). Also shown are the observed frost depths for reference only. The frost depth from the θ_{sim} case is similar to the frost depth from the θ_{high} because the simulated soil moisture is usually close to the effective porosity. Frost depth increases when θ_{low} is used, which coincides with other studies (Fox, 1992; Willis et al., 1961). The timing of the frozen ground (when it begins and ends) is identical in all three of the simulations. The consistent timing occurs because soil moisture is not used to calculate F and the same $F_{threshold}$ (which controls when frozen ground begins) was used for all three simulations. This result highlights a deficiency in the modeling framework. Specifically, soil moisture should be considered for determining the initiation of frozen ground because wet soils have a higher specific heat capacity and require more energy loss to cool and freeze the soil (Kurganova et al., 2007).

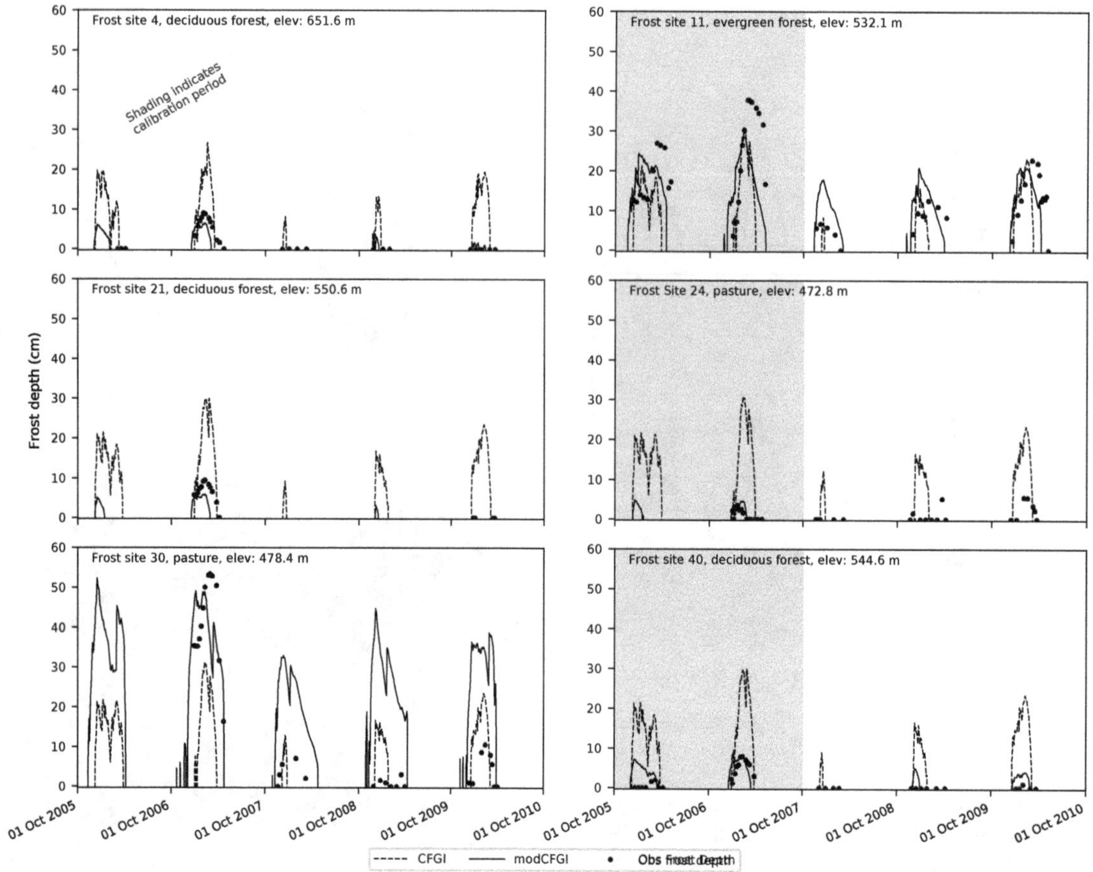

Figure 7. Observed frost depth compared against simulated (CFGI and modCFGI) frost depth at all six selected frozen ground test sites within the W-3 watershed.

5 Conclusions

The main purpose of this paper was to better estimate the spatial pattern of frozen ground for distributed watershed modeling by modifying an existing degree-day frozen ground model (CFGI), which uses a frost index value to determine whether the ground is frozen or not. The modifications to the CFGI model include (1) use of a radiation-derived temperature index (RTI) snow model instead of a standard temperature-index (TI) snow model, (2) use of a radiation-derived proxy temperature (T_{rad}) instead of air temperature (T_{a}) in the calculation of the frost index, (3) inclusion of ground cover (litter, debris, grass, etc.) as an insulator of the ground from air temperatures, and (4) an option to use a version of the modified Berggren equation to calculate frost depths based on the frost index values. The CFGI and modCFGI models were tested using the GSSHA hydrologic model over a five-year period within the W-3 watershed, which is part of Sleepers River experimental watershed in Vermont. The model results were compared against snow depth at eight sites, snow water equivalent at two sites, and frost depth at six sites. The primary conclusions of the paper are as follows:

1. The RTI snow model produces much more complex spatial patterns of snow depth than the TI snow model for the W-3 watershed. The TI model, which is based on SNOW-17 (Anderson, 2006), only produces spatial variation using elevation. The RTI model accounts for elevation, hillslope orientation, canopy shading, and longwave radiation from the canopy through the use of the radiation-derived proxy temperature. It also includes a simple sublimation method based on solar radiation. Thus, its snow depths exhibit spatial heterogeneity based on elevation, slope/aspect, and land cover, all of which are known to affect the large-scale distribution of observed snow depths (Fassnacht et al., 2017; Jost et al., 2007).

2. Both the RTI model and TI model produce accurate results for the eight snow depth sites at W-3. Two of the eight sites also measure snow water equivalent, where the RTI and TI model also show similarly accurate results. The eight test sites have similar topographic at-

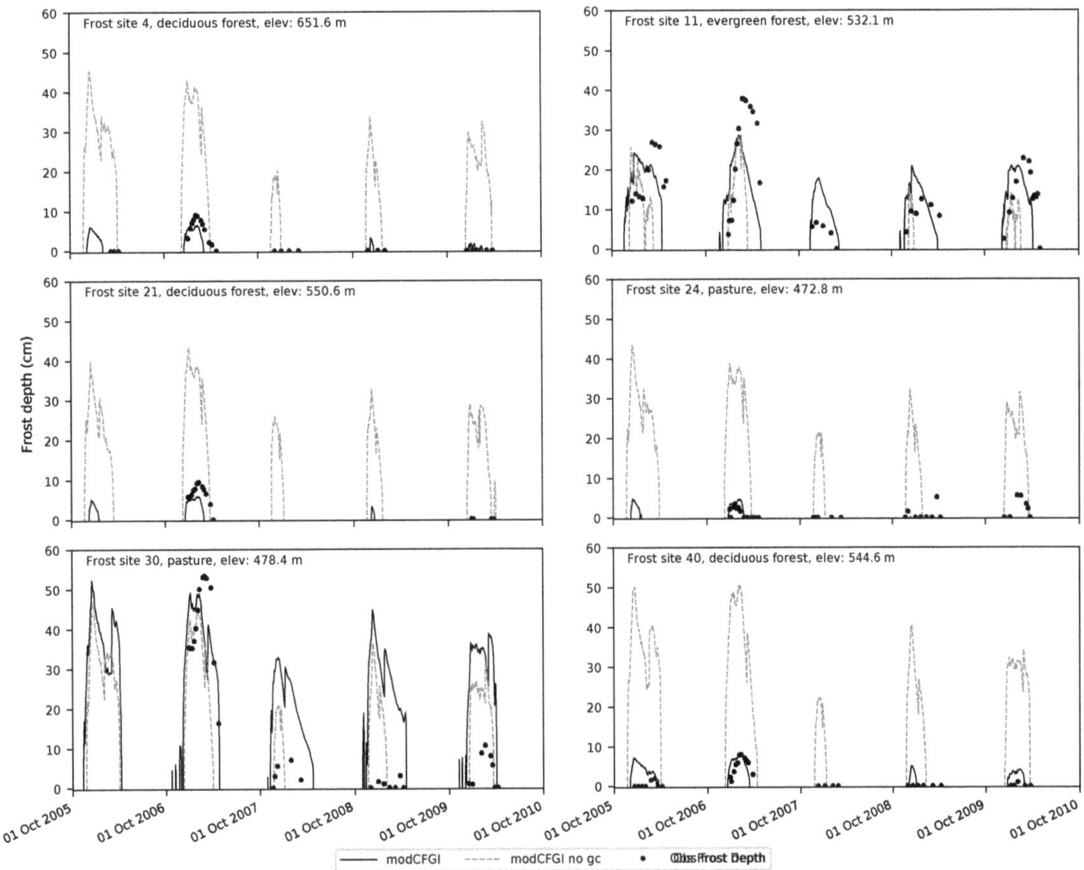

Figure 8. Observed frost depth compared against simulated (modCFGI with and without ground cover included) frost depth at all six selected frozen ground test sites within the W-3 watershed. The modCFGI model without ground cover is labeled as "modCFGI no gc".

tributes and primarily differ in their land covers, which include pasture, deciduous forest, and evergreen forest. Because the leaves have typically fallen prior to snow accumulation, all but the evergreen site behave similarly in snow accumulation and ablation.

3. The modCFGI frost model produces more complex spatial patterns of frost depth than the CFGI frost model for the W-3 watershed. The CFGI model uses elevation to infer the spatial variation of air temperature. It additionally uses the TI model for snow depth, which also depends on elevation. Thus, the simulated frost depths at W-3 primarily reflect the watershed elevations. In contrast, the modCFGI model uses the radiation-derived proxy temperature to infer the energy available to heat the ground and the RTI model to simulate snow depth. Furthermore, it accounts for the insulating effects of ground cover (in addition to snowpack), which also depends on the land cover. Thus, the frost depths simulated by the modCFGI model at W-3 depend on the local elevation, hillslope orientation, and land cover, all of which are known to affect the distribution of frozen ground (Fox, 1992; MacKinney, 1929; Wilcox et al., 1997; Willis et al., 1961).

4. The modCFGI model produces more accurate frost depths than the CFGI for all but one of the six test sites in the W-3 watershed. Overall, the modCFGI model more accurately captures the inter-annual variability in frost depth at a given site and variability of frost depth between sites. Although both the CFGI and modCFGI capture the majority of frozen ground events observed, the modCFGI model has 15.2 % better accuracy in capturing the presence of frozen ground, which is expected to be important for capturing runoff that is produced by frozen ground.

5. A key reason for the difference in performance between the two frost models is that the modCFGI model includes the insulation of the ground by ground cover while the CFGI model does not. When ground cover is removed from the modCFGI model its results for W-3 are less accurate and the variability in simulated frost depth between the sites is limited. Ground cover is likely important in this watershed because it is relatively

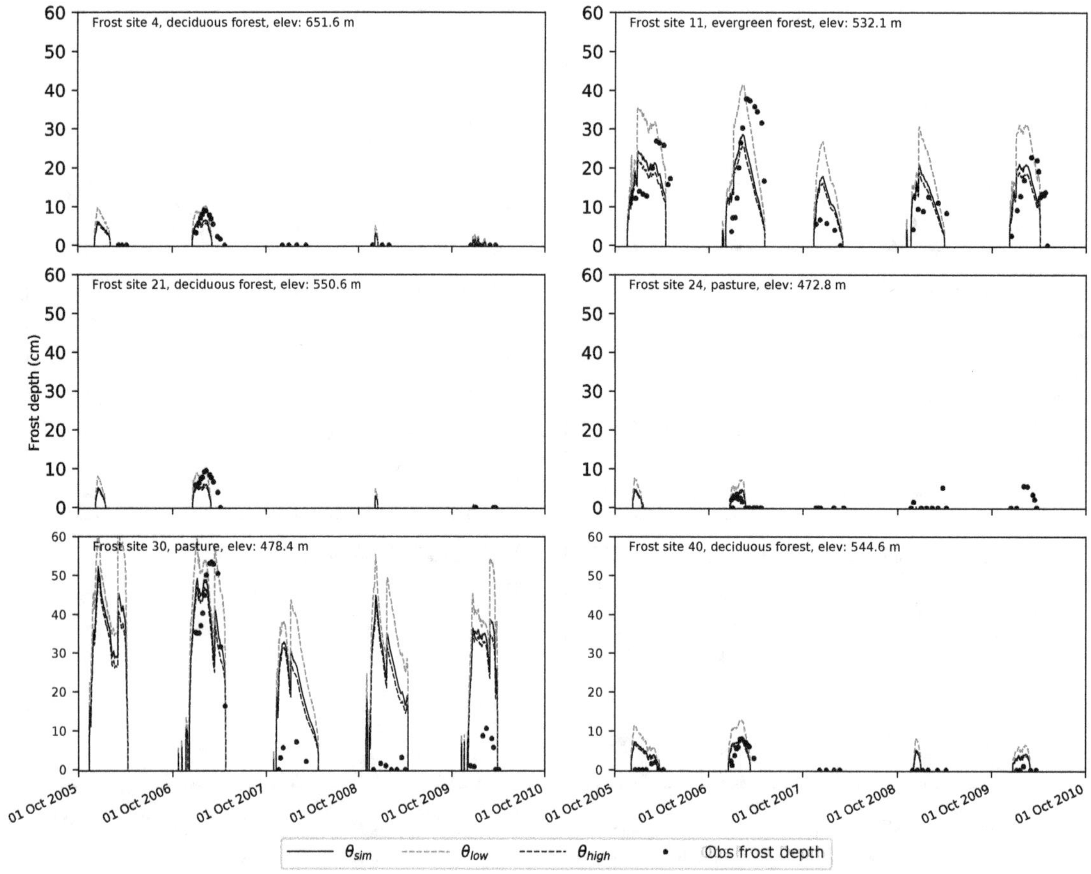

Figure 9. Simulated frost depths from the modCFGI model using simulated soil moisture (θ_{sim}), a constant high soil moisture (θ_{high}), and a constant low soil moisture (θ_{low}) at all six selected frozen ground test sites within the W-3 watershed.

thick and is also present at all stages of the winter while snowpack is not.

Overall, the modCFGI model provides improved spatial representation of frozen ground while requiring only cloud cover estimates as additional forcing data (more forcing data may be required if soil moisture is simulated to obtain frost depth). Limited data requirements should make modCFGI well suited for data sparse environments. Hydrologic models often need to account for the presence of frozen ground, which in data-sparse environments often means using simple degree-day approaches that typically vary frozen ground with elevation only (as was shown with the CFGI model). To calculate T_{rad} the modCFGI model does require cloud cover data, which are collected operationally at most airports within the US. If soil moisture is explicitly simulated within the hydrologic model the modCFGI model can also be used with the modified Berggren equation to simulate frost depth, which requires information on soil type and an estimate of the thermal conductivity of the soil.

Five main avenues are available for future research. First, the modCFGI model should be generalized to include the effects of wind (as it relates to the snowpack) and more completely consider the role of soil moisture. Soil moisture is not considered when calculating the frost index, so it does not impact the initiation or duration of frozen ground. This limitation results from using a degree-day approach and may be important in some cases (Kurganova et al., 2007; Willis et al., 1961). Second, the modCFGI model should be tested further. Additional testing should consider other areas where snow and frozen ground are known to affect runoff, such as the upper Midwest region of the US. Additional testing should also better characterize the insulation properties of ground cover under different management scenarios. Third, the calculation of T_{rad} is simple and applicable in data-sparse environments, but other approaches for adjusting a temperature value based on topography and land cover are available (Fox, 1992; Kang, 2005; Webster et al., 2017) and could be further tested. Fourth, future research should also determine the effects of spatial heterogeneity of snow and frost depth on runoff and streamflow at both the local and watershed scales. Similar to Campbell et al. (2010), the RTI and modCFGI models could be used in data-sparse watersheds to investigate how changes in historic and future climate affect snow, frozen ground, and runoff. Finally, although this paper focuses on the simulation of frost depth in the context of watershed modeling, the methods described could also be used for agriculture, overland mobility modeling, and infrastructure where snow and frost depth are major concerns.

Data availability. Hourly precipitation and temperature data used in this study are available by contacting the Vermont Field Office of the USGS (https://nh.water.usgs.gov/project/sleepers/index.htm; USGS, 2018a). Cloud cover, relative humidity, wind speed, and atmospheric pressure data used in this study are available from the National Centers for Environmental Information (NCEI, 2018; https://www.ncdc.noaa.gov/; last access: 7 November 2016). Land cover data (2006 National Land Cover Database; Fry et al., 2011) and elevation data (National Elevation Dataset; Gesch et al., 2002) used in this study are available from the USGS (https://nationalmap.gov/; USGS, 2018b). Soil data used in this study were obtained from the Digital General Soil Map of the United States (Soil Survey Staff, Natural Resources Conservation Service, United States Department of Agriculture, Web Soil Survey, available online at https://websoilsurvey.sc.egov.usda.gov/App/HomePage.htm; last access: 10 August 2016). Snow and frozen-ground data used in this study are available by contacting the Vermont Field Office of the USGS (https://nh.water.usgs.gov/project/sleepers/index.htm; USGS, 2018a).

Information regarding the GSSHA model is available at https://gsshawiki.com/Gridded_Surface_Subsurface_Hydrologic_Analysis (GSSHA, 2018). Additional information and output data sets are available by contacting the corresponding author.

Competing interests. The authors declare that they have no conflict of interest.

Acknowledgements. This project was funded by the Flood and Coastal Research Program at the U.S. Army Corps of Engineers, Engineering Research and Development Center, Coastal and Hydraulics Laboratory in Vicksburg, Mississippi. We thank the Vermont Field Office of the USGS (specifically James Shanley and Ann Chalmers) for their diligent efforts to collect, process, and analyze numerous hydrologic data for research. We also thank three anonymous reviewers and the editor who provided helpful comments that substantially improved this article.

Edited by: Markus Weiler

References

Aldrich Jr., H. P.: Frost penetration below highway and airfield pavements, Highway Research Board Bulletin, 135, 124–149, 1956.

Allen, R. G., Walter, I. A., Elliott, R., Howell, T., Itenfisu, D., and Jensen, M.: The ASCE Standardized Reference Evapotranspiration Equation, ASCE, New York, 2005.

Anderson, E. A.: National Weather Service River Forecast System – Snow Accumulation and Ablation Model, Technical Memorandum NWS Hydro-17, November 1973, 217 pp., Silver Spring, Maryland, 1973.

Anderson, E. A.: A Point Energy and Mass Balance Model of a Snow Cover, U.S. National Oceanic and Atmospheric Administration NOAA Technical Report NWS 9, Silver Spring, Maryland, 1976.

Anderson, E. A.: Snow Accumulation and Ablation Model – SNOW-17, NWSRFS User Documentation, U.S. National Weather Service, Silver Springs, MD, 2006.

Aquaveo: Watershed Modeling System Version 9.1.4, Provo, Utah, 2013.

Bayard, D., Stähli, M., Parriaux, A., and Flühler, H.: The influence of seasonally frozen soil on the snowmelt runoff at two Alpine sites in southern Switzerland, J. Hydrol., 309, 66–84, 2005.

Beck, P. S., Atzberger, C., Høgda, K. A., Johansen, B., and Skidmore, A. K.: Improved monitoring of vegetation dynamics at very high latitudes: A new method using MODIS NDVI, Remote Sens. Environ., 100, 321–334, 2006.

Bras, R. L.: Hydrology: An Introduction to Hydrologic Science, Addison-Wesley, Reading, Massachusetts, 643 pp., 1990.

Brown, J.: Influence of vegetation on permafrost, in: Permafrost International Conference, edited by: Woods, K. B., National Academy of Sciences-National Research Council, NRC Div. Bldg., Ottawa, Canada, 6 pp., 1966.

Campbell, J. L., Ollinger, S. V., Flerchinger, G. N., Wicklein, H., Hayhoe, K., and Bailey, A. S.: Past and projected future changes in snowpack and soil frost at the Hubbard Brook Experiment Forest, New Hampshire, USA, Hydrol. Process., 24, 2465–2480, 2010.

Carey, S. K. and Woo, M.-K.: Freezing of subarctic hillslopes, Wolf Creek Basin, Yukon, Canada, Arct. Antarct. Alp. Res., 37, 1–10, 2005.

Chen, R. S., Kang, E. S., Ji, X. B., Yang, Y., Zhang, Z. H., Qing, W. W., Bai, S. Y., Wang, L. D., Kong, Q. Z, Lei, Y. H., and Pei, Z. X.: Preliminary study of the hydrological processes in the alpine meadow and permafrost regions at the headwaters of Heihe River, Journal of Glaciology and Geocryology, 29, 387–396, 2007.

De Roo, A., Odijk, M., Schmuck, G., Koster, E., and Lucieer, A.: Assessing the effects of land use changes on floods in the Meuse and Oder catchment, Phys. Chem. Earth Pt. B, 26, 593–599, 2001.

DeWalle, D. R. and Rango, A.: Principles of Snow Hydrology, Cambridge University Press, Cambridge, England, 2008.

Diebold, C. H.: The effects of vegetation upon snow cover and frost penetration during the March 1936 floods, J. Forest., 36, 1131–1137, 1938.

Doherty, J., Brebber, L., and Whyte, P.: PEST: Model-Independent Parameter Estimation, Watermark Computing, Corinda, Australia, 122, 1994.

Downer, C. W.: Development of a Simple Soil Moisture Model in the Hydrologic Simulator GSSHA, ERDC-TN-SWWRP-07-8, U.S. Army Engineer Research and Development Center, Vicksburg, Mississippi, 2007.

Downer, C. W. and Ogden, F. L.: GSSHA: Model to simulate diverse stream flow producing processes, J. Hydrol. Eng., 9, 161–174, 2004.

Downer, C. W. and Ogden, F. L.: GSSHA Users' Manual. in: SR-06-1, U.S. Army Engineer Research and Development Center, Vicksburg, Mississippi, 2006.

Duffie, J. A. and Beckman, W. A.: Solar Engineering of Thermal Processes, 3, Wiley, New York, 1980.

Dun, S., Wu, J., McCool, D., Frankenberger, J., and Flanagan, D.: Improving frost-simulation subroutines of the Water Erosion Prediction Project (WEPP) model, T. ASABE, 53, 1399–1411, 2010.

Dunne, T. and Black, R. D.: Runoff processes during snowmelt, Water Resour. Res., 7, 1160–1172, 1971.

Dunne, T. and Black, R. D.: An experimental investigation of runoff production in permeable soils, Water Resour. Res., 6, 478–490, 1970a.

Dunne, T. and Black, R. D.: Partial area contributions to storm runoff in a small New England watershed, Water Resour. Res., 6, 1296–1311, 1970b.

Fahey, T. J. and Lang, G. E.: Concrete frost along an elevational gradient in New Hampshire, Can. J. Forest Res., 5, 700–705, 1975.

Farouki, O. T.: The thermal properties of soils in cold regions, Cold Reg. Sci. Technol., 5, 67–75, 1981.

Fassnacht, S. R., López-Moreno, J. I., Ma, C., Weber, A. N., Pfohl, A. K. D., Kampf, S. K., and Kappas, M.: Spatio-temporal snowmelt variability across the headwaters of the Southern Rocky Mountains, Front. Earth Sci., 11, 1–10, 2017.

Flerchinger, G. and Saxton, K. E.: Simultaneous heat and water model of a freezing snow-residue-soil system I, Theory and development, T. ASAE, 32, 565–0571, 1989.

Follum, M. L., Downer, C. W., and Niemann, J. D.: Simulating the spatial distribution of snow pack and snow melt runoff with different snow melt algorithms in a physics based watershed model, AGU Fall Meeting, American Geophysical Union, San Francisco, CA, 2014.

Follum, M. L., Downer, C. W., Niemann, J. D., Roylance, S. M., and Vuyovich, C. M.: A radiation-derived temperature-index snow routine for the GSSHA hydrologic model, J. Hydrol., 529, 723–736, 2015.

Fox, J. D.: Incorporating freeze-thaw calculations into a water balance model, Water Resour. Res., 28, 2229–2244, 1992.

Fry, J. A., Xian, G., Jin, S., Dewitz, J. A., Homer, C. G., Limin, Y., Barnews, C. A., Helold, N. D., and Wickham, J. D.: Completion of the 2006 national land cover database for the conterminous United States, Photogramm. Eng. Rem. S., 77, 858–864, 2011.

Gesch, D., Oimoen, M., Greenlee, S., Nelson, C., Steuck, M., and Tyler, D.: The national elevation dataset, Photogramm. Eng. Rem. S., 68, 5–32, 2002.

Green, W. and Ampt, G.: Studies on Soil Physics, J. Agr. Sci., 4, 1–24, 1911.

GSSHA (Gridded Surface Subsurface Hydrologic Analysis): available at: https://gsshawiki.com/Gridded_Surface_Subsurface_Hydrologic_Analysis, (last access: 25 April 2018), 2018.

Gustafson, J. R., Brooks, P. D., Molotch, N. P., and Veatch, W. C.: Estimating snow sublimation using natural chemical and isotropic tracers across a gradient of solar radiation, Water Resour. Res., 46, 1–14, 2010.

Henneman, H. E. and Stefan, H. G.: Albedo models for snow and ice on a freshwater lake, Cold Reg. Sci. Technol., 29, 31–48, 1999.

Jansson, P. E.: Coupled heat and mass transfer model for soil-plant-atmosphere systems, Royal Institute of Technology, Stockholm Sweden, 2001.

Jansson, P. E. and Karlberg, L.: Coupled heat and mass transfer model for soil-plant-atmosphere systems, Royal Institute of Technology, Stockholm, 454 pp., 2010.

Johansen, O.: Thermal Conductivity of Soils, CRREL-TL-637, Cold Regions Research and Engineering Laboratory, Hanover, NH, 1977.

Jost, G., Weiler, M., Gluns, D. R., and Alila, Y.: The influence of forest and topography on snow accumulation and melt at the watershed-scale, J. Hydrol., 347, 101–115, 2007.

Kang, D. H.: Distributed Snowmelt Modeling with GIS and CASC2D at California Gulch, Colorado, Colorado State University, Fort Collins, CO, 195 pp., 2005.

Kennedy, I. and Sharratt, B.: Model Comparisons to Simulate Soil Frost Depthwi, Soil Sci., 163, 636–645, 1998.

Koren, V., Schaake, J., Mitchell, K., Duan, Q. Y., Chen, F., and Baker, J. M.: A parameterization of snowpack and frozen ground intended for NCEP weather and climate models, J. Geophys. Res.-Atmos., 104, 19569–19585, 1999.

Kurganova, I., Teepe, R., and Loftfield, N.: Influence of freeze-thaw events on carbon dioxide emission from soils at different moisture and land use, Carbon Balance and Management, 2, 2–2, 2007.

Lin, C. and McCool, D.: Simulating snowmelt and soil frost depth by an energy budget approach, T. ASABE, 49, 1383–1394, 2006.

Lindstrom, G., Bishop, K., and Lofvenius, M. O.: Soil frost and runoff at Svartberget, northern Sweden – Measurements and model analysis, Hydrol. Process., 16, 3379–3392, 2002.

Liou, K. N.: An Introduction to Atmospheric Radiation, second edition, Academic Press, New York, 2002.

Liston, G. E. and Elder, K.: A distributed snow-evolution modeling system (SnowModel), J. Hydrometeorol., 7, 1259–1276, 2006.

MacKinney, A.: Effects of forest litter on soil temperature and soil freezing in autumn and winter, Ecology, 10, 312–321, 1929.

McNamara, J. P., Kane, D. P., and Hinzman, L. D.: Hydrograph separation in an Arctic watershed using mizing model and graphical techniques, Water Resour. Res., 33, 1707–1719, 1997.

Molnau, M. and Bissell, V. C.: A continuous frozen ground index for flood forecasting, Proceedings 51st Annual Meeting Western Snow Conference, Canadian Water Resources Association, Cambridge, Ontario, 109–119, 1983.

Monteith, J.: Evaporation and environment, Symposia of the Society for Experimental Biology, 4, Cambridge, England, 1965.

Monteith, J.: Evaporation and surface temperature, Q. J. Roy. Meteor. Soc., 107, 1–27, 1981.

Musselman, K., Molotch, N. P., and Brooks, P. D.: Effects of vegetation on snow accumulation and ablation in a mid-latitude subalpine forest, Hydrol. Process., 22, 2767–2776, 2008.

NCEI (National Centers for Environmental Information): Cloud cover, relative humidity, wind speed, and atmospheric pressure data, available at: https://www.ncdc.noaa.gov/, (last access: 7 November 2016), 2018.

Nikolaev, I. V., Leong, W. H., and Rosen, M. A.: Experimental investigation of soil thermal conductivity over a wide temperature range, Int. J. Thermophys., 34, 1110–1129, 2013.

Nyberg, L., Stähli, M., Mellander, P. E., and Bishop, K.: Soil frost effects on soil water and runoff dynamics along a boreal forest transact: 1. Field investigations, Hydrol. Process., 15, 909–926, 2001.

Ogden, F. L. and Saghafian, B.: Green and Ampt infiltration with redistribution, J. Irrig. Drain. E., 123, 386–393, 1997.

Pearson, G. A.: Factors controlling the distribution of forest types, Part I, Ecology, 1, 139–159, 1920.

Pomeroy, J.: Wind Transport of Snow, PhD Thesis, University of Saskatchewan, 1988.

Pomeroy, J., Gray, D., Hedstrom, N., and Janowicz, J.: Prediction of seasonal snow accumulation in cold climate forests, Hydrol. Process., 16, 3543–3558, 2002.

Prèvost, M., Barry, R., Stein, J., and Plamondon, A. P.: Snowmelt runoff modelling in a balsam fir forest with variable source are simulator (VSAS2), Water Resour. Res., 25, 1067–1077, 1990.

Rawls, W. J. and Brakensiek, D.: Prediction of soil water properties for hydrologic modelling, in: Watershed Management in the Eighties, edited by: Ward, T. J., ASCE, 293–299, 1985.

Rawls, W., Brakensiek, D., and Saxton, K.: Estimation of soil water properties, T. ASAE, 25, 1316–1320, 1982.

Rawls, W. J., Brakensiek, D. L., and Miller, N.: Green-Ampt infiltration parameters from soils data, J. Hydraul. Eng., 109, 62–70, 1983.

Rekolainen, S. and Posch, M.: Adapting the CREAMS model for Finnish conditions, Hydrol. Res., 24, 309–322, 1993.

Ricard, J. A., Tobiasson, W., and Greatorex, A.: The field assembled frost gage, United States Army Cold Regions Research and Engineering Laboratory, Hanover, New Hampshire, 1976.

Rinehart, A. J., Vivoni, E. R., and Brooks, P. D.: Effects of vegetation, albedo, and solar radiation sheltering on the distribution of snow in the Valles Caldera, New Mexico, Ecohydrology, 1, 253–270, 2008.

Sartz, R. S.: Effect of Forest Cover Removal on Depth of Soil Freezing and Overland Flow, Soil Sci. Soc Am. J., 37, 774–777, 1973.

Shanley, J. B.: Sleepers River, Vermont: A water, energy, and biogeochemical budgets program site, Fact Sheet-166-99, U.S. Department of the Interior, U.S. Geological Survey, 2000.

Shanley, J. B. and Chalmers, A.: The effect of frozen soil on snowmelt runoff at Sleepers River, Vermont, Hydrol. Process., 13, 1843–1857, 1999.

Shanley, J. B., Sundquist, E. T., and Kendall, C.: Water, energy, and biogeochemical budget research at Sleepers River Research Watershed, Vermont, 2331–1258, U.S. Geological Survey, Open-File Reports Section [distributor], 1995.

Sicart, J. E., Pomeroy, J. W., Essery, R. L. H, Hardy, J., Link, T., and Marks, D.: A sensitivity study of daytime net radiation during snowmelt to forest canopy and atmospheric conditions, J. Hydrometeorol., 5, 774–784, 2004.

Smith, J. A.: Precipitation, in: Handbook of Hydrology, edited by: Maidment, D. R., McGraw-Hill Inc., New York, 3.4, 1993.

Stähli, M.: Hydrological significance of soil frost for pre-alpine areas, J. Hydrol., 546, 90–102, 2017.

Stähli, M., Jansson, P.-E., and Lundin, L.-C.: Soil moisture redistribution and infiltration in frozen sandy soils, Water Resour. Res., 35, 95–103, 1999.

USGS: Sleepers River Research Watershed, Danville, Vermont, Vermont Field Office, available at: https://nh.water.usgs.gov/project/sleepers/index.htm, (last access: 23 April 2018), 2018a.

USGS: The National Map, available at: https://nationalmap.gov/, (last access: 23 April 2018), 2018b.

TVA: Heat and Mass Transfer Between a Water Surface and the Atmosphere, Tennessee Valley Authority, Norris, TN, 1972.

Van Der Knijff, J., Younis, J., and De Roo, A.: LISFLOOD: a GIS-based distributed model for river basin scale water balance and flood simulation, Int. J. Geogr. Inf. Sci., 24, 189–212, 2010.

Veatch, W., Brooks, P. D., Gustafson, J. R., and Molotch, N. P.: Quantifying the effects of forest canopy cover on net snow accumulation at a continental, mid-latitude site, Ecohydrology, 2, 115–128, 2009.

Vermette, S. and Christopher, S.: Using the rate of accumulated freezing and thawing degree days as a surrogate for determining freezing depth in a temperate forest soil, Middle States Geographer, 41, 68–73, 2008.

Vermette, S. and Kanack, J.: Modeling frost line soil penetration using freezing degree-day rates, day length, and sun angle, 69th Eastern Snow Conference, Frost Valley YMCA, Claryville, New York, USA, 2012.

Wang, Q., Adiku, S., Tenhunen, J., and Granier, A.: On the relationship of NDVI with leaf area index in a deciduous forest site, Remote Sens. Environ., 94, 244–255, 2005.

Webster, C., Rutter, N., and Jonas, T.: Improving representation of canopy temperatures for modeling subcanopy incoming longwave radiation to the snow surface, J. Geophys. Res.-Atmos., 122, 9154–9172, 2017.

Wilcox, B. P., Newman, B. D., Brandes, D., Davenport, D. W., and Reid, K.: Runoff from a semiarid ponderosa pine hillslope in New Mexico, Water Resour. Res., 33, 2301–2314, 1997.

Willis, W., Carlson, C., Alessi, J., and Haas, H.: Depth of freezing and spring run-off as related to fall soil-moisture level, Can. J. Soil Sci. Sci., 41, 115–123, 1961.

Woo, M.: Permafrost hydrology in North America, Atmos. Ocean., 24, 201–234, 1986.

Woo, M., Arain, M., Mollinga, M., and Yi, S.: A two-directional freeze and thaw algorithm for hydrologic and land surface modelling, Geophys. Res. Lett., 31, L12501, https://doi.org/10.1029/2004GL019475, 2004.

4

Intercomparison of different uncertainty sources in hydrological climate change projections for an alpine catchment (upper Clutha River, New Zealand)

Andreas M. Jobst[1], Daniel G. Kingston[1], Nicolas J. Cullen[1], and Josef Schmid[2]

[1]Department of Geography, University of Otago, P.O. Box 56, Dunedin, New Zealand
[2]Department of Geography, Ludwig Maximilian University (LMU), Munich, Germany

Correspondence: Andreas M. Jobst (andreas.jobst@otago.ac.nz)

Abstract. As climate change is projected to alter both temperature and precipitation, snow-controlled mid-latitude catchments are expected to experience substantial shifts in their seasonal regime, which will have direct implications for water management. In order to provide authoritative projections of climate change impacts, the uncertainty inherent to all components of the modelling chain needs to be accounted for. This study assesses the uncertainty in potential impacts of climate change on the hydro-climate of a headwater sub-catchment of New Zealand's largest catchment (the Clutha River) using a fully distributed hydrological model (WaSiM) and unique ensemble encompassing different uncertainty sources: general circulation model (GCM), emission scenario, bias correction and snow model. The inclusion of snow models is particularly important, given that (1) they are a rarely considered aspect of uncertainty in hydrological modelling studies, and (2) snow has a considerable influence on seasonal patterns of river flow in alpine catchments such as the Clutha. Projected changes in river flow for the 2050s and 2090s encompass substantial increases in streamflow from May to October, and a decline between December and March. The dominant drivers are changes in the seasonal distribution of precipitation (for the 2090s +29 to +84 % in winter) and substantial decreases in the seasonal snow storage due to temperature increase. A quantitative comparison of uncertainty identified GCM structure as the dominant contributor in the seasonal streamflow signal (44–57 %) followed by emission scenario (16–49 %), bias correction (4–22 %) and snow model (3–10 %). While these findings suggest that the role of the snow model is comparatively small, its contribution to the overall uncertainty was still found to be noticeable for winter and summer.

1 Introduction

Over recent decades climate change has had a considerable impact on the Earth's freshwater resources (Jiménez Cisneros et al., 2014), causing, amongst others, changes in the amount of runoff (Piao et al., 2010), the timing of peak discharge (Hidalgo et al., 2009), a reduction in glacier volume (Rosenzweig et al., 2007) and an increase in flood risk (Pall et al., 2011). Future impacts under mid- and late 21st century climate change are projected to intensify, affecting both the main processes and stores of the water cycle. The impacts include an increase of potential evapotranspiration (PET) over most land areas, a further shrinkage of glaciers and changes in the runoff regime of snowmelt-affected basins (Jiménez Cisneros et al., 2014). Thus, 21st century climate change is expected to have substantial implications for water users and operators alike, which makes robust projections of potential changes in the seasonality and magnitude of streamflow essential.

While general circulation model (GCM) land surface schemes can be used for climate change impact assessments (e.g. Haddeland et al., 2011; Gudmundsson et al., 2012), their coarse resolution makes them inadequate for modelling studies at the small scale and mesoscale. Thus, climate change impact studies typically use a cascade of models and processing steps to move between the scales of the lower resolution climate models and a separate higher resolution hydrological model (Maraun et al., 2010; Muerth et al., 2013).

As discussed by Muerth et al. (2013), the hydro-climatic model chain typically consists of the following components: emission scenario, GCM, regional climate model (RCM) or statistical downscaling, bias correction, and hydrological model. All of these components constitute a potential uncertainty source, and as such all need to be examined to provide a truly comprehensive understanding of the uncertainty associated with hydrological impact assessments (Teutschbein and Seibert, 2010). The uncertainty associated with the individual components of the model chain has been investigated by an increasing number of studies. Typically, GCM structure is identified as the dominant source of uncertainty (e.g. Graham et al., 2007; Prudhomme and Davies, 2009; Hagemann et al., 2011; Dobler et al., 2012). There is little agreement on the second most important source of uncertainty between the downscaling method (Wilby and Harris, 2006; Prudhomme and Davies, 2009; Dobler et al., 2012), the bias correction (Vormoor et al., 2015) or the emission scenario (Bennett et al., 2012). A common finding is that hydrological model uncertainty is less important than other uncertainty sources (i.e. GCM), but cannot be ignored (Prudhomme and Davies, 2009; Teng et al., 2012; Thompson et al., 2013; Velázquez et al., 2013). However, for certain hydrological indicators (e.g. high-flow events) hydrological models can be associated with a comparable uncertainty range to the driving climate projections (e.g. Ludwig et al., 2009; Muerth et al., 2012).

As an alternative to an ensemble of different hydrological models varying in their representation of spatial variation (i.e. lumped, semi-distributed, fully distributed) and process descriptions (i.e. stochastic, conceptual or physically oriented), some studies have explored uncertainty associated with particular routines within a single model. Examples include the sensitivity of climate change impacts on the PET method used (e.g. Kay and Davies, 2008; Thompson et al., 2014). However, in snowmelt-affected mid-latitude catchments PET-related uncertainty is often relatively small (e.g. Koedyk and Kingston, 2016), with uncertainty linked to snow-related processes being more important. For example, Troin et al. (2016) investigated the uncertainty introduced by the snowmelt routine in a hydrological model for three Canadian catchments. For a number of snow indicators (e.g. snow water equivalent, SWE), most of the uncertainty was found to be caused by natural climate variability. For temporal indices (e.g. duration of snowpack), however, the different snow models showed a greater variability. Troin et al. (2016) did not look at the implications of snow model uncertainty for river flow, but the greater uncertainty associated with temporal indices could be indicative of significant implications on the timing of snowmelt and therefore the annual streamflow regime. Thus, the choice of the snow model as a potential uncertainty source and its implications on streamflow needs to be explored further, particularly in alpine catchments.

The aim of this present study is to investigate the contribution of the snow model and three more commonly studied uncertainty sources (i.e. GCM, emission scenario and bias-correction method) to the climate change signal in hydrological projections. New Zealand's largest catchment, the Clutha, was selected for this purpose as its highly complex hydro-climate, including snow-affected headwaters, makes it a particularly interesting case study. To this end, the fully distributed hydrological model WaSiM (Schulla, 2012) was implemented for an alpine sub-catchment of the Clutha, with a total of 32 separate hydrological simulations produced. These comprised two emissions scenarios, four GCMs, two bias-correction methods and two snow models. Although previous New Zealand studies (including for the Clutha) have examined multiple GCM scenarios (e.g. Poyck et al., 2011; Gawith et al., 2012; Caruso et al., 2017), none have used an ensemble covering the present range of uncertainty sources. Furthermore, in using WaSiM this will be the first application of a fully distributed and grid-based hydrological model for this purpose in a medium- to large-scale New Zealand catchment. Consequently, this study will generate the most complete assessment of climate change impacts on streamflow and associated uncertainty for an alpine New Zealand catchment. Importantly, the study will also speak more widely to the issue of snowmelt uncertainty under climate change in alpine catchments.

2 Data and methods

2.1 The study area

The Clutha–Mata-Au is the largest catchment (20 586 km^2) in New Zealand and is situated in the lower half of the South Island, extending eastwards from the Southern Alps (Fig. 1a). It has the highest average streamflow of any river in New Zealand (approximately 570 $m^3 s^{-1}$) and drains 6 % of the South Island's water (Murray, 1975). The catchment is characterized by a highly variable hydro-climate ranging from very humid alpine headwaters dominated by seasonal snow accumulation and melt, to substantially drier areas in the central catchment. The Clutha catchment can thus be considered broadly representative of most of the South Island's hydrologic and climatic domain, and so an ideal candidate for investigating climate change impacts.

As described in Jobst (2017) WaSiM was implemented for the entire Clutha catchment as a tool for climate change impact modelling. Most of the upper and lower Clutha catchment are under extensive water management (particularly for hydro-electric dams and water abstractions), except for the north-western part (gauge Chards Rd in Fig. 1b), which is characterised by natural flow conditions. As the focus of this climate change impact study is on potential changes

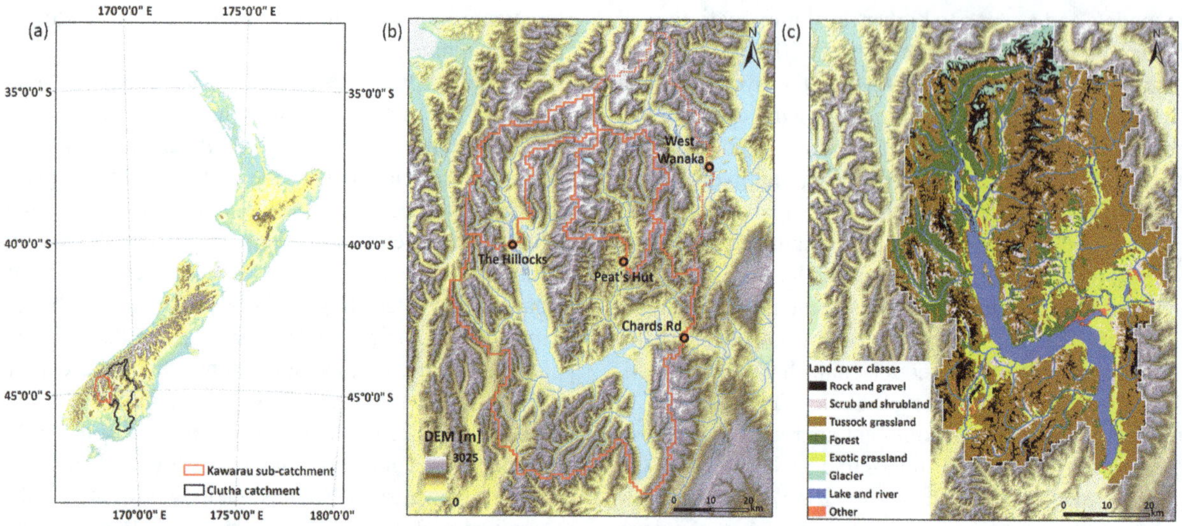

Figure 1. Maps showing **(a)** New Zealand with the Clutha catchment located in the lower South Island, **(b)** the sub-catchments of the gauges recording streamflow used in this study (note that the West Wanaka sub-catchment is outside the Kawarau sub-catchment) and **(c)** a land cover classification of the Kawarau sub-catchment based on New Zealand's Land Cover Database (LCDB v3.0) that was published by Landcare Research in 2012.

in natural streamflow and seasonal snow, the results focus mostly on the Kawarau River sub-catchment (see Jobst, 2017 for model simulations at the other key sites of the Clutha). The catchment area of the Kawarau River (at Chards Rd) is 4541 km^2 (22 % of entire Clutha basin) and with a mean discharge of 212 $m^3 s^{-1}$ is the largest component of the upper Clutha basin, comprising 36 % of flow at the catchment outlet. Snowmelt contributes approximately 20 % of annual flow. Streamflow at Chards Rd is also highly correlated with the headwater sub-catchment of Lake Wanaka (located to the east of the Kawarau sub-catchment), while the outflow of Lake Hāwea is controlled by a dam for hydropower generation further downstream.

Most of the Kawarau sub-catchment is covered by indigenous tussock grassland followed by low-producing exotic grassland and forest (Fig. 1c). The elevation of the sub-catchment ranges between 300 and 2800 m. The ice-covered area inside the Kawarau catchment is small and, based on 2001–2002 satellite imagery, amounts to 1.8 % or 84 km^2 (New Zealand's Land Cover Database v3.0 as published by Landcare Research in 2012). Chinn (2001) estimated the volume of ice in the Kawarau basin at 2.25 km^3 using the area-mean–depth relationship of the World Glacier Monitoring Service. Thus, based on Chinn's (2001) estimate the water stored in the glaciers of the Kawarau could only sustain its mean flow for $\sim$ 123 days, which highlights that glaciers in the catchment are less important from a hydrological perspective.

2.2 The WaSiM model of the Clutha

The fully distributed and physically oriented hydrological model WaSiM-Richards (version 9.06.10) was implemented at a spatial resolution of 1 km and at a daily time step. The main components of this implementation of WaSiM are described briefly here – for a more detailed description see Schulla (2012). The modelling of PET is solved by the Penman–Monteith approach, while actual evapotranspiration (ET) is a function of the simulated soil water content. Soil and groundwater processes are described by finite differencing of the 1-D Richards equation combined with a 2-D groundwater model. In addition, WaSiM's dynamic glacier model was used to describe the glacial processes for the ice-covered cells located in the upper catchment. The glacier model uses three degree-day factors to calculate melt from the three storage components snow, firn and ice. After each year the remaining snow of a cell forms a new firn layer and once the firn stack reaches seven layers (as recommended by Schulla, 2012) the lowest firn layer turns into ice.

WaSiM was parameterised using both remotely sensed data (i.e. MODIS-15A2-1km for leaf area index) and values obtained from the literature. Two versions of this WaSiM implementation were set up, one with a simple temperature index (Tindex) snowmelt routine (Schulla, 2012) and the other with the conceptual energy balance model of Anderson (1973). The Tindex model calculates the melting rate via a degree-day factor multiplied by the difference between actual temperature and the melting point temperature. The Anderson model is more complex as it computes four separate melt fractions, accounts for radiation by using a seasonal melt factor and also models the refreezing of liquid water stored in the snowpack when the actual temperature is below melting point.

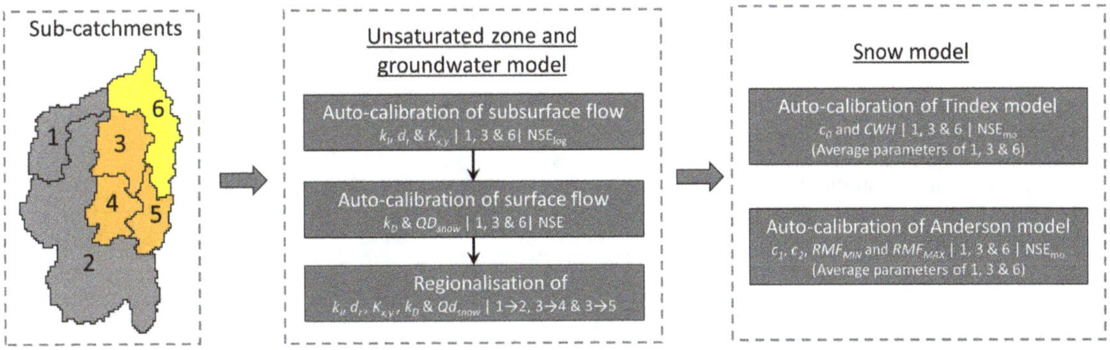

Figure 2. The calibration workflow that was used to calibrate WaSiM. For each calibration step (either manual or automatic) information is provided as follows: "calibrated parameters | calibrated sub-catchments | performance criterium used". The parameters controlling subsurface flow encompass the drainage density for interflow (d_r), the storage coefficient of interflow (k_I) and the groundwater conductivity in x or y-direction. Surface flow is controlled by the storage coefficient of surface runoff (k_D) and the fraction of snowmelt directly becoming surface runoff (QD_snow). The parameters of the snow models are shown in Table 1.

Station-based meteorological observations of mean daily air temperature (T_mean), precipitation, solar radiation, relative humidity and wind speed were interpolated (Jobst, 2017; Jobst et al., 2017) and served as input to WaSiM during the calibration (2008–2012) and validation (1992–2008) periods. The last four hydrological years of the reference period were chosen for calibration because of the higher density of weather stations compared to previous years and a better consistency of the streamflow records. Further, the relatively short calibration period constitutes a compromise between a reasonable processing time and a sufficient number of iterations (as part of the auto-calibration). Although the calibration period is rather short this means that the longer validation period allows for a robust assessment of the hydrological model's performance.

After a 2-year model spin-up the individual sub-models of WaSiM (unsaturated zone, groundwater, snow and glacier model) were calibrated iteratively using a combination of auto-calibration and manual parameter optimisation (Fig. 2; see Jobst, 2017 for a more detailed description of the calibration process). Particle swarm optimisation (PSO) (Kennedy and Eberhart, 1995) was used for auto-calibration due to its effective performance during the first iterations and fast operation (Jiang et al., 2010), allowing for an adequate compromise between processing time and efficiency. First the parameters controlling subsurface flow were calibrated using the Nash–Sutcliffe model coefficient of efficiency (NSE). Therefore, the NSE was based on daily logarithmic streamflow values ($\mathrm{NSE}_\mathrm{log}$) to account for the physical link between low-flow conditions and subsurface-flow components of WaSiM. The remaining parameters, controlling surface flow, were calibrated using regular daily streamflow values (standard NSE) to preserve the sensitivity of the NSE to flood peaks. A regionalisation based on spatial proximity and topographical similarity was carried out next (see map in Fig. 2) to parameterise sub-catchments that were ungauged or only had short records of streamflow (e.g. parameters of sub-catchment 3 $\rightarrow$ sub-catchment 4).

In the following step the two snow models were calibrated for three separate headwater sub-catchments (gauges: The Hillocks, Peat's Hut and West Wanaka as shown in Fig. 1b) against monthly streamflow (NSE_mo). The resulting parameter sets were then averaged resulting in a global parameter set for each of the two snow models respectively (Table 1).

As opposed to the other sub-models the glacier model was calibrated manually for the entire Clutha catchment using the annual volume estimates (1994–2010) of Willsman (2011) as calibration (1994–2001) and validation (2002–2010) source. The degree day factor (DDF) of ice (7.17 mm °C^{-1} d^{-1}) was based on the study of Anderson et al. (2006), while the DDF of snow (3.80 mm °C^{-1} d^{-1}) was calibrated manually. The DDF of firn (5.54 mm °C^{-1} d^{-1}) was calculated by averaging the DDFs of snow and ice, to ensure a physically sound relationship between the three parameters (DDF of ice > DDF of firn > DDF of snow).

With regards to the validation of streamflow, daily NSE values indicated a reasonable performance for the three tributaries (Dart, Shotover and Matukituki rivers), while monthly NSE values indicated a good performance at these sites (Table 2). For the Matukituki River the validation of both WaSiM versions revealed a substantially better simulation of monthly streamflow (NSE of 0.83 and 0.82, respectively) when compared to the TopNet-based modelling study of Gawith et al. (2012) (NSE of 0.68). For Chards Rd the validation of WaSiM-Anderson and WaSiM-Tindex revealed a strong performance at the daily and monthly timescale, with NSE values between 0.85 and 0.90 across all model versions, timescales and time periods (Table 2). The monthly hydrographs of WaSiM-Anderson and WaSiM-Tindex (Fig. 3c) further indicate a realistic representation of observed daily

Table 1. Parameters of the two snowmelt routines as determined by the calibration routine. T_0 and CWH are mutual parameters of both models. The snow accumulation parameters are based on the findings of Auer (1974).

Anderson and Tindex	
Threshold temperature for snowmelt (T_0)	0.00 (°C)
Water holding capacity of snow (CWH)	0.29 (–)
Tindex	
Temperature-dependent DDF (c_0)	1.91 (mm °C^{-1} d^{-1})
Anderson	
Temperature-dependent DDF (c_1)	0.63 (mm °C^{-1} d^{-1})
Wind-dependent DDF (c_2)	0.08 (mm (°C m s^{-1} d)$^{-1}$)
Minimum radiation melt factor (RMF_{MAX})	3.13 (mm °C^{-1} d^{-1})
Maximum radiation melt factor (RMF_{MIN})	0.36 (mm °C^{-1} d^{-1})
Snow accumulation	
Temperature, at which 50 % of precipitation are falling as snow ($T_{R/S}$)	3.00 (°C)
Half of the temperature-transition range from snow to rain (T_{trans})	3.00 (K)

runoff at Chards Rd. Obvious inaccuracies of both WaSiM versions are an underestimation of larger flow events during the melt period (e.g. December–January 1995, 1996 and 2000) and an overestimation during autumn (e.g. April–May 1994 and 1996). The likeliest explanation is that not enough snow is being accumulated from autumn to early winter and consequently the main melt peaks are under-simulated. This is largely supported by the monthly SWE simulations (Fig. 3b) and can be exemplified for the snow accumulation period of 1992 when WaSiM-Anderson results in a greater accumulation of SWE compared to WaSiM-Tindex. In the following melt period (November 1992 to February 1993) the observed peak is then approximated better by WaSiM-Anderson due to the larger SWE storage. Overall, inaccuracies are generally larger with WaSiM-Tindex as can be seen in December 1994 (Fig. 3c). Compared to WaSiM-Anderson the snowpack melts slower, which causes a greater underestimation of runoff during the main melt period. Overall the good performance of WaSiM is substantiated by the monthly NSE values which correspond to 0.89 and 0.88 for WaSiM-Anderson and WaSiM-Tindex, respectively.

The modelled glacier storage decreased from 540 to 265 mm during the reference period. For comparison Willsman (2011) found a decrease of 29 % for the entire Clutha catchment between 1994 and 2010, which is less than the decrease modelled here for the Kawarau (42 %). As there are no other studies that have quantified the glacier volume of the Kawarau the results of this study could not be assessed any further.

Due to the high annual precipitation totals the grid cells in the upper catchment could potentially be prone to large build-ups of snow that persist throughout the melting period and into the next hydrological year. As discussed by Freudiger et al. (2017) such snow towers can cause substantial errors in the modelled water balance and should therefore be made transparent. As indicated in Fig. 3b the snowpack of some cells persists across all of the melt seasons, with SWE values at the end of the melt season ranging between 7 and 46 mm (21 and 77 mm) for WaSiM-Anderson (WaSiM-Tindex). To further investigate the significance of this potential issue, all cells where the SWE value at the end of the hydrological year exceeded the annual snowfall (excluding glacier cells), were marked. The combined relative area of these cells is shown in Fig. 3a and does not exceed 1 % for WaSiM-Anderson, while for WaSiM-Tindex the affected area does not exceed 2 % except for the years 1997 and 1998. The mean exceedance (SWE / snowfall) for the individual years based on all affected cells varies from 1.3 to 1.8 for WaSiM-Anderson and from 1.4 to 2.3 for WaSiM-Tindex. Thus, for a number of cells along the main divide the modelled build-ups of snow could be classified as potential snow towers. However, the combined area of these cells is small, which means that the effect on the modelled water balance is negligible.

2.3 The model cascade

Most existing impact studies in the New Zealand domain (Poyck et al., 2011; Srinivasan et al., 2011; Zammit and Woods, 2011; Zemansky et al., 2012) have been based on statistically downscaled GCM simulations provided by the National Institute of Water and Atmospheric Research (NIWA) (Ministry for the Environment, 2008). More recently a small ensemble of four GCMs (CM2.1-GFDL, ECHAM5, HadCM3 and MK3.5-CSIRO) based on the A1B and A2 SRES emissions scenarios has been dynamically downscaled for the New Zealand domain using the HadRM3P RCM (Ackerley et al., 2012), and it is this ensemble of eight dy-

Table 2. Nash–Sutcliffe values based on daily (NSE), logarithmic (NSE_{log}) and monthly (NSE_{mo}) streamflow data during the calibration (Cal) and validation (Val) periods, calculated for the WaSiM-Anderson and WaSiM-Tindex (in brackets) simulations. Note the different validation periods for The Hillocks and Peat's Hut due to shorter records.

River	Gauge	Cal (1.4.2008–31.3.2012)			Val (1.4.1992–31.3.2008)		
		NSE	NSE_{log}	NSE_{mo}	NSE	NSE_{log}	NSE_{mo}
Dart	The Hillocks (1996–2012)	0.77 (0.77)	0.77 (0.78)	0.92 (0.92)	0.64 (0.65)	0.64 (0.68)	0.78 (0.79)
Shotover	Peat's Hut (1996–2012)	0.64 (0.65)	0.67 (0.70)	0.81 (0.82)	0.60 (0.62)	0.65 (0.70)	0.76 (0.79)
Kawarau	Chards Rd	0.87 (0.88)	0.88 (0.87)	0.89 (0.90)	0.87 (0.85)	0.86 (0.86)	0.89 (0.87)
Matukituki	West Wanaka	0.67 (0.67)	0.64 (0.65)	0.80 (0.80)	0.62 (0.62)	0.72 (0.72)	0.83 (0.82)

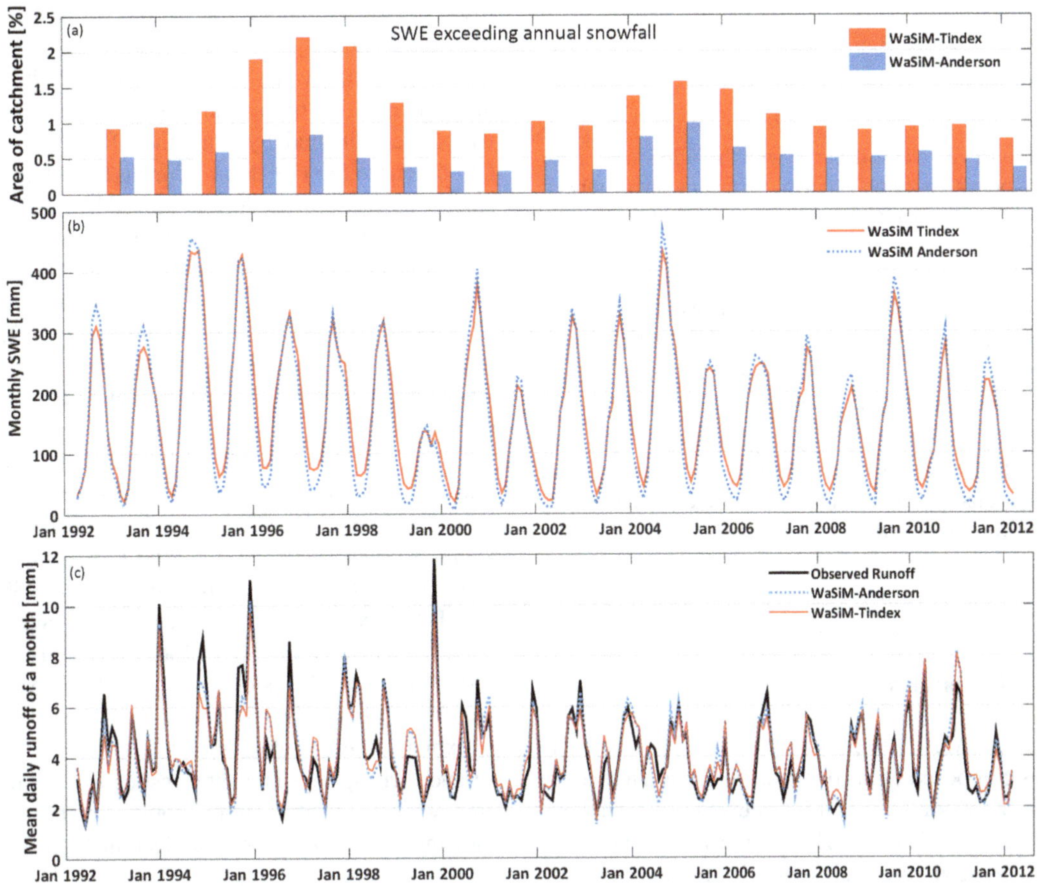

Figure 3. (a) Relative area of the Kawarau catchment where SWE exceeds annual snowfall for a hydrological year (e.g. 1 April 1992–31 March 1993) as an indicator for unrealistic build-up of snow (i.e. snow towers). **(b)** Modelled mean monthly snow water equivalent (SWE) based on the Tindex and Anderson simulations. **(c)** Observed and modelled (WaSiM-Anderson and WaSiM-Tindex) monthly runoff at Chards Rd for the reference period (1992–2012).

namically downscaled GCM simulations that forms the data set for the current study.

A model chain was constructed (Fig. 4) to process the raw RCM runs (from 1990 to 2099) and generate high-resolution climate change projections at the hydrological model scale. Two different bias-correction methods, linear transformation (LT; as described in Lenderink et al., 2007) and quantile mapping (QM; as described in Mpelasoka and Chiew, 2009), were used to correct the RCM data. Both methods have been successfully used by a number of studies (e.g. Boé et al., 2007; Chen et al., 2013; Gutjahr and Heinemann, 2013) and were selected here to allow for a direct comparison between a simple correction method based on additive or multiplicative correction terms (LT) and the more complex distribution-based QM approach.

To bridge the gap between the RCM grid (~27 km) and the hydrological model grid (1 km) an additional statistical downscaling step was required. The downscaling of precipi-

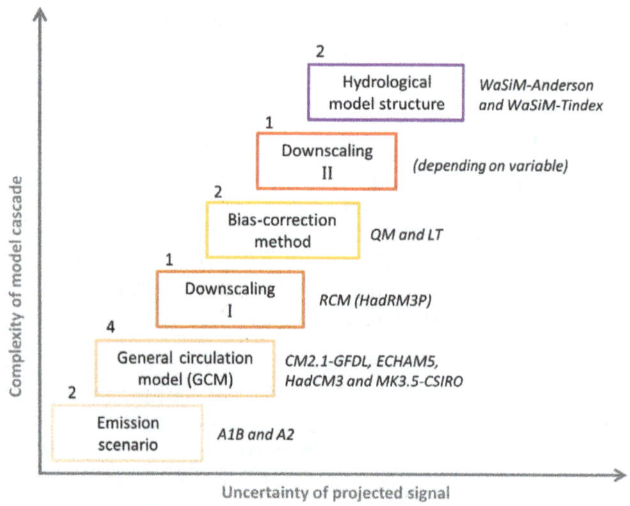

Figure 4. The model cascade is depicted with the individual members of each component listed on the right. Permuting the members of all components results in a total of 32 hydrological projections.

tation (and the remaining three variables) is based on the topographical scaling approach of Frueh et al. (2006), while maximum and minimum air temperatures are scaled via monthly lapse rate models (as described in Jobst et al., 2017 but excluding the thin plate spline layer). As part of the downscaling, additional processing steps were adopted from Marke (2011) to ensure the conservation of mass and energy when transforming the RCM data between the model scales.

3 Results

3.1 Baseline simulations

For the historic analysis, the ensemble was divided into four sub-ensembles composed of the two bias-correction methods and the two snow models (i.e. QM-Anderson, QM-Tindex, LT-Anderson and LT-Tindex). The regimes (i.e. monthly runoff values averaged across all years) of the eight RCM-driven simulations of each sub-ensemble were compared to the observed runoff regime (OBS) and the modelled runoff regime forced by the observed meteorology (MOD-METEO$_{OBS}$).

The skill in reproducing the observed historic regime varies substantially depending on both the bias-correction method and the snowmelt routine (Fig. 5). Overall QM-Anderson gave the most realistic approximation of the observed regime, although still with some overestimation in May (late autumn) followed by an underestimation during July and August (winter). QM-Tindex and LT-Anderson also underestimate the main peak; however, the general fit of their RCM members is still relatively close to the observed regime. The largest discrepancies occurred with LT-Tindex, with a substantially flatter regime, mainly due to too much flow being generated between May and September, leading to an underestimation of the main peak (November to January). Overall the LT method shows a lower skill in reproducing the observed regime, which is especially pronounced in combination with Tindex. This behaviour points to a high sensitivity of the modelled regime towards the bias-correction method and generally speaking the meteorological forcing.

The RCM-driven runs agree more closely with MOD-METEO$_{OBS}$ than with the observed regime, as monthly over- and underestimations of MOD-METEO$_{OBS}$ have propagated into the RCM-driven WaSiM simulations. This was expected as the RCM climate data have been tuned (i.e. bias-corrected) to the station-interpolated meteorology that was used to drive MOD-METEO$_{OBS}$.

Regarding the water balance (Table 3), the observed annual precipitation of the Kawarau sub-catchment (2007 mm) was underestimated by both the QM (1926 mm) and the LT sub-ensemble (1931 mm) during the reference period. A small part of that difference is caused by the shorter 360-day calendar of the RCM runs. Compared to MOD-METEO$_{OBS}$, ET was modelled almost identically by QM-Anderson and QM-Tindex, with slightly larger discrepancies (−1 %) under LT-Anderson and LT-Tindex. Regarding runoff, QM and LT resulted in an underestimation of −6 and −5 %, respectively, while the choice of the snow model had only a negligible impact.

In terms of the seasonal SWE volume, the QM-Anderson runs agree more closely with MOD-METEO$_{OBS}$ than the other sub-ensembles. The differences in the seasonal SWE volumes range from 3 % in spring to 6 % in autumn. Compared to MOD-METEO$_{OBS}$ the modelled SWE volumes of LT-Tindex were almost identical for summer and autumn but substantially lower for winter (−18 %) and spring (−16 %). Thus, the poor agreement between the observed runoff regime and the LT-Tindex runs (Fig. 5) can very likely be explained by too much melt being modelled between winter and early spring. The latter results in a reduced SWE volume, which is insufficient to supply streamflow with enough snowmelt during late spring and summer.

3.2 The climate change signals of precipitation and air temperature

The climate change signals of T_{mean} and precipitation that are presented in the following section correspond to the mean change of the spatio-temporal average between a future period (either 2050s or 2090s) and the reference period. For precipitation, the spread of the 2050s summer climate change signal (Fig. 6) is almost completely caused by the GCM structure. Both the emission scenario and bias-correction method have negligible effects on the extent of the signal range and median, with the latter showing a near-zero change in precipitation. Regarding the 2090s summer, the median change is more negative, while both the emission scenario and bias correction cause a slight increase in the uncertainty

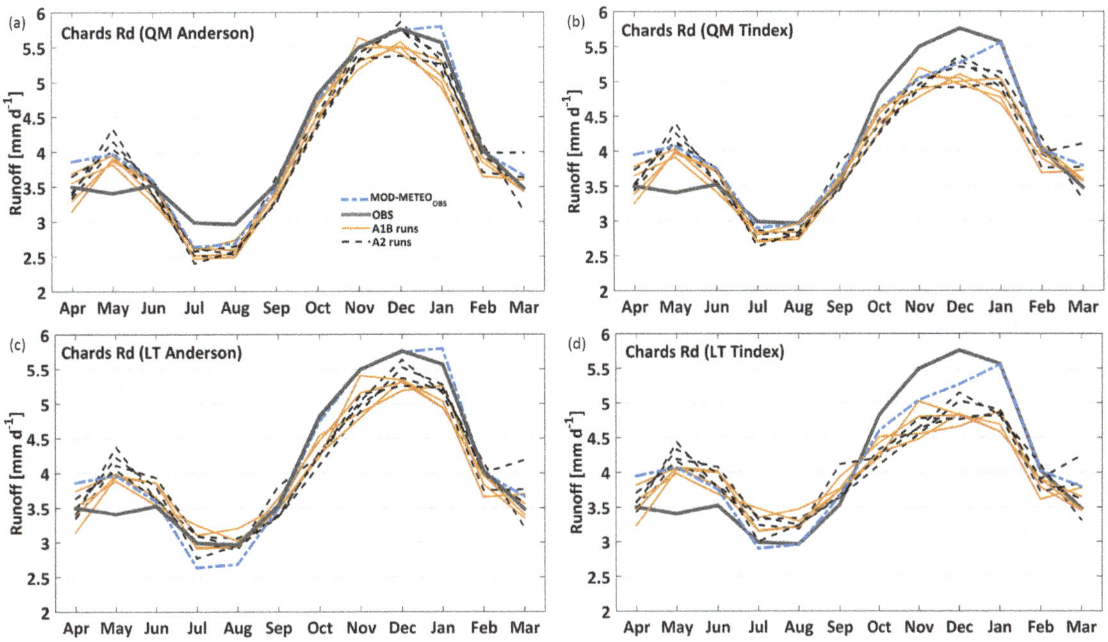

Figure 5. RCM-driven mean daily runoff simulations for the 1 April 1992–30 March 2011 period at Chards Rd. Simulations are compared with the observed regime (grey line) and the modelled (WaSiM forced with observed meteorology) regime (blue line) for the four sub-ensembles: **(a)** QM-Anderson, **(b)** QM-Tindex, **(c)** LT-Anderson and **(d)** LT-Tindex.

Table 3. The historic (1 April 1992–30 March 2011) water balance terms of the $MOD\text{-}METEO_{OBS}$ run compared with the corresponding (depending on snow model) RCM sub-ensemble means (QM and LT). The seasonal and annual SWE values are also shown.

Anderson	P (mm)	ET (mm)	Q (mm)	SWE (mm)				
				DJF	MAM	JJA	SON	YEAR
$MOD\text{-}METEO_{OBS}$	2007	471	1512	117	33	183	286	154
QM	1926	473	1425	123	31	176	277	152
LT	1931	466	1443	117	33	150	240	134
Tindex								
$MOD\text{-}METEO_{OBS}$	2007	473	1509	143	53	181	275	163
QM	1926	475	1422	143	44	172	262	154
LT	1931	468	1441	126	42	139	216	130

range. A different situation can be seen for the 2050s winter (Fig. 6), where the extent of the range is largely determined by the emission scenario. For the 2090s winter, all three components have a considerable impact on the uncertainty range. Here, the GCM spread is the largest of all seasons and future periods. It can also be seen that the precipitation signal is noticeably higher for the A2 sub-ensemble (mainly caused by ECHAM5-A2). In addition, the selection of the bias correction method considerably increases the extent of the whole ensemble, resulting in a total uncertainty range spanning 55.3 percentage points (i.e. a 28.5 to 83.8 % increase from the baseline).

For all seasons the uncertainty in the T_{mean} signals during the 2050s is predominantly caused by the GCM structure (Fig. 7). The selection of the emission scenario becomes a major source of uncertainty in the 2090s, with most of the A2 members projecting a stronger signal than their corresponding A1B members. However, this only holds for members stemming from the same GCM (e.g. ECHAM5-A1B and ECHAM5-A2), as can be seen for the 2090s winter, where an A1B member (MK3.5-CSIRO) has a greater warming signal than two of the A2 members (HadCM3 and CM2.1-GFDL).

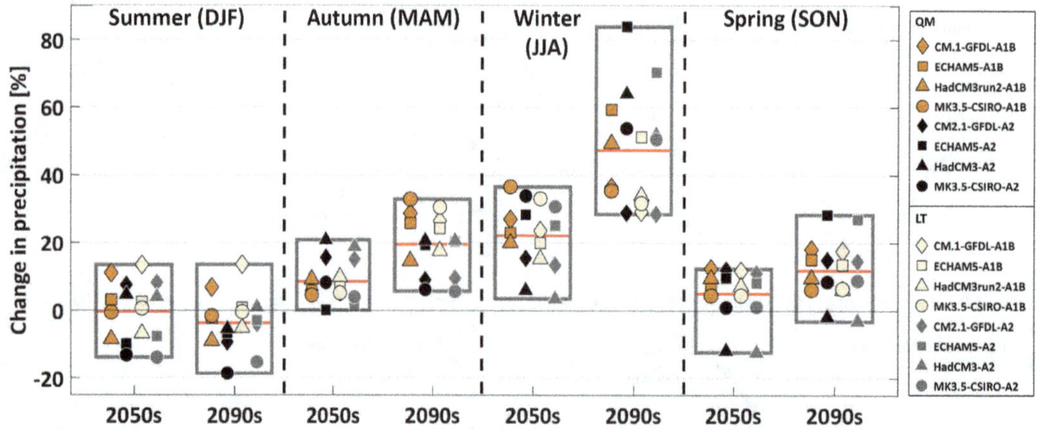

Figure 6. The uncertainty range of the precipitation signal (domain average) is shown for the entire ensemble and the four seasons. For each box the uncertainty range is broken down into the two emission scenarios and the two bias-correction methods. The red line represents the median of all 16 members.

3.3 The hydrological signals

3.3.1 Runoff

For both future periods the historic melt-driven December peak in the annual regime is projected to move earlier in the year (Fig. 8). In the 2050s, the highest monthly mean flow is projected to occur between October and November, with a further shift for the 2090s (to September and October). The most striking transformation for the 2090s is the dramatic enhancement of monthly flows during winter and spring, with uninterrupted increases from May to October.

In order to specifically compare the contribution of the snow model with the remaining sources of uncertainty, the seasonal signals in runoff are shown separately for WaSiM-Anderson and WaSiM-Tindex (Fig. 9). It can be seen that the influence of the snowmelt routine on seasonal flows is comparatively small for both periods and during all seasons. The most noticeable difference is an enhancement of the decrease during summer and a more pronounced increase during winter when using the Anderson model. Compared to the snow model the effect of the bias correction on the overall spread is more important. Positive signals were found to be enhanced by the QM method and vice versa for negative signals. Further, the influence of bias correction becomes visibly more important in the 2090s (except for autumn).

While both the bias correction and the snow model contribute to the overall spread, the GCM and the choice of the emission scenario appear to be the dominant sources of uncertainty. In most seasons the GCM range differs substantially depending on the underlying emission scenario. In the 2050s period this becomes especially apparent for the autumn season (Fig. 9a), during which the A2 runs show a much greater spread than the A1B runs. Regarding the 2090s period (Fig. 9b), differences between the A2 and the A1B runs become more pronounced and the most extreme signals are all represented by an A2 member.

Regarding the climate change signal derived from the overall ensemble mean, the largest changes were projected for winter and summer. For the 2090s (2050s) summer runoff is projected to decrease by −24 % (−10 %), with a substantial increase of 71 % (29 %) during winter. The overall ensemble spread becomes largest for the 2090s winter season, ranging from 40 to 116 %.

3.3.2 Snow water equivalent

During the historic period simulations of the monthly SWE storage varied considerably for the four sub-ensembles (as expected from the seasonal values in Table 3), with the lowest and highest volumes modelled by LT-Tindex and QM-Anderson, respectively. Despite these differences in the historic simulations, the relative changes are similar between the four sub-ensembles and thus the results are only shown for the QM-Anderson sub-ensemble (Fig. 10). A general observation is that the larger spread in the precipitation signal of the QM runs (Fig. 6) has clearly propagated into the uncertainty range of the monthly SWE volume, while the LT envelopes are visibly narrower during both time periods (particularly for the 2090s).

For the 2050s period (Fig. 10a), the A1B envelope predominantly lies within the upper and lower bounds of the A2 envelope, with the latter showing a substantially greater spread. The proportion of the two envelopes becomes reversed in the 2090s period, when the A1B envelope surpasses the A2 envelope in all months (Fig. 10b).

Although the 2090s envelopes of the four sub-ensembles have a relatively large overlapping area during winter and spring, all of the A2 members have a lower SWE volume than their A1B counterparts. It is, however, noticeable that the A1B member MK3.5-CSIRO has a lower SWE volume

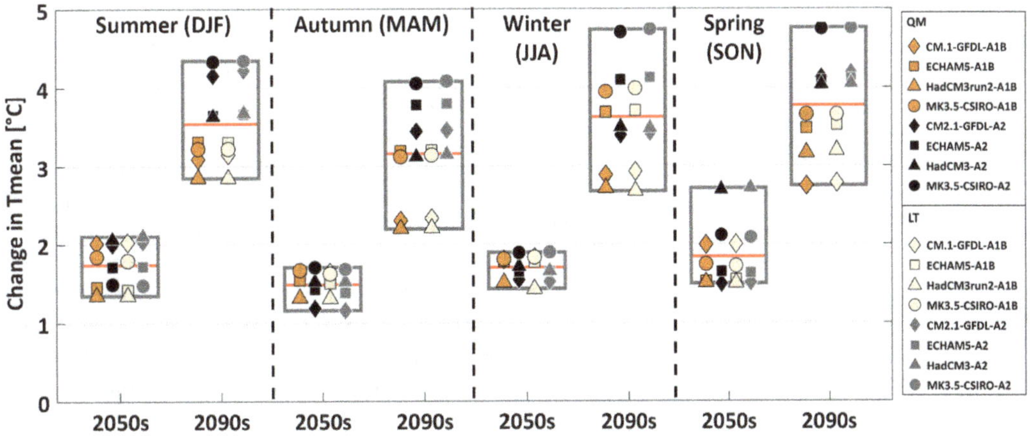

Figure 7. The uncertainty range of the T_{mean} signal is shown for the entire ensemble and the four seasons. For each box the uncertainty range is broken down into the two emission scenarios and the two bias-correction methods. The red line represents the median of all 16 members within a box.

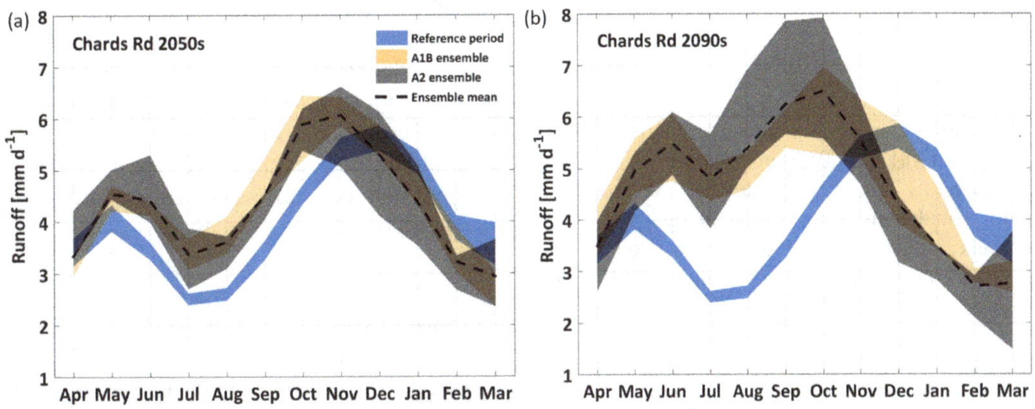

Figure 8. Mean daily modelled runoff (orange envelope = A1B and grey envelope = A2) at Chards Rd during **(a)** the 2050s and **(b)** 2090s is compared with the historic simulations (blue). All simulations are based on the QM-Anderson sub-ensemble.

than the ECHAM5 A2 member, which is associated with a greater warming signal during winter and spring (Fig. 7). A closer inspection of the corresponding climate signals revealed that the MK3.5-CSIRO-A1B signal can be primarily attributed to less precipitation (second smallest increase in winter precipitation of the ensemble), which combined with a relatively strong warming signal (strongest of all A1B members) would have resulted in less snowfall and therefore less snow accumulating. This indicates that despite the warming signal being a key driver of SWE changes in the future, the precipitation signal also plays an important role, which in this case led to an enhancement of the negative SWE signal.

As shown (Fig. 11) by the transient simulations of mean SWE from July to December (months with historically the highest SWE) there is no clear distinction between the A1B and A2 simulations until the last quarter of the 21st century when the median of the A2 runs stays consistently under the A1B runs. While the median SWE based on all 16 QM runs decreases substantially (60 %) during the 110-year period, years with large accumulations of SWE still occur in the second half of the 21st century for individual members and years. The negative trend was found to be more pronounced for the glaciers in the catchment, which are projected to lose 93 % of their volume (91 % based on A1B and 94 % based on A2) by 2099.

3.4 Quantifying the uncertainty in the seasonal runoff signal

In order to quantify the uncertainty induced by the individual components of the model chain compared to the overall uncertainty in runoff projections, the approach of Muerth et al. (2012) was adopted. First, the approach is exemplified for the uncertainty quantification of the winter runoff signal (Fig. 12). An uncertainty component (e.g. GCM) is selected and all possible permutations between the selected and the remaining three model components are computed, resulting in 32 combinations ($4 \times 2 \times 2 \times 2$). In the next step, the cur-

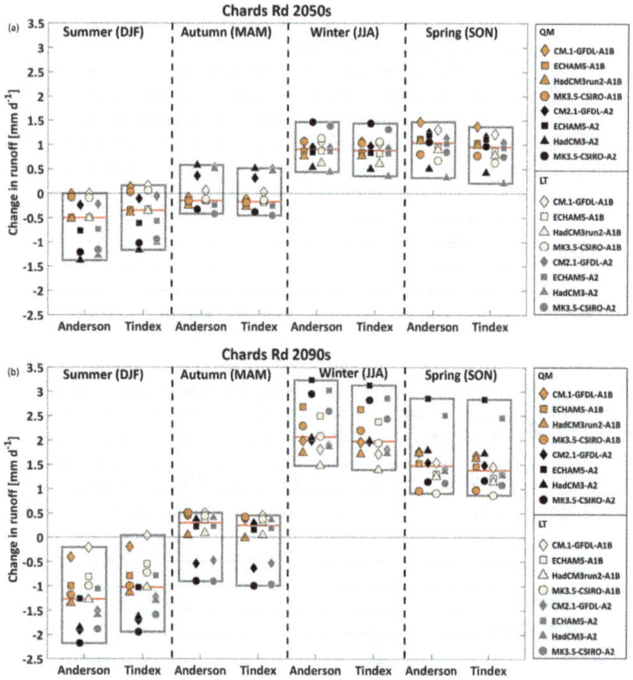

Figure 9. The uncertainty range of the seasonal runoff signal at Chards Rd is shown for **(a)** the 2050s and **(b)** the 2090s. For each season an uncertainty box is shown for WaSiM-Anderson and WaSiM-Tindex simulations, respectively. In each box (red line = median) the uncertainty range is broken down into the two emission scenarios and the two bias-correction methods (LT and QM).

rently selected component is varied (four GCMs = four circles), while the other three components (emission scenario, bias correction and hydrological model) are fixed to a certain combination of their members. As such, all of the four circles spanning the first bar (Fig. 12, left, first segment) have the emission scenario fixed to A1B (orange quarter), the bias-correction method fixed to QM (white quarter) and the snow model fixed to Anderson (blue quarter), while the fourth quarter, which corresponds to one of the four GCM members (CM2.1-GFDL, ECHAM5, HadCM3 and MK3.5-CSIRO), is varied. Thus, the effect of the GCM has been isolated by fixing the other components to one particular combination. This step is repeated for all the possible combinations between the three remaining components, translating to a total of eight combinations (eight bars). As each of the eight bars in the GCM segment (Fig. 12, first segment) contains four circles, all 32 possible permutations have been accounted for. The same procedure is then repeated for the remaining three uncertainty components. The mean contributions to the overall uncertainty (including the standard deviations) of the individual components are then displayed in a radar chart (Fig. 13a).

The uncertainty analysis (Fig. 13a–d) identified the GCM as the primary source of uncertainty across all seasons (44–57 % change in runoff). The selection of the emission scenario was the second largest contributor (16–49 %), except for winter when the choice of bias correction was greater (22 % vs. 16 %). The uncertainty induced by the emission scenario showed a pronounced seasonal variation and was found to be largest during summer (33 %) and autumn (49 %). A likely explanation for the latter is the significantly different T_{mean} signals under A1B and A2 (Fig. 7), which translate to different ET rates and consequently variable changes in runoff. This is supported by the fact that the most extreme decreases in summer and autumn runoff occurred under the A2 scenario (not shown here). The contribution of bias correction to the overall uncertainty ranged from 4 % in autumn to 22 % in winter and was higher (except for autumn) than the relative contribution of the snow model (3–10 %).

As described in Muerth et al. (2012) the standard deviation associated with the relative uncertainty contribution of an individual component indicates its degree of dependence on the other model components. Here the standard deviation was clearly largest for the emission scenario and the GCM. The standard deviations of both components also varied seasonally and were found to be largest during spring and autumn (Fig. 13b, d). Thus, it can be stated that during spring and autumn the uncertainty induced by the GCM (same holds for emission scenario) is associated with a relatively large dependence on the other variables.

4 Discussion

Before the climate change uncertainty assessment was carried out, the hydrological simulations were analysed during the reference period. Performance in reproducing the observed regime varied depending on both the selection of the snow model and the bias-correction method. The bias-correction method was expected to only have a minor effect on the simulated monthly runoff during the reference period, which made the observed sensitivity somewhat unexpected. A potential explanation could be related to the predominant air temperature during mature precipitation events along the main divide. At Brewster Glacier (located along the main divide just west of the Clutha catchment), Cullen and Conway (2015) found air temperature to be frequently around the rain–snow threshold during events with major solid precipitation, which led to the conclusion that the accumulation of snow in areas along the main divide is vulnerable to small variations in air temperature. The relatively large variability during the reference period was therefore seen as a first indicator for a potentially high sensitivity of the modelled snow storage and runoff to projected warming. The historic analysis also showed that the observed regime was captured more realistically by WaSiM-Anderson as opposed to WaSiM-Tindex. Studies targeting the controlling processes of snowmelt in the Southern Alps (Prowse and Owens, 1982; Sims and Orwin, 2011) identified net radiation

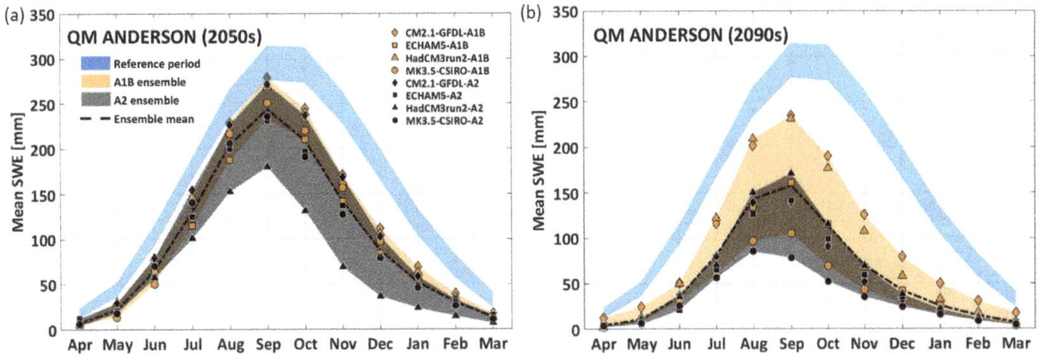

Figure 10. The monthly SWE storage simulated by the QM-Anderson sub-ensemble is depicted for **(a)** the 2050s and **(b)** the 2090s period. The individual members are augmented by the A1B envelope, the A2 envelope and the ensemble mean (dashed line). The blue envelope represents the range of SWE volume simulations during the baseline period.

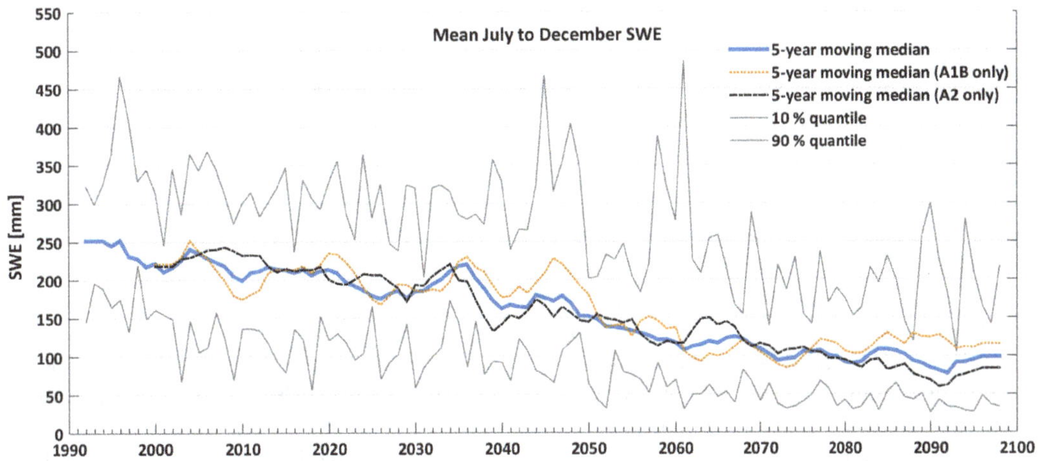

Figure 11. The mean SWE of the 6 months with the historically highest SWE (July to December) is shown for the 16 transient QM simulations (including QM-Anderson and QM-Tindex).

as an important driver of snowmelt (in addition to sensible heat). Thus, the better performance of the conceptual energy balance method (WaSiM-Anderson) compared to the Tindex model can likely be explained by the advances of accounting for individual melt fractions and using a seasonal radiation melt factor (see Anderson, 1973 or Schulla, 2012).

Overall the baseline analysis showed that the individual sub-ensembles performed differently and that the observed regime was not always enveloped by the corresponding range of simulations. This introduces some additional uncertainty into the projections that could not be quantified or accounted for in this study. A potential explanation is that neither snow model was able to accurately represent all of the spatio-temporal variation in the snowmelt process across the catchment, and that some driving processes (i.e. radiation-induced events) are also not represented adequately in either snow model. Either improved empirical relationships or a greater physical component to snowmelt modelling would be beneficial in this respect for future research. Accounting for snow redistribution by avalanches or wind as described in Freudiger et al. (2017) should also be investigated as this would potentially reduce some of the relatively large accumulations of snow that were found to persist over multiple years on some cells in the upper catchment during the reference period. Inaccuracies in the meteorological fields ($\mathrm{METEO_{OBS}}$) that were used for the bias correction could also have caused some of the seasonal over- and underestimations in the hydrological regime. As discussed in Jobst et al. (2017) the climate network in the upper Clutha is sparse, with very few sites located in medium to high elevations. Notwithstanding the improved representation of temperature provided by the Jobst et al. (2017) data set compared to other products, the remaining biases in this temperature field would have also propagated into the bias-corrected RCM fields and the corresponding hydrological baseline simulations.

For the two future periods (2050s and 2090s) the projections revealed substantial increases in runoff from May to October, and a decline between November and March. The dominant drivers behind this regime shift were changes in the seasonal distribution of precipitation (for the 2090s win-

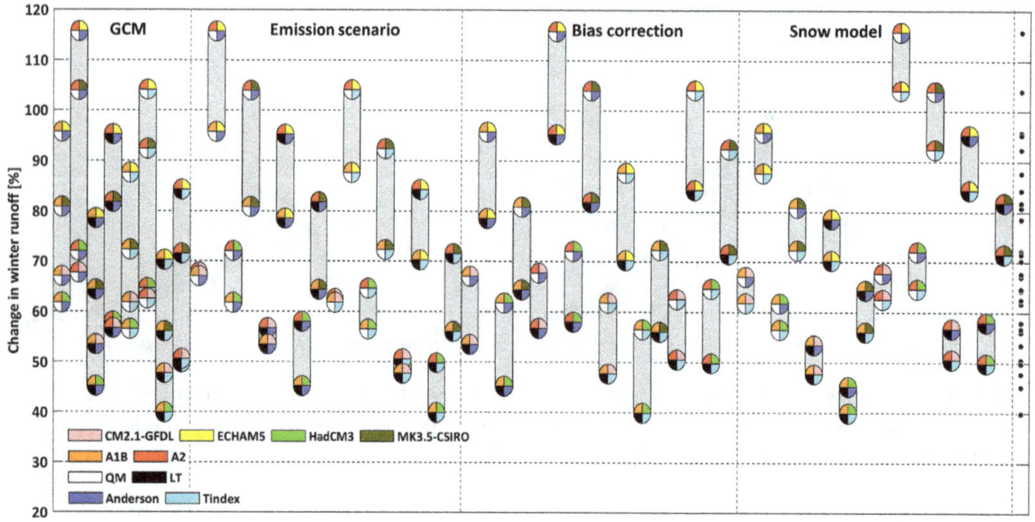

Figure 12. Uncertainty contributions are isolated for each of the four model components: GCM, emission scenario, bias-correction method and snow model (the approach is described in more detail in Section 3.4). The number of points in a bar corresponds to the number of members of the selected model component (e.g. four GCMs), while the number of bars is equal to the possible permutations based on the remaining three model components ($2 \times 2 \times 2 = 8$).

ter +29 to +84 %) and a rise in air temperature causing decreases in the seasonal snow storage. These findings are mostly consistent with previous New Zealand-based climate change impact assessments. In a New Zealand-wide study Hendrikx et al. (2012) also modelled substantial reductions in the peak snow accumulations along the Southern Alps, which they attributed to decreases in the fraction of solid precipitation due to increases in air temperature.

Using a semi-distributed hydrological model (TopNet) and ensemble of 12 CMIP-3 GCMs (including the 4 used herein), Poyck et al. (2011) and Gawith et al. (2012) found a similar ensemble-mean increase in winter streamflow in the 2090s (for Balclutha and the upper Clutha River), despite only relatively small decreases in summer river flow. In the upper Waitaki catchment (9490 km^2, located north-east of the Clutha and also with headwaters bordering the Main Divide of the Southern Alps), Caruso et al. (2017) found comparably large increases in lake inflows during winter (i.e. 76 % for August) and a noticeable decrease in summer (i.e. −13 % for February) using the same hydrological model and GCM ensemble as Poyck et al. (2011). An increase in winter precipitation was also identified as the main driver for the Waitaki.

Globally, similar changes in streamflow have been reported for many alpine catchments, for example in British Columbia (Mandal and Simonovic, 2017), Oregon (Chang and Jung, 2010) and the Austrian Alps (Laghari et al., 2012). In addition to increased winter precipitation, a reduction in solid precipitation is often reported to lead to an earlier melt peak and further enhanced winter flow (Kundzewicz, 2008). Here, a decrease in the proportion of solid precipitation combined with an intensification of snowmelt was also found to contribute to more flow being generated during winter and spring, but the main driver remained the increase in winter precipitation.

Analysis of the uncertainty in the hydrological projections for the upper Clutha (Figs. 12 and 13) showed that although the total spread of hydrological projections was large (i.e. increase of 40–116 % for winter), for most seasons (except autumn) the direction of change was found to be consistent amongst individual members (increases in winter and spring, decreases in summer). The main contributors to the spread in the projections for seasonal flow were as follows (in ascending order): snow model (3–10 %), bias-correction method (4–22 %), emission scenario (16–49 %) and GCM (44–57 %). As in this study, a large body of existing hydrological impact studies also identified GCM structure as the dominant source of uncertainty (e.g. Kingston and Taylor, 2010; Hughes et al., 2011; Teng et al., 2012; Thompson et al., 2014). It should be noted that the four GCMs constitute a subset of a total of 12 GCMs which had been previously selected by NIWA on the basis of a performance assessment for the South Pacific region (Ministry for the Environment, 2008). In terms of T_{mean} signal (A1B) the four GCMs had the 5th, 10th, 11th and 12th highest warming – hence, the A1B projections used in this study are at the lower end of the "full" GCM envelope.

A large part of the GCM-related uncertainty was found to be caused by the precipitation signal, which became especially uncertain during the winter season. This finding is in agreement with a number of studies targeting alpine catchments such as the Hindu Kush–Himalayan region (Palazzi et al., 2014; Lutz et al., 2016), the Pacific Northwest of the US (Jung et al., 2012), western Oregon (Chang and Jung, 2010) and the Southern Alps of New Zealand (Zammit and

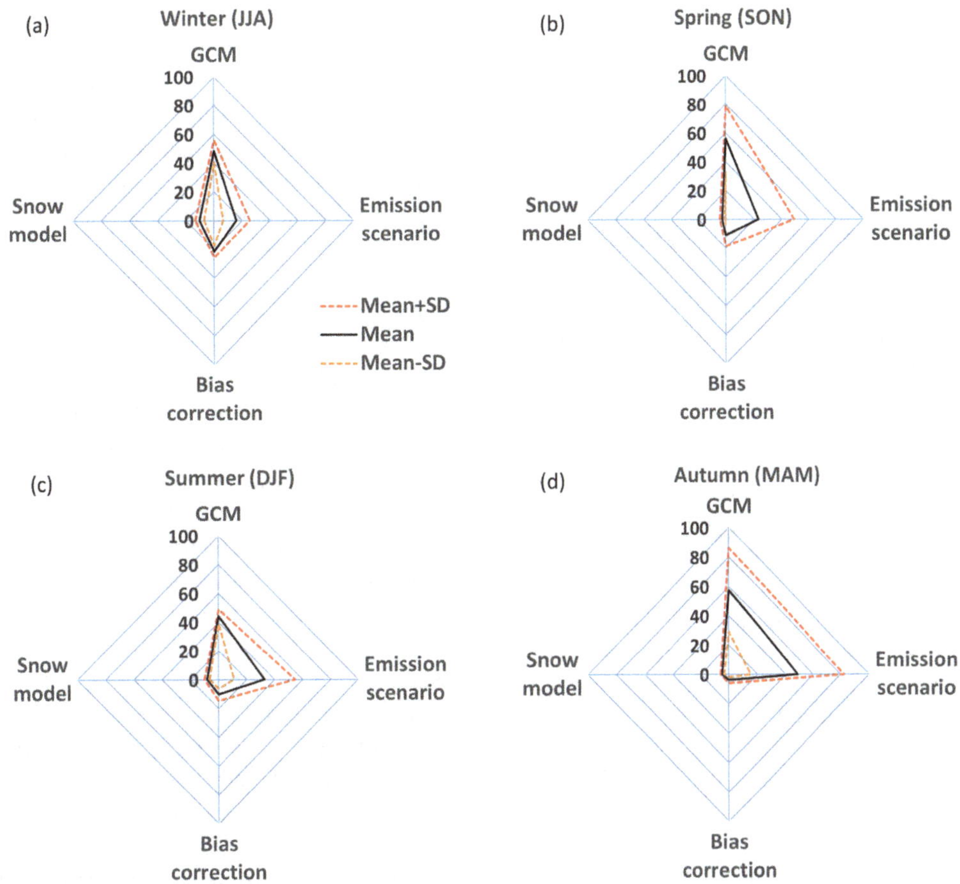

Figure 13. Relative contribution of the four uncertainty sources to the overall uncertainty of the seasonal runoff signal for **(a)** winter, **(b)** spring, **(c)** summer and **(d)** autumn. Based on the permutations (as exemplified for winter in Fig. 12) the radar charts show the mean contributions (%; mean length of bars shown in Fig. 12) of the four model components to the overall uncertainty as well as the standard deviations (%; standard deviation of bars shown in Fig. 12).

Woods, 2011). Hence, constraining and accounting for the uncertainty associated with the precipitation output of GCMs and RCMs remains a major research challenge in hydrological impact studies.

Emission uncertainty was identified as the second most important source during most seasons, while in winter bias correction was found to introduce a similar level of uncertainty. These findings generally agree with the study of Prudhomme and Davies (2009), in which emission scenario and downscaling (RCM vs. statistical method) uncertainty were of a comparable magnitude, but still considerably smaller when compared to GCM uncertainty. For alpine catchments in British Columbia the ranking order of uncertainty sources computed by Bennett et al. (2012) was also led by the GCM, followed by the emission scenario and, in third, hydrological parameter uncertainty.

As described in Kay and Davies (2008) and Thompson et al. (2014), different versions of the same hydrological model can be developed that differ in one particular routine (i.e. PET), allowing for a process specific uncertainty analysis. In the upper Clutha catchment, the high precipitation intensity in the headwaters combined with the relatively high proportion of snowmelt ($\sim 20\,\%$) means that the seasonal regime is largely controlled by the process of snowmelt rather than PET, which made the upper Clutha an appropriate candidate for the snow-model-specific uncertainty analysis. By using the two WaSiM versions that only differ in their snowmelt routine the contribution of that process to the overall uncertainty could be assessed in isolation.

As expected, the contribution of the snow model was highest for winter (10 %). However, interestingly, the contribution of the snow model was still relatively high during summer (8 %), a time of year when the influence of melt processes on streamflow were expected to be minor. This can likely be explained by the larger SWE volume that was modelled by Tindex (compared to Anderson) during the summer reference period (Table 3). Thus, the Tindex SWE storage had the potential to release more meltwater (compared to baseline) under the projected warming, which translated into an attenuation of the decrease in summer runoff. This is supported by the fact that the negative changes in summer runoff are consistently less pronounced for the Tindex model (not shown here). For autumn and spring, the snow model only added a small proportion (5 and 3 %, respectively) to the overall uncertainty. Considering that spring is (historically) the main melt period, projections were expected to vary more depending on the choice of the snow model. Hence, for the spring season the results suggest that under the projected warming the effect of the snow model can be considered negligible, especially when compared to the GCM and the emission scenario. At 10 % of overall uncertainty in winter, the effect of the snow model is noticeable but substantially smaller than the variation caused by the GCM output (48 %) (uncertainty of bias correction and emission scenario corresponds to 22 and 16 %, respectively).

The study of Troin et al. (2016), which focused on the direct output of the snow model (i.e. SWE or duration of snowpack), came to comparable conclusions in the sense that hydrological models are not the major source of uncertainty for SWE projections. In their study, natural variability had a far greater effect on the projections for the individual snow indicators as the snow model component, which was shown here in a similar way for GCM structure.

5 Conclusions

The implementation of WaSiM for the Clutha River constitutes the first application of a fully distributed and grid-based hydrological model for climate change impact assessment in a large-scale New Zealand catchment. The model chain that was built here to force WaSiM with RCM simulations can be regarded as an important contribution to the existing body of climate change impact studies targeting snowmelt-affected mid-latitude alpine catchments. The projections for the end of the 21st century encompass substantial increases in runoff for winter (71 %) and spring (35 %), while summer runoff is projected to decrease (−24 %). The key drivers behind the changes in the regime were found to be an increase in winter precipitation and a reduction in the SWE storage between winter and spring. The changes in the regime will likely impact the capacity of the two hydropower schemes (Clyde and Roxburgh) located downstream of Chards Rd, where production can be expected to increase from winter to early spring and decrease during the summer months.

Adopting the approach of Muerth et al. (2012), this study allowed the contribution of the individual uncertainty sources to be quantified in a more objective way as opposed to a mere visual interpretation of results. For the first time the role of the rarely investigated snowmelt routine was explored together with three of the key uncertainty sources in hydrological impact studies (i.e. GCM, emission scenario and bias-correction method). While all components contributed to the total ensemble uncertainty, the selection of the GCM introduced the biggest spread to the range of runoff projections during all seasons. When looking at the climate signals (Figs. 6 and 7) it becomes obvious that the uncertainty stemming from the precipitation signal (especially during winter) is the primary driver behind the large uncertainty in the hydrological projections and should therefore be the focus of future studies. In this context combining the limited number of RCM simulations with sophisticated statistical techniques (e.g. the use of probability density functions as described in Tait et al., 2016) could help to more fully explore the uncertainty range associated with the precipitation signal. Further, since the completion of this study additional RCM simulations based on RCP (Representative Concentration Pathways) scenarios and CMIP-5 GCMs have been generated for the New Zealand domain (Ministry for the Environment, 2016), which could be used to enlarge the existing ensemble of hydro-climatic projections for the Clutha.

The uncertainty linked to the snow model, which showed a pronounced seasonal variation (ranging from 3 % in spring to 10 % in winter), was found to be smaller when compared to the other components, but the findings of this study suggest that it should not be ignored as its effect was shown to be significant for both winter and summer runoff. Another important finding from this study is that the contribution of the snow model and the other model components to the overall uncertainty possesses a high inter-annual variability. While there was consistency regarding the main uncertainty source (i.e. GCM structure), the second largest contributor varied between emission scenario and bias correction for the individual seasons. Future work should investigate whether the selection of the snow model has a stronger impact in other regions or catchments of different size (i.e. small headwater sub-catchments). Model parameter uncertainty was not accounted for in this study but should be investigated as part of future work, which could help to understand and potentially improve misrepresentations in the historic runoff regime. The use of other hydrological indicators (i.e. low and high flow) should also be explored as the effect of the individual components of the model chain might differ for such alternative metrics.

Data availability. RCM data were taken from the CCII project (described at http://ccii.org.nz/research-aims/ra1/ra-1-in-progress/, last access: 5 June 2018). Meteorological data as used for the historical simulation and bias correction can be partly accessed from NIWA (https://cliflo.niwa.co.nz/, last access: 5 June 2018) while for some sites data need to be requested from MetService New Zealand (https://about.metservice.com/contact, last access: 5 June 2018). The hydrological projections generated in this study can be requested from the first author (andreas.jobst@gmail.com).

Competing interests. The authors declare that they have no conflict of interest.

Acknowledgements. We would like to thank a number of authorities who provided the various data sets used in this study. The National Institute of Water and Atmospheric Research (NIWA) for climate observations and the RCM simulations. Streamflow records were kindly provided by NIWA, Contact Energy Ltd. and the Otago Regional Council. The digital elevation model, soil data and land use information were obtained from Landcare Research New Zealand. This study was funded by a University of Otago Doctoral Scholarship and Publishing Bursary.

Edited by: Kerstin Stahl

References

Ackerley, D., Dean, S., Sood, A., and Mullan, A. B.: Regional climate modelling in New Zealand: Comparison to gridded and satellite observations, Weather and Climate, 32, 3–22, 2012.

Anderson, B., Lawson, W., Owens, I., and Goodsell, B.: Past and future mass balance of 'Ka Roimata o Hine Hukatere' Franz Josef Glacier, New Zealand, J. Glaciol., 52, 597–607, https://doi.org/10.3189/172756506781828449, 2006.

Anderson, E. A.: National Weather Service river forecast system – snow accumulation and ablation model, National Oceanographic and Atmospheric Administration (NOAA), Technical Memorandum, NWS-HYDRO-17, US Department of Commerce, Silver Spring, MD, 238 pp, 1973.

Auer, A. H.: The rain versus snow threshold temperatures, Weatherwise, 27, p. 67, https://doi.org/10.1080/00431672.1974.9931684, 1974.

Bennett, K. E., Werner, A. T., and Schnorbus, M.: Uncertainties in hydrologic and climate change impact analyses in headwater basins of British Columbia, J. Climate, 25, 5711–5730, https://doi.org/10.1175/JCLI-D-11-00417.1, 2012.

Boé, J., Terray, L., Habets, F., and Martin, E.: Statistical and dynamical downscaling of the Seine basin climate for hydro-meteorological studies, Int. J. Climatol., 27, 1643–1655, https://doi.org/10.1002/joc.1602, 2007.

Caruso, B. S., King, R., Newton, S., and Zammit, C.: Simulation of climate change effects on hydropower operations in mountain headwater lakes, New Zealand, River Res. Appl., 33, 147–161, https://doi.org/10.1002/rra.3056, 2017.

Chang, H. and Jung, I.: Spatial and temporal changes in runoff caused by climate change in a complex large river basin in Oregon, J. Hydrol., 388, 186–207, https://doi.org/10.1016/j.jhydrol.2010.04.040, 2010.

Chen, J., Brissette, F. P., Chaumont, D., and Braun, M.: Finding appropriate bias correction methods in downscaling precipitation for hydrologic impact studies over North America, Water Resour. Res., 49, 4187–4205, https://doi.org/10.1002/wrcr.20331, 2013.

Chinn, T. J.: Distribution of the glacial water resources of New Zealand, Journal of Hydrology New Zealand, 40, 139–187, 2001.

Cullen, N. J. and Conway, J. P.: A 22-month record of surface meteorology and energy balance from the ablation zone of Brewster Glacier, New Zealand, J. Glaciol., 61, 931–946, https://doi.org/10.3189/2015JoG15J004, 2015.

Dobler, C., Hagemann, S., Wilby, R. L., and Stötter, J.: Quantifying different sources of uncertainty in hydrological projections in an Alpine watershed, Hydrol. Earth Syst. Sci., 16, 4343–4360, https://doi.org/10.5194/hess-16-4343-2012, 2012.

Freudiger, D., Kohn, I., Seibert, J., Stahl, K., and Weiler, M.: Snow redistribution for the hydrological modeling of alpine catchments, WIREs Water, 4, e1232, https://doi.org/10.1002/wat2.1232, 2017.

Frueh, B., Schipper, J. W., Pfeiffer, A., and Wirth, V. A.: A pragmatic approach for downscaling precipitation in alpine-scale complex terrain, Meteorol. Z., 15, 631–646, 2006.

Gawith, D., Kingston, D. G., and McMillan, H.: The effects of climate change on runoff in the Lindis and Matukituki catchments, Otago, New Zealand, Journal of Hydrology New Zealand, 51, 121–135, 2012.

Graham, L. P., Hagemann, S., Jaun, S., and Beniston, M.: On interpreting hydrological change from Regional Climate Models, Climatic Change, 81, 97–122, 2007.

Gudmundsson, L., Tallaksen, L. M., Stahl, K., Clark, D. B., Dumont, E., Hagemann, S., Bertrand, N., Gerten, D., Heinke, J., Hanasaki, N., Voss, F., and Koirala, S.: Comparing large-scale hydrological model simulations to observed runoff percentiles in Europe, J. Hydrometeorol., 13, 604–620, https://doi.org/10.1175/JHM-D-11-083.1, 2012.

Gutjahr, O. and Heinemann, G.: Comparing precipitation bias correction methods for high resolution regional climate simulations using Cosmo-Clm: Effects on extreme values and climate change signal, Theor. Appl. Climatol., 114, 511–529, 2013.

Haddeland, I., Clark, D. B., Franssen, W., Ludwig, F., Voß, F., Arnell, N. W., Bertrand, N., Best, M., Folwell, S., Gerten, D., Gomes, S., Gosling, S. N., Hagemann, S., Hanasaki, N., Harding, R., Heinke, J., Kabat, P., Koirala, S., Oki, T., Polcher, J., Stacke, T., Viterbo, P., Weedon, G. P., and Yeh, P.: Multimodel estimate of the global terrestrial water balance: Setup and first results, J. Hydrometeorol., 12, 869–884, https://doi.org/10.1175/2011JHM1324.1, 2011.

Hagemann, S., Chen, C., Haerter, J. O., Heinke, J., Gerten, D., and Piani, C.: Impact of a statistical bias correction on the projected hydrological changes obtained from three GCMs and two hydrology models, J. Hydrometeorol., 12, 556–578, 2011.

Hendrikx, J., Hreinsson, E. Ö., Clark, M. P., and Mullan, A. B: The potential impact of climate change onseasonal snow in New Zealand: Part I – an analysis using 12 GCMs, Theor. Appl. Climatol., 110, 607–618, https://doi.org/10.1007/s00704-012-0711-1, 2012.

Hidalgo, H. G., Das, T., Dettinger, M. D., Cayan, D. R., Pierce, D. W., Barnett, T. P., Bala, G., Mirin, A., Wood, A. W., Bonfils, C., Santer, B. D., and Nozawa, T.: Detection and attribution of streamflow timing changes to climate change in the western United States, J. Climate, 22, 3838–3855, https://doi.org/10.1175/2009JCLI2470.1, 2009.

Hughes, D. A., Kingston, D. G., and Todd, M. C.: Uncertainty in water resources availability in the Okavango River basin as a result of climate change, Hydrol. Earth Syst. Sci., 15, 931–941, https://doi.org/10.5194/hess-15-931-2011, 2011.

Jiang, Y., Liu, C., Huang, C., and Wu, X.: Improved particle swarm algorithm for hydrological parameter optimization, Appl. Math. Comput., 217, 3207–3215, https://doi.org/10.1016/j.amc.2010.08.053, 2010.

Jiménez Cisneros, B. E., Oki, T., Arnell, N. W., Benito, G., Cogley, J. G., Döll, P., Jiang, T., and Mwakalila, S. S.: Freshwater resources, In: Climate Change 2014: Impacts, Adaptation, and Vulnerability. Part A: Global and Sectoral Aspects. Contribution of Working Group II to the Fifth Assessment Report of the Intergovernmental Panel on Climate Change, edited by: Field, C. B., Barros, V. R., Dokken, D. J., Mach, K. J., Mastrandrea, M. D., Bilir, T. E., Chatterjee, M., Ebi, K. L., Estrada, Y. O., Genova, R. C., Girma, B., Kissel, E. S., Levy, A. N., MacCracken, S., Mastrandrea, P. R., and White, L. L., Cambridge University Press, Cambridge, United Kingdom and New York, NY, USA, 229–269, 2014.

Jobst, A. M.: The potential impacts of climate change on the hydroclimate of the Clutha/Mata-Au catchment, PhD thesis, Department of Geography, University of Otago, Dunedin, New Zealand, 334 pp., 2017.

Jobst, A. M., Kingston, D. G., Cullen, N. J., and Sirguey, P.: Combining thin-plate spline interpolation with a lapse rate model to produce daily air temperature estimates in a data-sparse alpine catchment, Int. J. Climatol., 37, 214–229, https://doi.org/10.1002/joc.4699, 2017.

Jung, I., Moradkhani, H., and Chang, H.: Uncertainty assessment of climate change impacts for hydrologically distinct river basins, J. Hydrol., 466–467, 73–87, https://doi.org/10.1016/j.jhydrol.2012.08.002, 2012.

Kay, A. L. and Davies, H. N.: Calculating potential evaporation from climate model data: a source of uncertainty for hydrological climate change impacts, J. Hydrol., 358, 221–239, https://doi.org/10.1016/j.jhydrol.2008.06.005, 2008.

Kennedy, J. and Eberhart, R.: Particle swarm optimization, Paper presented at the IEEE International Conference on Neural Networks – Conference Proceedings, https://doi.org/10.1109/ICNN.1995.488968, 1995.

Kingston, D. G. and Taylor, R. G.: Sources of uncertainty in climate change impacts on river discharge and groundwater in a headwater catchment of the Upper Nile Basin, Uganda, Hydrol. Earth Syst. Sci., 14, 1297–1308, https://doi.org/10.5194/hess-14-1297-2010, 2010.

Koedyk, L. P. and Kingston, D. G.: Potential evapotranspiration method influence on climate change impacts on river flow: a mid-latitude case study, Hydrol. Res., 47, 951–963, https://doi.org/10.2166/nh.2016.152, 2016.

Kundzewicz, Z. W., Mata, L. J., Arnell, N. W., Döll, P., Jimenez, B., Miller, K., Oki, T., Sen, Z., and Shiklomanov, I.: The implications of projected climate change for freshwater resources and their management, Hydrolog. Sci. J., 53, 3–10, 2008.

Laghari, A. N., Vanham, D., and Rauch, W.: To what extent does climate change result in a shift in Alpine hydrology? A case study in the Austrian Alps, Hydrolog. Sci. J., 57, 103–117, https://doi.org/10.1080/02626667.2011.637040, 2012.

Lenderink, G., Buishand, A., and van Deursen, W.: Estimates of future discharges of the river Rhine using two scenario methodologies: direct versus delta approach, Hydrol. Earth Syst. Sci., 11, 1145-1159, https://doi.org/10.5194/hess-11-1145-2007, 2007.

Ludwig, R., May, I., Turcotte, R., Vescovi, L., Braun, M., Cyr, J.-F., Fortin, L.-G., Chaumont, D., Biner, S., Chartier, I., Caya, D., and Mauser, W.: The role of hydrological model complexity and uncertainty in climate change impact assessment, Adv. Geosci., 21, 63–71, https://doi.org/10.5194/adgeo-21-63-2009, 2009.

Lutz, A. F., Immerzeel, W. W., Kraaijenbrink, P. D. A., Shrestha, A. B, and Bierkens, M. F. P.: Climate Change Impacts on the Upper Indus Hydrology: Sources, Shifts and Extremes, PLOS ONE, 11, e0165630, https://doi.org/10.1371/journal.pone.0165630, 2016.

Mandal, S. and Simonovic, S. P.: Quantification of uncertainty in the assessment of future streamflow under changing climate conditions, Hydrol. Process., 31, 2076–2094, https://doi.org/10.1002/hyp.11174, 2017.

Maraun, D., Wetterhall, F., Ireson, A. M., Chandler, R. E., Kendon, E. J., Widmann, M., Brienen, S., Rust, H. W., Sauter, T., Themeßl, M., Venema, V. K. C., Chun, K. P., Goodess, C. M., Jones, R. G., Onof, C., Vrac, M., and Thiele-Eich, I.: Precipitation downscaling under climate change: Recent developments to bridge the gap between dynamical models and the end user, Rev. Geophys., 48, RG3003, https://doi.org/10.1029/2009rg000314, 2010.

Marke, T., Mauser, W., Pfeiffer, A., and Zängl, G.: A pragmatic approach for the downscaling and bias correction of regional climate simulations: evaluation in hydrological modeling, Geosci. Model Dev., 4, 759–770, https://doi.org/10.5194/gmd-4-759-2011, 2011.

Ministry for the Environment: Climate change effects and impacts assessment. A guidance manual for local government in New Zealand, 2nd Edn., prepared by: Mullan, B., Wratt, D., Dean, S., Hollis, M. (NIWA), Allan, S., Williams, T. (MWH NZ Ltd), and Kenny, G. (Earthwise Consulting Ltd), in consultation with Ministry for the Environment, NIWA Client Report No. WLG2007/62., 156 pp., 2008.

Ministry for the Environment: Climate Change Projections for New Zealand: Atmosphere projections based on simulations from the IPCC Fifth Assessment, prepared by: Mullan, B., Sood, A., and Stuart, S. (NIWA), Wellington, 127 pp., 2016.

Mpelasoka, F. S. and Chiew, F. H. S.: Influence of rainfall scenario construction methods on runoff projections, J. Hydrometeorol., 10, 1168–1183, 2009.

Muerth, M. J., St-Denis, G. B., Ludwig, R., and Caya, D.: Evaluation of different sources of uncertainty in climate change impact research using a hydro-climatic model ensemble, edited by: Seppelt, R., Voinov, A. A., Lange, S., and Bankamp, D., International Environmental Modelling and Software Society (iEMSs) 2012 International Congress on Environmental Modelling and Software Managing Resources of a Limited Planet: Pathways and Visions under Uncertainty, Sixth Biennial Meeting, Leipzig, Germany, 2012.

Muerth, M. J., Gauvin St-Denis, B., Ricard, S., Velázquez, J. A., Schmid, J., Minville, M., Caya, D., Chaumont, D., Ludwig, R., and Turcotte, R.: On the need for bias correction in regional climate scenarios to assess climate change impacts on river runoff, Hydrol. Earth Syst. Sci., 17, 1189–1204, https://doi.org/10.5194/hess-17-1189-2013, 2013.

Murray, D. L.: Regional hydrology of the Clutha River, Journal of Hydrology New Zealand, 14, 83–98, 1975.

Palazzi, E., von Hardenberg, J., Terzago, S., and Provenzale, A.: Precipitation in the Karakoram-Himalaya: a CMIP5 view, Clim. Dynam., 45, 21–45, https://doi.org/10.1007/s00382-014-2341-z, 2014.

Pall, P., Aina, T., Stone, D. A., Stott, P. A., Nozawa, T., Hilberts, A. G. J., Lohmann, D., and Allen, M. R.: Anthropogenic greenhouse gas contribution to flood risk in England and Wales in autumn 2000, Nature, 470, 382–385, https://doi.org/10.1038/nature09762, 2011.

Piao, S., Ciais, P., Huang, Y., Shen, Z., Peng, S., Li, J., Zhou, L., Liu, H., Ma, Y., Ding, Y., Friedlingstein, P., Liu, C., Tan, K., Yu, Y., Zhang, T., and Fang, J.: The impacts of climate change on water resources and agriculture in China, Nature, 467, 43–51, https://doi.org/10.1038/nature09364, 2010.

Poyck, S., Hendrikx, J., McMillan, H., Hreinsson, E. Ö., and Woods, R.: Combined snow and streamflow modelling to estimate impacts of climate change on water resources in the Clutha River, New Zealand, Journal of Hydrology New Zealand, 50, 293–311, https://doi.org/10.1007/s00704-012-0711-1, 2011.

Prowse, T. D. and Owens, I. F.: Energy balance over melting snow, Craigieburn Range, New Zealand, Journal of Hydrology New Zealand, 21, 133–147, 1982.

Prudhomme, C. and Davies, H.: Assessing uncertainties in climate change impact analyses on the river flow regimes in the UK. Part 2: Future climate, Climatic Change, 93, 197–222, 2009.

Rosenzweig, C., Casassa, G., Karoly, D. J., Imeson, A., Liu, C., Menzel, A., Rawlins, S., Root, T. L., Seguin, B., and Tryjanowski P.: Assessment of observed changes and responses in natural and managed systems, in: Climate Change 2007: Impacts, Adaptation and Vulnerability. Contribution of Working Group II to the Fourth Assessment Report of the Intergovernmental Panel on Climate Change, edited by: Parry, M. L., Canziani, O. F., Palutikof, J. P., van der Linden, P. J., and Hanson, C. E., Cambridge University Press, Cambridge, UK and New York, NY, USA, 79–131, 2007.

Schulla, J.: Model description WaSiM, Technical report, available at: http://wasim.ch (last access: 5 June 2018), 324 pp., 2012.

Sims, C. and Orwin, J. F.: Snowmelt generation on a hydrologically sensitive mountain range, Pisa Range, Central Otago, New Zealand, Journal of Hydrology New Zealand, 50, 383–405, https://doi.org/10.1002/joc.2216, 2011.

Srinivasan, M. S., Schmidt, J., Poyck, S., and Hreinsson, E.: Irrigation reliability under climate change scenarios: A modeling investigation in a river-based irrigation scheme in New Zealand, J. Am. Water Resour. As., 47, 1261–1274, https://doi.org/10.1111/j.1752-1688.2011.00568.x, 2011.

Tait, A. B., Sood, A., Mullan, B., Stuart, S., Bodeker, G., Kremser, S., and Lewis, J.: Updated climate change projections for New Zealand for use in impact studies, Synthesis report RA1, Climate Changes, Impacts and Implications (CCII) for New Zealand to 2100, MBIE contract C01X1225, 18 pp., 2016.

Teng, J., Vaze, J., Chiew, F. H. S., Wang, B., and Perraud, J. M.: Estimating the relative uncertainties sourced from GCMs and hydrological models in modeling climate change impact on runoff, J. Hydrometeorol., 13, 122–139, 2012.

Teutschbein, C. and Seibert, J.: Regional Climate Models for hydrological impact studies at the catchment scale: A review of recent modeling strategies, Geography Compass, 4, 834–860, 2010.

Thompson, J. R., Green, A. J., Kingston, D. G., and Gosling, S. N.: Assessment of uncertainty in river flow projections for the Mekong River using multiple GCMs and hydrological models, J. Hydrol., 486, 1–30, 2013.

Thompson, J. R., Green, A. J., and Kingston, D. G.: Potential evapotranspiration-related uncertainty in climate change impacts on river flow: An assessment for the Mekong River basin, J. Hydrol., 510, 259–279, https://doi.org/10.1016/j.jhydrol.2013.12.010, 2014.

Troin, M., Poulin, A., Baraer, M., and Brissette, F.: Comparing snow models under current and future climates: Uncertainties and implications for hydrological impact studies, J. Hydrol., 540, 588–602, https://doi.org/10.1016/j.jhydrol.2016.06.055, 2016.

Velázquez, J. A., Schmid, J., Ricard, S., Muerth, M. J., Gauvin St-Denis, B., Minville, M., Chaumont, D., Caya, D., Ludwig, R., and Turcotte, R.: An ensemble approach to assess hydrological models' contribution to uncertainties in the analysis of climate change impact on water resources, Hydrol. Earth Syst. Sci., 17, 565–578, https://doi.org/10.5194/hess-17-565-2013, 2013.

Vormoor, K., Lawrence, D., Heistermann, M., and Bronstert, A.: Climate change impacts on the seasonality and generation processes of floods – projections and uncertainties for catchments with mixed snowmelt/rainfall regimes, Hydrol. Earth Syst. Sci., 19, 913–931, https://doi.org/10.5194/hess-19-913-2015, 2015.

Wilby, R. L. and Harris, I.: A framework for assessing uncertainties in climate change impacts: Low-flow scenarios for the River Thames, UK, Water Resour. Res., 42, W02419, https://doi.org/10.1029/2005wr004065, 2006.

Willsman, A.: Annual glacier volumes in New Zealand 1995–2010, NIWA Client Report, No. CHC2011-050, available at: http://archive.stats.govt.nz/~/media/Statistics/browse-categories/environment/natural-resources/water-physical-stock-account-1995-2010/annual-glacier-ice-volumes-in-new-zealand-1995-2010.pdf (last access: 5 June 2018), 2011.

Zammit, C. and Woods, R.: Projected climate and river flow for the Waimakariri catchment for 2040s and 2090s, NIWA Client Report, No. CHCH2011-025, 2011.

Zemansky, G., Hong, Y. S. T., Rose, J., Song, S. H., and Thomas, J.: Assessing the effects of climate change on water resources: The Waimea Plains, Journal of Hydrology New Zealand, 51, 45–62, 2012.

5

Transferability of climate simulation uncertainty to hydrological impacts

Hui-Min Wang[1], **Jie Chen**[1], **Alex J. Cannon**[2], **Chong-Yu Xu**[1,3], **and Hua Chen**[1]

[1]State Key Laboratory of Water Resources and Hydropower Engineering Science, Wuhan University, Wuhan, 430072, China
[2]Climate Research Division, Environment and Climate Change Canada, Victoria BC, Canada
[3]Department of Geosciences, University of Oslo, Oslo, Norway

Correspondence: Jie Chen (jiechen@whu.edu.cn)

Abstract. Considering rapid increases in the number of climate model simulations being produced by modelling centres, it is often infeasible to use all of them in climate change impact studies. In order to thoughtfully select subsets of climate simulations from a large ensemble, several envelope-based methods have been proposed. The subsets are expected to cover a similar uncertainty envelope to the full ensemble in terms of climate variables. However, it is not a given that the uncertainty in hydrological impacts will be similarly well represented. Therefore, this study investigates the transferability of climate uncertainty related to the choice of climate simulations to hydrological impacts. Two envelope-based selection methods, K means clustering and the Katsavounidis–Kuo–Zhang (KKZ) method, are used to select subsets from an ensemble of 50 climate simulations over two watersheds with very different climates using 31 precipitation and temperature variables. Transferability is evaluated by comparing uncertainty coverage between climate variables and 17 hydrological variables simulated by a hydrological model. The importance of choosing climate variables properly when selecting subsets is investigated by including and excluding temperature variables. Results show that KKZ performs better than K means at selecting subsets of climate simulations for hydrological impacts, and the uncertainty coverage of climate variables is similar to that of hydrological variables. The subset of the first 10 simulations covers over 85 % of total uncertainty. As expected, temperature variables are important for the snow-related watershed, but less important for the rainfall-driven watershed. Overall, envelope-based selection of around 10 climate simulations, based on climate variables that characterize the physical processes controlling the hydrology of the watershed, is recommended for hydrological impact studies.

1 Introduction

In studies of climate change impacts on hydrology, multi-model ensembles (MMEs) formed by multiple global climate models (GCMs) and multiple emission scenarios have been widely used to drive hydrological models (Minville et al., 2008; Vaze and Teng, 2011; Chiew et al., 2009; Chen et al., 2011b). There are two strengths of using MMEs: (1) the MME mean typically performs better than any individual model in representing the mean of historical climate observations (Gleckler et al., 2008; Pierce et al., 2009; Pincus et al., 2008; Mehran et al., 2014); and (2) the spread of a MME can be used to estimate climate change impact uncertainties, for example those related to GCM structure, future greenhouse gas concentrations, and internal climate variability (Mendlik and Gobiet, 2016; Knutti et al., 2010; Chen et al., 2011b; Tebaldi and Knutti, 2007). While climate projection uncertainty and spread or coverage of a MME are not equivalent, the latter does provide an imperfect estimate of uncertainty and, for sake of simplicity, we use the terms interchangeably in the remainder of this study.

The number of GCM simulations available for impact studies is increasing rapidly. For instance, the Coupled Model Intercomparison Project Phase 3 (CMIP3) contains outputs from 25 different GCMs, whereas CMIP5 contains outputs from 61 GCMs (https://pcmdi.llnl.gov, last access: 19 June 2018), with each GCM contributing one or more simulation runs (Taylor et al., 2012). Although it is usually advised that as many climate simulations as possible be used in impact studies, the extraction, storage, and computational costs associated with a large MME may be prohibitive. In practice, it is not uncommon for impact studies to instead rely on a small subset of climate simulations, which are often selected manually, relying on expert judgement.

Several studies have considered more objective means of selecting subsets of climate simulations for impact studies based on different criteria. Generally, there are two main types of selection approaches. The past-performance approach selects a subset by weighting or selecting simulations according to their ability to represent the observed near-past climate (Gleckler et al., 2008; Perkins et al., 2007; Pincus et al., 2008). Climate model performance is generally evaluated based on agreement with observed climate conditions, which is often defined by a suite of climate metrics. For example, Perkins et al. (2007) ranked climate models based on probability density functions of observed temperature and precipitation. Similarly, Gleckler et al. (2008) evaluated the performances of 22 GCMs according to relative errors of some climatological fields, but stressed that a wider range of metrics might give more robust results. In general, the assumption that models with good performance over the near-past provide more realistic climate change signals is questionable (Knutti et al., 2010; Reifen and Toumi, 2009), although recent work on emergent constraints suggests that it may be possible to remove models that fail to represent certain key physical processes that dictate the evolution of long-term climate projections (Klein and Hall, 2015). In practice, however, the metrics commonly used to evaluate model performance are often manually defined based on the fields of interest, which leads to substantial subjectivity within the weighting process.

Another means of selecting climate simulations is the envelope-based approach, which tries to select a representative subset of climate simulations that covers as much of the full ensemble's range of future climate change signals as possible (Warszawski et al., 2014; Cannon, 2015; Logan et al., 2011). For instance, Cannon (2015) used two automated multivariate statistical algorithms, K means clustering and the Katsavounidis–Kuo–Zhang (KKZ) method (Katsavounidis et al., 1994), to select subsets of CMIP5 GCMs that bracket the overall range of changes in a suite of 27 climate extreme indices. The goal of the envelope-based approach coincides with the motivation behind the usage of a MME, namely to account for different sources of projection uncertainty, including structural uncertainty (Wilcke and Bärring, 2016; Tebaldi and Knutti, 2007).

Some studies have proposed selection methods that combine both near-past performance and climate change envelope coverage criteria (McSweeney et al., 2012; Lutz et al., 2016a; Giorgi and Mearns, 2002). For example, Lutz et al. (2016a) took both model historical skill and the range of projected changes in means and extremes into consideration through a three-step sequential selection procedure. Since many examples of such selection methods place emphasis on ranking or weighting climate model performance, they may inherit the potential flaws of the past-performance approach.

Regardless of underlying approaches, most selection methods are only conducted on climate variables that can be calculated directly from the MME simulation outputs. Even though subsets of simulations that account for most of the ensemble spread in climate variables can be identified, it is not guaranteed that the same level of spread coverage extends to hydrological impact variables due to the complexity and non-linearity of hydrological processes. For example, small perturbations in the frequency or intensity of temperature and precipitation regimes may have noticeable impacts on streamflow patterns and flood magnitudes (Muzik, 2001; Whitfield and Cannon, 2000). Consequently, whether the adequate coverage of climate simulation uncertainty is transferable to hydrological impacts should be evaluated before applying envelope-based selection methods in hydrological impact studies.

Chen et al. (2016) investigated the transferability of optimally selected climate simulations in the uncertainty quantification of hydrological impacts over a Canadian watershed. They concluded that the transferability of climate simulation uncertainty is limited to hydrological impacts. However, the selection methods used in their study were applied to just two climate variables, mean annual temperature and mean annual precipitation, which is a common strategy employed by practitioners who employ envelope-based approaches (Immerzeel et al., 2013; Warszawski et al., 2014). Hydrological responses are driven both by annual climate conditions and intra-annual climate processes, which may not be described by a small number of climate variables. The transferability of climate uncertainty may be diminished due to insufficient climate variables.

Accordingly, the main objective of this study is to investigate the transferability of climate simulation uncertainty to the assessment of hydrological impacts by using a large pool of climate variables, including seasonal means, annual means, and climate extremes. The case study is conducted over two watersheds with very different climate conditions, one of which is seasonally snow-covered and the other driven by summer monsoon rainfall with little winter snowfall. Two envelope-based approaches (K means clustering and the KKZ method) are used to select subsets of climate simulations based on different sets of climate variables. Transferability is evaluated by comparing the uncertainty coverage between the climate variables and 17 hydrological variables simulated by a hydrological model.

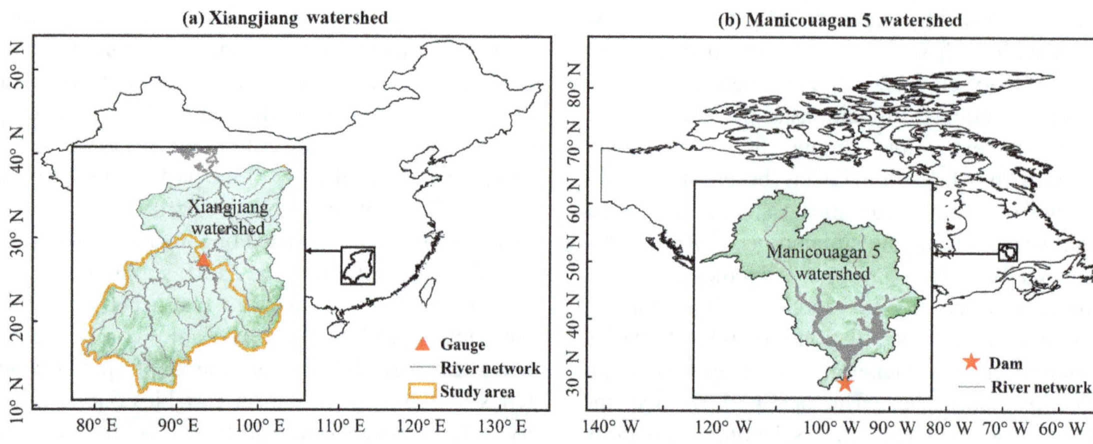

Figure 1. Location maps of the **(a)** Xiangjiang and **(b)** Manicouagan 5 watersheds. (The study area in the Xiangjiang watershed is one of its sub-basins, as the orange boundary shows.)

2 Study area and data

2.1 Study area

This study was conducted over two watersheds (the Xiangjiang and Manicouagan 5 watersheds) with different climate and hydrological characteristics (Fig. 1). The Xiangjiang watershed is a monsoon-climate and rainfall-dominated watershed located in southern central China, whereas Manicouagan 5 is a temperate-climate and seasonally snow-covered watershed located in central Quebec, Canada.

2.1.1 Xiangjiang watershed

The Xiangjiang watershed is one of the largest sub-basins of the Yangtze River watershed (Fig. 1a). The Xiangjiang River originates from the Haiyang Mountains in the Guangxi Autonomous Region and flows north to Dongting Lake in Hunan Province, which connects to the Yangtze River. The Xiangjiang River consists of several tributaries with a surface area of approximately 94 660 km^2, but only the catchment with an area of 52 150 km^2 above the Hengyang gauging station was used in this study. The catchment has a hilly topography ranging from a maximum elevation of 2042 m a.s.l. to a minimum elevation of 58 m a.s.l. at the Hengyang station. The Xiangjiang watershed is heavily influenced by a subtropical monsoon climate with hot and humid summers and mild and dry winters. The average annual precipitation over the catchment is about 1570 mm almost entirely in the form of rainfall. Around 61 % precipitation occurs from April to August, resulting in high flows during this period. The average daily maximum and minimum temperatures are around 22 and 15 °C, respectively. The average daily discharge at the Hengyang station is around 1400 m^3 s^{-1}. The peak discharge of the averaged daily hydrograph is about 4420 m^3 s^{-1}, mainly resulting from high-intensity rainfall.

2.1.2 Manicouagan 5 watershed

The Manicouagan 5 watershed, the largest sub-basin of the Manicouagan River watershed, is located in the centre of the province of Quebec, Canada (Fig. 1b). The Manicouagan 5 River discharges into the Manicouagan reservoir, an annular reservoir within the remnant of an ancient eroded impact crater, and ends at the Daniel Johnson Dam, which is the largest buttressed multiple arc dam in the world. The drainage area of the Manicouagan 5 River is about 24 610 km^2, which is mostly covered by forest and has a moderately hilly topography ranging from a maximum elevation of 952 m a.s.l. to a minimum elevation of 350 m a.s.l. (Chen et al., 2016). The Manicouagan 5 watershed has a continental subarctic climate dominated by long and cold winters. The annual precipitation is fairly evenly distributed within the year and averages about 912 mm, around 45 % of which is snowfall. The average daily maximum and minimum temperatures are around 2.4 and −7.8 °C, respectively. The average discharge of the Manicouagan 5 River is about 530 m^3 s^{-1}. The peak discharge of the averaged daily hydrograph is around 2200 m^3 s^{-1}, mainly resulting from snowmelt.

2.2 Data

Both observed and simulated daily meteorological (maximum and minimum temperatures and precipitation) data over both watersheds were used in this study. All the climate data from multiple stations or grids were averaged over the watersheds.

2.2.1 Climate simulations

Climate model simulation data used in this study were extracted from the CMIP5 archive (Taylor et al., 2012) for both the historical reference (1975–2004) and future projection (2070–2099) periods. Twenty-six GCMs from 15 insti-

Table 1. Basic information about the CMIP5 models.

Institution	Model name	Resolution (long. × lat.)
Commonwealth Scientific and Industrial Research Organization (CSIRO) and Bureau of Meteorology (BOM), Australia	ACCESS1.0 ACCESS1.3	1.875 × 1.25 1.875 × 1.25
Beijing Climate Center, China Meteorological Administration	BCC-CSM1.1 BCC-CSM1.1(m)	2.8 × 2.8 1.125 × 1.125
College of Global Change and Earth System Science, Beijing Normal University	BNU-ESM	2.8 × 2.8
Canadian Centre for Climate Modelling and Analysis	CanESM2	2.8 × 2.8
Centro Euro-Mediterraneo per I Cambiamenti Climatici	CMCC-CMS CMCC-CM CMCC-CESM	1.875 × 1.875 0.75 × 0.75 3.75 × 3.7
Centre National de Recherches Météorologiques/Centre Européen de Recherche et Formation Avancée en Calcul Scientifique	CNRM-CM5	1.4 × 1.4
Commonwealth Scientific and Industrial Research Organization in collaboration with Queensland Climate Change Centre of Excellence	CSIRO-Mk3.6.0	1.8 × 1.8
LASG, Institute of Atmospheric Physics, Chinese Academy of Sciences; and CESS, Tsinghua University	FGOALS-g2	1.875 × 1.25
NOAA Geophysical Fluid Dynamics Laboratory	GFDL-CM3 GFDL-ESM2G GFDL-ESM2M	2.5 × 2.0 2.5 × 2.0 2.5 × 2.0
Institute for Numerical Mathematics	INM-CM4	2.0 × 1.5
Institut Pierre-Simon Laplace	IPSL-CM5A-LR IPSL-CM5A-MR IPSL-CM5B-LR	3.75 × 1.9 2.5 × 1.25 3.75 × 1.9
Japan Agency for Marine-Earth Science and Technology, Atmosphere and Ocean Research Institute (The University of Tokyo), and National Institute for Environmental Studies	MIROC-ESM-CHEM MIROC-ESM	2.8 × 2.8 2.8 × 2.8
Atmosphere and Ocean Research Institute (The University of Tokyo), National Institute for Environmental Studies, and Japan Agency for Marine-Earth Science and Technology	MIROC5	1.4 × 1.4
Max Planck Institute for Meteorology (MPI-M)	MPI-ESM-LR MPI-ESM-MR	1.875 × 1.875 1.875 × 1.875
Meteorological Research Institute	MRI-ESM1 MRI-CGCM3	1.125 × 1.125 1.1 × 1.1

tutions were employed to represent climate modelling uncertainty (Table 1). Two Representative Concentration Pathways (RCP4.5 and RCP8.5) were used for each GCM to represent forcing scenario uncertainty, with the exception of CMCC-CESM, which only used RCP8.5, and MRI-ESM1, which only used RCP4.5. Only the first run of each GCM was used. On the whole, an ensemble of 50 climate simulations was used in this study.

2.2.2 Observations

Observed daily meteorological data used to downscale the GCM outputs and calibrate the hydrological model cover the 1975–2004 period for both watersheds. Meteorological data for the Manicouagan 5 watershed were obtained from the 10 km gridded dataset of Hutchinson et al. (2009), which was created by fitting spatially continuous functions of longitude, latitude, and elevation to daily station data using a trivariate thin-plate smoothing spline interpolation algorithm. Discharge data at the outlet of the Manicouagan 5 River were based on mass balance calculations at the Daniel Johnson Dam. Meteorological and discharge data for the Xiangjiang watershed were observed at 97 rain gauges, 8 temperature gauges, and 1 streamflow gauge in the catchment above the Hengyang station, which are the same as those used in Zeng et al. (2016) and Xu et al. (2013).

Table 2. Definitions of 31 climate variables. The final column indicates whether the change in a given variable is expressed in the form of relative difference (CT: change type).

Category	Index	Description	CT
ETCCDI extreme indices	TXx	Annual maximum value of daily max temperature	
	TXn	Annual minimum value of daily max temperature	
	TNx	Annual maximum value of daily min temperature	
	TNn	Annual minimum value of daily min temperature	
	TX10p	Percentage of days when daily max temperature < 10th percentile	
	TX90p	Percentage of days when daily max temperature > 90th percentile	
	TN10p	Percentage of days when daily min temperature < 10th percentile	
	TN90p	Percentage of days when daily min temperature > 90th percentile	
	WSDI	Warm spell duration index: annual count of days with at least 6 consecutive days when daily max temperature > 90th percentile	
	CSDI	Cold spell duration index: annual count of days with at least 6 consecutive days when daily min temperature < 10th percentile	
	DTR	Daily temperature range: monthly mean difference between daily max and min temperature	
Seasonal or annual mean indices	Tav	Annual average temperature	
	SpT	Seasonal average temperature in spring	
	SuT	Seasonal average temperature in summer	
	AuT	Seasonal average temperature in autumn	
	WiT	Seasonal average temperature in winter	
ETCCDI extreme indices	R1mm	Annual count of days when precipitation ≥ 1 mm	
	R10mm	Annual count of days when precipitation ≥ 10 mm	
	R20mm	Annual count of days when precipitation ≥ 20 mm	
	CDD	Consecutive dry days: maximum number of consecutive days when daily precipitation < 1 mm	
	CWD	Consecutive wet days: maximum number of consecutive days when daily precipitation ≥ 1 mm	
	Rx1day	Annual maximum 1-day precipitation	%
	Rx5day	Annual maximum consecutive 5-day precipitation	%
	SDII	Simple precipitation intensity index: mean precipitation amount on wet days (≥ 1 mm)	%
	R95pTOT	Annual total precipitation when daily precipitation > 95th quantile	%
	R99pTOT	Annual total precipitation when daily precipitation > 99th quantile	%
Seasonal or annual mean indices	ARav	Annual total precipitation	%
	SpR	Seasonal total precipitation in spring	%
	SuR	Seasonal total precipitation in summer	%
	AuR	Seasonal total precipitation in autumn	%
	WiR	Seasonal total precipitation in winter	%

3 Methodology

3.1 Subset selection of GCM simulations

Two automated envelope-based methods were used to select subsets of climate simulations. One is the K means clustering which finds cluster centroids that best characterize high-density regions of a multivariate space; the other is the KKZ method which recursively selects simulations that best span the spread of an ensemble (Cannon, 2015). Both selection methods operate on multivariate data, which means that they are sensitive to the choice and scaling of climate variables.

3.1.1 Climate variables

Since the hydrological response of a watershed not only depends on annual mean temperature and precipitation but is also sensitive to intra-annual climate variability (e.g. seasonal means or extremes), subset selection should be based on a set of climate variables that includes annual and seasonal means as well as extremes. The World Meteorological Organization's Expert Team on Climate Change Detection and Indices (ETCCDI) has recommended a set of core climate indices that can be easily derived from daily meteorological data series (http://etccdi.pacificclimate.org/list_27_indices.shtml, last access: 19 June 2018). The ETCCDI indices are designed to monitor changes in the frequency and intensity of climate extreme events and characterize the variability of extremes (Zhang et al., 2011). Here, we assume that the ETCCDI indices are sufficient to characterize climate extremes that lead to hydrological impacts.

Specifically, this study used a set of 31 climate variables as shown in Table 2 (21 ETCCDI extreme indices and 10 seasonal or annual mean indices), including 16 temperature variables and 15 precipitation variables. Since the focus of this study is on the capability of selected GCM subsets to cover uncertainty of climate change signals, changes in climate variables (relative change for precipitation and absolute change for temperature and duration) between the historical reference period (1975–2004) and the future projection period (2070–2099) were calculated for 50 climate simulations over the two study watersheds. Changes in each climate variable were standardized to zero mean and unit standard deviation to eliminate influences from different magnitudes and units between variables. These changes in climate variables are referred to as simulated climate change signals. Once changes were calculated, subsets could be selected based on the multivariate space formed by the climate variables.

3.1.2 *K* means clustering

The *K* means clustering algorithm divided the ensemble of 50 climate simulations into a user-specified number of clusters based on the objective of minimizing within-cluster sum of squared error (SSE) (Hartigan and Wong, 1979). Each cluster is represented by its centroid. The SSE is characterized by the Euclidean distances from simulations to their corresponding cluster centroids. Some studies have applied this method to select subsets of climate simulations (Logan et al., 2011; Cannon, 2015; Houle et al., 2012). The climate simulations closest to the centroids were selected as the subsets. Due to sensitivity of the *K* means clustering to initial cluster centroid positions, it was run 10 000 times with different initializations and the best solution with the lowest SSE was kept. A disadvantage of the *K* means clustering is that the selected climate simulations are not ordered. In other words, the selected simulations in a small subset may not be in a larger subset, which means that it is inconvenient for end-users to change the subset size for different applications.

3.1.3 KKZ method

The KKZ method was originally designed by Katsavounidis et al. (1994) to identify a set of optimal seed cases as initial centroids in the *K* means clustering, and was introduced by Cannon (2015) to the selection of climate simulations. This method prefers the peripheral simulations in the multivariate space. The specific procedure is as follows.

1. The climate simulation closest to the centroid of the whole ensemble is selected as the first simulation.
2. The simulation farthest from the first selected simulation is selected as the second simulation.
3. Subsequent simulations are selected as follows.
   - For each remaining simulation, its distances to every previously selected simulation are calculated.
   - Each remaining simulation is designated with the minimum distance among all distances calculated in step 3(1).
   - The simulation with the largest minimum distance, which is designated in step 3(2), is selected as the next selected simulation.

Compared to the *K* means clustering, the KKZ method is deterministic and ordered. However, it is more susceptible to selecting outliers than *K* means clustering. In addition, a random selection, repeated 100 times to minimize the influence of its stochastic nature, was conducted as a baseline to evaluate the *K* means clustering and KKZ method.

3.2 Generation of climate scenarios

GCM outputs are typically on a coarse spatial grid and contain systematic biases that preclude their direct use in hydrological modelling (Mpelasoka and Chiew, 2009; Chen et al., 2011a, b; Minville et al., 2008; Vaze and Teng, 2011). It is thus necessary to bias correct and downscale GCM outputs before running the hydrological model. The main objective of this study is to investigate the transferability of climate simulation uncertainty; hence, there is no need to use a complicated downscaling method. A commonly used change factor method, namely the daily scaling (DS) method proposed by Harrold and Jones (2003), was used in this study. This method assumes that climate change signals simulated by GCMs are credible and can be used to perturb observations to obtain future daily series. The DS method adjusts the observed daily series using the differences in distributions of simulated temperature/precipitation between the future period and the reference period. The specific steps are the following.

1. Distributions (represented by 100 quantiles in this study) of daily temperature and precipitation simulated by GCMs are calculated for both reference and future periods in each calendar month (i.e. January, February, etc.);
2. scaling factors are estimated as the differences (for temperatures) or ratios (for precipitation) in distributions of temperature and precipitation between reference and future periods for each calendar month; and
3. scaling factors are added (for temperatures) or multiplied (for precipitation) to corresponding distributions of observed daily temperature or precipitation for each calendar month.

The use of the DS method preserves the simulated climate change signal. It is based on differences in probability distributions between the reference and future periods, which are only caused by climate change signals. In addition, the consideration of quantile-dependent changes in the precipitation distribution is important in hydrological impact studies, because more runoff is generated in high-intensity precipitation events (Harrold and Jones, 2003; Chiew et al., 2009). The use of 100 quantiles in the DS method is the same as many other studies (Mpelasoka and Chiew, 2009; Chen et al., 2013). Lafon et al. (2013) showed that the empirical quantile mapping method based on 100 quantiles is more accurate than that based on 25, 50, or 75 quantiles. However, temporal sequencing in the future period is assumed to be the same as in the observed data. Changes in, for example, wet/dry spell lengths are not informed by the GCM simulations.

3.3 Hydrological response simulation

3.3.1 Hydrological modelling

The hydrological impacts were simulated by a six-parameter, lumped, conceptual hydrological model, GR4J-6. The GR4J-6 model consists of the GR4J rainfall–runoff model and the

Table 3. Definitions of 17 hydrological variables. The final column indicates whether the change in a given variable is expressed in the form of relative difference (CT: change type).

Category	Index	Description	CT
Water resource indicators (WRIs)	MD	Annual mean flow	%
	SpMD	Seasonal mean flow in spring	%
	SuMD	Seasonal mean flow in summer	%
	AuMD	Seasonal mean flow in autumn	%
	WiMD	Seasonal mean flow in winter	%
	tCMD	Centre of timing of annual flow	
Quantiles of daily flow	Q5	5th quantile of daily flow series	%
	Q50	50th quantile of daily flow series	%
	Q95	95th quantile of daily flow series	%
Indicators of hydrological alteration (IHAs)	Qx1day	Annual mean 1-day maximum flow	%
	Qx3day	Annual mean 3-day maximum flow	%
	Qx7day	Annual mean 7-day maximum flow	%
	tQx	Julian date of annual 1-day maximum flow	
	LPC	Number of low pulse periods when flow < 25th quantile in a year	
	HPC	Number of high pulse periods when flow > 75th quantile in a year	
	LPD	Mean duration of low pulses in a year	%
	HPD	Mean duration of high pulses in a year	%

CemaNeige snow accumulation and melt routines (Arsenault et al., 2015). The GR4J model is a reservoir-based four-parameter model developed on the basis of the GR3J model (Edijatno et al., 1999; Perrin et al., 2003). This model routes runoff through a production reservoir, two linear unit hydrographs and a non-linear routing reservoir. Four parameters have to be calibrated for this model. They are maximum capacity of the production reservoir, groundwater exchange coefficient, 1-day-ahead maximum capacity of the routine reservoir and time base of unit hydrograph. In an evaluation of the model, Perrin et al. (2003) found that GR4J outperformed 19 models over a large sample of catchments.

Due to its lack of snow accumulation and snowmelt algorithms, the GR4J model cannot be directly used in snow-related watersheds. Thus, the general snow accounting routine proposed by Valéry et al. (2014), CemaNeige, was added. In this routine, precipitation is divided into rainfall and snowfall depending on the daily range of temperatures, and the updating of snowpack and snowmelt is based on a degree-day approach that is controlled by two free parameters (cold content factor and snowmelt factor). In addition, the Oudin formulation (Oudin et al., 2005) was used to pre-process evapotranspiration for GR4J-6.

3.3.2 Hydrological variables

To examine the performance of subset selection in terms of hydrological response uncertainty, this study used a set of 17 hydrological variables based on water resource indicators (WRIs), indicators of hydrologic alteration (IHAs), and quantiles of daily flow series (Table 3). WRIs have been used in many hydrological impact studies to assess streamflow alteration due to natural and anthropogenic climate change (Eum et al., 2017; Shrestha et al., 2014; Chen et al., 2011b). IHAs are used to examine the temporal variation of key streamflow hydrograph components (Eum et al., 2017; Richter et al., 1996; Shrestha et al., 2014). Quantiles of daily flow series have been used to describe the characteristics of flow regimes (Mu et al., 2007; Wilby, 2005).

Similar to climate variables, changes in hydrological variables between the reference (1975–2004) and future (2070–2099) periods were calculated. To remove the influence of systematic biases between the observations and simulations, simulated runoff values instead of gauge observations were used as flow data in the reference period. The first year of each period was used to spin up the hydrological model and was excluded when calculating the hydrological variables. Once the projected changes in hydrological variables were calculated, the uncertainty coverage of subsets could be compared between climate variables and hydrological variables to evaluate the transferability of climate simulation uncertainty.

3.4 Data analysis

A criterion called the percentage of spread coverage (PSC) was used to measure the uncertainty coverage of a subset relative to the coverage of all simulations. For a given variable and subset, PSC was calculated by dividing the variable's range in the subset by the variable's range in all sim-

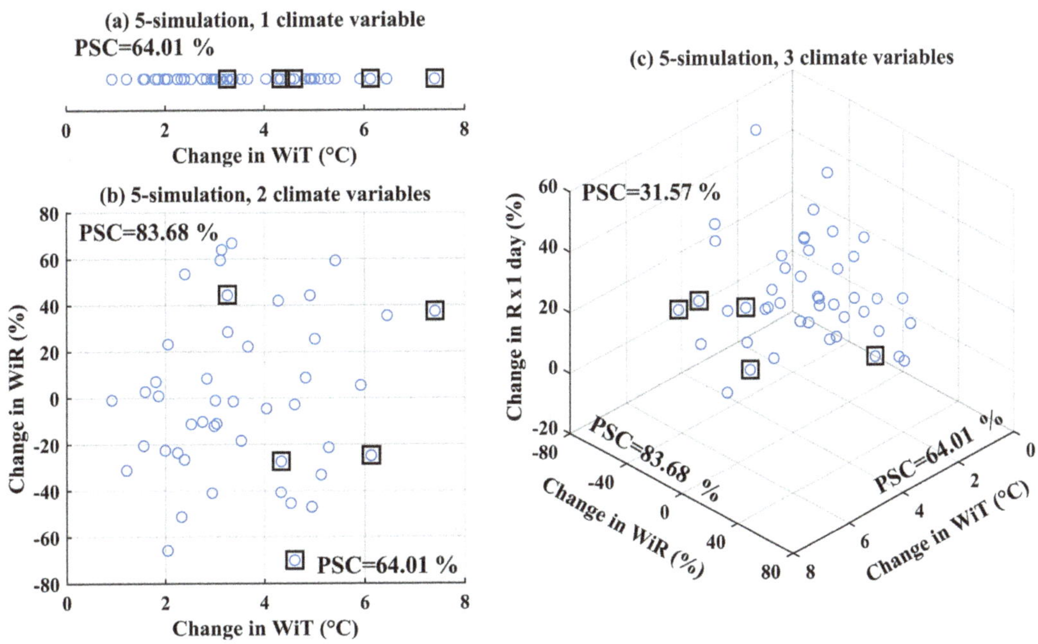

Figure 2. Examples of PSCs when selecting five climate simulations over the Xiangjiang watershed using the KKZ method. The PSCs of each variable are presented beside the corresponding axes.

ulations. A higher PSC means that the selected subset covers a larger uncertainty range. Figure 2 shows examples of PSC when five climate simulations are selected using the KKZ method. Since it is difficult to illustrate results in more than three dimensions, examples are limited to one, two, and three variables. In Fig. 2a, points represent the changes in "WiT" (seasonal average temperature in winter) for 50 GCM simulations. The larger squares represent the same variable for a subset of five climate simulations selected by KKZ. The PSC is calculated by dividing the temperature range of the selected subset, 4.15 °C, by that of the whole ensemble, 6.49 °C. Therefore, for this specific variable the PSC (uncertainty coverage) of the subset is 64.01 %. Similarly, every variable has a corresponding PSC associated with a subset of a given size; examples of "WiR" (seasonal total precipitation in winter) and "Rx1day" (annual maximum 1-day precipitation) are shown in Fig. 2b–c. For the random subset selection method, the reported PSC is the mean value of 100 PSCs, each calculated for a different random subset of the specified size.

4 Results

4.1 Calibration and validation of hydrological models

The basin-averaged daily minimum and maximum temperatures and precipitation, as shown in Table 4, were used to calibrate and validate the GR4J-6 model over the two watersheds. Model parameters were obtained by the shuffled complex evolution optimization (Duan et al., 1992) based on the objective to maximize Nash–Sutcliffe efficiency (NSE) (Nash and Sutcliffe, 1970). The optimally chosen sets of parameters yield NSE values between 0.87 and 0.93 over both watersheds for calibration and validation. The calibrated GR4J-6 also yields small relative errors (REs) of water balance between −0.3 and 5.4 % for calibration and validation (Table 4). These indicate the reasonable performance of GR4J-6 for both watersheds. Figure 3 presents the mean daily hydrographs (average discharge of each calendar day across all years) for the calibration and validation periods over the two watersheds. The calibrated GR4J-6 has good performance in most parts of the year, with the exception of overestimation of winter discharge for the Manicouagan 5 watershed. In addition, since snow accumulation and snowmelt processes are not important in the Xiangjiang watershed, the GR4J model (excluding snow module) was also calibrated in this watershed. Results showed that there was little difference between the calibrated GR4J and GR4J-6 models, and thus the presence of the CemaNeige snow module would not influence the performance of GR4J-6 in the rainfall-characterized Xiangjiang watershed (Table 4).

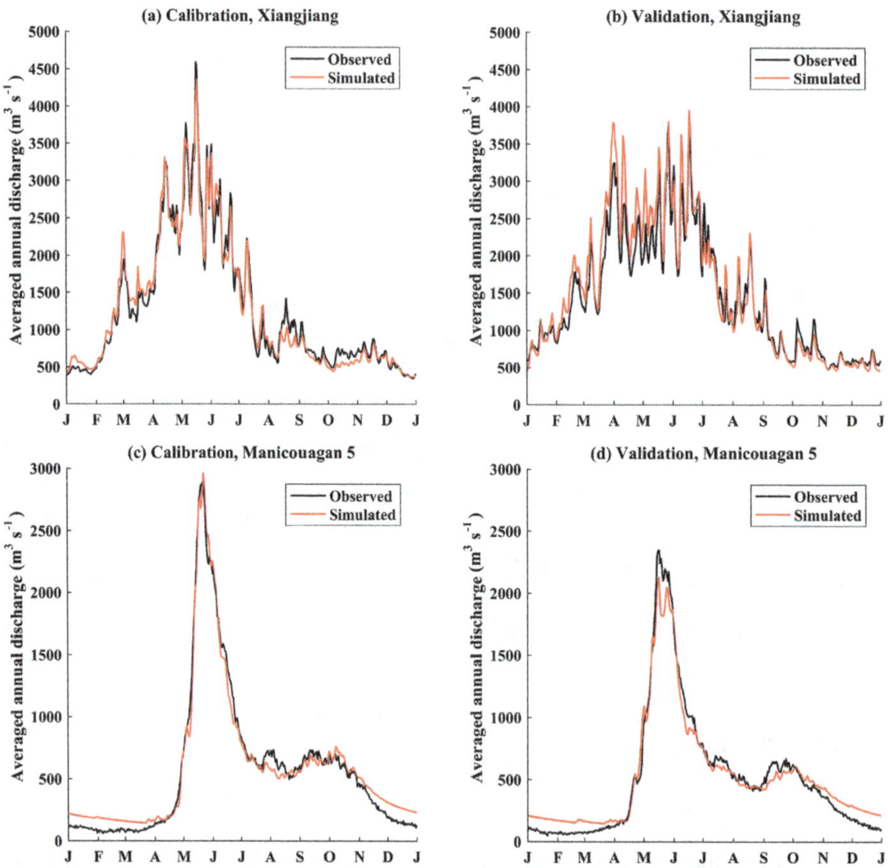

Figure 3. Observed and simulated mean hydrographs for **(a, c)** calibration and **(b, d)** validation periods over the **(a, b)** Xiangjiang and **(c, d)** Manicouagan 5 watersheds.

Table 4. Nash–Sutcliffe efficiency (NSE) and relative error (RE) of hydrological models in the calibration and validation over two watersheds.

Country	Watershed name	Area (km^2)	Hydrological model	Calibration period	NSE calibration	RE calibration	Validation period	NSE validation	RE validation
China	Xiangjiang	52 150	GR4J-6	1975–1987	0.912	−0.3 %	1988–2000	0.871	5.4 %
			GR4J	1975–1987	0.912	−0.2 %	1988–2000	0.872	5.5 %
Canada	Manicouagan 5	24 610	GR4J-6	1970–1979	0.926	3.8 %	1980–1989	0.881	2.7 %

4.2 Transferability of climate uncertainty

As an illustrative example, the uncertainty transferability from one climate variable to one hydrological variable in the Xiangjiang watershed is shown in Fig. 4. The larger squares represent the 5- and 10-climate simulation subsets selected by the KKZ method. The subfigures on the top display the PSC for "Rx5day" (maximum consecutive 5-day precipitation), whereas those at the bottom display the PSC for "Qx7day" (7-day maximum flow). The reason for choosing these two variables is that there is a generally accepted linkage between high-intensity precipitation and high flow in a rainfall-driven watershed. Although this particular choice of climate and hydrological variables is, in some ways, unfair because the overall selection process is based on a high-dimensional multivariate climate space, these subfigures still illustrate the process of uncertainty transferability from climate simulations to hydrological impacts. Here, the PSC of the climate variable increases from 66.45 to 87.37 % as the number of selected simulations goes from 5 to 10; at the same time, the PSC of the hydrological variable increases from 65.42 to 90.32 %. In this case, the uncertainty coverage of the subsets in terms of the climate variable is well translated to uncertainty coverage of the hydrological variable.

Figure 5 expands the example above from one to two dimensions. In this case, the subfigures on the top show a 2-D space formed by the changes in two climate variables, "ARav" (annual total precipitation) and "Rx1day" (maxi-

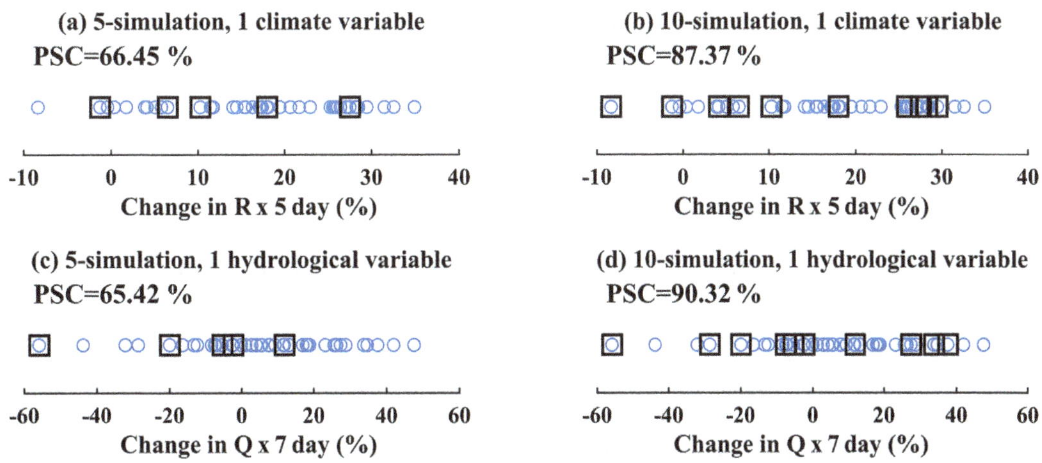

Figure 4. Examples of the transferability of climate uncertainty to hydrological impacts based on one variable when selecting **(a, c)** 5 and **(b, d)** 10 climate simulations over the Xiangjiang watershed using the KKZ method. The PSCs of each variable are presented in the top left corner.

mum 1-day precipitation), whereas those at the bottom show changes in two hydrological variables, "MD" (annual mean flow) and "HPD" (mean duration of high pulses). It should be noted that the subsets of climate simulations are the same as in the 1-D example above. As the number of selected simulations increases from 5 to 10, the mean PSC for the two climate variables increases from 50.83 to 88.72 %, while the mean PSC for the two hydrological variables increases from 59.46 to 94.05 %. The increases are mostly due to selection of outlying simulations in the top right corner of the plots (the sixth, eighth, and ninth selected simulations). There is strong consistency between locations of selected simulations in 2-D climate space and hydrology space. For example, the first, fourth, and tenth selected simulations are close to each other in both climate space (Fig. 5b) and hydrology space (Fig. 5d). Accordingly, the uncertainty coverage tends to translate well from climate variables to hydrological variables in this 2-D example. However, PSC increases are not consistent in all cases. For example, selection of the simulation on the left edge of Fig. 5b (the seventh selected simulation) substantially improves the PSC of "Rx1day", but it does not lie on the edge of Fig. 5d and hence does not contribute to improvements in PSC of either hydrology variable. This may be due to the non-linearity of the hydrological model or an imperfect explanatory relationship between the climate and hydrological variables.

The discussion above is limited to results for 5- and 10-simulation subsets for one watershed selected using the KKZ method. In the study as a whole, subset sizes from 1 to 50 simulations were evaluated in terms of transferability for two watersheds and two envelope-based methods, and PSCs for all 31 climate variables and 17 hydrological variables were calculated. Figure 6 shows distributions of climate and hydrological PSCs for 5-, 10-, 20-, 30-, and 40-simulation subsets. For the Xiangjiang watershed (Fig. 6a–b), PSCs for the climate variables are similar to those for the hydrological variables. For the Manicouagan 5 watershed (Fig. 6c–d), PSCs for the hydrological variables are slightly smaller than those for the climate variables. Overall, the tendency of the hydrological PSCs to increase with subset size is comparable to that of the climate PSCs in both watersheds. In other words, as the size of a subset becomes larger, the improvement in PSCs of the hydrological variables is similar to that of the climate variables. When comparing the two envelope-based methods, KKZ tends to outperform K means clustering.

Given the large number of climate and hydrological variables under consideration and the challenges inherent in communicating information about multi-dimensional data, two summary criteria were used to generalize subset coverage results in this study. The first criterion is the average PSC for all climate or hydrological variables. Following Cannon (2015), the second criterion is the percentage of variables that reach a 90 % PSC threshold (PSC90p).

Figure 7 presents the average PSC and PSC90p for climate variables (solid lines) and hydrological variables (dashed lines) when selected subsets contain K simulations ($K = 1$ to 50) over the two watersheds. Generally, the KKZ method performs better than K means clustering for both evaluation criteria and both watersheds, and the two automated envelope-based methods perform better than random selection. In the case of the Xiangjiang watershed (Fig. 7a), the 10-simulation KKZ subset reaches an average PSC of 90 % for climate variables, while K means and random selection require 23 and 29 simulations, respectively, to reach this threshold. For hydrological variables, the KKZ method still shows the best performance. To reach an average PSC of 90 %, KKZ and K means clustering require at least 13 and 26 simulations, respectively. In contrast to results for the climate variables, K means clustering even performs worse than random selection

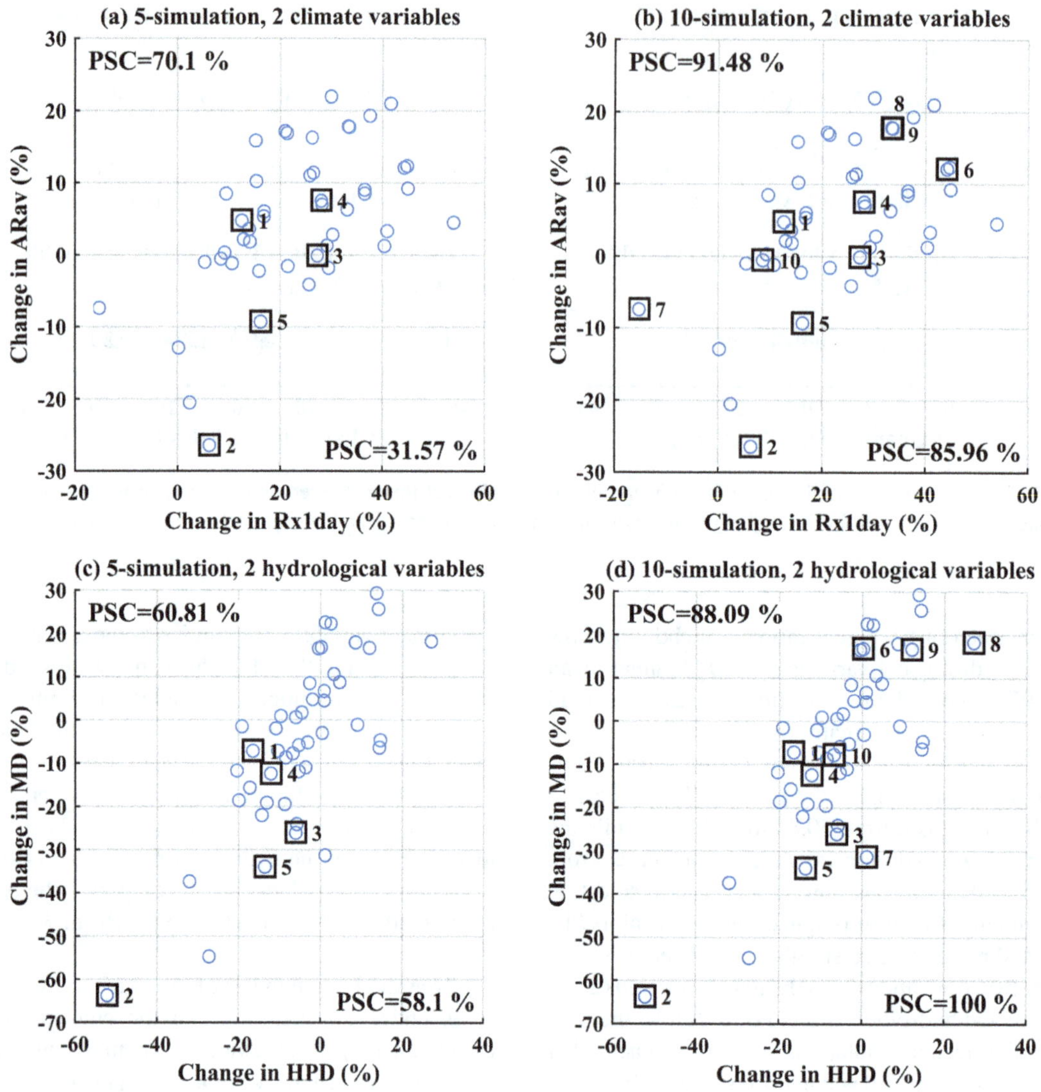

Figure 5. Examples of the transferability of climate uncertainty to hydrological impacts based on two variables when selecting **(a, c)** 5 and **(b, d)** 10 climate simulations over the Xiangjiang watershed using the KKZ method. The PSCs of each variable are presented beside the corresponding axes.

for the hydrological variables in some cases (e.g. $K = 27$ to 32). In the case of the Manicouagan 5 watershed (Fig. 7c), the KKZ method again outperforms K means clustering and random selection.

In addition, as more simulations are selected, the average PSC increases rapidly when the size of selected simulations is smaller than 10 for both watersheds, while the rate of increase slows when the number is larger than 10. For the KKZ method, a subset of 10 simulations covers more than 85 % of uncertainty for climate and hydrology variables in both watersheds; selecting more than 10 climate simulations leads to little change in uncertainty coverage. For both watersheds, a subset of 10 simulations selected by KKZ appears to be optimal for reducing computational costs while incurring the smallest possible loss of uncertainty information. In addition, the performance of the KKZ method is maintained for larger subsets, while the performance of K means clustering fluctuates. In other words, a larger subset selected by the K means clustering may not have a greater uncertainty coverage than a smaller subset. The recursive nature of the KKZ method effectively guarantees that average PSC will increase monotonically with subset size.

The focus of this study is the transferability of climate simulation uncertainty to uncertainty in hydrological impacts. For a given method this can be inferred from the difference in average PSC and PSC90p between climate and hydrological variables. For the Xiangjiang watershed, the average PSC of climate variables is close to that of hydrological variables for all selection methods (Fig. 7a). Especially for the KKZ method, differences in average PSC are less than 5 % (with

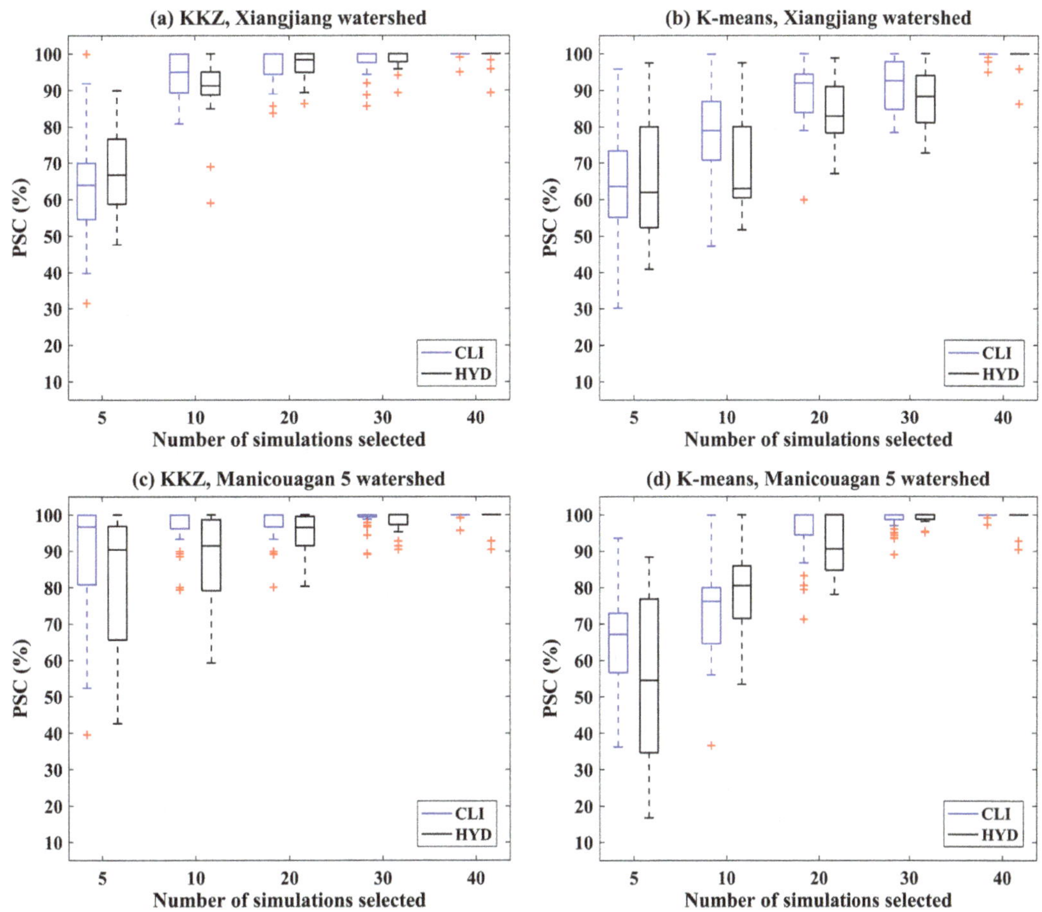

Figure 6. Boxplots of the PSCs of 31 climate variables (CLI) and 17 hydrological variables (HYD) when selecting different numbers of climate simulations over two watersheds using the KKZ method and K means clustering.

the exception of $K = 2$, 11, and 12). The differences in climate and hydrology uncertainty coverage are slightly larger when using the K means clustering and random selection methods. For PSC90p (Fig. 7b), transferability is somewhat less apparent due to the more rigorous 90 % PSC threshold. Although differences in PSC90p between climate and hydrological variables are sometimes large, especially for the K means clustering, the PSC90p of hydrological variables still exhibits similar overall tendency and behaviour to the climate variables. In general, subsets of climate simulations that are selected based on a large number of relevant climate variables are effective at transferring uncertainty coverage into the realm of hydrological impacts. However, this transferability is method-dependent; results are less variable and more consistent for KKZ than K means clustering.

Figure 7c–d present results for average PSC and PSC90p in the Manicouagan 5 watershed. On the whole, the selection methods behave similarly in terms of transferability to in the Xiangjiang watershed, but the uncertainty coverage of the subsets for the hydrological variables is reduced slightly. Degraded transferability is most apparent in larger differences in PSC90p between the climate and hydrological variables. As noted above, however, this criterion is much more stringent than average PSC.

4.3 Impact of temperature variables

The climate variables used in this study can be classified into two groups: temperature variables and precipitation variables. Each variable is given equal weight in the subset selection, regardless of inter-variable correlations, and all variables are assumed to exert the same influence on the hydrological variables. However, the impacts of climate variables on flow regimes may not be the same in watersheds with different hydroclimatic characteristics. For example, warmer temperatures lead to earlier spring floods in northern seasonally snow-covered watersheds (such as the Manicouagan 5 watershed) (Whitfield and Cannon, 2000; Chen et al., 2011b; Minville et al., 2008), whereas changes in temperature have little impact on the timing of floods in rainfall-dominated watersheds (such as the Xiangjiang watershed). Since the importance of temperature is different for the two study watersheds, a question is raised: can the transferability of climate uncertainty in the Xiangjiang watershed be improved

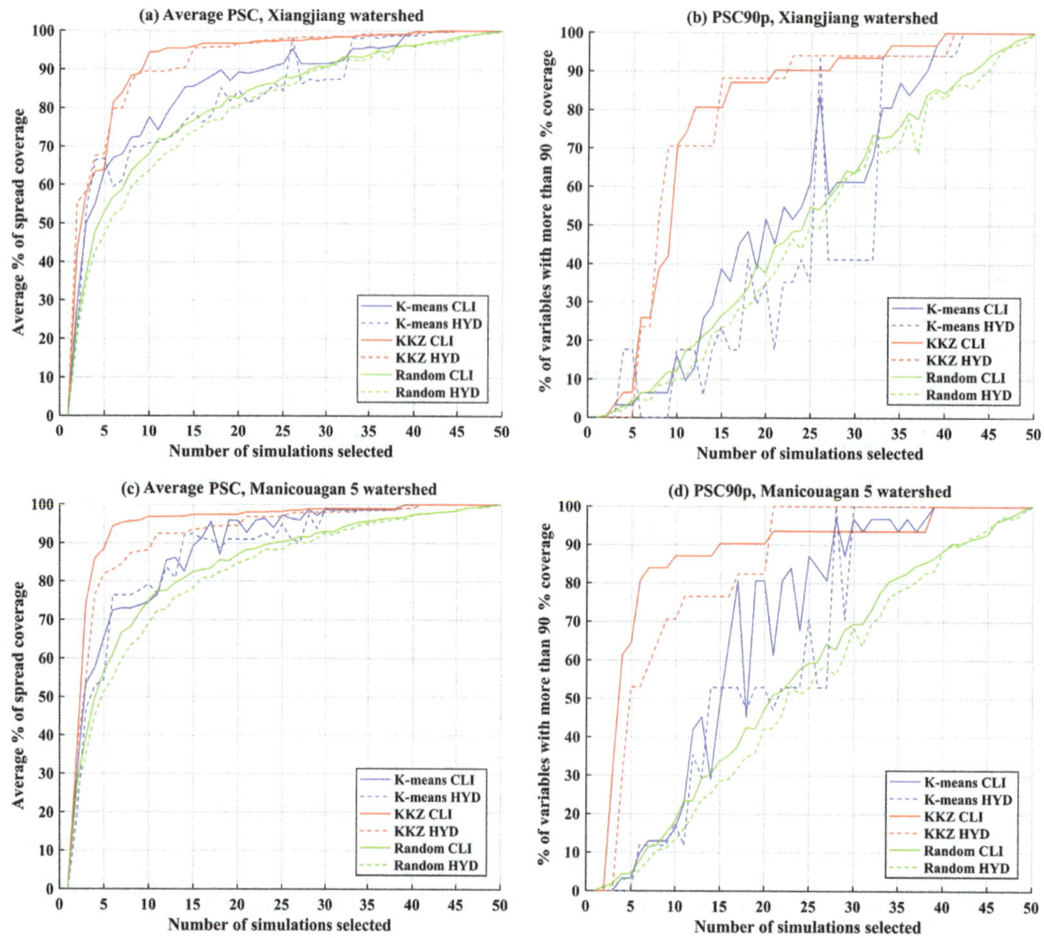

Figure 7. The **(a, c)** average PSC and **(b, d)** PSC90p for three different selection methods (K means, KKZ, and random selection) over the **(a, b)** Xiangjiang watershed and the **(b, d)** Manicouagan 5 watershed (CLI: climate variables; HYD: hydrological variables).

if irrelevant temperature variables are removed? To answer this question, temperature variables (the first 16 variables in Table 2) were removed and subset selection was conducted again using the 15 precipitation variables. The average PSC and PSC90p were then calculated to be compared with original results that include temperature variables. Results from the precipitation analysis are shown in Fig. 8.

For the Xiangjiang watershed (Fig. 8a–b), removing temperature variables from the subset selection leads to improved uncertainty coverage for the hydrological variables, especially for K means clustering. The K means clustering now performs better than random selection in most cases. For KKZ, average PSC for the hydrological variables reaches 90 % with a subset of only 6 simulations, whereas the same level of coverage required 13 simulations when considering both temperature and precipitation. However, the effect of removing temperature variables is the opposite for the Manicouagan 5 watershed (Fig. 8c–d). Here, uncertainty coverage for the hydrological variables is reduced when temperature variables are not considered. The contrasting effects are consistent with the processes that generate runoff in the two watersheds. As mentioned above, the Manicouagan 5 watershed is seasonally snow-covered – snow accumulation and snowmelt are the dominant processes that contribute to runoff generation – and hence it is sensitive to changes in temperature. However, temperature variables are not relevant in the rainfall-dominated Xiangjiang watershed. The different impacts of temperature variables in the two watersheds highlight the necessity of carefully choosing climate variables for subset selection based on physical process knowledge.

4.4 Transferability of the multi-model mean

In addition to the overall spread in the projected climate change signal, policymakers are also concerned with the MME mean when communicating hydrological climate change impacts. Therefore, the selection methods were also evaluated in terms of their ability to preserve the multi-model mean of the full MME. It bears noting that the CMIP5 MME considered in this study is an ensemble of opportunity. Models are not statistically independent, for example

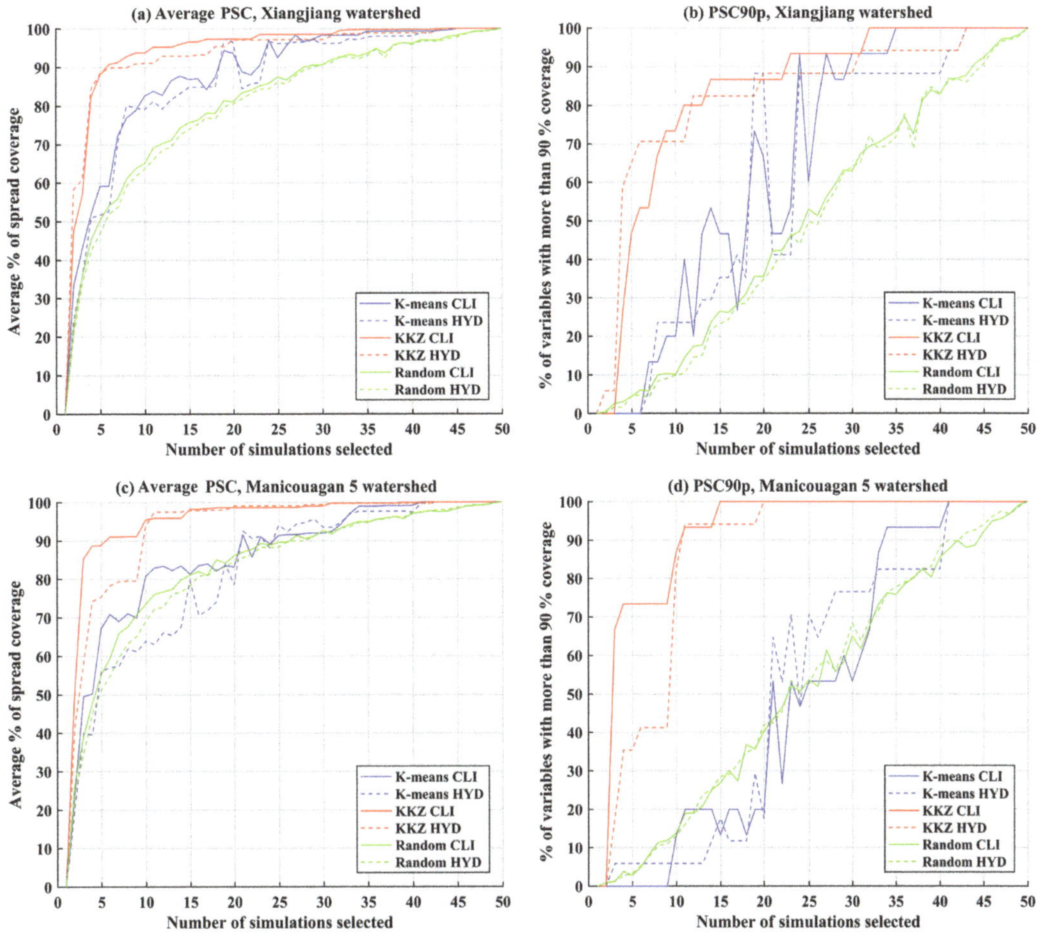

Figure 8. The **(a, c)** average PSC and **(b, d)** PSC90p for three different selection methods (K means, KKZ, and random selection) over two watersheds when temperature variables are excluded in the process of simulation selection (CLI: climate variables; HYD: hydrological variables).

due to shared physical parameterizations, and multiple simulations may be contributed by the same model. Also, the two envelope-based methods make very different assumptions about the underlying nature of the statistical distribution of the ensemble. The KKZ method is not biased towards high-density regions of the multivariate space, preferring uniform coverage, whereas the K means method, which assumes a mixture of multivariate normal clusters with equal variance, will tend to select simulations that lie in regions populated by a large number of simulations. These characteristics will have implications for preservation of the MME mean.

In order to generalize the MME mean over multiple variables, standardized changes in each variable were averaged across variables and selected simulations to obtain a dimensionless criterion (referred to as averaged standardized mean change). For different sizes of subsets selected by the three selection methods, corresponding climate and hydrological averaged standardized mean changes were calculated and compared with values for the whole ensemble. Because projected changes are preprocessed by standardizing to zero mean and unit standard deviation, the averaged standardized mean change in the whole ensemble is zero by construction. Therefore, if the averaged standardized mean change in a subset is close to zero, the MME mean change simulated by that subset is similar to that simulated by the entire ensemble. Figure 9 shows the averaged standardized mean changes in climate and hydrological variables when K simulations ($K = 1$ to 50) are selected for the two watersheds. When averaged over a large number of random trials, mean values will, by definition, lie close to zero for the random selection method; thus, the envelopes of results across all 100 random selections are presented as blue and pink shaded areas in each subfigure for climate and hydrological variables, respectively. Figure 9a–b present results for subsets when temperature variables are included in the selection process, whereas Fig. 9c–d present results when temperature variables are excluded.

Overall, when gauged against the range of variability in the 100 random selections, subsets selected by both statis-

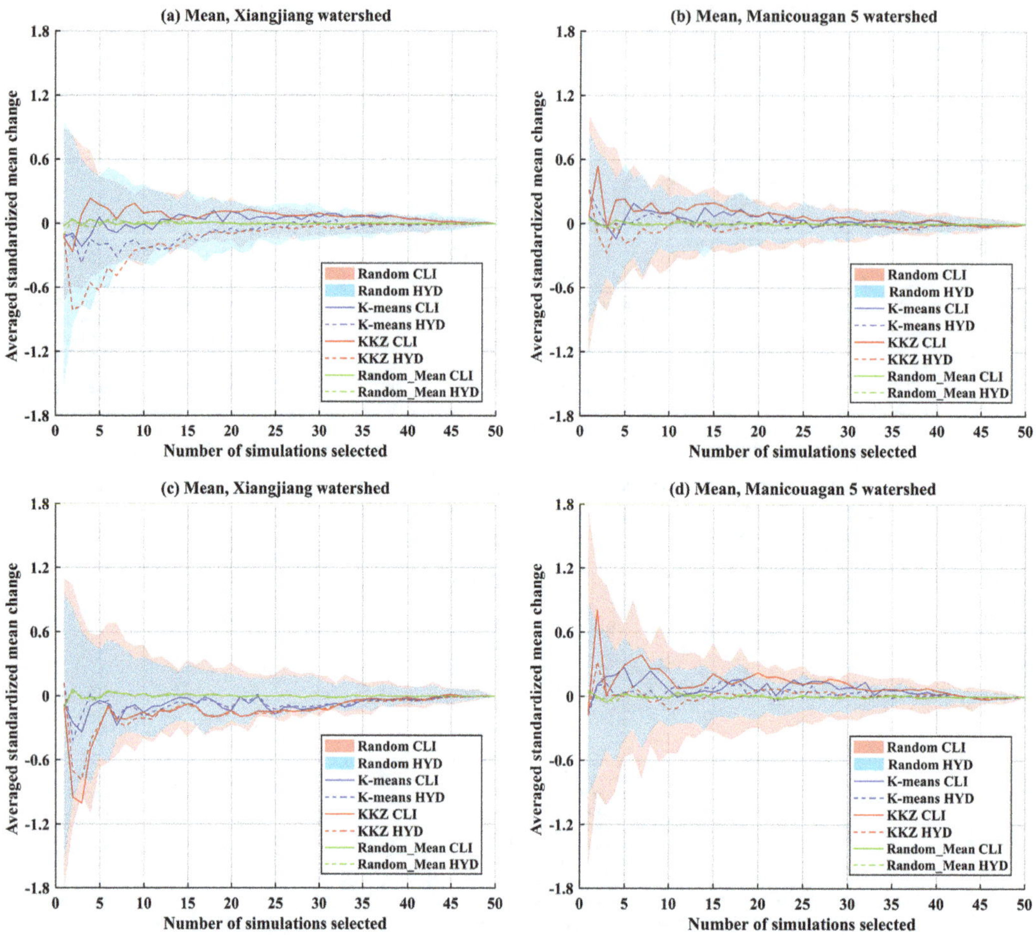

Figure 9. Averaged standardized mean change in climate (CLI) and hydrological (HYD) variables of subsets selected by three selection methods (K means, KKZ and random selection) over the Xiangjiang and Manicouagan 5 watersheds when temperature variables are **(a, b)** included or **(c, d)** excluded in the process of selection. The pink and blue panels are the envelopes resulting from 100 random selections.

tical methods perform well in reproducing the MME mean of the entire ensemble, with K means clustering performing slightly better than the KKZ method. When looked at in more detail, in the Xiangjiang watershed, the averaged standardized mean changes of subsets in climate variables tend to differ from those in hydrological variables when temperature variables are included (Fig. 9a). For example, when five simulations are selected using the KKZ method, the averaged standardized mean change for climate variables is 0.18, whereas it is −0.63 for hydrological variables. Subsets selected by the KKZ method often have higher means than the whole ensemble for climate variables, while they have lower values for hydrological variables. In other words, a subset with positive changes in climate variables gives negative changes in hydrological variables, which means that selected subsets have poor transferability in terms of MME mean. However, when temperature variables are not included in the selection process, the transferability of the multi-model mean is improved (Fig. 9c). In the Manicouagan 5 watershed, by contrast, differences between average changes in climate variables and hydrological variables are slightly smaller when temperature variables are included (Fig. 9b, d). Again, this highlights the importance of selecting the appropriate climate variables when conducting ensemble subset selection.

5 Discussion

In order to recommend a practical subset of climate simulations for end-users who deal with the assessment of climate change impacts on hydrology, various selection methods have been proposed based on different criteria (Mendlik and Gobiet, 2016; Cannon, 2015; Gleckler et al., 2008; Lutz et al., 2016a; McSweeney et al., 2012; Warszawski et al., 2014; Perkins et al., 2007). Even though these methods usually perform well in terms of the climate variables to which they are applied, their performance in terms of hydrological impacts needs to be verified. In normal usage, for example, envelope-based methods may only consider changes in mean temperature and annual precipitation (Immerzeel et al., 2013; Murdock and Spittlehouse, 2011; Warszawski et al., 2014),

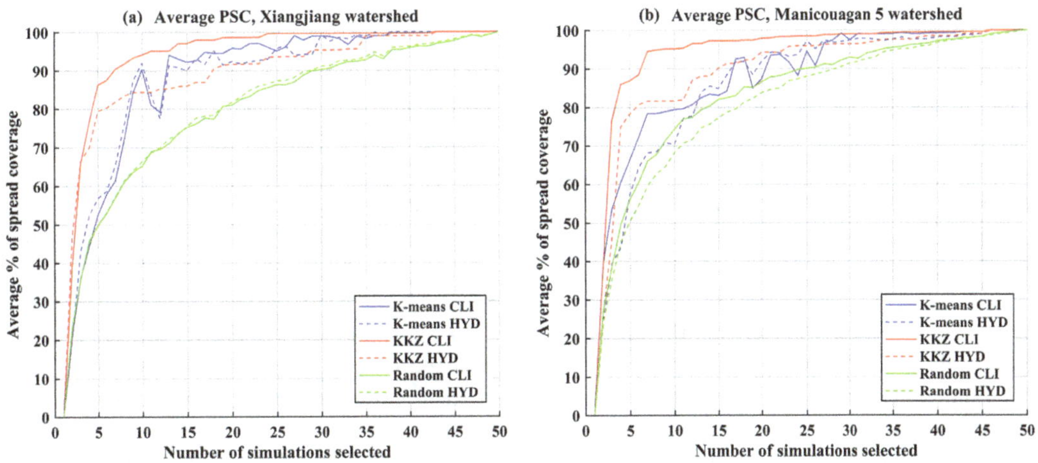

Figure 10. The average PSC for three different selection methods (K means, KKZ and random selection) over two watersheds when using QM methods (temperature variables are excluded in the Xiangjiang watershed, while they are included in the Manicouagan 5 watershed).

which will have a strong influence on both the overall measurement of climate uncertainty and subset selection results. By not considering relevant climate variables, there may be a loss of information when transferring climate uncertainty to hydrological uncertainty (Chen et al., 2016). When one considers the fact that it is often hard to determine a one-to-one correspondence between climate and hydrological variables, it may be reasonable to use a large suite of climate variables.

Therefore, this study investigated the transferability of climate simulation uncertainty to the hydrological world by K means clustering and KKZ method using a large number of climate and hydrological variables, including both seasonal and annual means and extremes. Multiple variables, when selected carefully, can improve the transferability of climate simulation uncertainty to hydrological impacts. Although the introduction of multiple climate variables may lead to redundant information, and it may be unnecessary for impact studies with different aims (e.g. water balance or hydrological drought) to consider so many climate extreme indices, this general approach can nonetheless give a more useful and reasonable selection for the purpose of covering an overall range of future climate change and its hydrological impacts. This is crucial for hydrological modellers as they usually spend a lot of computational costs in running a large number of climate simulations with a complicated hydrological model. The results of this study show that the climate simulation uncertainty is transferable in the envelope-based selection based on multiple climate variables, and the subset of around 10 climate simulations can cover the majority of uncertainty. Thus, the selected 10 climate simulations can be directly used to drive a hydrological model for impact studies instead of using all climate simulations. In addition, depending on the choice of climate variables and climate model ensemble, it may not be necessary to extract, store, and compute climate variables from all climate model simulations in the ensemble of interest. For example, pre-computed ETCCDI climate extreme indices for GCMs in CMIP3 and CMIP5 are publicly available from http://climate-modelling.canada.ca/climatemodeldata/climdex/ (last access: 19 June 2018).

In order to further investigate the impact of variable selection on transferable uncertainty, this study also compares uncertainty transferability in a rainfall-dominated watershed and a seasonally snow-covered watershed when including and excluding temperature variables. The different impacts of temperature variables over two watersheds indicate that climate variables, if not chosen with consideration of runoff-generating processes, can affect the performance of the subset selection algorithms. In the rainfall-dominated Xiangjiang watershed, inclusion of temperature variables, which play little role in generating runoff, leads to a small loss of performance, whereas in the snow-related Manicouagan 5 watershed, exclusion of temperature variables results in a large loss of performance. This is reflected in results for both ensemble spread and MME mean. Thus, it is important to choose proper climate variables that characterize the physical processes controlling the hydrology of the watershed for subset selection. Both watersheds in this study are located in humid regions with moderately hilly topography. The proper climate variables for watersheds in other climate regions may need to be modified accordingly. For example, for high-mountain regions, precipitation may be influenced by complex topography and snowmelt often has the greatest contribution to runoff, and thus it is recommended that climate variables related to orographic precipitation and the evolution of snowpack be included in the selection process (Cannon et al., 2017; Immerzeel et al., 2012); for arid regions, Hortonian overland flow may be the predominant runoff mechanism, and thus precipitation variables that are capable of describing the intensity of precipitation events may need to be stressed (Pilgrim et al., 1988). End-users can also choose the group of climate variables according to their knowledge of the climate and hydrological characteristics of the watershed of interest and then select the representative subset of climate simulations to save computational costs in the hydrology world, although it is likely that some level of site and study-specific analysis will be required in other climate regions. However, the judgement on relevant climate variables in this study is somewhat subjective. Some automated variable selection procedure may provide a more objective selection of relevant climate variables, such as redundancy analysis or multivariate sparse group lasso (Li et al., 2015).

In terms of selection methods, the results of this study reveal two strengths of the KKZ method over *K* means clustering. First, the KKZ method selects simulations on the boundaries of the climate simulation ensemble and, as a result, it is better able to cover overall climate uncertainty, as measured by average PSC and PSC90p, of the ensemble than *K* means clustering. Second, uncertainty coverage of the KKZ method for climate variables increases monotonically as more climate simulations are selected, whereas the *K* means clustering is unstable. This is because climate simulations are added incrementally, in a recursive fashion, by the KKZ method as subset size increases, whereas *K* means clustering needs to be run independently for each subset. Consequently, *K* means clustering produces a disordered sequence of solutions. The results of this study show that these two strengths of the KKZ method are retained for hydrological impacts. Therefore, in the aspect of overall uncertainty coverage, the KKZ method outperforms *K* means clustering. Performance in terms of MME mean was also evaluated in this study. Results show that the subsets selected by *K* means clustering produce a more similar MME mean to the whole ensemble, although differences between the two methods are small. This result is expected because *K* means clustering selects representative simulations for each cluster according to their closeness to the cluster centroid, which is the multivariate mean.

The two envelope-based methods in this study are from a single branch of selection methods whose purpose is to cover the spread (uncertainty) in projected changes of an ensemble. The model weighting approach is another common way to select model simulations, usually based on historical model performance, measures of statistical independence, and other evaluation metrics. Some studies have investigated the impact of weighting GCMs on the projection of climate conditions or hydrological impacts (Chen et al., 2017; Christensen et al., 2010). They concluded that weighting methods have little influence on the ensemble mean and uncertainty, and it is more appropriate to consider GCMs as being equiprobable.

Some studies have argued that certain GCMs may not be independent of one another because of shared code or parameterization schemes (Evans et al., 2013; Knutti et al., 2010; Mendlik and Gobiet, 2016). In an ensemble of opportunity like CMIP5, this dependence may lead to high-density regions of climate variable space and hence influence the selection of models by methods like *K* means clustering. On the other hand, the KKZ method is designed to select simulations that lie on the edges of the ensemble. If these simulations are outliers because their projections are not credible, for example due to poor process representation, then their selection may not be warranted. Therefore, previously removing any obviously dependent or ill-behaving GCMs through model weighting methods may improve the rationality of these two equal-weighting selection methods in regional impact studies.

In this study, only one downscaling method was used to generate climate scenarios in the scale of a watershed. In order to consider different downscaling methods, the quantile mapping (QM) approach (Maurer and Pierce, 2014; Piani et al., 2010) was additionally examined. Figure 10 presents the results of average PSC in this case. Compared with the results of the DS method, the overall character of the results is roughly the same. Thus, the choice of a downscaling method may have little influence on the conclusions of this study. In addition, only two emission scenarios (RCP4.5 and RCP8.5) were considered. RCP4.5 is the medium stabilization scenario and RCP8.5 represents the very high radiative forcing scenario. The mitigation scenario, RCP2.6, was not used because recent analyses suggested that this RCP will be very difficult to achieve with current emission trajectories (Arora et al., 2011; Rozenberg et al., 2015). RCP6.0 is a scenario with radiative forcing that is bracketed by RCP4.5 and RCP8.5 and was not simulated by as many modelling centres as RCP4.5 and RCP8.5. Thus, RCP4.5 and RCP8.5 were used to include a range of realistic projections (Lutz et al., 2016b). Due to unknown future emission scenarios, two concentration pathways were used in an undifferentiated manner to cover uncertainty resulting from emission scenarios. However, the two emission scenarios do generate different climate simulations. Thus, each pathway was also separately input into the subset selection to examine the transferability when only one scenario was considered. Figure 11 shows the average PSC for either emission scenario over the Xiangjiang watershed. The main character of the results is the same as the original research, where two RCPs were considered and selection of five or six climate simulations by the KKZ method can cover an adequate uncertainty range. Therefore, the specific choice of emission scenarios can be decided by end-users according to their own needs.

6 Conclusions

In this study, the transferability of climate simulation uncertainty to climate change impacts on hydrology was investigated over two watersheds with different climate and hydrological regimes based on multiple climate variables. The main conclusions are summarized as follows.

1. In terms of uncertainty coverage, both the KKZ method and *K* means clustering are effective at selecting subsets

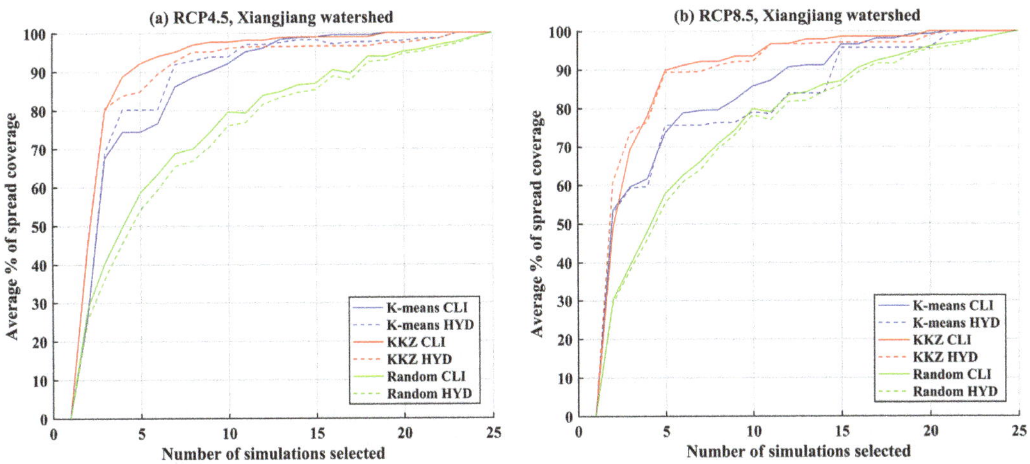

Figure 11. The average PSC for three different selection methods (*K* means, KKZ and random selection) over the Xiangjiang watershed when only one emission scenario (RCP4.5 or RCP8.5) is considered.

of climate simulations that represent the range of the climate change signal. However, when it comes to hydrological impacts, the KKZ method always performs better than random selection, while *K* means clustering sometimes performs worse than random selection.

2. Both *K* means clustering and the KKZ method are capable of reproducing the MME mean of the whole ensemble, although *K* means clustering performs slightly better than the KKZ method in some cases.

3. The uncertainty of climate simulations based on multiple climate variables can be transferred to the assessment of hydrological impact uncertainty. In other words, selected subsets can generate similar uncertainty coverage in terms of both climate simulation and hydrological impacts.

4. In order to cover an adequate range of climate simulation and hydrological impact uncertainty with fewer computational costs, selection of about 10 simulations from the ensemble of 50 simulations is required. Little improvement is gained when the number of simulations is increased beyond 10.

5. The choice of climate variables affects the transferability of climate uncertainty to hydrological uncertainty. Thus, the climate and hydrological regimes of a watershed should be considered when choosing variables used to select climate model simulations for hydrological impact studies.

Data availability. The observation data in the Xiangjiang and Manicouagan 5 watersheds were provided by the Changjiang Water Resources Commission and Hydro-Québec, respectively. These data are not publicly available because of governmental restrictions, but can be accessed by contacting the corresponding author. The climate simulation data can be accessed from the CMIP5 archive (https://esgf-node.llnl.gov/projects/esgf-llnl/, last access: 19 June 2018).

Competing interests. The authors declare that they have no conflict of interest.

Acknowledgements. This work was partially supported by the National Natural Science Foundation of China (grant nos. 51779176, 51339004, and 51539009), the Thousand Youth Talents Plan from the Organization Department of CCP Central Committee (Wuhan University, China), and the Research Council of Norway (FRINATEK Project 274310). The authors would like to acknowledge the contribution of the World Climate Research Program Working Group on Coupled Modelling, and all climate modelling groups listed in Table 1 for making available their respective model outputs. The authors would also like to acknowledge Hydro-Québec and the Changjiang Water Resources Commission for providing observation data in the Manicouagan 5 and Xiangjiang watersheds, respectively.

Edited by: Kerstin Stahl

References

Arora, V. K., Scinocca, J. F., Boer, G. J., Christian, J. R., Denman, K. L., Flato, G. M., Kharin, V. V., Lee, W. G., and Merryfield, W. J.: Carbon emission limits required to satisfy future representative concentration pathways of greenhouse gases, Geophys. Res. Lett., 38, https://doi.org/10.1029/2010GL046270, 2011.

Arsenault, R., Gatien, P., Renaud, B., Brissette, F., and Martel, J.-L.: A comparative analysis of 9 multi-model averaging approaches in hydrological continuous streamflow simulation, J. Hydrol.,

529, 754–767, https://doi.org/10.1016/j.jhydrol.2015.09.001, 2015.

Cannon, A. J.: Selecting GCM Scenarios that Span the Range of Changes in a Multimodel Ensemble: Application to CMIP5 Climate Extremes Indices, J. Climate, 28, 1260–1267, https://doi.org/10.1175/jcli-d-14-00636.1, 2015.

Cannon, F., Carvalho, L. M. V., Jones, C., Norris, J., Bookhagen, B., and Kiladis, G. N.: Effects of topographic smoothing on the simulation of winter precipitation in High Mountain Asia, J. Geophys. Res.-Atmos., 122, 1456–1474, https://doi.org/10.1002/2016jd026038, 2017.

Chen, J., Brissette, F. P., and Leconte, R.: Uncertainty of downscaling method in quantifying the impact of climate change on hydrology, J. Hydrol., 401, 190–202, https://doi.org/10.1016/j.jhydrol.2011.02.020, 2011a.

Chen, J., Brissette, F. P., Poulin, A., and Leconte, R.: Overall uncertainty study of the hydrological impacts of climate change for a Canadian watershed, Water Resour. Res., 47, W12509, https://doi.org/10.1029/2011wr010602, 2011b.

Chen, J., Brissette, F. P., Chaumont, D., and Braun, M.: Performance and uncertainty evaluation of empirical downscaling methods in quantifying the climate change impacts on hydrology over two North American river basins, J. Hydrol., 479, 200–214, https://doi.org/10.1016/j.jhydrol.2012.11.062, 2013.

Chen, J., Brissette, F. P., and Lucas-Picher, P.: Transferability of optimally-selected climate models in the quantification of climate change impacts on hydrology, Clim. Dynam., 47, 3359–3372, https://doi.org/10.1007/s00382-016-3030-x, 2016.

Chen, J., Brissette, F. P., Lucas-Picher, P., and Caya, D.: Impacts of weighting climate models for hydro-meteorological climate change studies, J. Hydrol., 549, 534–546, https://doi.org/10.1016/j.jhydrol.2017.04.025, 2017.

Chiew, F. H. S., Teng, J., Vaze, J., Post, D. A., Perraud, J. M., Kirono, D. G. C., and Viney, N. R.: Estimating climate change impact on runoff across southeast Australia: Method, results, and implications of the modeling method, Water Resour. Res., 45, https://doi.org/10.1029/2008wr007338, 2009.

Christensen, J. H., Kjellström, E., Giorgi, F., Lenderink, G., and Rummukainen, M.: Weight assignment in regional climate models, Clim. Res., 44, 179–194, https://doi.org/10.3354/cr00916, 2010.

Duan, Q., Sorooshian, S., and Gupta, V.: Effective and efficient global optimization for conceptual rainfall-runoff models, Water Resour. Res., 28, 1015–1031, https://doi.org/10.1029/91WR02985, 1992.

Edijatno, De Oliveira Nascimento, N., Yang, X., Makhlouf, Z., and Michel, C.: GR3J: a daily watershed model with three free parameters, Hydrol. Sci. J., 44, 263–277, https://doi.org/10.1080/02626669909492221, 1999.

Eum, H.-I., Dibike, Y., and Prowse, T.: Climate-induced alteration of hydrologic indicators in the Athabasca River Basin, Alberta, Canada, J. Hydrol., 544, 327–342, https://doi.org/10.1016/j.jhydrol.2016.11.034, 2017.

Evans, J. P., Ji, F., Abramowitz, G., and Ekstrom, M.: Optimally choosing small ensemble members to produce robust climate simulations, Environ. Res. Lett., 8, 044050, https://doi.org/10.1088/1748-9326/8/4/044050, 2013.

Giorgi, F. and Mearns, L. O.: Calculation of Average, Uncertainty Range, and Reliability of Regional Climate Changes from AOGCM Simulations via the "Reliability Ensemble Averaging" (REA) Method, J. Climate, 15, 1141–1158, https://doi.org/10.1175/1520-0442(2002)015<1141:coaura>2.0.co;2, 2002.

Gleckler, P. J., Taylor, K. E., and Doutriaux, C.: Performance metrics for climate models, J. Geophys. Res., 113, D06104, https://doi.org/10.1029/2007jd008972, 2008.

Harrold, T. I. and Jones, R. N.: Generation of rainfall scenarios using daily patterns of change from GCMs, in: Water Resources Systems – Water Availability and Global Change, edited by: Franks, S., Blöschl, G., Kumagai, M., Musiake, K., and Rosbjerg, D., 280, IAHS Press, 165–172, 2003.

Hartigan, J. A. and Wong, M. A.: Algorithm AS 136: A *K* means Clustering Algorithm, Journal of the Royal Statistical Society. Series C (Applied Statistics), 28, 100–108, https://doi.org/10.2307/2346830, 1979.

Houle, D., Bouffard, A., Duchesne, L., Logan, T., and Harvey, R.: Projections of Future Soil Temperature and Water Content for Three Southern Quebec Forested Sites, J. Climate, 25, 7690–7701, https://doi.org/10.1175/jcli-d-11-00440.1, 2012.

Hutchinson, M. F., McKenney, D. W., Lawrence, K., Pedlar, J. H., Hopkinson, R. F., Milewska, E., and Papadopol, P.: Development and Testing of Canada-Wide Interpolated Spatial Models of Daily Minimum–Maximum Temperature and Precipitation for 1961–2003, J. Appl. Meteorol. Climatol., 48, 725–741, https://doi.org/10.1175/2008jamc1979.1, 2009.

Immerzeel, W. W., van Beek, L. P., Konz, M., Shrestha, A. B., and Bierkens, M. F.: Hydrological response to climate change in a glacierized catchment in the Himalayas, Clim. Change, 110, 721–736, https://doi.org/10.1007/s10584-011-0143-4, 2012.

Immerzeel, W. W., Pellicciotti, F., and Bierkens, M. F. P.: Rising river flows throughout the twenty-first century in two Himalayan glacierized watersheds, Nat. Geosci., 6, 742–745, https://doi.org/10.1038/ngeo1896, 2013.

Katsavounidis, I., Jay Kuo, C. C., and Zhen, Z.: A new initialization technique for generalized Lloyd iteration, IEEE Signal Processing Letters, 1, 144–146, https://doi.org/10.1109/97.329844, 1994.

Klein, S. A. and Hall, A.: Emergent Constraints for Cloud Feedbacks, Current Climate Change Reports, 1, 276–287, https://doi.org/10.1007/s40641-015-0027-1, 2015.

Knutti, R., Furrer, R., Tebaldi, C., Cermak, J., and Meehl, G. A.: Challenges in Combining Projections from Multiple Climate Models, J. Climate, 23, 2739–2758, https://doi.org/10.1175/2009jcli3361.1, 2010.

Li, Y., Nan, B., and Zhu, J.: Multivariate sparse group lasso for the multivariate multiple linear regression with an arbitrary group structure, Biometrics, 71, 354–363, https://doi.org/10.1111/biom.12292, 2015.

Lafon, T., Dadson, S., Buys, G., and Prudhomme, C.: Bias correction of daily precipitation simulated by a regional climate model: a comparison of methods, Int. J. Climatol., 33, 1367–1381, https://doi.org/10.1002/joc.3518, 2013.

Logan, T., Charron, I., Chaumont, D., and Houle, D.: Atlas of climate scenarios for Québec forests, uranos for Ministère des Ressources naturelles and de la Faune du Québec Techical Report ISBN:978-2-923292-12-0, 1-132, 2011.

Lutz, A. F., ter Maat, H. W., Biemans, H., Shrestha, A. B., Wester, P., and Immerzeel, W. W.: Selecting representative climate models for climate change impact studies: an advanced envelope-based selection approach, Int. J. Climatol., 36, 3988–4005, https://doi.org/10.1002/joc.4608, 2016a.

Lutz, A. F., Immerzeel, W. W., Kraaijenbrink, P. D., Shrestha, A. B., and Bierkens, M. F.: Climate Change Impacts on the Upper Indus Hydrology: Sources, Shifts and Extremes, PLoS One, 11, e0165630, https://doi.org/10.1371/journal.pone.0165630, 2016b.

Maurer, E. P. and Pierce, D. W.: Bias correction can modify climate model simulated precipitation changes without adverse effect on the ensemble mean, Hydrol. Earth Syst. Sci., 18, 915–925, https://doi.org/10.5194/hess-18-915-2014, 2014.

McSweeney, C. F., Jones, R. G., and Booth, B. B. B.: Selecting Ensemble Members to Provide Regional Climate Change Information, J. Climate, 25, 7100–7121, https://doi.org/10.1175/jcli-d-11-00526.1, 2012.

Mehran, A., AghaKouchak, A., and Phillips, T. J.: Evaluation of CMIP5 continental precipitation simulations relative to satellite-based gauge-adjusted observations, J. Geophys. Res.- Atmos., 119, 1695–1707, https://doi.org/10.1002/2013jd021152, 2014.

Mendlik, T. and Gobiet, A.: Selecting climate simulations for impact studies based on multivariate patterns of climate change, Clim. Change, 135, 381–393, https://doi.org/10.1007/s10584-015-1582-0, 2016.

Minville, M., Brissette, F., and Leconte, R.: Uncertainty of the impact of climate change on the hydrology of a nordic watershed, J. Hydrol., 358, 70–83, https://doi.org/10.1016/j.jhydrol.2008.05.033, 2008.

Mpelasoka, F. S., and Chiew, F. H. S.: Influence of Rainfall Scenario Construction Methods on Runoff Projections, J. Hydrometeorol., 10, 1168–1183, https://doi.org/10.1175/2009jhm1045.1, 2009.

Mu, X., Zhang, L., McVicar, T. R., Chille, B., and Gau, P.: Analysis of the impact of conservation measures on stream flow regime in catchments of the Loess Plateau, China, Hydrol. Proc., 21, 2124–2134, https://doi.org/10.1002/hyp.6391, 2007.

Murdock, T. and Spittlehouse, D.: Selecting and using climate change scenarios for British Columbia, Pacific Climate Impacts Consortium, University of Victoria, Victoria, BC, 1–39, 2011.

Muzik, I.: Sensitivity of Hydrologic Systems to Climate Change, Can. Water Resour. J., 26, 233–252, https://doi.org/10.4296/cwrj2602233, 2001.

Nash, J. E. and Sutcliffe, J. V.: River flow forecasting through conceptual models Part I – A discussion of principles, J. Hydrol., 10, 282–290, https://doi.org/10.1016/0022-1694(70)90255-6, 1970.

Oudin, L., Hervieu, F., Michel, C., Perrin, C., Andréassian, V., Anctil, F., and Loumagne, C.: Which potential evapotranspiration input for a lumped rainfall-runoff model?, J. Hydrol., 303, 290–306, https://doi.org/10.1016/j.jhydrol.2004.08.026, 2005.

Perkins, S. E., Pitman, A. J., Holbrook, N. J., and McAneney, J.: Evaluation of the AR4 Climate Models' Simulated Daily Maximum Temperature, Minimum Temperature, and Precipitation over Australia Using Probability Density Functions, J. Climate, 20, 4356–4376, https://doi.org/10.1175/jcli4253.1, 2007.

Perrin, C., Michel, C., and Andréassian, V.: Improvement of a parsimonious model for streamflow simulation, J. Hydrol., 279, 275–289, https://doi.org/10.1016/s0022-1694(03)00225-7, 2003.

Piani, C., Weedon, G. P., Best, M., Gomes, S. M., Viterbo, P., Hagemann, S., and Haerter, J. O.: Statistical bias correction of global simulated daily precipitation and temperature for the application of hydrological models, J. Hydrol., 395, 199–215, https://doi.org/10.1016/j.jhydrol.2010.10.024, 2010.

Pierce, D. W., Barnett, T. P., Santer, B. D., and Gleckler, P. J.: Selecting global climate models for regional climate change studies, P. Natl. Acad. Sci., 106, 8441–8446, 2009.

Pilgrim, D. H., Chapman, T. G., and Doran, D. G.: Problems of rainfall-runoff modelling in arid and semiarid regions, Hydrol. Sci. J., 33, 379–400, https://doi.org/10.1080/02626668809491261, 1988.

Pincus, R., Batstone, C. P., Hofmann, R. J. P., Taylor, K. E., and Glecker, P. J.: Evaluating the present-day simulation of clouds, precipitation, and radiation in climate models, J. Geophys. Res., 113, https://doi.org/10.1029/2007jd009334, 2008.

Reifen, C. and Toumi, R.: Climate projections: Past performance no guarantee of future skill?, Geophys. Res. Lett., 36, https://doi.org/10.1029/2009gl038082, 2009.

Richter, B. D., Baumgartner, J. V., Powell, J., and Braun, D. P.: A Method for Assessing Hydrologic Alteration within Ecosystems, Conserv. Biol., 10, 1163–1174, https://doi.org/10.1046/j.1523-1739.1996.10041163.x, 1996.

Rozenberg, J., Davis, S. J., Narloch, U., and Hallegatte, S.: Climate constraints on the carbon intensity of economic growth, Environ. Res. Lett., 10, 95006, https://doi.org/10.1088/1748-9326/10/9/095006, 2015.

Shrestha, R. R., Peters, D. L., and Schnorbus, M. A.: Evaluating the ability of a hydrologic model to replicate hydro-ecologically relevant indicators, Hydrol. Proc., 28, 4294–4310, https://doi.org/10.1002/hyp.9997, 2014.

Taylor, K. E., Stouffer, R. J., and Meehl, G. A.: An Overview of CMIP5 and the Experiment Design, Bull. Am. Meteorol. Soc., 93, 485–498, https://doi.org/10.1175/bams-d-11-00094.1, 2012.

Tebaldi, C. and Knutti, R.: The use of the multi-model ensemble in probabilistic climate projections, Philosophical Transactions of the Royal Society A: Mathematical, Phys. Eng. Sci., 365, 2053–2075, https://doi.org/10.1098/rsta.2007.2076, 2007.

Valéry, A., Andréassian, V., and Perrin, C.: As simple as possible but not simpler: What is useful in a temperature-based snow-accounting routine? Part 2 – Sensitivity analysis of the Cemaneige snow accounting routine on 380 catchments, J. Hydrol., 517, 1176–1187, https://doi.org/10.1016/j.jhydrol.2014.04.058, 2014.

Vaze, J. and Teng, J.: Future climate and runoff projections across New South Wales, Australia: results and practical applications, Hydrol. Proc., 25, 18–35, https://doi.org/10.1002/hyp.7812, 2011.

Warszawski, L., Frieler, K., Huber, V., Piontek, F., Serdeczny, O., and Schewe, J.: The Inter-Sectoral Impact Model Intercomparison Project (ISI-MIP): project framework, P. Natl. Acad. Sci., 111, 3228–3232, https://doi.org/10.1073/pnas.1312330110, 2014.

Whitfield, P. H., and Cannon, A. J.: Recent Variations in Climate and Hydrology in Canada, Can. Water Resour. J., 25, 19–65, https://doi.org/10.4296/cwrj2501019, 2000.

Wilby, R. L.: Uncertainty in water resource model parameters used for climate change impact assessment, Hydrol. Proc., 19, 3201–3219, https://doi.org/10.1002/hyp.5819, 2005.

Wilcke, R. A. I. and Bärring, L.: Selecting regional climate scenarios for impact modelling studies, Environ. Modell. Softw., 78, 191–201, https://doi.org/10.1016/j.envsoft.2016.01.002, 2016.

Xu, H., Xu, C.-Y., Chen, H., Zhang, Z., and Li, L.: Assessing the influence of rain gauge density and distribution on hydrological model performance in a humid region of China, J. Hydrol., 505, 1–12, https://doi.org/10.1016/j.jhydrol.2013.09.004, 2013.

Zeng, Q., Chen, H., Xu, C.-Y., Jie, M.-X., and Hou, Y.-K.: Feasibility and uncertainty of using conceptual rainfall-runoff models in design flood estimation, Hydrol. Res., 47, 701–717, https://doi.org/10.2166/nh.2015.069, 2016.

Zhang, X., Alexander, L., Hegerl, G. C., Jones, P., Tank, A. K., Peterson, T. C., Trewin, B., and Zwiers, F. W.: Indices for monitoring changes in extremes based on daily temperature and precipitation data, Wiley Interdisciplinary Reviews: Climate Change, 2, 851–870, https://doi.org/10.1002/wcc.147, 2011.

6

Projected cryospheric and hydrological impacts of 21st century climate change in the Ötztal Alps (Austria) simulated using a physically based approach

Florian Hanzer[1,2], **Kristian Förster**[1,3], **Johanna Nemec**[4], **and Ulrich Strasser**[1]

[1]Department of Geography, University of Innsbruck, Innsbruck, Austria
[2]Wegener Center for Climate and Global Change, University of Graz, Graz, Austria
[3]Institute of Hydrology and Water Resources Management, Leibniz Universität Hannover, Hannover, Germany
[4]ENVEO IT GmbH, Innsbruck, Austria

Correspondence: Florian Hanzer (florian.hanzer@uibk.ac.at)

Abstract. A physically based hydroclimatological model (AMUNDSEN) is used to assess future climate change impacts on the cryosphere and hydrology of the Ötztal Alps (Austria) until 2100. The model is run in 100 m spatial and 3 h temporal resolution using in total 31 downscaled, bias-corrected, and temporally disaggregated EURO-CORDEX climate projections for the representative concentration pathways (RCPs) 2.6, 4.5, and 8.5 scenarios as forcing data, making this – to date – the most detailed study for this region in terms of process representation and range of considered climate projections. Changes in snow coverage, glacierization, and hydrological regimes are discussed both for a larger area encompassing the Ötztal Alps ($1850\,km^2$, 862–3770 m a.s.l.) as well as for seven catchments in the area with varying size ($11–165\,km^2$) and glacierization (24–77 %).

Results show generally declining snow amounts with moderate decreases (0–20 % depending on the emission scenario) of mean annual snow water equivalent in high elevations (> 2500 m a.s.l.) until the end of the century. The largest decreases, amounting to up to 25–80 %, are projected to occur in elevations below 1500 m a.s.l. Glaciers in the region will continue to retreat strongly, leaving only 4–20 % of the initial (as of 2006) ice volume left by 2100. Total and summer (JJA) runoff will change little during the early 21st century (2011–2040) with simulated decreases (compared to 1997–2006) of up to 11 % (total) and 13 % (summer) depending on catchment and scenario, whereas runoff volumes decrease by up to 39 % (total) and 47 % (summer) towards the end of the century (2071–2100), accompanied by a shift in peak flows from July towards June.

1 Introduction

Current and future climate change is expected to significantly alter the mountain cryosphere of the European Alps. Rising temperatures are expected to result in more precipitation falling as rain rather than snow, a delayed onset of the snow-covered period, and an earlier onset of snowmelt, regardless of the considered emission scenarios (e.g., Frei et al., 2018; Gobiet et al., 2014; Marty et al., 2017). Possibly increasing winter precipitation amounts are expected to partly compensate for these temperature-induced effects of reduced average snowpack depths only in very high-elevated regions (e.g., Schmucki et al., 2015b).

The European mountain glaciers have already lost substantial parts of their volume and area during the past decades. Due to their delayed response to changed climatic conditions, they are still out of balance with the current climate and would continue to recede throughout the century even without any further climatic changes (e.g., Marzeion et al., 2014; Jouvet et al., 2011; Zekollari et al., 2014). Further rising temperatures will only amplify this process, resulting in a total glacier volume reduction in the European Alps of 65–100 % by 2100 according to several global-scale studies (Bliss et al., 2014; Huss and Hock, 2015; Marzeion et al., 2012; Radić et al., 2013).

Consequently, the runoff regimes of snowmelt-dominated and ice melt-dominated Alpine catchments will also undergo significant changes. Meltwater contribution of the seasonal snow cover will be reduced, resulting in increased winter low flows, declining summer runoff, and shifts of the peak flows towards earlier periods of the year (Barnett et al., 2005; Horton et al., 2006; Stewart, 2009). Since these projected impacts are mainly triggered by increasing temperatures (shift from snowfall to rainfall, earlier onset of snowmelt), the likelihood of occurrence is very high. In lower-elevated catchments on the other hand, projected changes in precipitation exhibit a larger impact on changes in runoff (Horton et al., 2006). Besides a projected future increase in winter precipitation on which most current climate projections agree on, future changes in precipitation over the Alpine region are however highly uncertain (Gobiet et al., 2014; Smiatek et al., 2016). Consequently, future trends in total annual streamflow volume for purely snowmelt-dominated catchments are also uncertain, as changes in precipitation might (over)compensate for increased evaporation rates due to higher temperatures. While in glacierized catchments on the other hand runoff volumes will eventually decrease due to the retreating glaciers and subsequently reduced ice melt runoff volumes, prolonged periods of glacier mass loss can lead to increased glacier runoff volumes in the short term to midterm, depending on if increased melt rates are able to overcompensate for the loss in glacier area (e.g., Jansson et al., 2003; Beniston, 2003; Collins, 2008; Bliss et al., 2014). Detecting the occurrence of this moment of peak water is of high interest, e. g., for hydropower production and planning (Schaefli, 2015).

While there are numerous studies on climate change impacts on the future of glaciers and hydrology for various parts of the European Alps, particularly Switzerland (e.g., Addor et al., 2014; Bosshard et al., 2013; Farinotti et al., 2011; Fatichi et al., 2015; Finger et al., 2012; Horton et al., 2006; Huss et al., 2008, 2014; Kobierska et al., 2013; Milano et al., 2015), few such studies exist for Austria. Kuhn and Batlogg (1998) used hypothetical temperature change scenarios to simulate future runoff for nine Austrian catchments with varying glacierization using a simple conceptual water balance model while assuming constant glacier areas. The same model was applied by Kuhn and Olefs (2007) for three catchments in the Ötztal Alps while accounting for changed glacier areas using an approach taking into account observed mean annual ice thickness changes with an additional constant surface lowering per degree of temperature increase. Tecklenburg et al. (2012) investigated climate change impacts on the Ötztaler Ache catchment using the conceptual semi-distributed model HBV-D REG and one realization of the A1B climate scenario, however without accounting for glacier geometry changes but rather investigating only the two extremes of either constant glacier areas or entirely ice-free areas through the entire simulation period, respectively. Weber et al. (2010) used the fully distributed physically based model PROMET with the subgrid-scale extension SURGES for simulating glacier processes (Prasch et al., 2011) to calculate future hydrological changes in the Upper Danube basin using a single regional climate model (RCM) realization based on the A1B scenario, and Wijngaard et al. (2016) applied the two conceptual hydrological models HBV and HQsim for the simulation of future hydrology in two catchments of the Ötztal Alps while updating glacier extents in 10-year intervals using precalculated ice thickness distributions.

Most of the cited studies rely on air temperature and precipitation as meteorological forcing data, applying simple temperature index methods for calculating snow and ice melt. However, the degree-day factors are calibrated for past conditions and their transferability in space and time is uncertain. Several studies hence have pointed out that more physical methods should be favored over classical temperature index melt calculations in climate change impact studies (e.g., Farinotti et al., 2011; Huss et al., 2009; Radić et al., 2013; Viviroli et al., 2011). Some studies have for example applied enhanced temperature index methods that also take solar radiation into account for melt calculation (e.g., Addor et al., 2014; Bosshard et al., 2013; Fatichi et al., 2015; Finger et al., 2012), addressing the fact that glacier melt rates are especially sensitive to variations in solar radiation (e.g., Huss et al., 2009; Ohmura et al., 2007). Only very few studies (e.g., Kobierska et al., 2013; Weber et al., 2010) however have applied full energy balance melt models for climate change impact assessment. While their superiority to more empirical methods is undisputed under the premise of in situ recordings of the required meteorological variables at the point scale, it remains challenging to provide adequate meteorological forcing data for their application in distributed mode. Nevertheless, due to their physical basis energy balance models are in principle better suited to account for changed climatic conditions than conceptual models (e.g., Klemeš, 1990; Walter et al., 2005; Pomeroy et al., 2007).

In this study we apply the fully distributed energy and mass balance model AMUNDSEN (Strasser, 2008) to assess future climate change impacts on the cryosphere and hydrology of the Ötztal Alps (Austria). This region has a long tradition of hydrological and glaciological research; however – as outlined above – only scattered information about future climate change impacts exists to date, with most studies applying simplified modeling approaches and lacking a comprehensive consideration of state-of-the-art climate scenarios. The aim of this study is to expand on this knowledge by applying a model with a comparatively high level of process representation and considering the entire set of the currently available EURO-CORDEX climate scenario simulations for the representative concentration pathways (RCPs) 2.6, 4.5, and 8.5. We analyze the simulation results with regard to changes in snow cover, glacier extent and volume, and hydrology, and we discuss the associated uncertainties.

2 Study site and data

The study site (Fig. 1) is situated in the Ötztal Alps (Austria), a heavily glacierized central Alpine mountain range stretching east–west at the main ridge, with Austria in the north and Italy in the south. The Ötztal Alps are characterized by a warm and dry climate, with average annual precipitation sums being as low as 660 mm at station Vent (1890 m a.s.l.). In the study, we focus on the headwaters of the three north–south-trending valleys Kaunertal, Pitztal, and Ötztal (located from west to east in Fig. 1), which are tributaries to the Inn river in the north. Elevations in the study site range from 862 to 3770 m a.s.l., with a mean elevation of approx. 2400 m a.s.l. For the analysis of the cryospheric impacts in terms of changes in snow and glacierization, we take the entire area shown in Fig. 1 into account (1850 km^2), while the hydrological investigations are carried out for the seven gauged catchments highlighted in the figure and listed in Table 1, with a total area of 379 km^2. The runoff regimes are characterized by a strong seasonality due to being fed mostly by snow and ice melt (glacial to nival regime types). Several of the catchments contribute to the Gepatsch hydropower reservoir situated in the south of the Kaunertal valley, whose natural catchment area of 106 km^2 is extended by a further 171 km^2 by diversion from the adjacent Pitztal and Radurschltal valleys. In the year 1997, 206 glaciers with a combined area of 150.7 km^2 were located in the area as determined from the second Austrian glacier inventory (Kuhn et al., 2015).

For the model simulations, a digital elevation model (DEM) resampled to 100 m resolution was used. Initial ice thickness distributions for all glaciers in the region were calculated based on the glacier outlines and surface elevations of the year 1997 using the methodology by Huss and Farinotti (2012). The resulting ice thicknesses as shown in Fig. 1 were validated against ground-penetrating radar measurements for 11 glaciers in the region, with deviations ranging between −12 and 26 % and an average deviation of −3.2 % (Seiser et al., 2012).

Meteorological stations with long-term data record availability in the region were used to drive the model in 3-hourly temporal resolution for the historical simulations. For the scenario simulations until 2100, we used the EURO-CORDEX climate change projections (Jacob et al., 2013) as climatic forcing, while considering the scenarios RCP2.6, RCP4.5, and RCP8.5. The latter is a scenario assuming no implementation of climate mitigation policies, resulting in considerably and steadily increasing emissions and concentrations of greenhouse gases over time. RCP4.5 is an intermediate scenario consistent with peaking emissions around the mid-century and a strong decline afterwards, resulting in stabilizing concentrations by the end of the century. Finally, the intervention scenario RCP2.6 is at the very low end of the spectrum in terms of future emissions and radiative forcing, assuming a peak in CO_2 concentrations in the middle of the century and a slow decline afterwards along with negative emissions toward the end of the century.

The entire range of (as of July 2016) publicly available EURO-CORDEX simulations for the high-resolution ($0.11° \approx 12.5$ km) EUR-11 domain that contained all required meteorological variables (minimum, maximum and mean 2 m air temperature, precipitation, relative or specific humidity, global radiation, and wind speed) in daily temporal resolution was selected, amounting to a total of 14 general circulation model (GCM)–RCM combinations (Table 2). All of these 14 model combinations include simulations for both the RCP4.5 and RCP8.5 scenarios, whereas only three realizations for the RCP2.6 scenario were available. Hence, in total 31 different sets of climate projections were available for the glaciohydrological simulations. The scenario period of the models is 2006–2100, with the exception of the three HadGEM-driven models which terminate in November 2099.

3 Methods

3.1 Model

For the glaciohydrological simulations in this study, we used the fully distributed hydroclimatological model AMUNDSEN (Strasser, 2008). AMUNDSEN has specifically been designed as a scenario-capable model for the application in high-mountain regions and has been set up and extensively validated for historical conditions in the study site in a recent study (Hanzer et al., 2016). In the following, the most important model components are briefly discussed – for a more detailed model description, we refer to, e.g., Hanzer et al. (2014, 2016), Marke et al. (2015), Pellicciotti et al. (2005), Strasser (2004, 2008), and Strasser et al. (2008).

The model is capable of operating in temporal resolutions of 1–3 h, while the spatial resolution is generally arbitrary but typically chosen on the order of 10–100 m for the application in high-mountain regions in order to accurately capture the complex topography and the underlying processes in these regions. As meteorological variables, AMUNDSEN requires point measurements or gridded inputs of air temperature, precipitation, relative humidity, global radiation, and wind speed. For the application using point data, the model includes a meteorological preprocessor for the spatial interpolation of scattered point measurements using a combination of lapse rates and inverse distance weighting. Lapse rates are either calculated from the point data in each time step or are pre-supplied as average monthly values. A radiation model based on the method by Corripio (2002) is used to calculate

Table 1. Area, elevation, and glacierization (as of 1997) of the investigated catchments. The catchment IDs correspond to the labels in Fig. 1. Indentations indicate subcatchments.

ID	Catchment	Area (km^2)	z_{min} (m a.s.l.)	z_{max} (m a.s.l.)	z_{mean} (m a.s.l.)	Glacierization 1997 (%)
1	Venter Ache	165.3	1880	3762	2887	35.4
2	– Rofenache	98.6	1893	3762	2889	38.3
3	—- Vernagtbach	11.5	2638	3622	3120	77.0
4	Gurgler Ache	72.4	1885	3533	2785	31.8
5	Gepatschalm	53.9	1898	3524	2824	39.7
6	Taschachbach	60.5	1791	3761	2753	24.0
7	Pitze	27.0	1812	3548	2835	48.2
	Total	379.1	1791	3762	2833	34.4

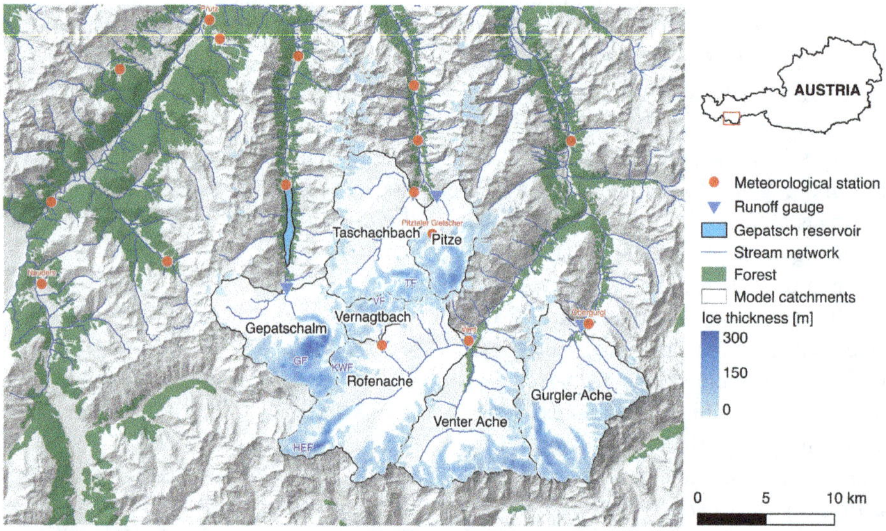

Figure 1. Location and topography of the study site including forested areas, catchment boundaries, the locations of the meteorological stations and runoff gauges, and the calculated initial ice thickness distribution for the year 1997. Selected meteorological stations and glaciers explicitly mentioned in the article are labeled in red and blue, respectively. Glacier abbreviations refer to Hintereisferner (HEF), Kesselwandferner (KWF), Vernagtferner (VF), Taschachferner (TF), and Gepatschferner (GF).

potential clear-sky solar radiation while taking into account terrain slope and orientation, hill shading, and transmission losses and gains due to scattering, absorption, and reflections, and it uses measured point values of global radiation to derive cloud factor fields and subsequently actual global radiation fields. Similarly, incoming longwave radiation is calculated using parameterizations for radiation received from the clear sky, clouds, and surrounding terrain. The precipitation phase (rain or snow) is determined as a function of the wet-bulb temperature. For solid precipitation, different correction methods are implemented in order to account for the undercatch of precipitation gauges when measuring snow accumulation. Hanzer et al. (2016) showed that a combination of a station-based snow correction factor (SCF) taking into account wind speed and air temperature with a subsequent constant post-interpolation SCF of 1.15 yielded plausible long-term precipitation amounts for the study area, while an additional redistribution of the interpolated snowfall fields using an approach based on topographic openness (Yokoyama et al., 2002) distinctly improved the spatial snow accumulation patterns. Accumulated snow on the surface is subdivided into three layers called new snow, old snow, and firn. Transitions between new snow and old snow occur depending on snow density (which is calculated based on approaches by Anderson (1976) and Jordan (1991)), while remaining snow amounts at the end of the ablation season (30 September) are transferred to the firn layer. In glacierized catchments, an additional layer is used to track the evolution of ice amounts. A linear densification of the firn layer is employed, while firn is converted to ice once reaching a threshold density of $900\,kg\,m^{-3}$. Snow albedo is parameterized using an aging curve approach with exponential decay down to a specified minimum value, while firn and ice albedo is kept constant for this application (with values of 0.4 and 0.2, respectively).

Table 2. EURO-CORDEX scenario simulations used in this study.

ID	RCM	GCM	RCPs
1	CCLM4-8-17	CNRM-CM5	4.5, 8.5
2	CCLM4-8-17	EC-EARTH	4.5, 8.5
3	CCLM4-8-17	HadGEM2-ES	4.5, 8.5
4	CCLM4-8-17	MPI-ESM-LR	4.5, 8.5
5	HIRHAM5	EC-EARTH	2.6, 4.5, 8.5
6	RACMO22E	EC-EARTH	4.5, 8.5
7	RACMO22E	HadGEM2-ES	4.5, 8.5
8	RCA4	CNRM-CM5	4.5, 8.5
9	RCA4	EC-EARTH	2.6, 4.5, 8.5
10	RCA4	CM5A-MR	4.5, 8.5
11	RCA4	HadGEM2-ES	4.5, 8.5
12	RCA4	MPI-ESM-LR	4.5, 8.5
13	REMO2009	MPI-ESM-LR	2.6, 4.5, 8.5
14	WRF331F	CM5A-MR	4.5, 8.5

In forested areas the interpolated meteorological fields are modified in order to capture sub-canopy conditions, resulting in reduced shortwave radiation, precipitation, and wind speed, increased longwave radiation and humidity, and an attenuation of the diurnal temperature cycle. Additionally the effects of the forest snow processes of interception, sublimation, and melt unload are accounted for (Strasser et al., 2011). Snow and ice melt is calculated using an energy balance approach taking into account short- and longwave radiation fluxes, latent and sensible heat fluxes, and advective energy, as well as the ground heat flux. The delay of the onset of snowmelt in cold snowpacks is accounted for by a parameterization of the cold content of the snowpack, whereas melting snow may persist in the snowpack in the form of liquid water up to a certain amount (and possibly refreeze again) before actual outflow occurs (Hanzer et al., 2016). Evapotranspiration in snow-free areas is calculated following Allen et al. (1998). Finally, runoff at predefined catchment outlets is calculated using a linear reservoir approach for snow on glaciers, firn, bare ice, snow outside of glaciers, and soil (Hanzer et al., 2016). Apart from the parameters of this linear reservoir model which have to be calibrated individually for each catchment, most parameters in the model have a physical meaning, and in general no site-specific calibration beyond the parameters for the runoff module and the precipitation correction method is performed.

The setup and extensive validation of the model for the study region under historical conditions has been described in Hanzer et al. (2016). Essentially the same model setup was used for this study, aside from (i) the newly implemented glacier retreat parameterization as described in the following section, (ii) reducing the spatial and temporal model resolution from 50 m and 1 h to 100 m and 3 h due to performance reasons, and (iii) the fact that only a subset of the meteorological stations that were utilized in the simulations presented in the aforementioned study was available for this study due to the constraints of a sufficiently long measurement period required for the bias correction of the RCM data (see Sect. 3.3).

3.2 Glacier geometry change

The model setup as described in Hanzer et al. (2016) already incorporated spatially distributed glacier thicknesses; however it only accounted for the climatic forcing on the glaciers without any adjustment of glacier geometry. Due to the generally small vertical ice flow contribution for the majority of the glaciers in the study region (Helfricht et al., 2014) this is a reasonable approach for shorter simulation periods. However, when performing simulations on multidecadal scales the effect of ice flow dynamics must be considered in glaciohydrological models. Otherwise, systematic errors in the simulated ice distributions (and subsequently in glacier runoff) are introduced as the glaciers thicken in their accumulation areas and retreat too quickly in the tongue parts.

For this study, we implemented the Δh method (Huss et al., 2010) to periodically update the simulated glacier geometries. This approach is particularly suited for spatially distributed models operating on a regional scale, as it does not necessarily require glacier-specific parameterizations. Rather, it uses simple assumptions to translate the climatic forcing (i.e., the surface mass balance as computed by the mass balance model) into a geometric response in terms of the glacier thickness distribution.

Essentially, the Δh parameterization scales the spatial distribution of the annual glacier surface mass balance such that the changes in glacier surface elevation match patterns observed in the past. This is accomplished by applying a prescribed function to each glacier in regular time intervals, which adjusts the simulated surface elevation change as a function of the normalized glacier elevation (assumed to be a proxy for the central flow line). To derive the Δh function, two glacier surface DEMs are required, preferably covering multidecadal periods to reduce possible errors due to DEM uncertainty.

In principle, a separate Δh function could be derived for each glacier; however Huss et al. (2010) showed that glaciers with similar characteristics show similar geometric responses and derived three Δh functions for glaciers of different size classes: large valley glaciers ($A \geq 20\,\mathrm{km}^2$), medium valley glaciers ($5\,\mathrm{km}^2 \leq A < 20\,\mathrm{km}^2$), and small glaciers ($A < 5\,\mathrm{km}^2$):

$$\Delta h = \begin{cases} (h_r - 0.02)^6 + 0.12\,(h_r - 0.02) & A \geq 20\,\mathrm{km}^2 \\ (h_r - 0.05)^4 + 0.19\,(h_r - 0.05) + 0.01 & 5\,\mathrm{km}^2 \leq A < 20\,\mathrm{km}^2 \\ (h_r - 0.30)^2 + 0.60\,(h_r - 0.30) + 0.09 & A < 5\,\mathrm{km}^2, \end{cases} \quad (1)$$

where h_r corresponds to the normalized surface elevation between 0 and 1.

While these parameterizations have been derived for the Swiss Alps, they are assumed to be applicable for all mountain glaciers (Huss et al., 2010) and have already been ap-

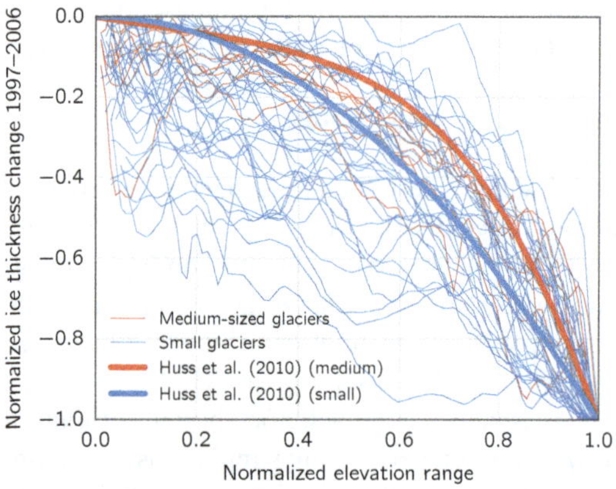

Figure 2. Normalized observed ice thickness change for 42 small ($A < 5\,\mathrm{km}^2$, thin blue lines) and 8 medium-sized ($5\,\mathrm{km}^2 \leq A < 20\,\mathrm{km}^2$, thin red lines) glaciers in the Ötztal Alps during the period 1997–2006 and the corresponding parameterizations from Huss et al. (2010) (thick lines).

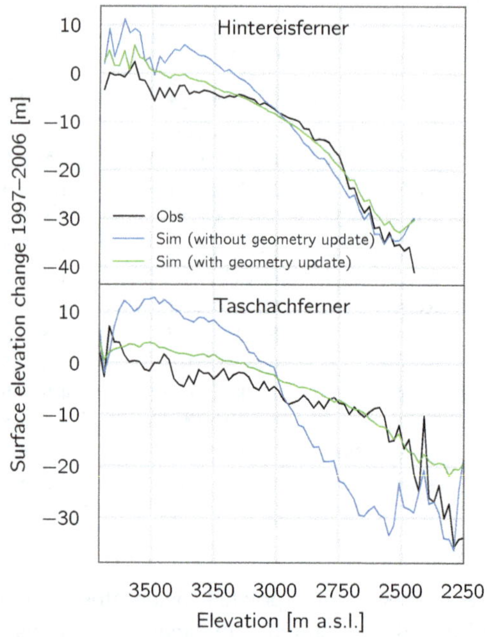

Figure 3. Observed and simulated glacier surface elevation change during the period 1997–2006 as a function of elevation for the glaciers Hintereisferner and Taschachferner (see Fig. 1 for their geographical setting). "Without geometry update" refers to the simulated surface mass balances without accounting for ice flow, while "with geometry update" refers to the simulation results obtained after implementing the Δh parameterization.

plied in other regions of the world (e.g., Huss and Hock, 2015; Ragettli et al., 2013). While for our study region DEMs and glacier outlines exist for the years 1997 and 2006, we chose to adopt these generalized parameterizations, as (i) they have been derived over much longer time periods, hence likely being more robust, and (ii) match the average observed Δh patterns for our study area in the period 1997–2006 well (Fig. 2).

In AMUNDSEN, the glacier geometry update is performed at the end of each glaciological year (30 September): for each glacier, the total volume change in terms of the surface mass balance is used to scale the thickness change prescribed by the Δh function (determined by Eq. 1), which is applied to each glacier pixel and the respective normalized elevations. Following Huss et al. (2010), the maximum surface lowering is limited to the surface mass balance lowering at the glacier terminus, and the glacier borders (pixels with ice thicknesses $< 10\,\mathrm{m}$) are not modified by the geometry update but rather change their thickness only due to the respective surface mass balance change.

Figure 3 demonstrates the effect of the glacier geometry update as implemented in AMUNDSEN exemplarily for the Hintereisferner and Taschachferner glaciers. While the average change in surface elevation between 1997 and 2006 is reproduced reasonably well for both glaciers in the run without updating the glacier geometries, the increase in surface elevation at the uppermost glacier parts is severely overestimated for both glaciers, whereas mass losses in the middle to lower parts of the glaciers are too high. Adding the glacier retreat module to the model distinctly improves the spatial distribution of mass gain or loss for both glaciers.

3.3 Spatial downscaling of RCM data

While state-of-the-art RCM simulations already feature comparatively high spatial resolutions, usually the performance of the raw RCM data is still inadequate for directly using it in hydrological impact studies, both due to mismatches in space that are still present as well as due to systematic model errors (biases) introduced by the RCMs. A common approach for impact studies is hence to use ensembles of GCM–RCM chains in combination with statistical bias correction methods (Teutschbein and Seibert, 2012).

In this study, we use quantile mapping (QM) (e.g., Déqué, 2007; Themeßl et al., 2011a), an empirical bias correction method that corrects the distribution functions of the RCM variables to fit the distribution functions of the observations. QM has regularly been shown to outperform other statistical bias correction methods (e.g., Teutschbein and Seibert, 2012, 2013), is applicable in mountain regions (e.g., Finger et al., 2012; Themeßl et al., 2011a, b), and allows for correcting climate variables other than temperature and precipitation (e.g., Finger et al., 2012; Wilcke et al., 2013). Limitations of QM are, among others, that it is generally not trend-preserving with respect to both means and extremes and that – like in most statistical bias correction methods – spatial, temporal, and intervariable correlations are not explicitly accounted for (Cannon et al., 2015; Maurer and Pierce, 2014). However, QM has been shown to perform well under changed climatic conditions (Teutschbein and Seibert, 2013) and can preserve intervariable dependency structures (Wilcke et al., 2013).

In the study, we used the QM methodology by Gudmundsson et al. (2012) as implemented in the R package *qmap*. Due to strong season-dependent biases in the RCM data, we performed the bias correction separately for DJF, MAM, JJA, and SON. Only stations and variables with a minimum amount of 20 years of data within the period 1971–2005 were considered for the bias correction procedure, with the exception of global radiation where two stations with only approx. 10 years of data were also included due to an otherwise insufficient number of stations. Given these constraints, in total 16 precipitation time series were available, as well as 13 for air temperature, 6 for wind speed, 5 for relative humidity, and 3 for global radiation.

In order to avoid the possibility of unrealistic temperature values in terms of corrected $T_{min} > T_{max}$, following Thrasher et al. (2012) we did not correct T_{min} and T_{max} independently, but rather corrected T_{max} and the diurnal temperature range (DTR) ($T_{max} - T_{min}$) with a subsequent calculation of corrected T_{min}.

For the EURO-CORDEX realizations for which only specific rather than relative humidity was available (IDs 1–4 and 13 in Table 2), mean daily specific humidity q (kg kg^{-1}) was first converted to relative humidity RH (%) using the vapor pressure formulations by Sonntag (1990).

With regard to the selection of RCM grid points for the downscaling to the point scale, we followed the approach by Hofer et al. (2017) (who however used linear regressions rather than QM) to find the optimum scale (OS) for each station and target variable: for each station and variable, spatial averages of the closest 1×1, 2×2, ..., 10×10 RCM grid points were calculated and subsequently used for bias correction. The OS for a given station and variable was then defined as the spatial window which minimizes the deviations between the cumulative distribution functions of the corrected and observed data in terms of the mean absolute error (MAE). Histograms of the resulting OS values for the different variables are shown in Fig. 4. While for all variables an OS of 1 (corresponding to using only the closest RCM grid point without spatial averaging) is the most common value, for all variables except global radiation all OS values from 1 to 10 are found – for precipitation, the maximum possible OS of 10 is even the second most common value.

RCM outputs from models using fixed 360- or 365-day calendars were linearly rescaled to Gregorian calendar dates (e.g., for a 360-day calendar each day from 1 to 360 was mapped to a respective day from 1 to 365 (or 366) in the Gregorian calendar). Resulting days with missing values were filled by duplicating the values from the preceding day.

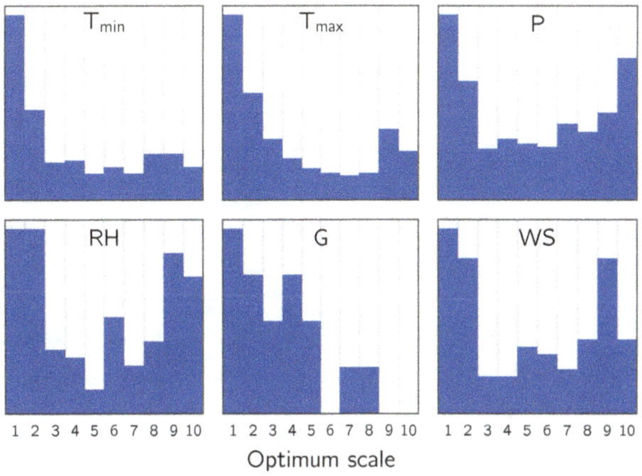

Figure 4. Histograms of the optimum scales for bias correction found using the approach by Hofer et al. (2017) for the variables minimum temperature (T_{min}), maximum temperature (T_{max}), precipitation (P), relative humidity (RH), global radiation (G), and wind speed (WS). Histograms are calculated using the derived optimum scales for all available stations for a given variable and all 14 GCM–RCM combinations (Table 2).

Figure 5 shows the mean deviation (MD) and the standard deviation ratio (SDR) of the bias-corrected RCM data vs. the daily observations for the stations in the study area in the historical period 1971–2005. The model IDs (x-axis) correspond to the IDs in Table 2. While some models tend to perform slightly better than others on average, both measures are in general close to their optimum values (0 and 1, respectively) for all realizations, indicating that the corrected RCM outputs adequately represent the observed climate in terms of mean and variability. The higher deviations in the variability of relative humidity for some models are not a result of the bias correction per se, but rather of the required conversion of specific to relative humidity for these models.

3.4 Temporal downscaling of RCM data

For the calculation of the snow and ice surface energy balance in AMUNDSEN, 1–3-hourly meteorological input time series are required in order to capture the diurnal variability of the contributing energy fluxes. As the EURO-CORDEX simulations were however only available in daily temporal resolution, an additional processing step was necessary. This step was performed with the open-source temporal disaggregation tool MELODIST (Förster et al., 2016). MELODIST provides simple, empirical disaggregation routines for the sub-daily disaggregation of daily point-scale data for the variables air temperature, precipitation, relative humidity, solar radiation, and wind speed. For the application in this study several new disaggregation methods were added to MELODIST, now published as version 0.1.2 (Hanzer et al., 2017).

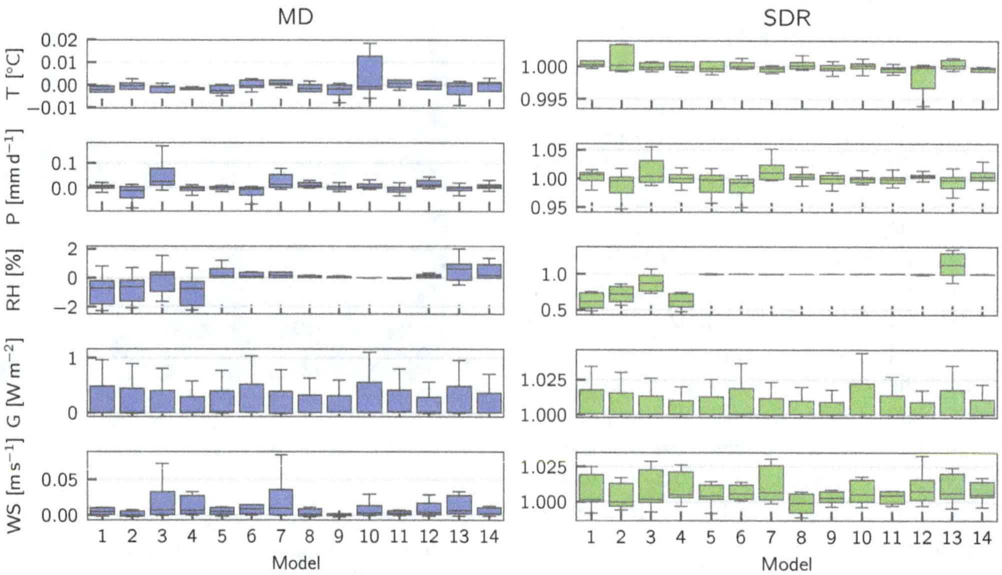

Figure 5. Mean deviation (MD) and standard deviation ratio (SDR) of simulated (i.e., downscaled and bias-corrected RCM data) vs. observed meteorological data for the historical period 1971–2005. The individual boxes represent the downscaled station meteorology for a given GCM–RCM combination (model IDs correspond to the ones in Table 2). Units given for the different variables apply only to MD (SDR is unitless).

1. For air temperature, a disaggregation method conserving the daily mean temperature and the DTR ($T_{max} - T_{min}$) while applying a sinusoidal temperature course in the same way as the default method (which preserves T_{min} and T_{max} but not T_{mean}) was implemented. However, most stations showed no clear sinusoidal temperature course; hence the application of this method still resulted in a (positive) temperature bias for most stations. Therefore, a new method based on mean values depending on the calendar month and hour of the day derived from hourly recordings was implemented. In a preprocessing step, for each month and station an average temperature course (normalized to a [0, 1] range) is calculated. Then, daily values are disaggregated on the basis of these temperature courses again by either preserving T_{min} and T_{max}, or T_{mean} and the DTR ($T_{max} - T_{min}$).

2. With regard to solar radiation, a similar method was implemented, which uses monthly varying average diurnal radiation courses derived from observations to scale the daily mean radiation. In addition, the methods for deriving daily radiation values from sunshine duration or the diurnal temperature range have been updated to allow monthly or seasonally varying conversion factors.

3. For relative humidity, an additional disaggregation method based on Waichler and Wigmosta (2003) was implemented. Hourly humidity values are generated using (month, hour, dry/wet day) categorical mean values with the additional option to preserve the daily mean humidity.

Figure 6 exemplarily shows the density functions for the observed hourly values (black lines) as well as for the various disaggregation methods for station Obergurgl (1938 m a.s.l.) and the variables air temperature, relative humidity, global radiation, and wind speed. Based on these comparisons and the results of the multilevel validation of the glaciohydrological simulation results when driven with the disaggregated values using multiple configurations, we decided on the following disaggregation methods: for all stations, air temperature was disaggregated using the "mean course" method while preserving daily minimum and maximum temperatures, daily precipitation amounts were distributed uniformly across the day, relative humidity was disaggregated using (month, hour, dry/wet day) categorical mean values while preserving daily mean humidity, solar radiation was disaggregated using the mean course method while preserving daily mean radiation, and wind speed was distributed uniformly over the day. Not all of these disaggregation methods performed best with regard to the reproduction of hourly values at the station scale; however this method combination yielded the best overall glaciohydrological model performance according to the multilevel validation procedure described in Hanzer et al. (2016). This can be attributed to the fact that the disaggregation is performed independently for each station and variable.

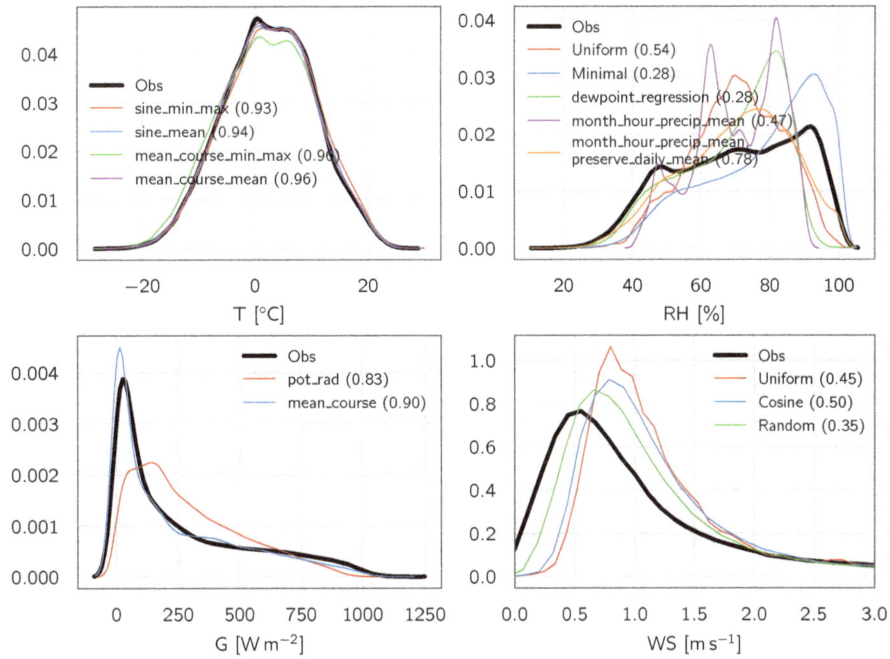

Figure 6. Density functions of the observed hourly values (black lines) for station Obergurgl (period 1999–2013) and the respective aggregated and disaggregated values using different disaggregation methods for the variables temperature, relative humidity, global radiation, and wind speed. Numbers in parentheses indicate the R^2 of the regression against the observations.

4 Results and discussion

In the following sections, first we discuss the future climatic changes in the study area as projected by the EURO-CORDEX scenarios. Then, we analyze the subsequent changes in snow, glaciers, and hydrology according to the AMUNDSEN simulations for the scenario period. These simulations were carried out by initializing the model in 1997 (using the glacier inventory and ice thickness data) and running it until September 2006 using the observed meteorological data. Afterwards, we switched to the EURO-CORDEX scenario period and carried out the scenario simulations until 2100 (with the exception of the three HadGEM-driven models for which the simulations could only be carried out until the year 2098).

We did not perform the entire range of glaciohydrological simulations for a longer historical period due to restrictions of the current model setup: (i) the initial ice thickness distributions were calculated for the year 1997 and are hence not applicable for earlier periods, and (ii) the generally negative mass balances of the glaciers in the Ötztal Alps during the last decades were interrupted by a short period of glacier advance between the mid-1970s and the early 1980s (Abermann et al., 2009), which cannot adequately be accounted for using the currently implemented Δh parameterization for the glacier geometry update. However, for discussing solely the changes in snow conditions, we also carried out model runs for the period 1971–2005 (hence obtaining transient runs from 1971 to 2100) using the historical RCM simulations as forcing, while neglecting the effect of glaciers in this case.

The validation of the model setup for the study area has been described in detail in Hanzer et al. (2016). As essentially the same model setup was used for the present study, in this section only the results of the scenario simulations are discussed. A brief discussion on how model performance is impacted by the few modifications in model setup can be found in the discussion of uncertainties (Sect. 4.5).

4.1 Future climate

Figure 7 shows the range of the projected seasonal and overall changes (spatial averages, relative to the baseline period 1971–2000) of the five meteorological variables for the study area as provided by the climate simulations. These changes are calculated based on areal means of the respective variables as calculated by the meteorological preprocessor in AMUNDSEN (i.e., after bias correction). While in the figure the changes are plotted for three time slices throughout the 21st century, for the following short discussion of the changes we focus on the period 2071–2100. If not otherwise stated, values refer to multi-model means in the following.

Temperature is projected to increase by all models and for all seasons, with average values of 1.1 (RCP2.6) to 3.8 °C (RCP8.5) in the annual mean and a maximum spread of 2.2 °C between individual models for a given scenario. The highest warming is projected for the winter season, with

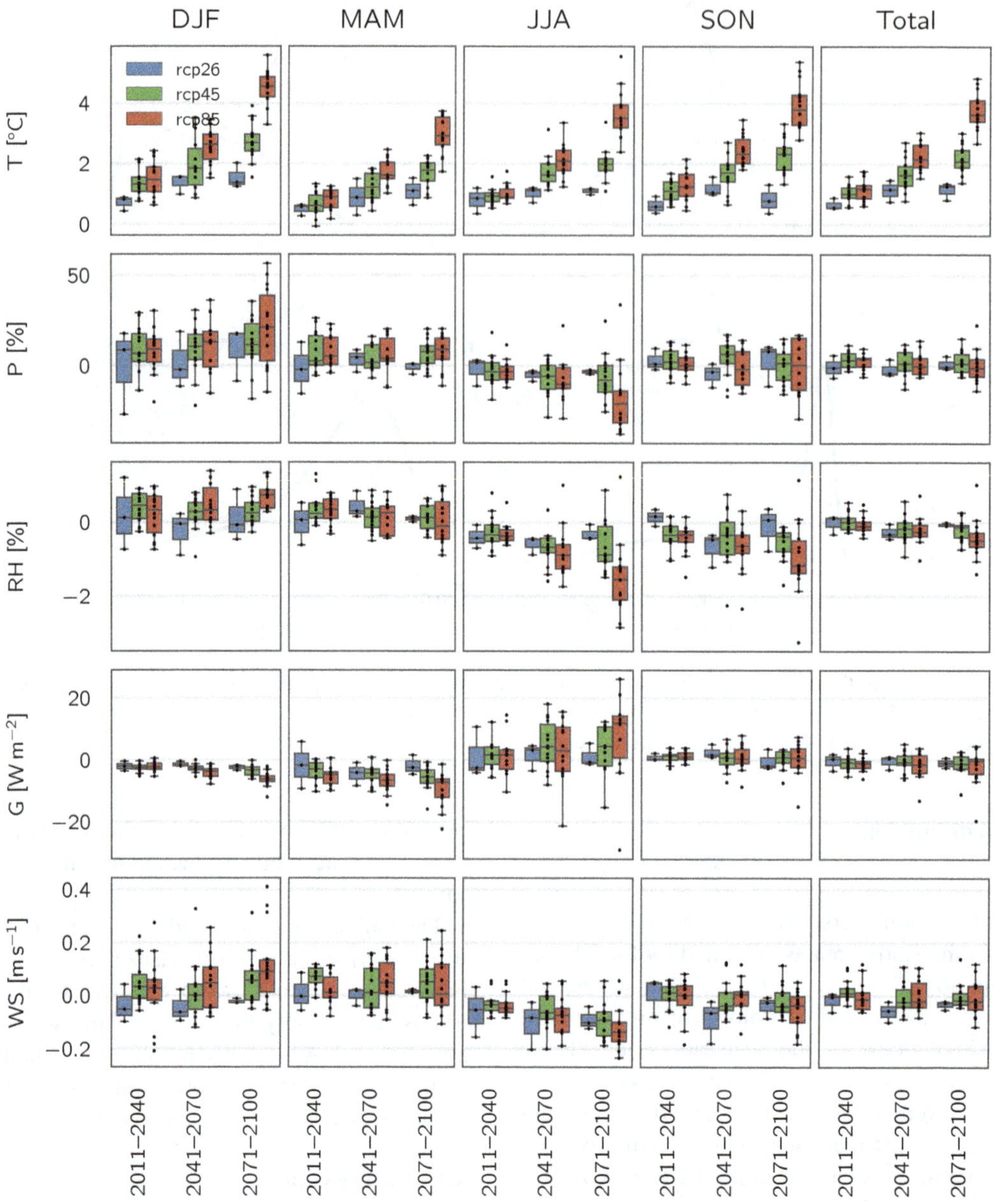

Figure 7. Projected seasonal changes for temperature (T), precipitation (P), relative humidity (RH), global radiation (G), and wind speed (WS) according to the selected EURO-CORDEX scenario simulations (spatial averages over the study area; changes calculated relative to 1971–2000). Small dots represent the individual GCM–RCM realizations.

up to 4.5 °C in the RCP8.5 scenario, while the smallest increases are projected for spring.

For precipitation, no clear general trend can be discerned with regard to annual sums. Both decreases (up to −14 %) and increases (up to 24 %) are projected by individual models, while the multi-model averages are close to zero for all three scenarios. However, with the exception of some outliers, there is a general consensus between the model towards a precipitation shift from summer towards winter. Projected multi-model average increases in winter precipitation are between 8 % (RCP2.6) and 21 % (RCP8.5), with individual models projecting increases of up to 57 %.

Changes in relative humidity largely follow the trend of precipitation changes, with decreases in summer (up to −1.5 %) and smaller increases in winter (up to 0.7 %). At least for the RCP8.5 scenario slight decreases in the annual mean are projected.

For global radiation, the disagreement between the individual models is considerable. The multi-model means indicate a decrease in overall global radiation for all RCPs, with

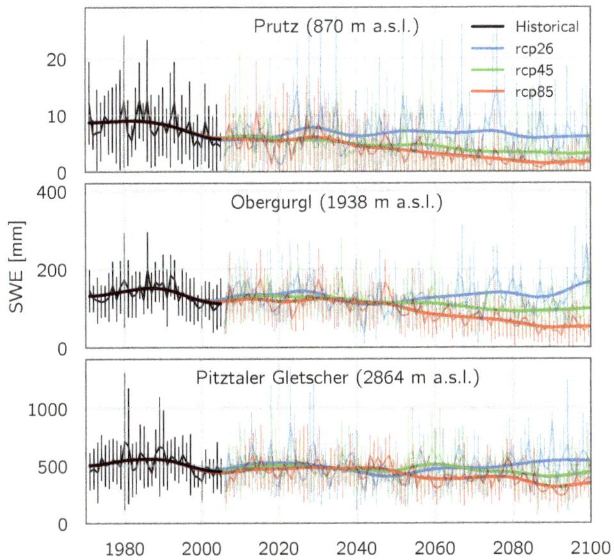

Figure 8. Evolution of mean annual SWE for the stations Prutz, Obergurgl, and Pitztaler Gletscher as simulated using the EURO-CORDEX scenarios for the RCPs2.6, 4.5, and 8.5. Thin lines and error bars indicate the multi-model mean and standard deviation, respectively, while the thick lines show the 5-year Gaussian low-pass filtered multi-model mean.

values between $-1.2\,\mathrm{W\,m^{-2}}$ for RCP4.5 and $-2.4\,\mathrm{W\,m^{-2}}$ for RCP2.6, however with individual model results ranging between -20 and $4\,\mathrm{W\,m^{-2}}$. The spread between the models is even larger when looking at seasonal changes. While all models agree on a decrease of global radiation in winter and (with one exception) spring, projections for summer are uncertain. Here, most models tend towards an increase in radiation for the RCP4.5 and RCP8.5 scenarios; however individual results range between -29 and $26\,\mathrm{W\,m^{-2}}$.

Finally, wind speed projections also show quite a large spread between individual models. Multi-model means indicate a very small overall decrease in wind speed, with seasonal increases of up to $0.12\,\mathrm{m\,s^{-1}}$ during winter and decreases of up to $-0.12\,\mathrm{m\,s^{-1}}$ during summer.

4.2 Changes in snow

Changes in the amount and duration of the simulated seasonal snow cover in the study area were analyzed both temporally and spatially. As a reference for computing the changes, we performed AMUNDSEN simulation runs using the historical periods of the EURO-CORDEX simulations for the period 1970–2005. Through combination with the scenario runs we hence obtained transient simulations for the period 1970–2100.

Figure 8 shows the evolution of the simulated mean annual snow water equivalent (SWE) for three meteorological stations covering an approx. 2000 m elevation range: Prutz (870 m a.s.l.), Obergurgl (1938 m a.s.l.), and Pitztaler

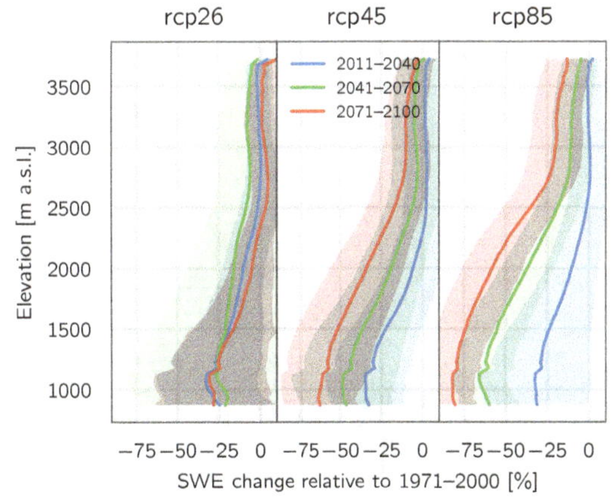

Figure 9. Elevation-dependent projected change in SWE (multi-model means; shadings indicate ±1 SD (one standard deviation)) relative to the period 1971–2000 for the RCPs 2.6, 4.5, and 8.5.

Gletscher (2864 m a.s.l.). At the highest-elevated station, Pitztaler Gletscher, the mean snow amounts remain relatively unchanged for all three RCPs until the middle of the century. For RCP2.6, snow amounts are then slowly increasing during the last third of the century, reaching an average level of 517 mm (multi-model mean), almost as high as the 530 mm in the historical period 1971–2000. For the other two scenarios comparatively small decreases in average SWE are simulated in this period, amounting to 427 mm (−18 %) for RCP4.5 and 357 mm (−31 %) for RCP8.5. For the station Obergurgl, similarly no strong changes in average snow amounts are simulated within the first half of the century, whereas afterwards the curves of the three scenarios begin to diverge more strongly than for Pitztaler Gletscher. For RCP2.6 again a general trend of increasing SWE after 2050 is projected, while snow amounts in the RCP4.5 scenario decrease to 94 mm (−31 %) in the period 2071–2100, and 59 mm (−57 %) for the RCP8.5 scenario. The largest relative changes are found for the lowest-elevated station, Prutz, where the curves for the three scenarios begin to divert by around 2020. RCP2.6 snow amounts then stay relatively constant during the remainder of the century at 6 mm (−31 %), while RCP4.5 and RCP8.5 snow amounts continue to decrease strongly, amounting to −63 and −80 %, respectively, at the end of the century.

Similar results of stronger declining snow amounts for lower elevations can be seen in Fig. 9, where the multi-model mean change in SWE relative to the baseline period 1971–2000 is plotted against elevation. For RCP2.6, with the exception of the very lowest-elevated parts of the study area, the strongest decreases in snow amounts generally are projected for the period 2041–2070, whereas the largest average snow amounts are found at the end of the century. Generally, however, the projected decreases are comparatively small, with maximum values of approx. 12 % for elevations above 2500 m a.s.l., up to 25 % for elevations be-

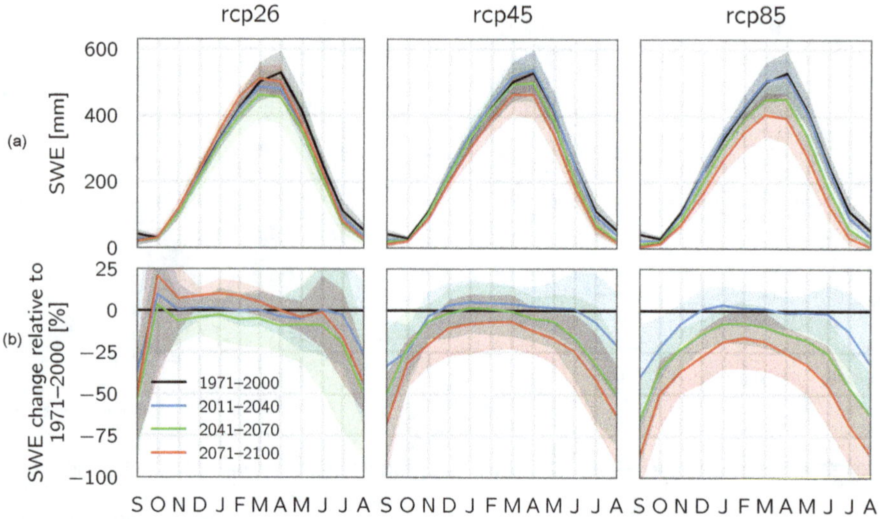

Figure 10. Mean monthly SWE (multi-model means; shadings indicate ± one standard deviation) over the entire study area (Fig. 1) for the three RCPs **(a)** and relative changes compared to 1971–2000 **(b)**.

tween 1500 and 2500 m a.s.l., and slightly stronger below. For both RCP4.5 and RCP8.5, average snow amounts stay virtually constant (even slightly increasing for RCP4.5) during the period 2011–2040 in elevations above 2500 m a.s.l., whereas for the lower-elevated parts, decreases during this period amount to up to 37 %. However, for both scenarios strong decreases are projected for the remainder of the century, with the largest changes in the lowest-elevated areas amounting to up to −64 and −83 % for RCP4.5 and RCP8.5, respectively.

The analysis of seasonal changes (Fig. 10) reveals that the strongest relative changes are projected for the summer months, while simulated changes for winter are comparatively small, amounting to maximum decreases of approx. 30 % for RCP8.5. For all three scenarios, a gradual shift in the timing of peak SWE amounts from April towards March is simulated.

These projected changes in snow coverage generally show similar patterns to studies for other Alpine sites (e.g., Beniston et al., 2003; Marty et al., 2017; Schmucki et al., 2015a; Steger et al., 2012), however with a tendency to lower relative decreases in average snow amounts. This is likely due to the comparatively strong increases in winter precipitation especially for the warmest (RCP8.5) scenario (Fig. 7). A comprehensive study recently done for Switzerland (Marty et al., 2017) for example projected more dramatic changes in snow amounts especially in high elevations (> 3000 m a.s.l.) until the end of the century for the A2 scenario (similar warming as in the RCP8.5 scenario), however projecting only very minor winter precipitation increases. Especially in high elevations however, increases in precipitation amounts are more likely to compensate for temperature increases, as temperatures in these elevations are usually still below melting conditions for most parts of the year.

4.3 Changes in glaciers

According to the simulation results, all glaciers in the study region will continue to retreat significantly throughout the 21st century. This is illustrated in Fig. 11, where the evolution of glacier volume and area (relative to the respective values of 2006) is plotted for all 31 model combinations as well as the respective multi-model means. In general, the evolution of glacier area and volume is very similar, with only a slightly stronger decline of glacier volume compared to the area. Looking at the multi-model mean values for the three emission scenarios, the largest changes in glacierization already occur within the first half of the century, where – relatively independent of the emission scenario – the glaciers lose between approx. 60 % (RCP2.6) and 65 % (RCP8.5) of their volume from the beginning of the century. After 2050, the slope of the curves decreases and differences between the three scenarios become more prominent. In the RCP2.6 scenario, glacier volume decreases to approx. 25 % until 2080, whereas the glaciers stabilize afterwards and only slightly further retreat until 2100. In the RCP8.5 scenario, glaciers decrease more rapidly than in the RCP4.5 case between 2050 and 2080, whereas afterwards the two curves again follow a similar course. For both scenarios, the glaciers have mostly vanished by the end of the century, with only small remains of approx. 10 % (RCP4.5) and 4 % (RCP8.5) still left in terms of their initial volume. Results for the evolution of glacier area are similar.

Looking beyond the multi-model mean, it becomes clear that there is a remarkable spread between the individual model runs, most strikingly for the RCP2.6 scenario, where the glacier volume decrease is between 58 and 97 % until 2100 for the three model runs. For the RCP4.5 and RCP8.5 scenarios the range between the individual model realiza-

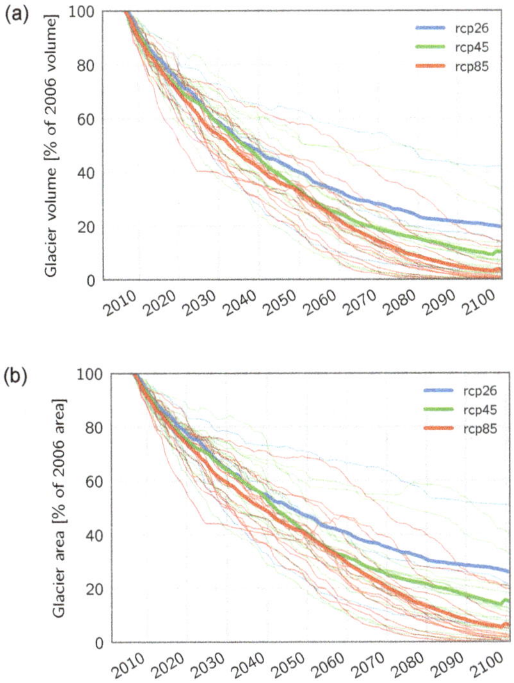

Figure 11. Simulated total glacier volume **(a)** and area (**b**, both shown as changes relative to the year 2006) for all glaciers in the study area and all 31 available climate scenarios. Thin lines indicate individual model results; thick lines show the multi-model mean. The small spikes in the multi-model means for RCP4.5 and RCP8.5 at the very end of the century are due to the fact that three models could only be run until 2098.

tions is also considerable, with individual realizations resulting in practically ice-free conditions as early as 2070, while in other realizations up to 33 % of glacier volume is preserved until 2100.

The comparatively small influence of the different emission scenarios on the other side is visualized in Fig. 12, where the ice thickness evolution of a single glacier (Hintereisferner) along its centerline is plotted in 10-year intervals for a single GCM–RCM combination (REMO2009 driven by MPI-ESM-LR) and the RCPs2.6, 4.5, and 8.5. Here again a strong loss of glacier volume can be seen in the first decades of the century. Regardless of the emission scenario, by 2040 the glacier retreats by approx. 1.4 km in length and loses approx. half of its volume (49–54 %). Only after 2060 does the RCP8.5 scenario generate a markedly stronger loss in ice volume compared to the other two scenarios, resulting in an almost complete disappearance of the glacier by 2100. The results for RCP2.6 and RCP4.5 largely follow the same evolution until 2090. Only in the last decade of the century does the RCP4.5 scenario produce an accelerated retreat of the glacier, whereas the RCP2.6 glacier surface stays approximately constant. However, also for these two less extreme scenarios, only 25 % (RCP2.6) and 13 % (RCP4.5) of the original ice volume is left by the end of the century.

Figure 13 shows the evolution of the absolute and relative glacierization as well as the absolute glacier volume and mean ice thickness for each catchment in the RCP4.5 scenario, illustrating that different catchments partly show different responses to the climatic forcing. For some of them (most notably Taschachbach) the average ice thickness stays approximately the same during the entire simulation period due to the similar rates of volume reduction and area reduction. For the Gepatschalm catchment, which includes the largest and thickest glacier in the study area (Gepatschferner), this is also the case for the first third of the century, whereas afterwards the rate of volume decline increases. In other catchments such as Vernagtbach and Gurgler Ache the reverse effect is observed during the last part of the century: glacier area retreats more quickly than the volume decreases (visible by the increase in mean ice thickness in this period).

The spatial evolution of glacier coverage within the study area is shown in Fig. 14, where the simulated multi-model mean glacier coverage is displayed in 10-year intervals for the three emission scenarios. The rate of glacier retreat is mainly dependent on elevation and glacier thickness. In all three scenarios, most of today's small glaciers have disappeared by 2050 according to the model. However, for the largest glaciers considerable differences resulting from the different emission scenarios appear. Gepatschferner, for example, loses 54 % of its 1997 area (17.1 km^2) by 2100 in the RCP2.6 scenario, while in the RCP4.5 and RCP8.5 scenarios 71 and 95 %, respectively, of the glacier area disappears.

4.4 Changes in hydrology

Figure 15 exemplarily shows the simulated seasonal runoff cycle for present-day conditions as well as the projected values for the future periods for four of the study catchments. While the reactions of the catchments are slightly different depending on their characteristics, the general pattern of change is the same for all catchments and emission scenarios: summer runoff strongly decreases with simultaneously increasing spring runoff, indicating a shift from glacial/glacio-nival to nivo-glacial runoff regimes. While in the RCP2.6 scenario the month of peak runoff remains unchanged, in the RCP4.5 and RCP8.5 scenarios the peak gradually shifts from July towards June for all catchments, with the exception of Vernagtbach in the RCP4.5 scenario. For the Rofenache, Gepatschalm, and Pitze catchments, this shift already occurs in the period 2041–2070 in both scenarios. Only for Vernagtbach, the most glacierized catchment, in the RCP8.5 scenario does the shift occur only towards the end of the century, as in the prior period the catchment is still glacierized to a considerable degree. Monthly peak runoff slightly increases for the Pitze catchment in the period 2011–2040; however total annual runoff volumes stay approximately constant due to lower August runoff volumes. In the other three catchments, both monthly peak runoff and total annual runoff volumes

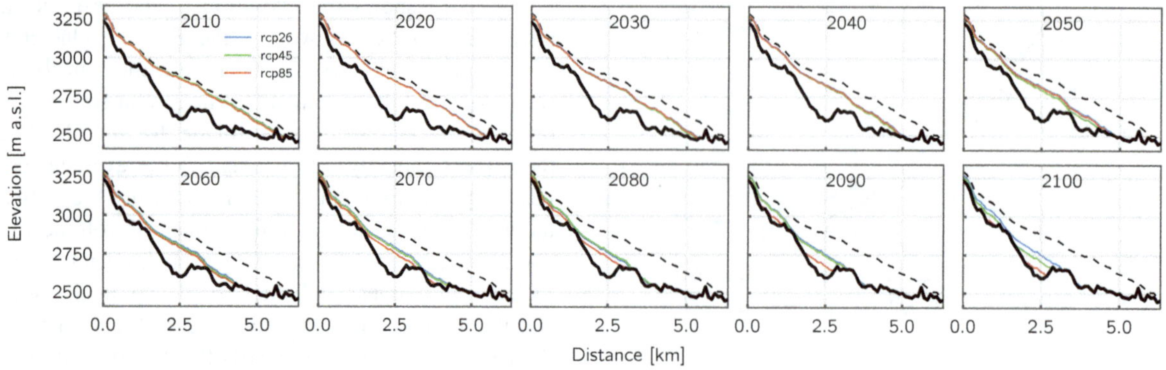

Figure 12. Bedrock topography (thick black lines) and simulated glacier thickness (thin colored lines) along the centerline of Hintereisferner as simulated for a single GCM–RCM combination (REMO2009 driven by MPI-ESM-LR) and the three RCPs, shown in 10-year intervals. Dashed black lines indicate the initial ice thickness as of the year 1997.

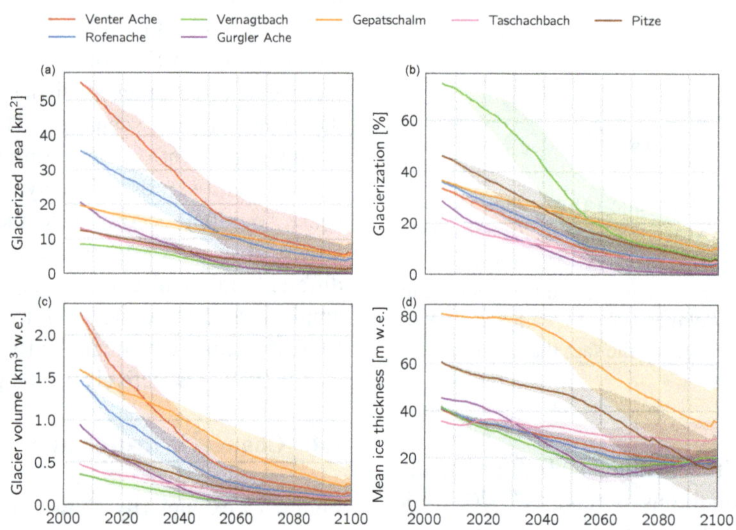

Figure 13. Evolution of the absolute **(a)** and relative **(b)** glacierized area for each catchment in the RCP4.5 scenario, as well as the absolute glacier volume **(c)** and the mean ice thickness **(d)**. Note that the latter is calculated only over the glacierized fraction of the catchment and not the entire catchment area (i.e., it is the result of dividing **c** by **a**). Lines and shadings represent multi-model means ±1 SD.

do not exceed the levels simulated for the historical period in all scenarios. With regard to ice runoff (dashed lines), Gepatschalm is the only catchment where ice runoff does not decrease monotonically over time, but rather increases in the period 2041–2070 compared to 2011–2040 before it strongly decreases towards the end of the century. This mid-century increase in glacier runoff is likely due to the average ice thickness for this catchment that is by far the largest and the slower decline in glacierization compared to the other catchments.

A more detailed analysis of the changes in monthly runoff is shown in Fig. 16, where the average monthly changes in runoff relative to the historical period 1998–2013 for the catchments in the study region throughout the 21st century are displayed. With respect to the different emission scenarios, again the differences between them are very small within the first considered period (2011–2040). In this period, also no clear trend of changes in the seasonality of the runoff regimes can be discerned, apart from slight increases during the winter months and slight decreases during summer. In the following period, 2041–2071, a clearer pattern begins to emerge. While May to June runoff stays approximately unchanged, runoff strongly decreases during August and September in all three scenarios, amounting to up to −55 %. From October to November, a transition period between decreasing and increasing runoff occurs, while during winter and spring increases of up to 50–125 % (depending on emission scenario) are projected (although still amounting to very small absolute values in comparison to summer as can be seen in Fig. 15). During the last part of the century (2071–2100), summer runoff continues to decrease, although not as strongly as between the first two periods. Winter runoff

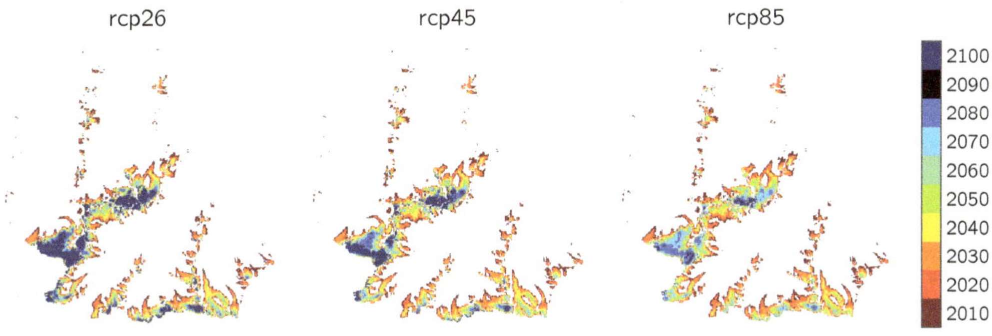

Figure 14. Spatially distributed simulated glacier coverage (multi-model mean) shown in 10-year intervals for the three RCPs.

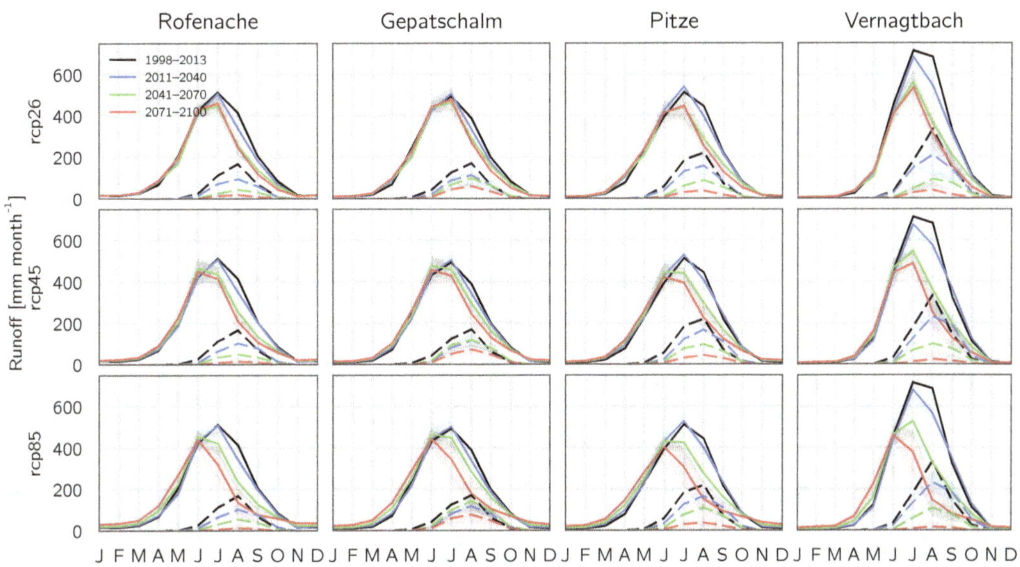

Figure 15. Average monthly runoff (multi-model mean ±1 SD indicated as shaded bands) as simulated for the early, middle, and late 21st century for four catchments and the three emission scenarios. Dashed lines indicate bare ice melt runoff.

also increases, most notably for the RCP8.5 scenario, where November to March runoff is at least approximately doubled for all catchments compared to the reference period. In this period, different catchment characteristics can also be more clearly distinguished. The largest changes are simulated for Vernagtbach, the smallest, highest-elevated, and (initially) most glacierized catchment. Here, increases between 43 and 225 % for November to May runoff are simulated for the RCP8.5 scenario, while the corresponding runoff decreases by 70–75 % during August and September due to the by then almost completely melted glacier (Vernagtferner). However, while in the figure the relative (positive) changes during winter appear similar or larger than the (negative) changes during summer for all catchments, it has to be emphasized that this still corresponds to a strong decrease in total annual runoff considering the respective absolute values (for example as seen in Fig. 15, the absolute JJA runoff volumes are approx. 2 orders of magnitude larger than the respective DJF values).

The simulated future streamflow composition of the components ice melt, snowmelt, and rainfall at the end of the scenario period (2071–2100) is shown in Fig. 17. While the ice melt fractions at the end of the century are negligible for all but the most glacierized catchments, snowmelt is still the major contributor to total runoff for all seasons. Rainfall runoff contributions are comparatively low in general, which is however partly due to the model structure – as rain falling on snow contributes to the liquid water storage of the snowpack up to a certain amount, these rainfall amounts are part of the snowmelt contribution in the runoff concentration scheme. Only rain falling on bare ground, ice, firn, or already saturated snow is part of the rainfall runoff contribution in this case.

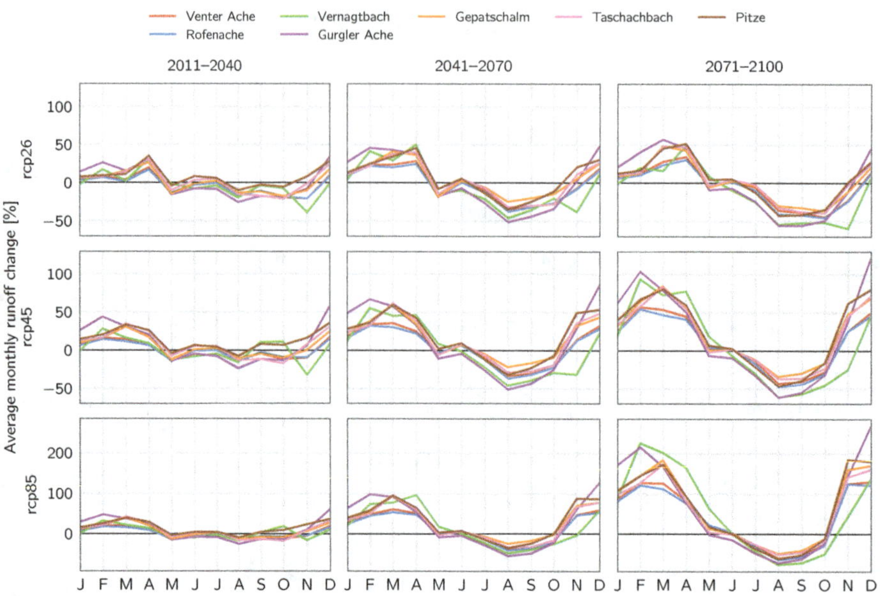

Figure 16. Average changes in monthly runoff relative to the historical period 1998–2013 over the course of the 21st century for the study catchments as simulated for the three emission scenarios (lines are multi-model means; note the differently scaled y-axes for RCP8.5).

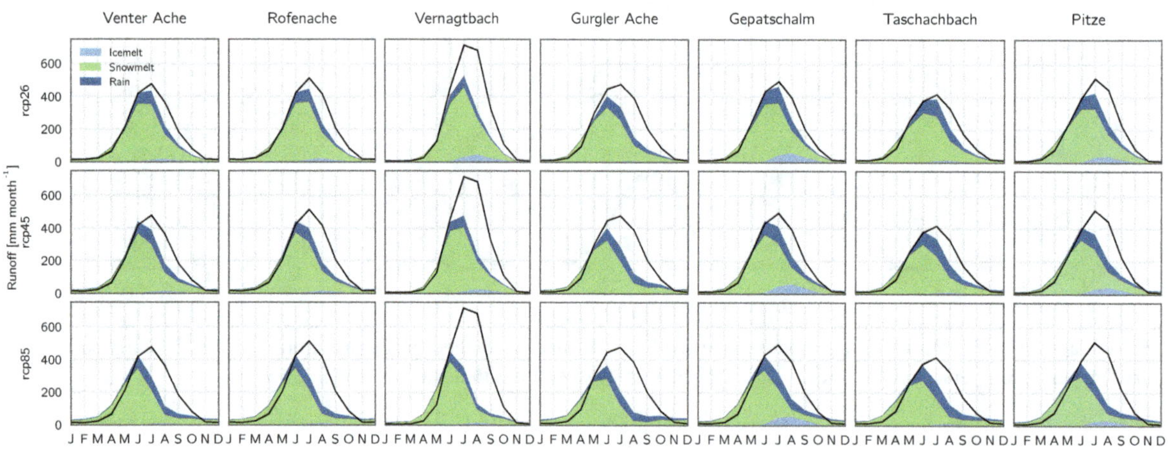

Figure 17. Average monthly runoff and composition (multi-model mean) for the period 2071–2100. The black lines show the average runoff in the historical period 1997–2013.

4.5 Uncertainty

Huss et al. (2014) assessed the influence of various model assumptions in projections of glacier evolution and runoff in high-mountain catchments. Their results indicate that major uncertainty sources are especially (i) the winter snow accumulation in terms of both volume and spatial distribution, (ii) the approach to account for glacier geometry changes, (iii) the initial glacier ice volume distribution, and (iv) the individual climate projections.

While the model has been shown to capture (i) the amount and distribution of winter snow accumulation well for the study area in a recent study (Hanzer et al., 2016), the model setup was slightly altered for the present study. The most substantial changes result from the requirement of a sufficiently long historical measurement period of the meteorological stations. As a consequence of this, only a subset of the stations utilized in the aforementioned study could be used, and from these stations only daily instead of hourly recordings were available, requiring an additional processing step in terms of temporal disaggregation. In addition, at several of these stations the daily and (sub-)hourly measurements are performed with different instruments, resulting in considerable differences in the recorded meteorological variables as illustrated in Fig. 18 for three stations. Especially the latter impacted model performance, resulting in

Table 3. NSE, BE, and PBIAS of observed vs. simulated runoff in the period 1998–2006 for the model runs using original hourly meteorological data and disaggregated daily data, respectively.

		Original			Disaggregated		
ID	Catchment	NSE	BE	PBIAS (%)	NSE	BE	PBIAS (%)
1	Venter Ache	0.92	0.60	8.71	0.79	−0.09	21.44
2	Rofenache	0.93	0.70	8.27	0.78	0.12	23.54
3	Vernagtbach	0.92	0.70	−1.12	0.88	0.57	13.19
4	Gurgler Ache	0.91	0.59	1.96	0.83	0.22	18.87
5	Gepatschalm	0.94	0.68	3.09	0.88	0.41	11.95
6	Taschachbach	0.92	0.54	0.01	0.84	0.08	19.19
7	Pitze	0.94	0.72	2.30	0.88	0.45	16.46

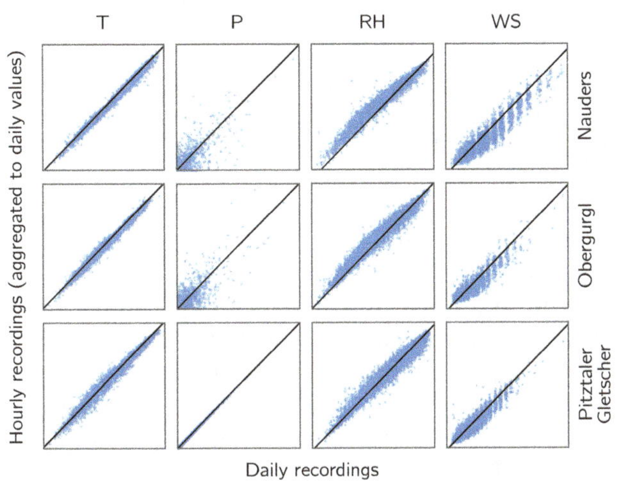

Figure 18. Scatter plots of daily vs. aggregated hourly recordings of air temperature (T), precipitation (P), relative humidity (RH), and wind speed (WS) for three stations in the study area.

a tendency to slightly overestimate winter snow accumulation and spring runoff volumes. Reducing the spatial and temporal model resolution from 50 m and 1 h to 100 m and 3 h on the other hand only resulted in minor impacts on model performance. Figure 19 shows the average observed monthly runoff for the seven study catchments and the period 1998–2006 as well as the corresponding simulation results for the 50 m/1 h run using original hourly meteorological recordings and the 100 m/3 h run using disaggregated daily data, highlighting the tendency to overestimate runoff particularly during spring for the latter run. Table 3 shows the corresponding skill scores derived from these two simulation runs: the Nash–Sutcliffe efficiency (NSE), the benchmark efficiency (BE) (Schaefli and Gupta, 2007), and the percent bias (PBIAS). However, as for the scenario simulations, mainly changes rather than absolute values are analyzed; these partial model biases likely do not affect the main conclusions of our study. Related to this issue is the question whether the implemented precipitation correction methods are still applicable in the future, especially concerning the empirically derived fixed snow correction factor of 15 % as a threshold-based adjustment. Utilizing the RCM-simulated snowfall fractions and precipitation fields instead of downscaling to point locations could reduce this uncertainty. While this was beyond the scope of this study, in the future promising advances in downscaling methods such as quasi-dynamical approaches (e.g., Gutmann et al., 2016) could help to bridge this gap between statistical and dynamical methods and allow more realistic small-scale precipitation fields to be derived.

Overestimations of winter snow accumulation and, subsequently, glacier mass balances might partly also have consequences for (ii) the approach to account for glacier geometry changes. The Δh parameterization as implemented is able to account only for conditions of glacier retreat, while in the case of positive mass balances no update of the glacier extents is performed. As Fig. 20 shows, 14–24 % of all simulated specific mass balances are positive in our simulations. However, previous studies have shown that the Δh parameterization performed significantly better than alternative approaches such as accumulation area ratio-based methods and similarly well as complex ice flow models (Huss et al., 2010). Hence, we believe that this approach is an adequate tradeoff for the application on a regional scale where the application of process-based ice flow models is not feasible.

We also assessed the influence of (iii) the initial glacier volume distribution by scaling the original initial ice thickness distribution by factors of 0.7 and 1.3, respectively (i.e., 30 % decrease/increase) and re-running the entire set of RCP4.5 simulations. However, as Fig. 21b–c exemplarily show for selected catchments, this does not significantly change the model results with regard to the evolution of glaciers and runoff volume. Similar results are also obtained for all other catchments.

With regard to (iv), many studies (e.g., Addor et al., 2014; Bosshard et al., 2013; Horton et al., 2006; Huss et al., 2014) have shown that the largest uncertainty in hydrological impact studies usually is due to the chosen climate models and

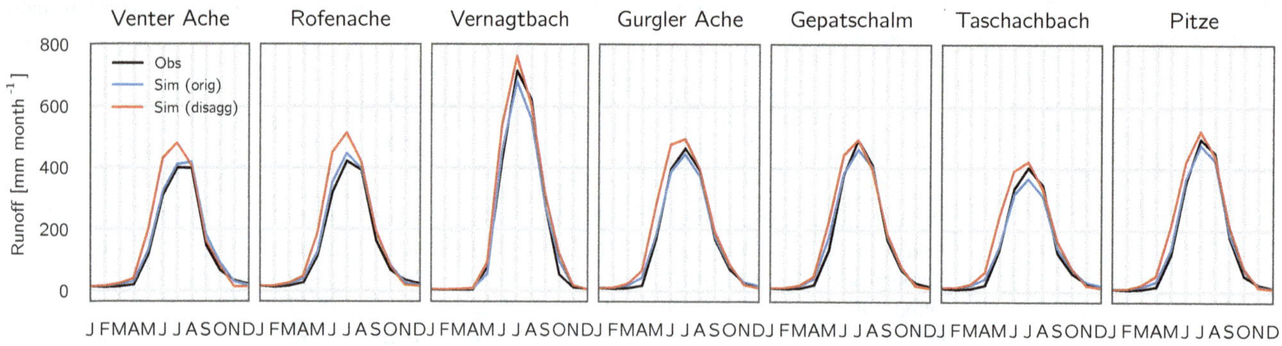

Figure 19. Seasonal cycle of observed (black lines) and simulated monthly runoff for the study catchments and the period 1998–2006. Blue lines correspond to the simulation results for the "reference run" (50 m spatial resolution, 1 h temporal resolution, original hourly meteorological data as forcing), while the red lines correspond to the simulation run using 100 m spatial and 3 h temporal resolution as well as disaggregated daily data as forcing.

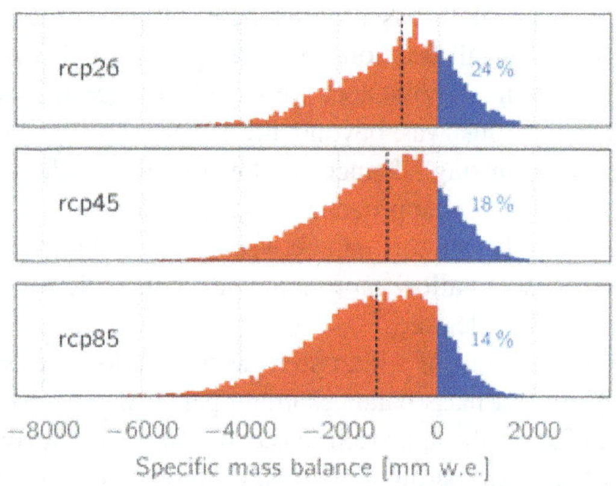

Figure 20. Histograms of the specific mass balance values as simulated for all (initially) 206 study area glaciers, all simulation years (2006–2100), and all GCM–RCM combinations. Dashed vertical lines refer to the median mass balance, and percentages indicate the total fractions of positive mass balances.

that the spread between individual climate model simulations is often larger than the spread between different emission scenarios within a single climate model. In the case of GCM–RCM chains, GCMs tend to have a larger impact on the hydrological model results than the RCMs. We addressed these uncertainties by utilizing the entire range of available EURO-CORDEX simulations, resulting in a total of 14 GCM–RCM chains with five different GCMs (each driving 2–4 RCMs) and six different RCMs for the RCP4.5 and RCP8.5 scenarios. The results for the RCP2.6 scenario have to be interpreted with care, as for this scenario only three GCM–RCM combinations (using two different GCMs) were available. Hence, especially when looking at multi-model averages, the RCP2.6 results are not directly comparable to the other two scenarios.

Besides the climate models themselves, the spatial and temporal downscaling approaches from the coarse daily-resolution grids to the 3-hourly resolution point-scale time series also add an uncertainty component to the results, as they are statistically derived from past conditions for which it is uncertain if they will persist in the future.

The influence of the climate models on the hydrological results is illustrated in Fig. 21a. Here, average monthly runoff for the period 2071–2100 as simulated by all 31 individual GCM–RCM realizations is shown exemplarily for the Pitze catchment. While the influence of the emission scenarios is visible to a part, there is significant overlap between the realizations of the different emission scenarios. For example, the realizations resulting in both the lowest and the highest August streamflow volumes are both driven by the RCP8.5 scenario. Similar results were already shown for the simulated glacier volume and area in Fig. 11.

5 Conclusions

In this study, we have forced a fully distributed physically based hydroclimatological model with the most state-of-the-art climate projections available. This is the most detailed study on cryospheric–hydrological climate change impacts in the Ötztal Alps to date and to our knowledge also in high-elevation glacierized catchments in Austria in general.

While some uncertainty in the results is due to the model configuration, the largest uncertainty can be traced back to the climate projections. This leads to a considerable range in the projected snow coverage, glacier extents, and hydrological regimes. Snow cover is projected to decrease by up to 80 % in elevations below 1500 m a.s.l., while only comparatively moderate decreases (up to 25 %) are found for high-elevated areas (> 2500 m a.s.l.) due to strongly increasing winter precipitation which partly compensates for the increased warming. Glaciers will continue to recede strongly throughout the century. Until 2050, glacier volume will decline by approx. 60–65 % largely independent of the emission scenario, whereas by the end of the century 80–96 %

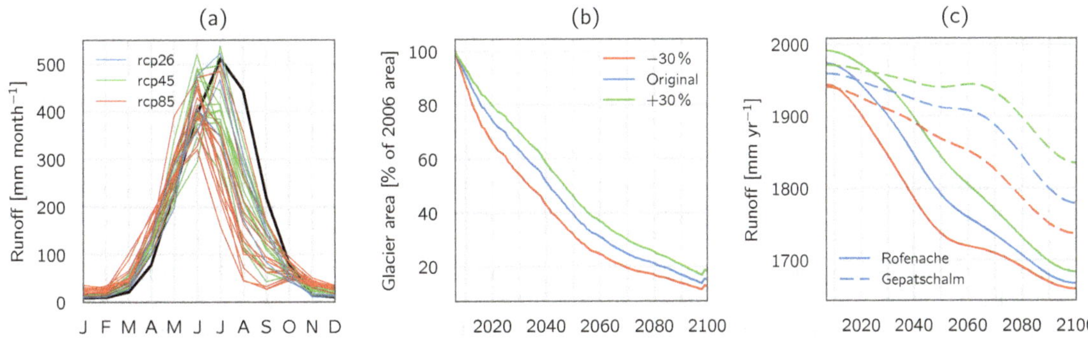

Figure 21. **(a)** Average monthly runoff (period 2071–2100) for the Pitze catchment as simulated for all 31 individual GCM–RCM realizations, **(b)** evolution of the total glacierized area for the entire Ötztal Alps study site as simulated by altering the initial ice thickness distribution by ±30 % (multi-model mean values for the RCP4.5 scenario), and **(c)** corresponding runoff evolution (10-year Gaussian low-pass filtered) for the Rofenache and Gepatschalm catchments.

of the original ice volume will be lost. Consequently, glacier runoff will diminish proportionally and summer runoff will strongly decrease in all investigated catchments by up to 55 %, resulting in a shift of the annual runoff peak from July towards June. Winter runoff volumes on the other hand will increase, however to absolute values that are still low. While the total annual runoff volumes stay approximately constant during the early 21st century compared to present-day levels, they gradually decrease throughout the rest of the century. Only for some catchments and scenarios do runoff volumes slightly exceed present-day levels, indicating that the peak water period of maximum runoff is currently under way or has already passed in this region.

Data availability. Most of the meteorological and hydrological recordings used are not publicly available. Exceptions are the meteorological data for station Vent (https://doi.pangaea.de/10.1594/PANGAEA.806582, https://doi.pangaea.de/10.1594/PANGAEA.876595) and the meteorological and runoff data for Vernagtbach (https://doi.pangaea.de/10.1594/PANGAEA.775113, https://doi.pangaea.de/10.1594/PANGAEA.829530). EURO-CORDEX scenario data can be acquired from the respective data servers (see http://www.euro-cordex.net). Glacier outlines for the Ötztal Alps are accessible at https://doi.pangaea.de/10.1594/PANGAEA.844988; the derived glacier thicknesses are available from the authors upon request. The digital elevation model is available at https://www.data.gv.at/katalog/dataset/land-tirol_tirolgelnde.

Competing interests. The authors declare that they have no conflict of interest.

Acknowledgements. This work was carried out within the framework of the projects "W01 MUSICALS II – Multiscale Snow/Ice Melt Discharge Simulation into Alpine Reservoirs", carried out in the research programme of alpS – Centre for Climate Change Adaptation in Innsbruck, and "HydroGeM3", financed by the Austrian Academy of Sciences. The computational results presented have been achieved in part using the Vienna Scientific Cluster (VSC). The authors want to thank the COMET research programme of the Austrian Research Promotion Agency (FFG), the TIWAG – Tiroler Wasserkraft AG, and the Austrian Academy of Sciences. Meteorological and hydrological data were provided by the Zentralanstalt für Meteorologie und Geodynamik (ZAMG), the Hydrographic Service of Tyrol, the TIWAG – Tiroler Wasserkraft AG, the Commission for Glaciology of the Bavarian Academy of Sciences and Humanities, and the Autonomous Province of Bozen/Bolzano. We would also like to thank Matthias Huttenlau for managing the MUSICALS II project and for many helpful discussions and Johannes Schöber for valuable comments during the preparation of the manuscript. Finally, we wish to thank two anonymous reviewers for their helpful comments that helped to improve the manuscript.

Edited by: Luis Samaniego

References

Abermann, J., Lambrecht, A., Fischer, A., and Kuhn, M.: Quantifying changes and trends in glacier area and volume in the Austrian Ötztal Alps (1969–1997–2006), The Cryosphere, 3, 205–215, https://doi.org/10.5194/tc-3-205-2009, 2009.

Addor, N., Rössler, O., Köplin, N., Huss, M., Weingartner, R., and Seibert, J.: Robust changes and sources of uncertainty in the projected hydrological regimes of Swiss catchments, Water Resour. Res., 50, 7541–7562, 2014.

Allen, R. G., Pereira, L. S., Raes, D., and Smith, M.: Crop evapotranspiration: guidelines for computing crop water requirements, FAO Irrigation and Drainage Paper 56, Food and Agriculture Organization, Rome, 1998.

Anderson, E. A.: A point energy and mass balance model of a snow cover, Tech. Rep. NWS 19, NOAA, Silver Spring, MD, 1976.

Barnett, T. P., Adam, J. C., and Lettenmaier, D. P.: Potential impacts of a warming climate on water availability in snow-dominated regions, Nature, 438, 303–309, 2005.

Beniston, M.: Climatic Change in Mountain Regions: A Review of Possible Impacts, Climatic Change, 59, 5–31, 2003.

Beniston, M., Keller, F., Koffi, B., and Goyette, S.: Estimates of snow accumulation and volume in the Swiss Alps under changing climatic conditions, Theor. Appl. Climatol., 76, 125–140, 2003.

Bliss, A., Hock, R., and Radić, V.: Global response of glacier runoff to twenty-first century climate change, J. Geophys. Res.-Ea. Surf., 119, 717–730, 2014.

Bosshard, T., Carambia, M., Goergen, K., Kotlarski, S., Krahe, P., Zappa, M., and Schär, C.: Quantifying uncertainty sources in an ensemble of hydrological climate-impact projections, Water Resour. Res., 49, 1523–1536, 2013.

Cannon, A. J., Sobie, S. R., Murdock, T. Q., Cannon, A. J., Sobie, S. R., and Murdock, T. Q.: Bias Correction of GCM Precipitation by Quantile Mapping: How Well Do Methods Preserve Changes in Quantiles and Extremes?, J. Climate, 28, 6938–6959, https://doi.org/10.1175/JCLI-D-14-00754.1, 2015.

Collins, D. N.: Climatic warming, glacier recession and runoff from Alpine basins after the Little Ice Age maximum, Ann. Glaciol., 48, 119–124, 2008.

Corripio, J.: Modelling the energy balance of high altitude glacierised basins in the Central Andes, PhD thesis, University of Edinburgh, Edinburgh, 2002.

Déqué, M.: Frequency of precipitation and temperature extremes over France in an anthropogenic scenario: Model results and statistical correction according to observed values, Global Planet. Change, 57, 16–26, 2007.

Farinotti, D., Usselmann, S., Huss, M., Bauder, A., and Funk, M.: Runoff evolution in the Swiss Alps: projections for selected high-alpine catchments based on ENSEMBLES scenarios, Hydrol. Process., 26, 1909–1924, 2011.

Fatichi, S., Rimkus, S., Burlando, P., Bordoy, R., and Molnar, P.: High-resolution distributed analysis of climate and anthropogenic changes on the hydrology of an Alpine catchment, J. Hydrol., 525, 362–382, 2015.

Finger, D., Heinrich, G., Gobiet, A., and Bauder, A.: Projections of future water resources and their uncertainty in a glacierized catchment in the Swiss Alps and the subsequent effects on hydropower production during the 21st century, Water Resour. Res., 48, W02521, https://doi.org/10.1029/2011WR010733, 2012.

Förster, K., Hanzer, F., Winter, B., Marke, T., and Strasser, U.: An open-source MEteoroLOgical observation time series DISaggregation Tool (MELODIST v0.1.1), Geosci. Model Dev., 9, 2315–2333, https://doi.org/10.5194/gmd-9-2315-2016, 2016.

Frei, P., Kotlarski, S., Liniger, M. A., and Schär, C.: Future snowfall in the Alps: projections based on the EURO-CORDEX regional climate models, The Cryosphere, 12, 1–24, https://doi.org/10.5194/tc-12-1-2018, 2018.

Gobiet, A., Kotlarski, S., Beniston, M., and Heinrich, G.: 21st century climate change in the European Alps – A review, Sci. Total Environ., 493, 1138–1151, 2014.

Gudmundsson, L., Bremnes, J. B., Haugen, J. E., and Engen-Skaugen, T.: Technical Note: Downscaling RCM precipitation to the station scale using statistical transformations – a comparison of methods, Hydrol. Earth Syst. Sci., 16, 3383–3390, https://doi.org/10.5194/hess-16-3383-2012, 2012.

Gutmann, E., Barstad, I., Clark, M., Arnold, J., and Rasmussen, R.: The Intermediate Complexity Atmospheric Research Model (ICAR), J. Hydrometeorol., 17, 957–973, https://doi.org/10.1175/JHM-D-15-0155.1, 2016.

Hanzer, F., Marke, T., and Strasser, U.: Distributed, explicit modeling of technical snow production for a ski area in the Schladming region (Austrian Alps), Cold Regions Science and Technology, 108, 113–124, 2014.

Hanzer, F., Helfricht, K., Marke, T., and Strasser, U.: Multilevel spatiotemporal validation of snow/ice mass balance and runoff modeling in glacierized catchments, The Cryosphere, 10, 1859–1881, https://doi.org/10.5194/tc-10-1859-2016, 2016.

Hanzer, F., Förster, K., Winter, B., and Marke, T.: melodist: MELODIST 0.1.2, https://doi.org/10.5281/zenodo.1123277, 2017.

Helfricht, K., Kuhn, M., Keuschnig, M., and Heilig, A.: Lidar snow cover studies on glaciers in the Ötztal Alps (Austria): comparison with snow depths calculated from GPR measurements, The Cryosphere, 8, 41–57, https://doi.org/10.5194/tc-8-41-2014, 2014.

Hofer, M., Nemec, J., Cullen, N. J., and Weber, M.: Evaluating predictor strategies for regression-based downscaling with a focus on glacierized mountain environments, J. Appl. Meteorol. Clim., 56, 1707–1729, https://doi.org/10.1175/JAMC-D-16-0215.1, 2017.

Horton, P., Schaefli, B., Mezghani, A., Hingray, B., and Musy, A.: Assessment of climate-change impacts on alpine discharge regimes with climate model uncertainty, Hydrol. Process., 20, 2091–2109, 2006.

Huss, M. and Farinotti, D.: Distributed ice thickness and volume of all glaciers around the globe, J. Geophys. Res.-Ea. Surf., 117, F04010, https://doi.org/10.1029/2012JF002523, 2012.

Huss, M. and Hock, R.: A new model for global glacier change and sea-level rise, Front. Earth Sci., 3, 54, https://doi.org/10.3389/feart.2015.00054, 2015.

Huss, M., Farinotti, D., Bauder, A., and Funk, M.: Modelling runoff from highly glacierized alpine drainage basins in a changing climate, Hydrol. Process., 22, 3888–3902, 2008.

Huss, M., Funk, M., and Ohmura, A.: Strong Alpine glacier melt in the 1940s due to enhanced solar radiation, Geophys. Res. Lett., 36, L23501, https://doi.org/10.1029/2009GL040789, 2009.

Huss, M., Jouvet, G., Farinotti, D., and Bauder, A.: Future high-mountain hydrology: a new parameterization of glacier retreat, Hydrol. Earth Syst. Sci., 14, 815–829, https://doi.org/10.5194/hess-14-815-2010, 2010.

Huss, M., Zemp, M., Joerg, P. C., and Salzmann, N.: High uncertainty in 21st century runoff projections from glacierized basins, J. Hydrol., 510, 35–48, 2014.

Jacob, D., Petersen, J., Eggert, B., Alias, A., Christensen, O. B., Bouwer, L. M., Braun, A., Colette, A., Déqué, M., Georgievski, G., Georgopoulou, E., Gobiet, A., Menut, L., Nikulin, G., Haensler, A., Hempelmann, N., Jones, C., Keuler, K., Kovats, S., Kröner, N., Kotlarski, S., Kriegsmann, A., Martin, E., van Meijgaard, E., Moseley, C., Pfeifer, S., Preuschmann, S., Radermacher, C., Radtke, K., Rechid, D., Rounsevell, M., Samuelsson, P., Somot, S., Soussana, J.-F., Teichmann, C., Valentini, R., Vautard, R., Weber, B., and Yiou, P.: EURO-CORDEX: new high-resolution climate change projections for European impact research, Reg. Environ. Change, 14, 563–578, 2013.

Jansson, P., Hock, R., and Schneider, T.: The concept of glacier storage: a review, J. Hydrol., 282, 116–129, 2003.

Jordan, R.: A one-dimensional temperature model for a snow cover: Technical documentation for SNTHERM 89, Tech. rep., Hanover, NH, 1991.

Jouvet, G., Huss, M., Funk, M., and Blatter, H.: Modelling the retreat of Grosser Aletschgletscher, Switzerland, in a changing climate, J. Glaciol., 57, 1033–1045, 2011.

Klemeš, V.: The modelling of mountain hydrology: the ultimate challenge, in: Hydrology of Mountainous Areas, IAHS Publication 190, edited by: Molnar, L., IAHS Press, Wallingford, UK, 29–43, 1990.

Kobierska, F., Jonas, T., Zappa, M., Bavay, M., Magnusson, J., and Bernasconi, S. M.: Future runoff from a partly glacierized watershed in Central Switzerland: A two-model approach, Adv. Water Resour., 55, 204–214, 2013.

Kuhn, M. and Batlogg, N.: Glacier runoff in Alpine headwaters in a changing climate, in: Hydrology, Water Resources and Ecology in Headwaters, IAHS Publ. no. 248, Proceedings of the HeadWater'98 conference, April 1998, Meran/Merano, Italy, 79–88, 1998.

Kuhn, M. and Olefs, M.: Auswirkung von Klimaänderungen auf das Abflussverhalten von vergletscherten Einzugsgebieten im Hinblick auf Speicherkraftwerke, in: StartClim2007E, Institute of Meteorology and Geophysics, University of Innsbruck, Vienna, Austria, 2007.

Kuhn, M., Lambrecht, A., and Abermann, J.: The Austrian glacier inventory GI 2, 1998, in: ArcGIS (shapefile) format, PANGAEA, 2015.

Marke, T., Strasser, U., Hanzer, F., Stötter, J., Wilcke, R. A. I., and Gobiet, A.: Scenarios of Future Snow Conditions in Styria (Austrian Alps), J. Hydrometeorol., 16, 261–277, 2015.

Marty, C., Schlögl, S., Bavay, M., and Lehning, M.: How much can we save? Impact of different emission scenarios on future snow cover in the Alps, The Cryosphere, 11, 517–529, https://doi.org/10.5194/tc-11-517-2017, 2017.

Marzeion, B., Jarosch, A. H., and Hofer, M.: Past and future sea-level change from the surface mass balance of glaciers, The Cryosphere, 6, 1295–1322, https://doi.org/10.5194/tc-6-1295-2012, 2012.

Marzeion, B., Jarosch, A. H., and Gregory, J. M.: Feedbacks and mechanisms affecting the global sensitivity of glaciers to climate change, The Cryosphere, 8, 59–71, https://doi.org/10.5194/tc-8-59-2014, 2014.

Maurer, E. P. and Pierce, D. W.: Bias correction can modify climate model simulated precipitation changes without adverse effect on the ensemble mean, Hydrol. Earth Syst. Sci., 18, 915–925, https://doi.org/10.5194/hess-18-915-2014, 2014.

Milano, M., Reynard, E., Bosshard, N., and Weingartner, R.: Simulating future trends in hydrological regimes in Western Switzerland, J. Hydrol., 4, 748–761, 2015.

Ohmura, A., Bauder, A., Müller, H., and Kappenberger, G.: Long-term change of mass balance and the role of radiation, Ann. Glaciol., 46, 367–374, 2007.

Pellicciotti, F., Brock, B., Strasser, U., Burlando, P., Funk, M., and Corripio, J.: An enhanced temperature-index glacier melt model including the shortwave radiation balance: Development and testing for Haut Glacier d'Arolla, Switzerland, Journal of Glaciology, 51, 573–587, 2005.

Pomeroy, J. W., Gray, D. M., Brown, T., Hedstrom, N. R., Quinton, W. L., Granger, R. J., and Carey, S. K.: The cold regions hydrological model: a platform for basing process representation and model structure on physical evidence, Hydrol. Process., 21, 2650–2667, 2007.

Prasch, M., Marke, T., Strasser, U., and Mauser, W.: Large scale integrated hydrological modelling of the impact of climate change on the water balance with DANUBIA, Adv. Sci. Res., 7, 61–70, https://doi.org/10.5194/asr-7-61-2011, 2011.

Radić, V., Bliss, A., Beedlow, A. C., Hock, R., Miles, E., and Cogley, J. G.: Regional and global projections of twenty-first century glacier mass changes in response to climate scenarios from global climate models, Clim. Dynam., 42, 37–58, 2013.

Ragettli, S., Pellicciotti, F., Bordoy, R., and Immerzeel, W. W.: Sources of uncertainty in modeling the glaciohydrological response of a Karakoram watershed to climate change, Water Resour. Res., 49, 6048–6066, 2013.

Schaefli, B.: Projecting hydropower production under future climates: a guide for decision-makers and modelers to interpret and design climate change impact assessments, Wiley Interdisciplin. Rev.: Water, 2, 271–289, 2015.

Schaefli, B. and Gupta, H. V.: Do Nash values have value?, Hydrol. Process., 21, 2075–2080, 2007.

Schmucki, E., Marty, C., Fierz, C., and Lehning, M.: Simulations of 21st century snow response to climate change in Switzerland from a set of RCMs, Int. J. Climatol., 35, 3262–3273, 2015a.

Schmucki, E., Marty, C., Fierz, C., Weingartner, R., and Lehning, M.: Impact of climate change in Switzerland on socioeconomic snow indices, Theor. Appl. Climatol., 127, 875–889, 2015b.

Seiser, B., Helfricht, K., Huss, M., Fischer, A., Veulliet, E., Schönlaub, H., Kuhn, M., and Strasser, U.: MUSICALS – (MUltiscale Snow/ICemelt Discharge Simulation into ALpine ReservoirS): Validation of a method to estimate ice volume and ice thickness distribution in the Ötztal Alps, Tyrol, Austria, Geophys. Res. Abstr., Vienna, Austria, 2012.

Smiatek, G., Kunstmann, H., and Senatore, A.: EURO-CORDEX regional climate model analysis for the Greater Alpine Region: Performance and expected future change, J. Geophys. Res.-Atmos., 121, 7710–7728, 2016.

Sonntag, D.: Important new values of the physical constants of 1986, vapour pressure formulations based on the ITS-90, and psychrometer formulae, Z. Meteorol., 70, 340–344, 1990.

Steger, C., Kotlarski, S., Jonas, T., and Schär, C.: Alpine snow cover in a changing climate: a regional climate model perspective, Clim. Dynam., 41, 735–754, 2012.

Stewart, I. T.: Changes in snowpack and snowmelt runoff for key mountain regions, Hydrol. Process., 23, 78–94, 2009.

Strasser, U.: Spatial and temporal variability of meteorological variables at Haut Glacier d'Arolla (Switzerland) during the ablation season 2001: Measurements and simulations, J. Geophys. Res., 109, D03103, https://doi.org/10.1029/2003JD003973, 2004.

Strasser, U.: Modelling of the mountain snow cover in the Berchtesgaden National Park, Tech. Rep. 55, Berchtesgaden National Park, Berchtesgaden, 2008.

Strasser, U., Bernhardt, M., Weber, M., Liston, G. E., and Mauser, W.: Is snow sublimation important in the alpine water balance?, The Cryosphere, 2, 53–66, https://doi.org/10.5194/tc-2-53-2008, 2008.

Strasser, U., Warscher, M., and Liston, G. E.: Modeling Snow-Canopy Processes on an Idealized Mountain, J. Hydrometeorol., 12, 663–677, 2011.

Tecklenburg, C., Francke, T., Kormann, C., and Bronstert, A.: Modeling of water balance response to an extreme future scenario in the Ötztal catchment, Austria, Adv. Geosci., 32, 63–68, 2012.

Teutschbein, C. and Seibert, J.: Bias correction of regional climate model simulations for hydrological climate-change impact studies: Review and evaluation of different methods, J. Hydrol., 456-457, 12–29, 2012.

Teutschbein, C. and Seibert, J.: Is bias correction of regional climate model (RCM) simulations possible for non-stationary conditions?, Hydrol. Earth Syst. Sci., 17, 5061–5077, https://doi.org/10.5194/hess-17-5061-2013, 2013.

Themeßl, M., Gobiet, A., and Leuprecht, A.: Empirical-statistical downscaling and error correction of daily precipitation from regional climate models, Int. J. Climatol., 31, 1530–1544, 2011a.

Themeßl, M. J., Gobiet, A., and Heinrich, G.: Empirical-statistical downscaling and error correction of regional climate models and its impact on the climate change signal, Climatic Change, 112, 449–468, 2011b.

Thrasher, B., Maurer, E. P., McKellar, C., and Duffy, P. B.: Technical Note: Bias correcting climate model simulated daily temperature extremes with quantile mapping, Hydrol. Earth Syst. Sci., 16, 3309–3314, https://doi.org/10.5194/hess-16-3309-2012, 2012.

Viviroli, D., Archer, D. R., Buytaert, W., Fowler, H. J., Greenwood, G. B., Hamlet, A. F., Huang, Y., Koboltschnig, G., Litaor, M. I., López-Moreno, J. I., Lorentz, S., Schädler, B., Schreier, H., Schwaiger, K., Vuille, M., and Woods, R.: Climate change and mountain water resources: overview and recommendations for research, management and policy, Hydrol. Earth Syst. Sci., 15, 471–504, https://doi.org/10.5194/hess-15-471-2011, 2011.

Waichler, S. R. and Wigmosta, M. S.: Development of Hourly Meteorological Values From Daily Data and Significance to Hydrological Modeling at H. J. Andrews Experimental Forest, J. Hydrometeorol., 4, 251–263, 2003.

Walter, M. T., Brooks, E. S., McCool, D. K., King, L. G., Molnau, M., and Boll, J.: Process-based snowmelt modeling: does it require more input data than temperature-index modeling?, J. Hydrol., 300, 65–75, 2005.

Weber, M., Braun, L., Mauser, W., and Prasch, M.: Contribution of rain, snow- and icemelt in the Upper Danube discharge today and in the future, Geogr. Fis. Dinam. Quat., 33, 221–230, 2010.

Wijngaard, R. R., Helfricht, K., Schneeberger, K., Huttenlau, M., Schneider, K., and Bierkens, M. F. P.: Hydrological response of the Ötztal glacierized catchments to climate change, Nord. Hydrol., 47, 979–995, 2016.

Wilcke, R. A. I., Mendlik, T., and Gobiet, A.: Multi-variable error correction of regional climate models, Climatic Change, 120, 871–887, 2013.

Yokoyama, R., Shirasawa, M., and Pike, R. J.: Visualizing topography by openness: a new application of image processing to digital elevation models, Photogram. Eng. Remote Sens., 68, 257–266, 2002.

Zekollari, H., Fürst, J. J., and Huybrechts, P.: Modelling the evolution of Vadret da Morteratsch, Switzerland, since the Little Ice Age and into the future, J. Glaciol., 60, 1155–1168, 2014.

7

The CAMELS-CL dataset: catchment attributes and meteorology for large sample studies – Chile dataset

Camila Alvarez-Garreton[1,2], Pablo A. Mendoza[3,4], Juan Pablo Boisier[1,5], Nans Addor[6], Mauricio Galleguillos[1,7], Mauricio Zambrano-Bigiarini[1,8], Antonio Lara[1,2], Cristóbal Puelma[1,7], Gonzalo Cortes[9], Rene Garreaud[1,5], James McPhee[3,4], and Alvaro Ayala[10,11]

[1]Center for Climate and Resilience Research (CR2), Santiago, Chile
[2]Instituto de Conservación, Biodiversidad y Territorio, Universidad Austral de Chile, Valdivia, Chile
[3]Department of Civil Engineering, Universidad de Chile, Santiago, Chile
[4]Advanced Mining Technology Center, Universidad de Chile, Santiago, Chile
[5]Department of Geophysics, Universidad de Chile, Santiago, Chile
[6]Climatic Research Unit, School of Environmental Sciences, University of East Anglia, Norwich, UK
[7]Faculty of Agronomic Sciences, Universidad de Chile, Santiago, Chile
[8]Department of Civil Engineering, Faculty of Engineering and Sciences, Universidad de La Frontera, Temuco, Chile
[9]Department of Civil and Environmental Engineering, University of California, Los Angeles, California, USA
[10]Laboratory of Hydraulics, Hydrology and Glaciology (VAW), ETH Zurich, Zurich, Switzerland
[11]Swiss Federal Institute for Forest, Snow and Landscape Research (WSL), Birmensdorf, Switzerland

Correspondence: Camila Alvarez-Garreton (camila.alvarez@uach.cl)

Abstract. We introduce the first catchment dataset for large sample studies in Chile. This dataset includes 516 catchments; it covers particularly wide latitude (17.8 to 55.0° S) and elevation (0 to 6993 m a.s.l.) ranges, and it relies on multiple data sources (including ground data, remote-sensed products and reanalyses) to characterise the hydroclimatic conditions and landscape of a region where in situ measurements are scarce. For each catchment, the dataset provides boundaries, daily streamflow records and basin-averaged daily time series of precipitation (from one national and three global datasets), maximum, minimum and mean temperatures, potential evapotranspiration (PET; from two datasets), and snow water equivalent. We calculated hydroclimatological indices using these time series, and leveraged diverse data sources to extract topographic, geological and land cover features. Relying on publicly available reservoirs and water rights data for the country, we estimated the degree of anthropic intervention within the catchments. To facilitate the use of this dataset and promote common standards in large sample studies, we computed most catchment attributes introduced by Addor et al. (2017) in their Catchment Attributes and MEteorology for Large-sample Studies (CAMELS) dataset, and added several others.

We used the dataset presented here (named CAMELS-CL) to characterise regional variations in hydroclimatic conditions over Chile and to explore how basin behaviour is influenced by catchment attributes and water extractions. Further, CAMELS-CL enabled us to analyse biases and uncertainties in basin-wide precipitation and PET. The characterisation of catchment water balances revealed large discrepancies between precipitation products in arid regions and a systematic precipitation underestimation in headwater mountain catchments (high elevations and steep slopes) over humid regions. We evaluated PET products based on ground data and found a fairly good performance of both products in humid regions ($r > 0.91$) and lower correlation ($r < 0.76$) in hyper-arid regions. Further, the satellite-based PET showed a consistent overestimation of observation-based PET. Finally, we explored local anomalies in catchment response by analysing the relationship between hydrological signatures and an attribute characterising the level of anthropic interventions. We showed that larger anthropic interventions are correlated with lower than normal annual flows, runoff ratios, elasticity of runoff with respect to precipitation, and flashiness of runoff, especially in arid catchments.

CAMELS-CL provides unprecedented information on catchments in a region largely underrepresented in large sample studies. This effort is part of an international initiative to create multi-national large sample datasets freely available for the community.

---

1 Introduction

Large sample hydrology has been recognised as a fundamental framework to advance hydrological science (e.g. Andréassian et al., 2006; Ehret et al., 2014). The insights provided by studying a large set of catchments complement the findings from intensive place-based studies, where more detailed analyses are conducted over a small number of catchments. A common approach in large sample studies is to explore interrelationships between catchment attributes describing landscape, climate and hydrologic behaviour, typically obtained from topography, soil types, geology, land cover and hydrometeorological datasets (e.g. Oudin et al., 2008; Sawicz et al., 2011; Gupta et al., 2014; Newman et al., 2015; Addor et al., 2017). Accounting for catchment attributes in a comprehensive dataset serves various purposes. For example, comparative hydrology and catchment classification studies use these attributes to explore catchment (dis)similarities (e.g. McDonnell and Woods, 2004; Wagener et al., 2007; Sawicz et al., 2011; Berghuijs et al., 2014). Likewise, regionalisation studies incorporate catchment attributes to identify (hydro-climatically and physically) similar catchments that can be used to transfer model information from gauged to ungauged locations (Blöschl et al., 2013; Sawicz et al., 2011) – a fundamental motivation of the Predictions in Ungauged Basins (PUB) initiative (Sivapalan et al., 2003). In summary, the main goal of large sample applications is to learn from diversity in order to define generalizable rules that can help to improve the predictability of the water cycle. This is addressed by disentangling the interplay between landscape, climate and hydrologic behaviour, which provides insights into hydrological systems and into suitable model structures to represent them.

As highlighted by Gupta et al. (2014), a key challenge in large sample hydrology is data accessibility, which is particularly critical in data-scarce regions such as South America (see Fig. 2 in Gupta et al., 2014). Although there is a tendency for large sample datasets to be shared worldwide (see examples in Gupta et al., 2014), available hydrometeorological records from different countries typically use different formats and come from different providers. Moreover, they are rarely spatially aggregated to the catchment scale, which makes it difficult for researchers and practitioners to use them for basin-oriented applications.

In this paper, we introduce a unique dataset that includes 516 catchments in Chile, and show how this dataset serves to improve our understanding of hydrological systems and their predictability through the assessment of (1) the uncertainties in two key meteorological variables (precipitation and PET) and (2) the impacts of anthropic intervention on catchment response.

The dataset built here consists of catchment boundaries in shapefile format, hydro-meteorological time series, and a suite of catchment attributes based on climate, hydrology, topography, geology, land cover, and water use. To facilitate and encourage the use of this dataset, and to promote common standards and formats in large sample studies, we compute five (out of six) classes of catchment attributes (location and topography, geology, land cover characteristics, climatic indices and hydrological signatures) used in Addor et al. (2017, referred to as A17 hereafter). A17 introduced the Catchment Attributes and MEteorology for Large-sample Studies dataset (CAMELS dataset), which encompasses meteorological and streamflow datasets collated by Newman et al. (2015) and provides quantitative estimates of a wide range of attributes for 671 catchments in the contiguous United States (CONUS). The CAMELS dataset has already been used in a myriad of applications, including assessment of streamflow skill elasticity to initial conditions and climate prediction (Wood et al., 2016), snow data assimilation for seasonal streamflow prediction (Huang et al., 2017), continental-scale hydrologic parameter estimation (Mizukami et al., 2017), and climate change impacts on the hydrology of the CONUS (Melsen et al., 2018), among others. Following this nomenclature, we name our dataset CAMELS-CL, which stands for CAMELS dataset in Chile. We add an attribute class not covered by A17: the degree of human intervention in each catchment. This novel information is valuable since anthropogenic activities may have major impacts on catchment behaviour, but human influence is often difficult to quantify, especially for hundreds of catchments.

We characterise hydrological systems in Chile by analysing the spatial distribution of catchment attributes provided in CAMELS-CL. Subsequently, we apply CAMELS-CL to assess uncertainties in precipitation and potential evapotranspiration estimates, and to quantify anthropic impacts on catchment response. To this end, we compared the different precipitation products and evaluated them based on the observed water balance. This analysis includes one national dataset (CR2MET) and three widely used global datasets (CHIRPS, MSWEP and TMPA); thus, the results may have implications beyond the domain covered by CAMELS-CL. Then, we assess PET products based on an independent set of PET point values calculated from meteorological records. Finally, we analyse human influence on catchment behaviour by relating hydrological signatures to a human intervention attribute calculated from water extraction information.

The paper is structured as follows: Sect. 2 describes the study area; Sect. 3 describes the collected datasets (Sect. 3.1) and provides a description of the derived catchment attributes with a discussion of their spatial distribution (Sect. 3.2); Sect. 4 presents the precipitation (Sect. 4.1) and potential evapotranspiration (Sect. 4.2) uncertainty analyses; Sect. 5 presents the analysis of human influence on catchment behaviour; and Sect. 6 summarises the main conclusions of the paper.

2 Study area

The area covered by CAMELS-CL corresponds to continental Chile, a territory with a distinct geographical configuration that spans 4300 km along a north–south axis over the west of South America (17.8° S to 55.0° S). The country lies on the Nazca and Antarctic tectonic plates. The tectonic activity in the Quaternary (early Pleistocene) led to the formation of the three main physiographic characteristics of the territory (from west to east): the coastal range, the intermediate depression, and the Andes Cordillera. Featuring altitudes well above 3000 m a.s.l., with summits up to 7000 m a.s.l. (e.g. Aconcagua mountain or Ojos del Salado volcano), the Andes act as an effective barrier for atmospheric flows, leading to particularly large precipitation amounts at high elevations (Garreaud, 2009) and to a noticeable contrast between the rainfall regimes of southern Chile (wet) and Argentinean Patagonia (dry).

Chile has 16 administrative regions (Fig. 1) split into four macro-zones defined by the Chilean Water Directorate (DGA), based on hydrological, climatic and topographic features (DGA, 2016a): North (from Arica and Parinacota to Coquimbo regions); Central (from Valparaiso to Maule regions); South (from Bio-Bio to Los Lagos regions); and Austral (from Aysén to Magallanes regions). To provide a more detailed description, we divided the North macro-zone into Far North (from Arica and Parinacota to Antofagasta regions) and Near North (from Atacama to Coquimbo regions), and the Austral macro-zone into the Austral Zone (Aysen region) and Southern Patagonia (Magallanes region). The resulting six macro-zones are presented in Fig. 1.

The country includes five primary climatic regimes according to the Köppen–Geiger climate classification (Kottek et al., 2006; Sarricolea et al., 2017). The Far North is dominated by a cold desert climate (BW_k) and tundra (ET) along the Andes range. The Near North is characterised by a cold desert climate in the Atacama region and a cold semi-arid climate (BS_k) in the Coquimbo region. The Central Zone is dominated by a sub-humid Mediterranean climate (C_{sb}). The Southern Zone includes a humid Mediterranean climate in the Bio-Bio and Araucanía regions, and a temperate rain–oceanic climate (C_{fb}) in the Los Rios and Los Lagos regions.

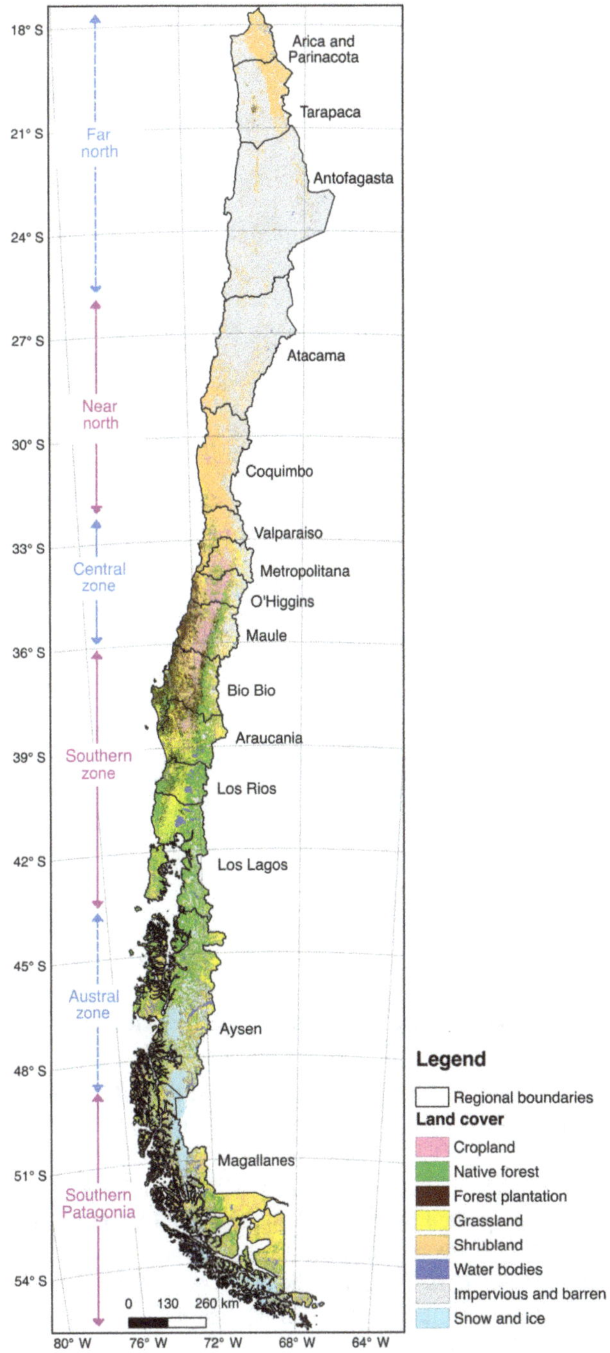

Figure 1. Chilean regional boundaries and names, and the six defined macro-zones (blue and magenta arrows). The background colour corresponds to the main land cover classes, obtained from Zhao et al. (2016).

The Austral and Southern Patagonia zones are dominated by rain–cool oceanic (C_{fc}) and cold steppe (BS_k) climates.

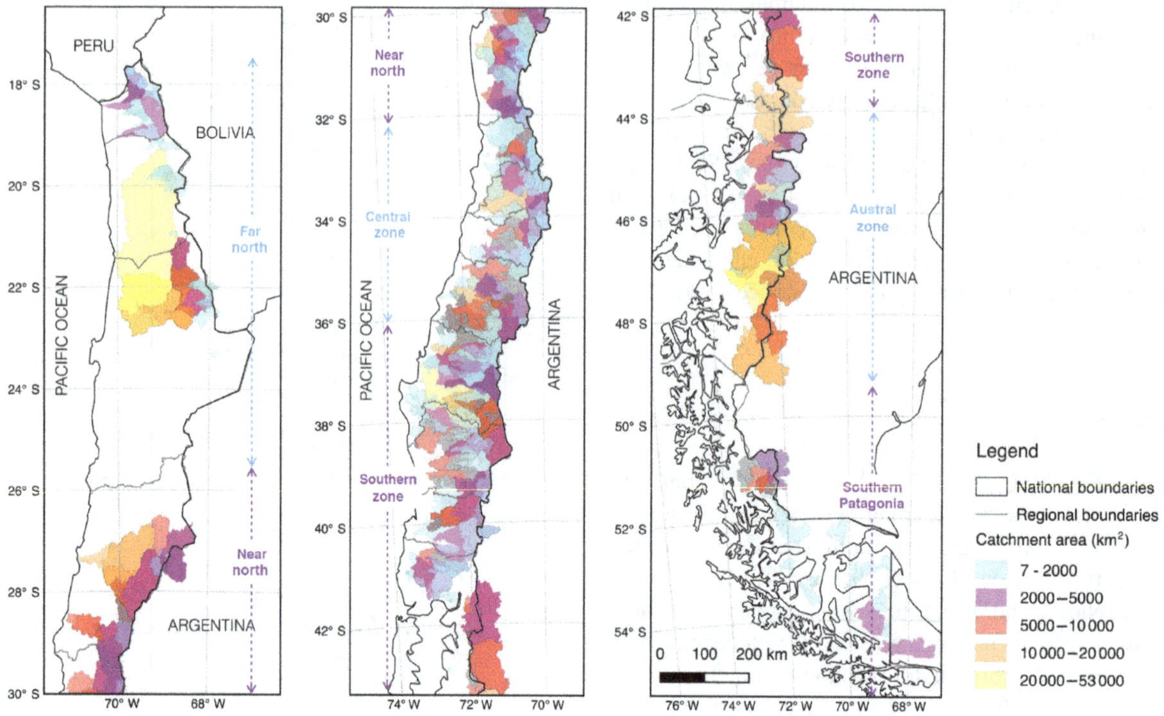

Figure 2. Catchment boundaries and contributing areas (km^2) of the 516 watersheds included in this study. The six defined macro-zones are indicated in blue and purple arrows.

3 CAMELS-CL dataset

3.1 Input data

3.1.1 Topography and catchment boundaries

The first step in the development of CAMELS-CL was the delimitation of catchment boundaries (Fig. 2). An official database for the Chilean hydrographic network was developed by the Instituto Geográfico Militar in 1984 (IGM, 1984) and updated by the DGA in 2014 (DGA and CIREN, 2014). This network delineation followed the Strahler hierarchy (Strahler, 1957), using the 30 m ASTER GDEM elevation data (Tachikawa et al., 2011). The DGA network includes 102 catchments, 491 sub-catchments and 1481 sub-sub-catchments, and has been largely used by government agencies, the private sector and the general public. However, a key limitation of this hydrographic network is that – given the methodology used for its implementation – the existing streamflow gauges do not correspond to catchment outlets. Furthermore, DGA catchment boundaries are truncated at the administrative national border, even though some catchments contribute runoff from the Bolivian and Argentinian territories. Since any hydrologic application within a controlled basin requires the total contributing area associated with streamflow measurements, and there is no official catchment boundary database, different studies have applied their own basin delineations, making it difficult to compare results.

To overcome this limitation, we created our own catchment boundary database for CAMELS-CL, defining the basin outlets at the location of 516 selected streamflow gauges (Sect. 3.1.5) and following only topographic-driven limits (not the administrative national border). A key challenge for this task is the mismatch between some station geographic coordinates reported by the DGA and the actual river network location – detected through the inspection of Google Earth imagery (Google, 2016). For some of those cases, expert advice was obtained from DGA technicians regarding gauge locations, while, for others, ancillary information (e.g. gauge name, road maps, Google Earth imagery) was used to determine the most likely location.

Basin delineation was performed in Quantum GIS (QGIS Development Team, 2015) by using watershed delineation packages from the Geographic Resources Analysis Support System (GRASS) (Neteler et al., 2012) and 30 m ASTER GDEM as input elevation data. Given the location of streamflow gauges, several catchments collected in this dataset are nested. We report this by using a logical hierarchy matrix indicating which basins are contained within another catchment of the dataset. The hierarchy matrix can be used to filter independent catchments, which is required for some applications such as hydrological modelling of large basins, catchment classification and parameter regionalisation.

The main topographic properties, including area, median, mean, minimum and maximum elevation, and mean slope, were computed for each catchment from ASTER GDEM 30 m raster data, clipped by the catchment boundary polygons and processed with the R raster package (Hijmans, 2016). An important limitation of this dataset is that its spatial resolution is relatively coarse, which can lead to errors when delineating catchments over very flat regions (such as the Far North; see Sect. 3.2.1).

3.1.2 Geology

Catchment-scale geological characteristics were retrieved from the Global Lithological Map database (GLiM) produced by Hartmann and Moosdorf (2012). GLiM is a compilation of national datasets into a unified global map. In the case of Chile, GLiM uses the map produced by the Servicio Nacional de Geología y Minería (Sernageomin, 2004), which has a resolution of 1 : 1 000 000 and is the most complete and commonly used map for the country. For each catchment, we reported the most and second-most frequent geological classes, as well as the fraction of the catchment they cover. We also extracted the fraction of the catchment described as "carbonate sedimentary rocks", as it is a useful indicator of the presence of karstic systems.

3.1.3 Land cover

We used the 30 m resolution land cover map provided by Zhao et al. (2016), which integrates multi-seasonal Landsat 8 imagery acquired during 2013 and 2014, complemented with Moderate Resolution Imaging Spectroradiometer (MODIS) Enhanced Vegetation Index data, and high-resolution imagery on Google Earth. The classification scheme adopted by Zhao et al. (2016) was designed with Chilean geographers and biodiversity researchers, based on the FROM-GLC project (Gong et al., 2013), which is similar to the Land Cover Classification System (Di Gregorio and Jansen, 2005). This classification scheme is compatible with other land cover classification systems such as FAO or IGBP, with minor ancillary data. It consists of 10 main (level 1) classes (Fig. 1): croplands; forests, grasslands; shrublands; wetlands; water bodies; impervious surfaces; barren lands; and snow and ice. Some classes were refined in level 2 (e.g. forests were separated into native forest and exotic forest plantation) and level 3 subclasses (a total of 30 and 35 subclasses, respectively). For CAMELS-CL, we used the R raster package (Hijmans, 2016) to clip the land cover map within each catchment boundary polygon, and compute the fractional area associated with each class or subclass (as described in Table 3).

3.1.4 Glaciers

Glaciers in Chile can be found at several locations, varying from small ice bodies at high-elevation sites in the Atacama region to alpine glaciers in the Central Zone and the large Patagonian ice fields in the Austral and Southern Patagonia regions (Pellicciotti et al., 2014). Even though the land cover map from Zhao et al. (2016) identifies areas of snow and ice, we included a global glacier inventory for calculating the degree of glacierisation of the selected catchments. Glacier inventories have the advantages of using geomorphologic glacier-delineation techniques and the recognition of debris-covered areas that cannot be identified by land cover classification schemes. In this study, we used the latest version of the Randolph Glacier Inventory (RGI 6.0; RGI Consortium, 2017). RGI 6.0 is a globally complete inventory of glacier outlines, and it is widely used in regional and global studies on land surface fluxes, climatology and meteorology (e.g. Huss and Hock, 2015; Marzeion et al., 2012; Mernild et al., 2017). We preferred to use RGI 6.0 rather than the Chilean glacier inventory from the DGA (DGA, 2014) because there are portions of some catchments lying in Argentinean territory (Fig. 2). RGI 6.0 was clipped within each catchment and two attributes were computed: the total glacierised area (km^2) and the percentage of glacierised area in the catchment (%).

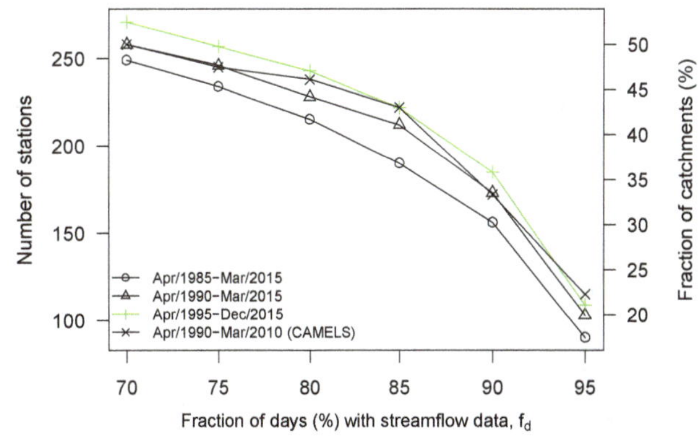

Figure 3. Number of stations (left y axis) with at least f_d % of days with daily streamflow records, for different periods. The right y axis shows the percentage of catchments (out of 516) that meets the criterion. The period used in the CAMELS dataset (Addor et al., 2017) was included as a reference.

3.1.5 Streamflow

We compiled daily streamflow records for gauges maintained by the DGA, available from the CR2 Climate Explorer (http://explorador.cr2.cl/, last access: March 2018). From the 809 gauges included there, we selected those currently operational (independently of their data period) or suspended after 31 December 1980 with a record period longer than 10 years. We also discarded gauges located in artificial channels, ending up with 516 selected gauges. The record lengths from the selected gauges range from 192 to 366 667 days, with a mean (median) of 10 979 (9909) days. Figure 3 illustrates the availability of daily streamflow records for different time pe-

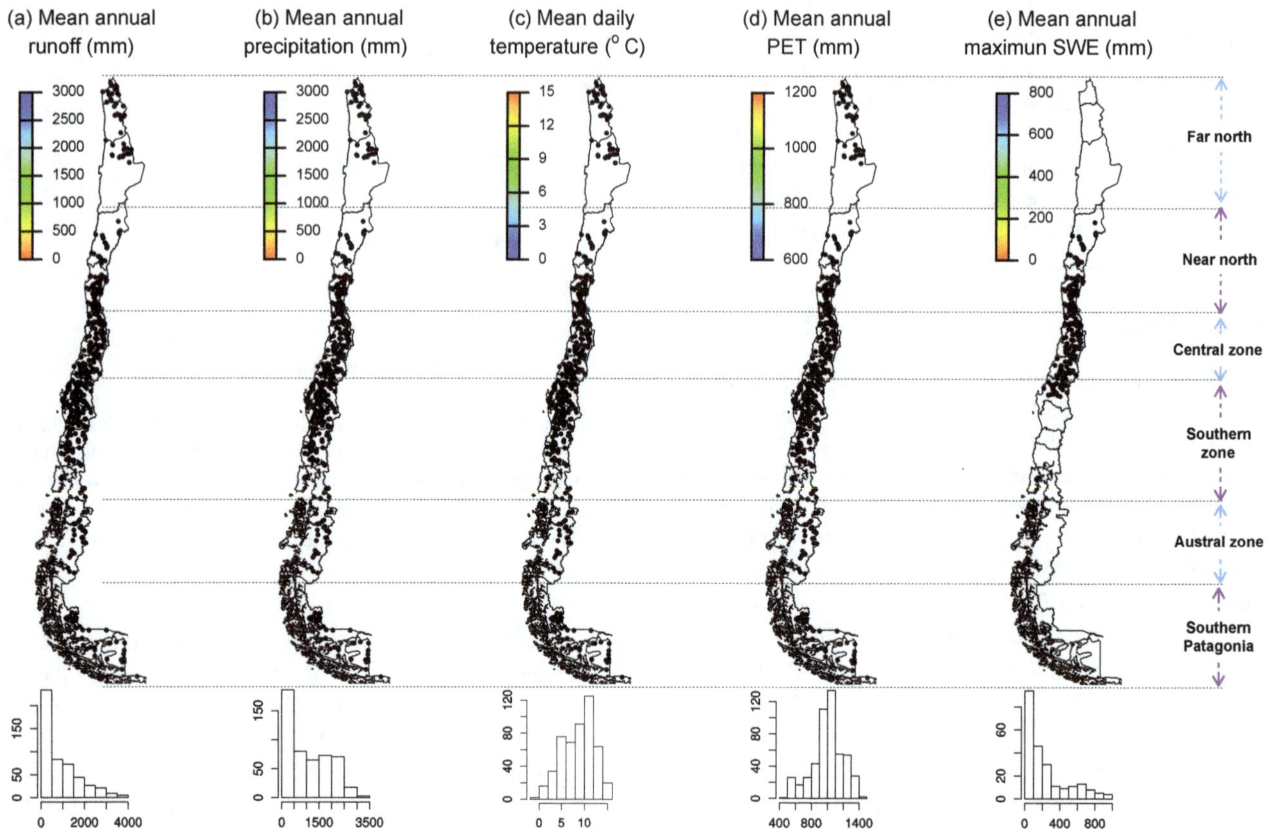

Figure 4. Mean annual hydro-meteorological variables, calculated for the complete recording period of each variable. **(b)** and **(d)** were generated with $\text{precip}_{\text{cr2met}}$ and pet_{har} products, respectively. The SWE product in **(e)** covers only the Near North and Central zones. The histograms indicate the number of catchments (out of 516) in each bin. The points represent the locations of catchment outlets.

riods (represented by different colours), which is mainly due to the different record periods of the selected gauges. Further, sparse missing daily records may be found in streamflow time series due to specific extreme events, where the station might have not worked properly or might have been broken. Note that hydrological year is considered from 1 April to 31 March. As expected, the number of stations decreases with longer data availability. For example, if only stations with at most 5 % of missing data were selected, this would lead to a subset of 90 to 115 stations (which corresponds to 18 % and 22 % of the total number of catchments within the database, respectively), depending on the time period selected. When considering all stations with at most 30 % of missing data, then 249 to 258 stations (48 % and 50 % of catchments, respectively) would meet this criterion, depending on the period (Fig. 3). Figure 4a presents the mean annual discharge for each station (computed for the entire record period).

3.1.6 Precipitation

In most cases, precipitation is the main driver of hydrological systems. However, the geographical distribution of this variable is highly uncertain, even in densely monitored regions (Tian and Peters-Lidard, 2010; Woldemeskel et al., 2013). This limitation is aggravated in regions with difficult accessibility, where only a sparse network of meteorological stations is available. In order to account for robust precipitation estimates and to characterise the uncertainty of this variable, we processed catchment-scale precipitation from four different products, whose main characteristics are summarised in Table 1. Daily precipitation fields provided by each product were clipped and averaged within the catchment boundaries, resulting in four daily time series for each catchment, named $\text{precip}_{\text{cr2met}}$, $\text{precip}_{\text{chirps}}$, $\text{precip}_{\text{mswep}}$ and $\text{precip}_{\text{tmpa}}$.

The $\text{precip}_{\text{cr2met}}$ time series was derived from CR2MET, a spatially distributed daily precipitation product developed for Chile, which is currently used by DGA to update the national water balance (DGA, 2017). The CR2MET product is partly based on a statistical downscaling of ERA-Interim reanalysis data (Balsamo et al., 2015). The method builds on multiple linear regression models used to transfer precipitation, moisture fluxes and other variables from ERA-Interim onto 0.05° resolution precipitation estimates. The statistical models, which also consider a number of topographic parameters,

Table 1. Precipitation products.

Name	Description	Spatial resolution	Temporal resolution	Period of record
$precip_{cr2met}$	Obtained from the CR2METv1.3 dataset (DGA, 2017)	0.05° lat–lon	daily	1979–2016
$precip_{tmpa}$	Obtained from the TMPA 3B42v7 dataset (Huffman et al., 2007, 2010)	0.25° lat–lon	daily	1998–2016
$precip_{chirps}$	Obtained from the Climate Hazards Group InfraRed Precipitation with Station data (CHIRPS) version 2 dataset (Funk et al., 2015)	0.05° lat–lon	daily	1981–2016
$precip_{mswep}$	Obtained from the Multi-Source Weighted-Ensemble Precipitation (MSWEP) v1.1 dataset (Beck et al., 2017)	0.25° lat–lon	daily	1979–2016

were calibrated using a large network of quality-controlled rain-gauge records. Depending on the distance of a given grid cell to neighbouring stations, the final product was obtained from merging downscaled precipitation and spatially interpolated in situ observations. Further information about formulation, quality control and product assessments can be found in DGA (2017).

The three satellite-based precipitation products used in CAMELS-CL were selected following the inter-comparison reported by Zambrano-Bigiarini et al. (2017) for the entire Chilean territory. The $precip_{chirps}$ time series was computed from the Climate Hazards Group InfraRed Precipitation with Station data version 2 (CHIRPS, Funk et al., 2015), a long-term (1981 to near present), quasi global (50° N to 50° S) daily dataset available at a spatial resolution of 0.05°, designed to monitor agricultural drought and global environmental changes over land. CHIRPS uses the Tropical Rainfall Measuring Mission Multi-Satellite Precipitation Analysis version 7 (TRMM 3B42v7) in order to calibrate global cold cloud duration rainfall estimates (Funk et al., 2015). CHIRPS also incorporates surface rain-gauge data in order to reduce estimation biases, based on public and private monthly data. Originally, this dataset spanned from 50° N to 50° S, but since November 2012 data have not been produced south of 46° S. More information can be found in Funk et al. (2015).

Variable $precip_{mswep}$ was computed from Multi-Source Weighted-Ensemble Precipitation (MSWEP, Beck et al., 2017) data version 1.1, a fully global precipitation dataset released in June 2016, with 3-hourly temporal and 0.25° spatial resolutions, specifically produced for hydrological modelling applications. MSWEP was designed to improve the performance of satellite products in representing precipitation over mountainous, tropical, and snowmelt-driven regions. The algorithm used in MSWEP merges observed rain-gauge data, satellite observations and reanalysis data to provide reliable precipitation estimates over the entire globe. In this paper, we used daily data from MSWEP version 1.1, but newer versions (already available) will be included in CAMELS-CL after validation with ground measurements in Chile.

Finally, $precip_{tmpa}$ was computed from the Tropical Rainfall Measuring Mission (Huffman et al., 2007) Multi-satellite Precipitation Analysis (TMPA) dataset, which provides quasi global (50° N to 50° S) precipitation estimates at 0.25° spatial resolution. TMPA integrates infrared and passive microwave data from a wide variety of satellite-borne precipitation-related sensors. In this study, we used TRMM research product 3B42v7, which makes use of Global Precipitation Climatology Project (GPCP; Adler et al., 2003) and Climate Assessment and Monitoring System (CAMS, Ropelewski et al., 1984) data to rescale its estimations on a monthly basis.

3.1.7 Temperature

Daily time series of minimum, maximum and mean temperatures for each catchment were also derived from the CR2MET dataset (DGA, 2017). Daily minimum and maximum temperatures in CR2MET ($CR2MET/T_{max}$ and $CR2MET/T_{min}$, respectively) were mapped for the period 1979–2016 using a slightly different approach than the one used for precipitation (Sect. 3.1.6). In this case, the method used land surface temperature (LST) estimates from MODIS satellite retrievals, in addition to near-surface temperature provided by ERA-Interim. Multivariate regression models for both $CR2MET/T_{max}$ and $CR2MET/T_{min}$ were developed using LST as part of the explanatory variables and local temperature records in Chile as target data. Given the data gaps and relatively short period available for LST, the final product was derived for the whole period (1979–2016) by fitting the ERA-Interim data to the preliminary (incomplete) MODIS-based product. To get mean daily temperatures ($CR2MET/T_{mean}$), the long-term $CR2MET/T_{max}$ and $CR2MET/T_{min}$ were used to adjust the ERA-Interim 3-hourly near-surface temperature. The adjusted 3-hourly data were then averaged to derive $CR2MET/T_{mean}$. Gridded daily mean, minimum and maximum temperatures from CR2MET (0.05° lat–lon resolution) were clipped to obtain basin-averaged daily time series for CAMELS-CL, named T_{mean}, T_{min} and T_{max}, respectively.

3.1.8 Potential evapotranspiration

We processed catchment-scale PET from two different sources. The first PET product uses the formulae proposed by

Hargreaves and Samani (1985), which is solely based on surface temperature data (see Hargreaves and Allen, 2003, for further details). We used CR2MET/T_{max} and CR2MET/T_{min} (described in Sect. 3.1.7) to generate a gridded PET estimate (PET_{har}).

The second PET data included in CAMELS-CL (PET_{mod}) are those provided by the MODIS PET product (MOD16 collection 5; Mu et al., 2005), which is processed from different sources of information, including leaf area index and fractional photosynthetically active radiation, FPAR/LAI (MOD15A2; Myneni et al., 2002), land cover type 2 (MOD12Q1; Friedl et al., 2002), albedo (MCD43B2 and MCD43B3; Jin et al., 2003; Lucht et al., 2000), and daily meteorological reanalysis data from NASA's MERRA GMAO (GEOS-5). MOD16 is calculated following the Penman–Monteith approach (Howell and Evett, 2001), and the final product is available at an 8-day temporal resolution for the period 2000–2014, on a $1 \times 1\,km^2$ grid. As for other gridded variables, the PET_{har} and PET_{mod} products were clipped and averaged within basin boundaries to generate daily (called pet_{har}) and 8-day (called pet_{mod}) catchment-scale time series, respectively.

3.1.9 Snow water equivalent

We processed daily snow water equivalent (SWE) data using the 180 m resolution SWE product generated by Cortés and Margulis (2017), which covers the Near North and Central zones (25–37° S). Cortés and Margulis (2017) obtained SWE ensemble estimates from forward modelling "prior" values, which were conditioned through the assimilation of historical fractional snow-covered area (fSCA) data from Landsat TM, ETM+ and OLI sensors. The "posterior" SWE and fSCA estimates were probabilistically conditioned on the observed depletion record from Landsat, the uncertainty in fSCA observations, and the forward model state uncertainty. The fSCA retrieval was obtained with a spectral unmixing algorithm (Cortés et al., 2014). The forward models for prior ensemble generation were the SSiB3 land surface model (Yang et al., 1997) and a snow depletion curve model (SDC; Liston, 2004). Detailed assessments of this reanalysis framework were performed for the Sierra Nevada using in situ sensor data (Margulis et al., 2016) and for the Andes (Cortés et al., 2016) using snow survey points, site years of peak annual snow pillow and snow course SWE observations. Validation results showed unbiased posterior SWE estimates with a correlation coefficient of 0.73, an RMSE of 0.29 m and a mean error of less than 0.01 m using snow pillow and snow course peak SWE. Results using snow survey data showed similar unbiased estimates, with a correlation coefficient of 0.50, an RMSE of 0.29 m and a mean error of less than 0.01 m. The daily SWE gridded product generated by Cortés and Margulis (2017) was clipped and averaged within the catchment boundaries to obtain daily time series for each catchment.

3.1.10 Water rights and reservoir information

A public reservoir dataset (http://www.ide.cl/descarga/capas/item/embalses-2016.html, last access: September 2017) was processed to identify the presence of dams within catchments. We also compiled and processed granted water rights available from the National Water Atlas (DGA, 2016a). This water allocation dataset includes information about the source (surface or groundwater), the type of right (i.e. consumptive or non-consumptive), use (i.e. industrial, irrigation, domestic and drinking water, hydroelectric power, pisciculture, mining, and classified as "other uses"), annual allocated flow (expressed in units of volume per time or as "shares"), and temporal allocation (i.e. permanent and continuous, permanent and discontinuous, permanent and alternated, eventual and continuous, eventual and discontinuous, or eventual and alternated). A detailed explanation of this water right classification can be found in Carey (2014). A key limitation of this dataset is the lack of information on the actual use of granted rights (Larraín, 2006). Additionally, some water right records have incomplete information (e.g. missing coordinates, water volume assigned and temporal allocation).

Figure 5 illustrates water allocation in central–southern Chile (30–43° S), showing surface and groundwater rights (all types). It is clear that groundwater rights dominate in central Chile (31–36° S), especially in low-elevation areas, compared to surface water rights. On the other hand, more surface water rights are granted in southern Chile, especially within high-elevation areas towards the Andes.

3.2 Derived catchment attributes

We computed 70 catchment attributes grouped into six classes (Table 2). To motivate the use of common standards in the development of large sample catchment datasets, we included most of the attributes presented by A17. A comparative summary between CAMELS and CAMELS-CL attributes is presented in Table 2, from which one can note that the attributes from classes climatic indices and hydrological signatures were fully adopted from A17. The attributes from the class soil characteristics were not computed at this stage since there is no publicly available national dataset. Given the differences in input datasets, some of the attributes from the classes location and topography, geologic characteristics, and land cover characteristics in A17 were not computed here. On the other hand, new attributes were derived for the classes' location and topography (Sect. 3.2.1), land cover characteristics (Sect. 3.2.3) and hydrological signatures (Sect. 3.2.5). Further, a new class was added to describe the degree of intervention within the catchments (Sect. 3.2.6).

A complete list of catchment attributes included in CAMELS-CL, their description and the corresponding data sources are presented in Table 3. To ensure the reproducibility of our results, the reference to the explicit formulation of

Table 2. Summary of attributes computed in CAMELS and CAMELS-CL.

Attribute class	CAMELS (A17)	CAMELS-CL
Location and topography	9 attributes	6 attributes adopted from A17 11 additional attributes
Geology	7 attributes	5 attributes adopted from A17
Soils characteristics	11 attributes	not computed
Land cover characteristics	8 attributes	3 attributes adopted from A17 13 additional attributes
Climatic indices	11 attributes	11 attributes adopted from A17 1 additional attribute
Hydrological signatures	13 attributes	13 attributes adopted from A17 1 additional attribute
Intervention degree	not computed	6 attributes
Total number of attributes	*59*	*38 adopted from A17* *32 introduced*

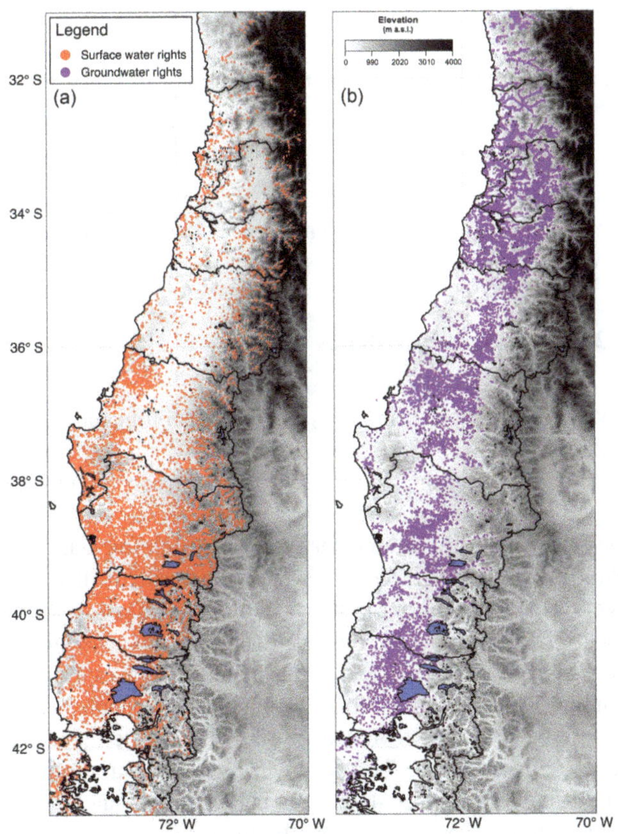

Figure 5. Surface **(a)** and ground **(b)** water rights (consumptive, non-consumptive, permanent, eventual and alternated) granted by the Chilean Water Directorate (DGA) for a portion of the country. Background colours represent topography (greyscale) and the main water bodies (highlighted in blue).

climatic indices and hydrological signatures is provided in Table 3. Discussions on the spatial distribution of catchment attributes, separated by class, are presented in the following sub-sections.

3.2.1 Location and topography

Figure 6 shows 6 (out of 14) location and topography attributes. Figure 6a presents the elevation of catchment outlets, illustrating two main elevation gradients: (i) a north to south (N–S) decrease, starting with high-elevation basins in the Far North macro-zone – which corresponds to the southern portion of the Altiplano Plateau (18–22° S) (Allmendinger et al., 1997) – towards lower elevations in the southern macro-zones; and (ii) an east to west (E–W) gradient, dominated by high elevations in the Andes (located along the eastern border) decreasing towards sea level at the western border. This gauge elevation attribute can be used to classify catchments based on their location with respect to the coast or the Andes. We proposed the attribute location_type (see Table 3 and Fig. 6f) with three categories: coastal (or low elevation), foothills, and Altiplano catchments, defined by gauge elevations lower than 50 m a.s.l., between 1000 and 1200 m a.s.l., and above 3500 m a.s.l., respectively.

Figure 6b reveals smoother N–S and E–W gradients of basin-averaged elevations, compared to gauge elevation gradients (Fig. 6a). This is because the mean elevation calculated for downstream catchments includes nested catchments (located at higher altitudes). From the complete set of catchments, 178 have a mean elevation greater than 2000 m a.s.l. The spatial distribution of mean catchment slopes follows different patterns depending on the macro-zone (Fig. 6c). The Far North – dominated by the flat Altiplano Plateau – exhibits relatively small variations in mean slopes, with rel-

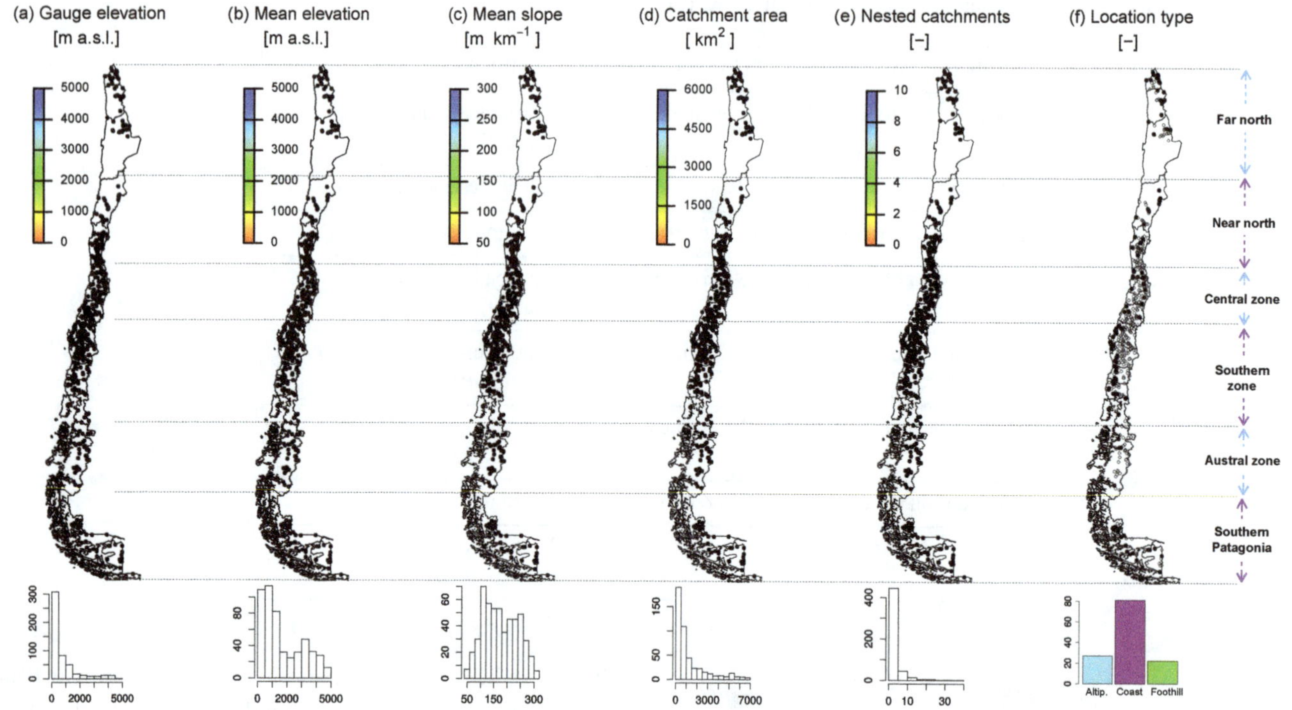

Figure 6. Location and topography. For visualisation purposes, catchment areas **(d)** are shown up to their 90th percentile. The histograms indicate the number of catchments (out of 516) in each bin.

atively low values. From the Near North to Austral zones, the mean slope shows a spatial distribution similar to that from mean elevation, with an E–W gradient dominated by high slopes in the Andes and flatter areas towards the sea. In Southern Patagonia, such an E–W gradient is reversed given the relative position of the Andes.

The spatial distribution of basin areas shows a general increase from east to west (Fig. 6d), which is consistent with smaller headwater catchments at the Andes and larger downstream catchments towards the sea. Some exceptions to this E–W distribution pattern are catchments located near the eastern border, featuring either a N–S drainage direction or a portion of their total contributing area in Argentina (beyond the eastern national border). Additional exceptions to such E–W distribution are small inner sub-catchments near the western border or small headwater catchments originated at the Chilean Coastal Range, which runs from north to south along the Pacific coast and reaches up to 3000 m a.s.l. in the Antofagasta region (Figueroa and Moffat, 2000).

Given that all catchments were delineated using available streamflow gauge locations as outlets (Sect. 3.1.1), the contributing area (Fig. 6d) is not necessarily correlated with the number of nested basins within each catchment (Fig. 6e). For example, some small catchments might be highly instrumented (i.e. with many controlled nested basins, because of – for example – water allocation priorities or having a high population density like those in Region Metropolitana, which concentrates more than 40 % of the country's population), while large but poorly instrumented catchments might not have inner basins defined.

3.2.2 Geological attributes

Overall, the most common dominant geological classes within CAMELS-CL catchments are acid plutonic rocks (24 %), acid volcanic rocks (20 %) and pyroclastic (14 %). In the Far North zone, there is a strong presence of Pyroclastics, Siliclastic sedimentary rocks and Intermediate volcanic rocks (Fig. 7a and b), which can result in the connection of groundwater systems through fractured volcanic rocks (DGA, 1986). This means that there might be differences between surface catchment boundaries (based on topography) and the extension of groundwater systems, which should be considered when analysing basin-scale hydrological response. Figure 7a also indicates that strong geological differences may exist between neighbouring catchments. Furthermore, one can see generally high geological variability within the catchments. Indeed, the dominant geological class covers less than half of the contributing area in most catchments, as indicated by the histogram in Fig. 7c. The presence of carbonate sedimentary rocks is particularly low (Fig. 7e), with only 24 catchments having at least 10 % of this type of rock. This suggests low formation of karst, a subsurface characteristic featuring large fissures and voids, which

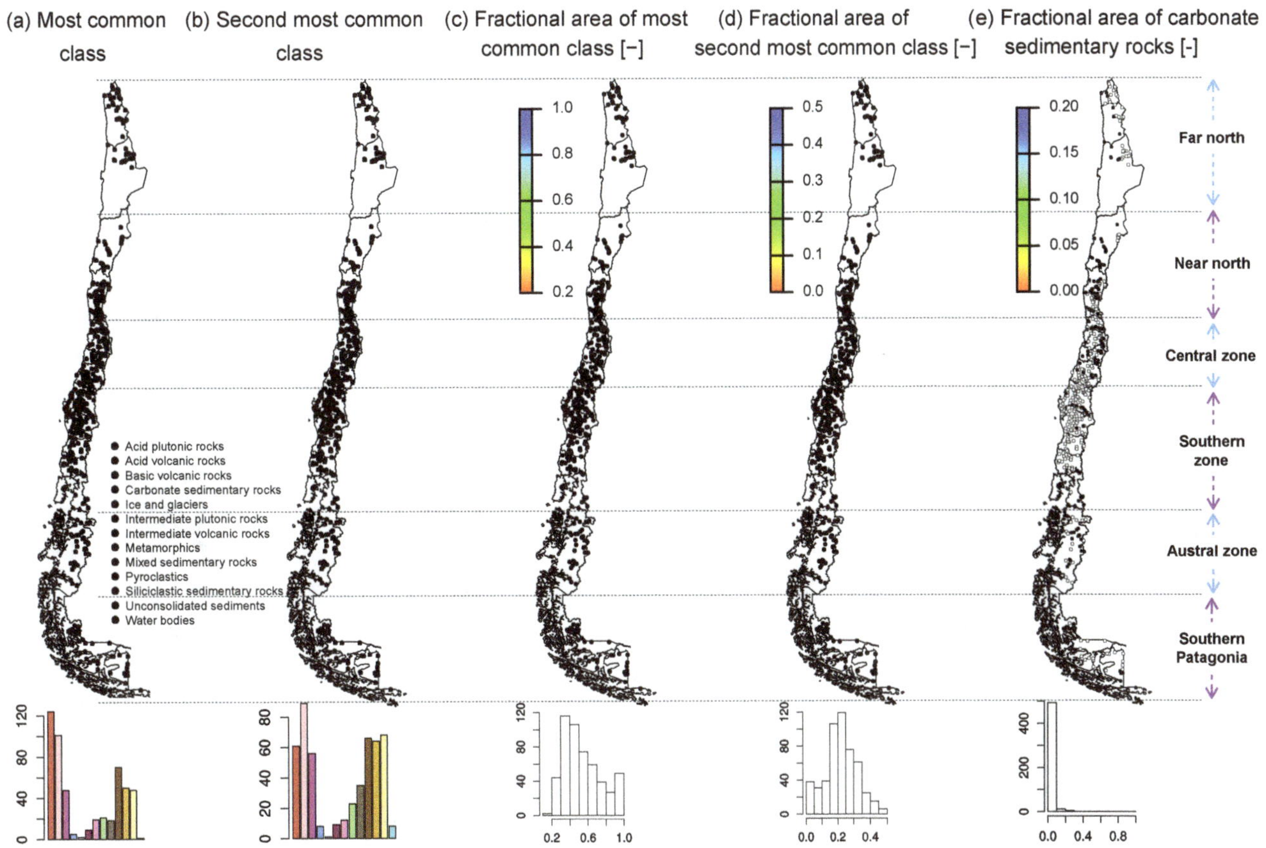

Figure 7. Geology attributes. The histograms indicate the number of catchments (out of 516) in each bin.

results in fast infiltration rates and preferential permeability channels (La Moreaux et al., 1984).

3.2.3 Land cover attributes

As summarised in Table 2, five land cover attributes in A17 were not computed since the land cover map used here (from Zhao et al., 2016) does not provide information about leaf area index, green vegetation fraction, or depth. Instead, we included land cover attributes based on the catchment area encompassed by the main classes of the land cover dataset (Table 3). The first nine land cover attributes described in Table 3 were computed as the percentage of the catchment area covered by level 1 and 2 land cover classes defined by Zhao et al. (2016). We also computed a forest plantation index to quantify the ratio between forest exotic plantation (mainly *Pinus radiata* and *Eucalyptus* spp.) and native forest within a catchment, which is critical information for forest hydrology and ecosystem studies (e.g. Lara et al., 2009).

Considering that several catchments (almost 50; Fig. 1) extend beyond the Chilean territory, a key limitation of the land cover attributes derived from Zhao et al. (2016) is the lack of information outside the national boundary. To address this, we generated an attribute indicating the fractional catchment area contained within the land cover map, serving also as a quality flag for basin-averaged land cover characteristics.

Another limitation of the land cover map and derived attributes is that there is no characterisation of inter-annual variability (the map was constructed by using imagery from 2013 and 2014). This can be particularly important for land cover types that are sensitive to climatic variations, such as altiplanic wetlands, which largely influence the hydrology in the Far North. This limitation is also critical for classes featuring drastic changes within time, such as forest plantation and cropland.

Figure 8 illustrates a subset of the land cover attributes listed in Table 3. Figure 8a shows the forested (native forests and forest plantation types) catchment area, which prevails in the Southern Zone, Austral Zone and Southern Patagonia. In forested catchments, exotic forest plantations dominate the coastal areas of the Central and Southern zones, with forest plantation indices up to 1 (Fig. 8b). Such a distribution illustrates the extensive land use change experienced in southern–central Chile over the last 50 years, where native forests have been progressively converted into agricultural lands and forest plantations (Armesto et al., 2010; Miranda et al., 2015). This conversion has had significant impacts on forest ecosys-

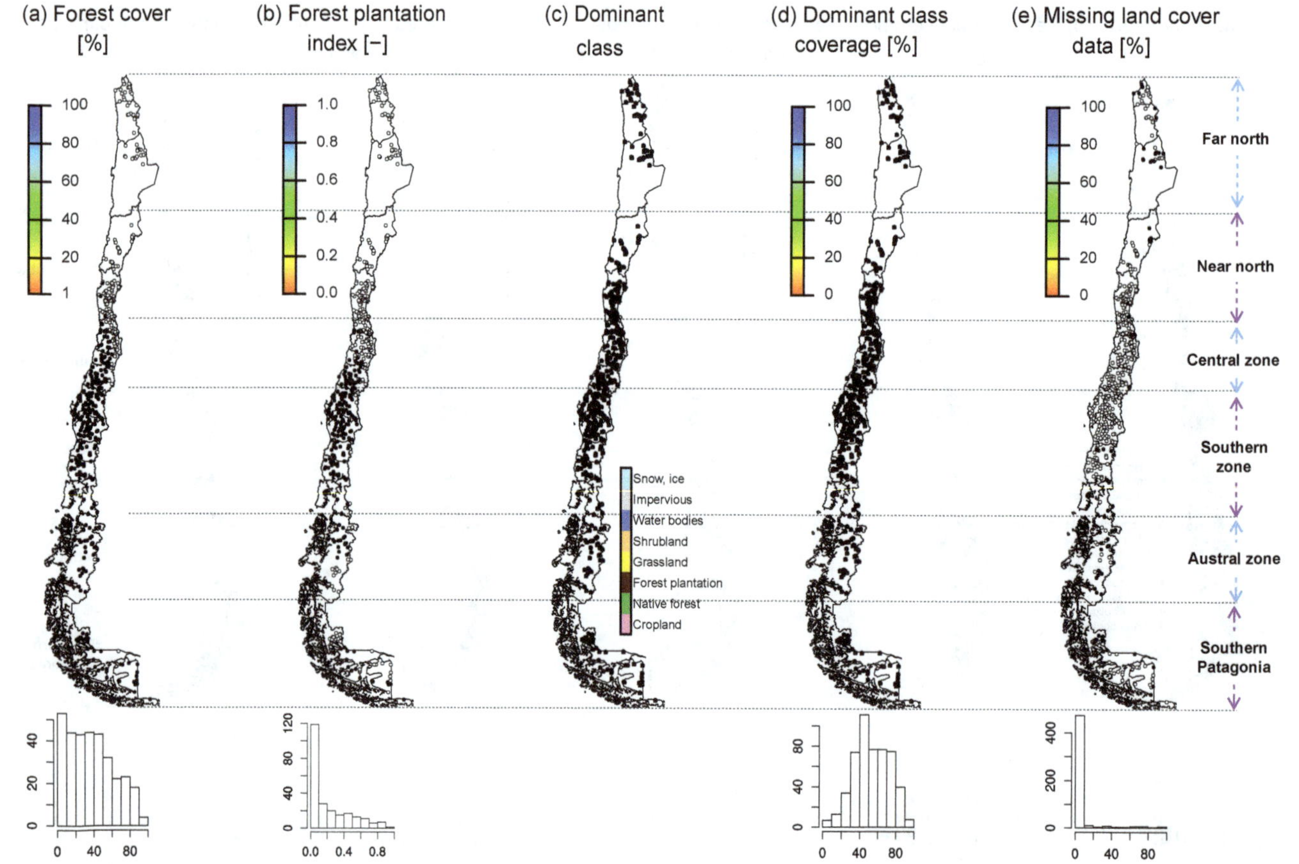

Figure 8. Land cover characteristics. Values below the colour bar lower limits are shown in blank (i.e. blank points in panel **a** represent no forest cover, and in **e** they represent no missing land cover data within those catchments). The histograms indicate the number of catchments (out of 516) in each bin.

tem services such as water provision (Jones et al., 2017; Lara et al., 2009).

Figure 8c and d show that the Far and Near North zones have more homogeneous land cover types, with shrublands and impervious lands occupying more than 60 % of the catchment areas. In southern areas, the coverage of the dominant classes decreases substantially, transitioning towards a mosaic of different land cover types. Missing land cover data are presented in Fig. 8e, which should be accounted for if the land cover attributes of the affected catchments (i.e. the ones with portions in Argentina, as shown in Fig. 2) are used for applications such as catchment classification or parameter regionalisation.

Because of the glaciological contributions to the water balance within the domain (Mernild et al., 2017; Le Quesne et al., 2009), we added two attributes (Table 3) based on information from the glacier inventory described in Sect. 3.1.4. We found that 255 catchments (48 % of the total) have some degree of glacierisation, reaching up to 62 % in the Geike River catchment in Southern Patagonia. The catchments with the largest degree of glacierisation (more than 15 %) are located in the Austral and Southern Patagonia regions, followed by the Olivares and Volcan river catchments (about 14 %) in the Central Zone. The glaciers included in CAMELS-CL span 7321 km^2, corresponding to almost a quarter of the glacierised area in the southern Andes (RGI Consortium, 2017).

3.2.4 Climatic indices

To allow direct comparisons between CAMELS (A17) and CAMELS-CL, climatic indices were computed for the same period as in A17, i.e. water years 1990 to 2009, corresponding to 1 April 1990 to 31 March 2010 for Chile. If these indices are required for different periods, the formulae provided in the references from Table 3 can be used with the raw hydro-meteorological time series (available from the CAMELS-CL website). The complete spatial and temporal coverages of meteorological variables allow the estimation of climatic indices for all 516 catchments – in contrast to hydrological signatures, computed for a subset of catchments (Sect. 3.2.5). Precipitation and PET-based attributes were calculated for all precipitation products (Sect. 3.1.6) using the daily PET product (pet_{har}, Sect. 3.1.8).

Table 3. Summary of catchment attributes. Climate indices and hydrological signatures were computed for the period 1 January 1990–31 December 2010. Index i refers to the precipitation product ($i = 1, 2, 3, 4$ for $\text{precip}_{\text{cr2met}}$, $\text{precip}_{\text{tmpa}}$, $\text{precip}_{\text{chirps}}$ and $\text{precip}_{\text{mswep}}$, respectively).

Attribute class	Attribute name	Description	Unit	Data source	Reference
Location and topography	gauge_id	catchment identifier (corresponds to the station code provided by DGA)	–	Gauges information collected from http://explorador.cr2.cl (last access: March 2018)	Sect. 3.1
	gauge_name	gauge name (based on DGA records)	–		
	gauge_lat	gauge latitude (based on DGA records)	° S		
	gauge_lon	gauge longitude (based on DGA records)	° W		
	gauge_record_start	start date of streamflow records	–		
	gauge_record_end	end date of streamflow records	–		
	gauge_n_obs	number of days with valid streamflow records	day		
	area	catchment area	km^2	ASTER GDEM 30 m raster data (Tachikawa et al., 2011)	
	elev_gauge	gauge elevation (catchment outlet) obtained from the 30 m ASTER GDEM elevation data and the location provided by DGA	m a.s.l.		
	elev_mean	catchment mean elevation	m a.s.l.		
	elev_med	catchment median elevation	m a.s.l.		
	elev_max	catchment maximum elevation	m a.s.l.		
	elev_min	catchment minimum elevation	m a.s.l.		
	slope_mean	catchment mean slope	$m\,km^{-1}$		
	nested_inner	number of inner catchments contained within gauge_id catchment (the gauge_id for the inner catchments can be obtained from the hierarchy matrix described in Sect. 3.1.1)	–	–	
	nested_outer	number of catchments containing gauge_id catchment (the gauge_id for the outer catchments can be obtained from the hierarchy matrix described in Sect. 3.1.1)	–		
	location_type	classification in “coastal (or low elevation)”, “foothill” and “altiplano” catchments, based on gauge elevations (gauge_elev) below 50 m a.s.l., between 1000 and 1200 m a.s.l., and above 3500 m a.s.l., respectively.	–	–	Sect. 3.2
Geological characteristics	geol_class_1st	most common geologic class in the catchment	–	Global Lithological Map database (GLiM) (Hartmann and Moosdorf, 2012)	Table 6 in A17
	geol_class_1st_frac	fraction of the catchment area associated with its most common geologic class	–		
	geol_class_2nd	second most common geologic class in the catchment	–		
	geol_class_2nd_frac	fraction of the catchment area associated with its second most common geologic class	–		
	carb_rocks_frac	fraction of the catchment area characterised as “carbonate sedimentary rocks”	–		
Land cover characteristics	crop_frac	Percentage of the catchment covered by croplands, level 1. Includes five types of level 2 classes: rice fields; greenhouse farming; other croplands; orchards; and bare croplands	%	30 m resolution land cover map provided by Zhao et al. (2016)	Sects. 3.1.3 and 3.2.3
	nf_frac	Percentage of the catchment covered by forest (level 1) classified as natural broadleaf (level 2) or natural conifer (level 2).	%		
	fp_frac	Percentage of the catchment covered by forest (level 1) classified as broadleaf plantations (level 2) or conifer plantations (level 2).	%		
	grass_frac	Percentage of the catchment covered by grasslands, level 1. Includes three types of level 2 classes: pastures; other grasslands; and withered grasslands.	%		
	shrub_frac	Percentage of the catchment covered by shrublands, level 1. Includes five types of level 2 classes: shrublands; shrubs and sparse trees mosaic; succulents; shrub plantations; and withered shrublands.	%		
	wet_frac	Percentage of the catchment covered by wetlands and water bodies (level 1). Includes six types of level 2 classes: marshlands; mudflats; other wetlands; lakes; reservoirs/ponds; rivers; and ocean.	%		
	imp_frac	Percentage of the catchment covered by impervious surfaces (level 1). Urbanised areas are usually contained in this class.	%		

Table 3. Continued.

Attribute class	Attribute name	Description	Unit	Data source	Reference
	barren_frac	Percentage of the catchment covered by barren lands (level 1). Includes three types of level 2 classes: dry salt flats; sandy areas; and bare exposed rocks	%		
	snow_frac	Percentage of the catchment covered by snow and ice, level 1. Includes two types of level 2 classes: snow and ice.	%		
	fp_nf_index	forest plantation index: calculated as the ratio between fp_frac and the total forested area (fp_frac + nf_frac)	–		
	forest_frac	fraction of the catchment covered by forests, including native forest and forest plantation (fp_frac + nf_frac)	%		
	dom_land_cover	dominant land cover class	–		
	dom_land_cover_frac	fraction of the basin associated with dominant land cover class	%		
	land_cover_missing	percentage of the basin not covered by the land cover map	%		
	glaciers_area	glacierised area within the catchment	km^2	Randolph Glacier Inventory v. 6.0 (RGI Consortium, 2017)	Sects. 3.1.4 and 3.2.3
	glaciers_frac	percentage of the catchment covered by glaciers.	%		
Climatic indices (computed for 1 April 1990 to 31 March 2010)	p_mean_i	mean daily precipitation of product *i*	mm day^{-1}	Precipitation, temperature and potential evapotranspiration products introduced in Sects. 3.1.6, 3.1.7 and 3.1.8, respectively.	Table 2 in A17
	p_mean_spread	coefficient of variation of basin-averaged mean annual precipitation from different products (standard deviation of p_mean_i from the four precipitation products, normalised by multi-product mean)	–		
	pet_mean	mean daily PET of pet_{har} product	mm day^{-1}		
	aridity_i	aridity, calculated as the ratio of mean daily PET (pet_mean) to mean daily precipitation (p_mean_i)	–		
	p_seasonality_i	seasonality and timing of precipitation (product *i*) estimated using sine curves to represent the annual temperature and precipitation cycles; positive (negative) values indicate that precipitation peaks in summer (winter); values close to 0 indicate uniform precipitation throughout the year)	–		
	frac_snow_i	fraction of precipitation (product *i*) falling as snow (i.e. on days colder than 0 °C)	–		
	high_prec_freq_i	frequency of high-precipitation days ($\geq$ 5 times mean daily precipitation) for product *i*	days yr^{-1}		
	high_prec_dur_i	average duration of high-precipitation events (number of consecutive days $\geq$ 5 times mean daily precipitation), for product *i*	days		
	high_prec_timing_i	season during which most high-precipitation days ($\geq$ 5 times mean daily precipitation) occur	season		
	low_prec_freq_i	frequency of dry days ($<$ 1 mm day^{-1}), for product *i*	day yr^{-1}		
	low_prec_dur_i	average duration of dry periods (number of consecutive days $<$ 1 mm day^{-1}), for product *i*	days		
	low_prec_timing_i	season during which most dry days ($<$ 1 mm day^{-1}) occur, for product *i*	season		
Hydrological signatures (computed for 1 April 1990 to 31 March 2010)	q_mean	mean daily discharge	mm day^{-1}	Streamflow records collected from http://explorador.cr2.cl (last access: March 2018)	Table 3 in A17
	runoff_ratio_i	runoff ratio (ratio of mean daily discharge to mean daily precipitation), for product *i*	–		
	stream_elas_i	streamflow precipitation elasticity (sensitivity of streamflow to changes in precipitation at the annual timescale, using the mean daily discharge as reference and precipitation product *i*)	–		
	slope_fdc	slope of the flow duration curve, FDC (between the log-transformed 33rd and 66th streamflow percentiles)	–		
	baseflow_index	baseflow index (ratio of mean daily baseflow to mean daily discharge, hydrograph separation performed using the Ladson et al. (2013) digital filter with α set to 0.925)	–		
	hdf_mean	mean half-flow date (date on which the cumulative discharge since 1 April reaches half of the annual discharge)	day of the year		
	Q_5	5 % flow quantile (low flows)	mm day^{-1}		
	Q_{95}	95 % flow quantile (high flows)	mm day^{-1}		
	high_q_freq	frequency of high-flow days ($>$ 9 times the median daily flow)	day yr^{-1}		

Table 3. Continued.

Attribute class	Attribute name	Description	Unit	Data source	Reference
	high_q_dur	average duration of high-flow events (number of consecutive days > 9 times the median daily flow)	days		
	low_q_freq	frequency of low-flow days (< 0.2 times the mean daily flow)	day yr^{-1}		
	low_q_dur	average duration of low-flow events (number of consecutive days < 0.2 times the mean daily flow)	days		
	zero_q_freq	percentage of days with $Q = 0$	%		
	swe_ratio	ratio of peak of snow water equivalent to mean annual discharge	–	SWE product developed by Cortés and Margulis (2017)	
Intervention	sur_rights_n	total number of granted surface rights within the catchment	–	Water Atlas developed by the DGA (2016a)	Sects. 3.1.10 and 3.2.6
	sur_rights_flow	annual flow calculated for consumptive permanent continuous surface water rights	$m^3 s^{-1}$		
	interv_degree	intervention degree defined as the annual flow of surface water rights (consumptive permanent continuous), normalised by mean annual streamflow.	–		
	gw_rights_n	total number of granted groundwater rights within the catchment	–		
	gw_rights_flow	annual flow calculated for consumptive permanent continuous groundwater rights	$m^3 s^{-1}$		
	large_dam	0 if there is no dam within the catchment, 1 if there is at least one dam classified as "large"	–	http://www.ide.cl/descarga/capas/item/embalses-2016.html (last access: September 2017)	

The climatic attributes presented in Figs. 4 and 9 reveal basic features of Chilean climatology, described in more detail by Miller (1976) and Garreaud et al. (2017), among others. Mean annual precipitation ranges from less than 10 mm yr^{-1} in the Atacama Desert (northern Chile) to more than 3000 mm yr^{-1} in western Patagonia (Fig. 4b). Such a marked precipitation gradient reflects the relative influence of the subtropical, semi-permanent south-eastern Pacific anticyclone, and the frequent incursion of frontal systems at higher latitudes. The frequency of high-precipitation events also increases southward, with a maximum in southern–central Chile (Fig. 9d). The Andean domain in the Far North (Chilean Altiplano), receiving about 300 mm yr^{-1} above 4000 m a.s.l., is influenced by the monsoonal regime developing over the interior of the continent. In addition to the N–S gradient, precipitation increases strongly in the west–east direction due to the orographic enhancement of air masses over the windward slope of the Andes Cordillera (a factor of 2–3 between lowlands and windward slopes; Viale and Garreaud, 2014). PET has a more restricted range than precipitation (400–1400 mm yr^{-1}; Fig. 4d); therefore, the aridity index (PET/P, Fig. 9c) is higher in northern Chile (> 1.0) compared to that of southern regions (< 1.0). A positive precipitation seasonality (Fig. 9a) in northern Chile indicates precipitation peaks during summer (DJF), following the monsoonal precipitation regime governing in this region (Fig. 9f). In contrast, the negative seasonality values obtained for all macro-zones, except the Far North and Southern Patagonia, illustrate the increased storm frequency and high-precipitation events in most of the country during the winter (JJA) (Fig. 9a). Seasonality values close to zero indicate uniform precipitation throughout the year in Southern Patagonia (Fig. 9a and f). The zero-temperature isotherm during winter storms ranges between 1500 and 4000 m a.s.l., so most of the precipitation is liquid along the coast and interior valleys (Fig. 9b), while a larger fraction of solid precipitation is obtained in high-elevation basins.

3.2.5 Hydrological signatures

Hydrological signatures were computed for the period 1 April 1990 to 31 March 2010, as in Sect. 3.2.4. To exclude the effects of anthropic intervention on hydrologic response, we selected 94 catchments with valid daily streamflow records in at least 85 % of the period, based on the following criteria: interv_degree lower than 0.1 (i.e. less than 10 % of the annual streamflow allocated to surface rights), large_dam equal to zero (absence of large dams within the catchment), imp_frac lower than 5 % (negligible urban areas), copr_frac lower than 20 % (negligible irrigation effects) and fp_frac lower than 20 % (negligible forest plantation effects). Further, we excluded glacier-dominated catchments by selecting glacier_frac lower than 5 %. It should be noted that, despite calculating hydrological indices for a subset of catchments, raw daily time series for all 516 catchments are

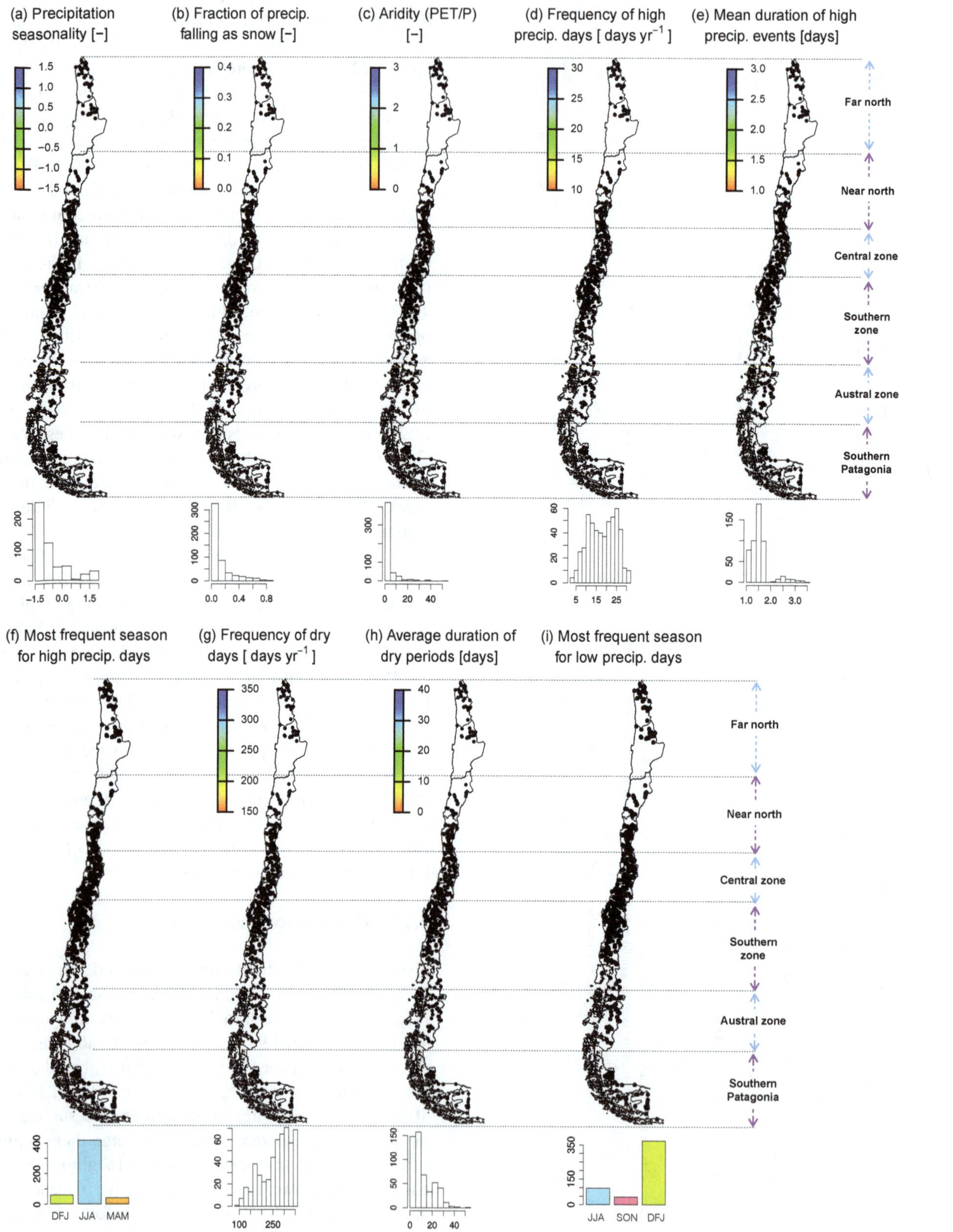

Figure 9. Climatic indices (calculated from the $precip_{cr2met}$ product). The histograms indicate the number of catchments (out of 516) in each bin.

included in the CAMELS-CL database. These time series and the formulae provided in Table 3 may be used if the signatures are required for different time periods.

Figure 10 illustrates the spatial distribution of 12 (out of 14) hydrological signatures (Table 3), revealing the leading patterns of catchment responses. Both mean daily flow and runoff ratio increase from the Far North to the Southern Zone, showing strong correlations with mean annual precipitation (Fig. 4b) and the aridity index (Fig. 9c). Further, the runoff ratio shows a positive west to east gradient (i.e. increase towards the Andes), reaching values above 1 in mountain catchments in the Southern Zone. These non-behavioural catchments are further analysed in Sect. 4.1. Mean half-flow dates (Fig. 10c) present a similar west to east gradient, with higher values in steep (Fig. 6c) snow-dominated (Fig. 9b) basins in central Chile – where the most frequent season for low-precipitation days is DJF (Fig. 9i).

The mid-segment slope of the flow duration curve (FDC, Fig. 10d) – a signature that quantifies flashiness of runoff – shows that slow basin-averaged responses occur in the Far North and part of the Near North, in spatial correspondence to a high baseflow index (Fig. 10e) and low discharge precipitation elasticity (Fig. 10f). Such behaviour is expected in this region due to substantial subsurface and groundwater contributions to total runoff. Although flashiness of runoff and discharge elasticity to precipitation (baseflow index) are relatively higher (lower) and show some correlation towards the south, no clear spatial gradients are observed within the domain spanning from central Chile to Southern Patagonia. Figure 10f shows negative elasticity values (-0.13 and -0.03) in two catchments located in the Austral Zone and Southern Patagonia. Such values indicate a negative annual runoff anomaly in response to the positive annual precipitation anomaly, which is not expected in near-natural catchments. We attribute this behaviour to two main factors. First, there is a numerical problem in the formula used to calculate streamflow elasticity (Eq. 7 in Sankarasubramanian et al., 2001, adapted in A17, Table 3) when annual precipitation of a single year approaches the long-term mean, causing the elasticity to approach infinity (Sankarasubramanian et al., 2001). A second factor is the use of incomplete streamflow daily records. Since the elasticity is computed from concurrent daily streamflow and precipitation, its calculation in catchments with missing streamflow records can be problematic. This can be particularly important in snow-dominated catchments (delayed runoff response to precipitation) and in catchments with a weak precipitation seasonality (i.e. precipitation falling during the whole year, Fig. 9a close to zero), which is the case for catchments in the Austral Zone and Southern Patagonia.

The examination of signatures related to extreme (high or low) streamflow conditions exposes some interesting features. Although no clear spatial relationship is observed between high-flow signatures (Fig. 10g–i), similar spatial distributions of low-flow frequencies (Fig. 10j) and mean low-flow durations (Fig. 10k) are obtained across the country. Q_{95} (Fig. 10i) and Q_5 (Fig. 10l) provide generally similar patterns to those of mean daily discharge (Fig. 10a), with positive increases from the Far North to Southern zones, and a positive west to east gradient. The comparison between the signatures displayed in Fig. 10g–l and climatic indices in Fig. 9d–i highlights the complex relationship between climate and hydrologic catchment behaviour. For example, the spatial structure in the frequency of low-/high-precipitation days is not reflected in the spatial distribution of high-/low-flow frequencies. A similar disjunction is observed between the duration of low-precipitation (Fig. 9h) and low-flow (Fig. 10k) events, whereas those catchments with low-duration of high-precipitation events also provide low durations in high-flow events.

Sharp variations in hydrological signatures (Fig. 10) – in contrast to generally smooth patterns in climate indices (Fig. 9) – are the result of complex, non-linear processes across a range of spatiotemporal scales, enhanced by heterogeneities in topography, soils, vegetation, geology and other landscape properties. Careful attention should be paid to such interactions and to the uncertainties involved in the calculation of hydrological signatures, in particular when attempting to extrapolate hydrological behaviour from gauged to ungauged basins based on climatic similarities alone (Westerberg et al., 2016; Westerberg and McMillan, 2015).

3.2.6 Human intervention

Figure 11 summarises water rights records used to characterise human intervention degree within the catchments. We can see that the number of surface rights (Fig. 11a) increases from north to south, while the number of groundwater rights (Fig. 11d) increases from east to west. Although these values do not provide information about allocated volumes, they show how many water rights holders interact to coordinate water use within a particular catchment. The CAMELS-CL database provides information about each water right within a catchment (not only the attributes with synthesised information), in case more detailed analyses are required.

In terms of allocated surface and groundwater flows (Fig. 11b and e, respectively), we only considered consumptive permanent water rights. Further, we considered only water rights recorded as volume per time, since water rights expressed as "shares" (6 % of the national water rights database) were not provided with their corresponding conversion into volume units (DGA, 2016b). It should be noted that shares rights are the oldest (allocated prior to the 2005 water code reform), thus probably representing a majority of the rights within the Central Zone (the region that concentrates the oldest rights).

The above limitations may lead to an underestimation of the allocated flow for (at least) the following reasons: (i) non-consumptive rights may have their restitution points outside the catchment boundaries (however, they were not consid-

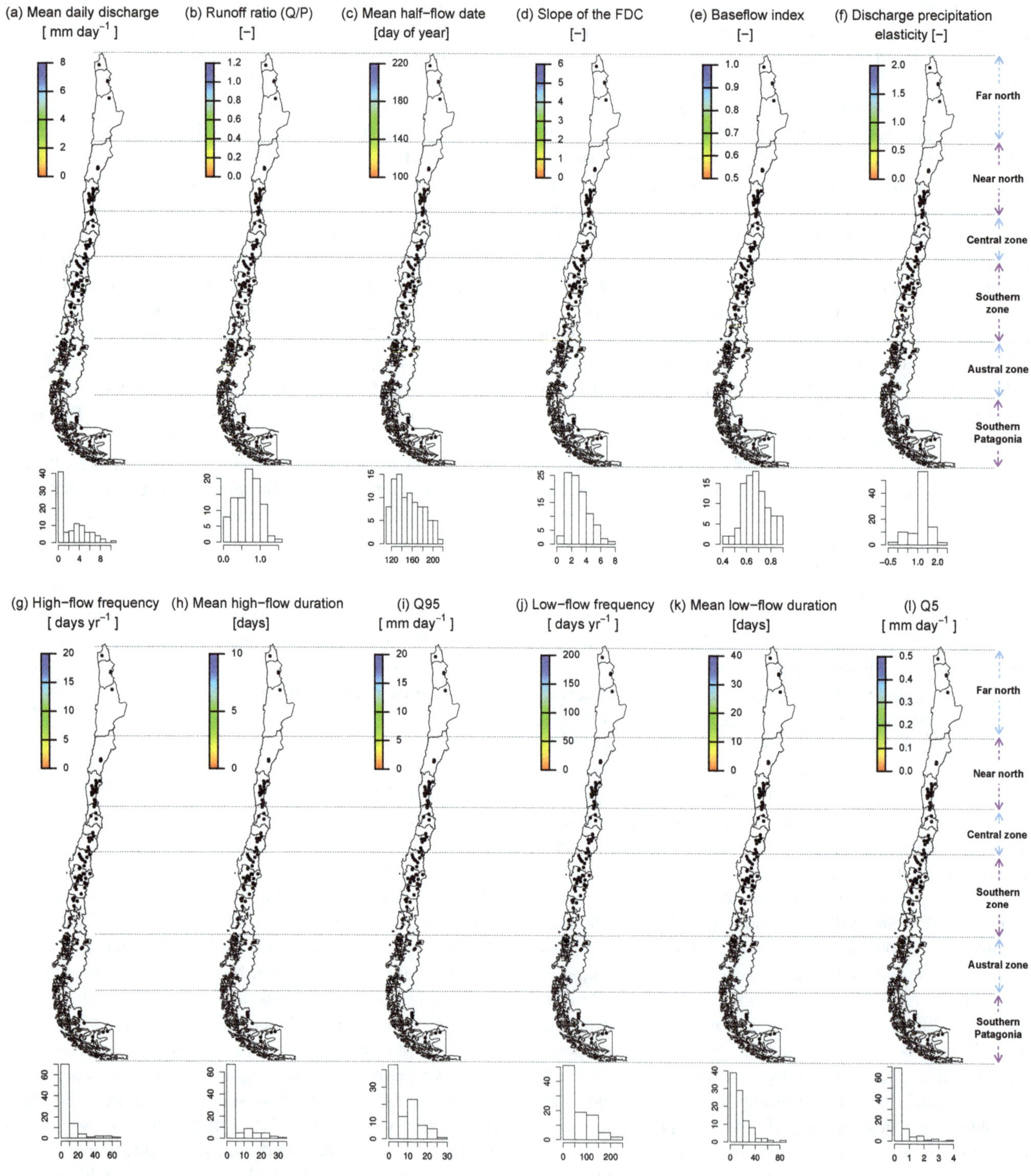

Figure 10. Hydrologic signatures for 94 near-natural catchments. The histograms indicate the number of catchments (out of 94) in each bin.

ered for the allocated flow calculation); (ii) share rights are disregarded; (iii) there is missing information, and therefore some rights may be omitted (Sect. 3.1.10). On the other hand, allocation estimates may differ considerably from the actual extraction within a catchment. Possible reasons for this are the sub/over use of a granted allocated flow and unauthorised extractions of surface and groundwater.

Despite the limitations of the water use dataset and the attributes presented in Fig. 11, water rights information is critical to quantify human intervention, and it has not been offi-

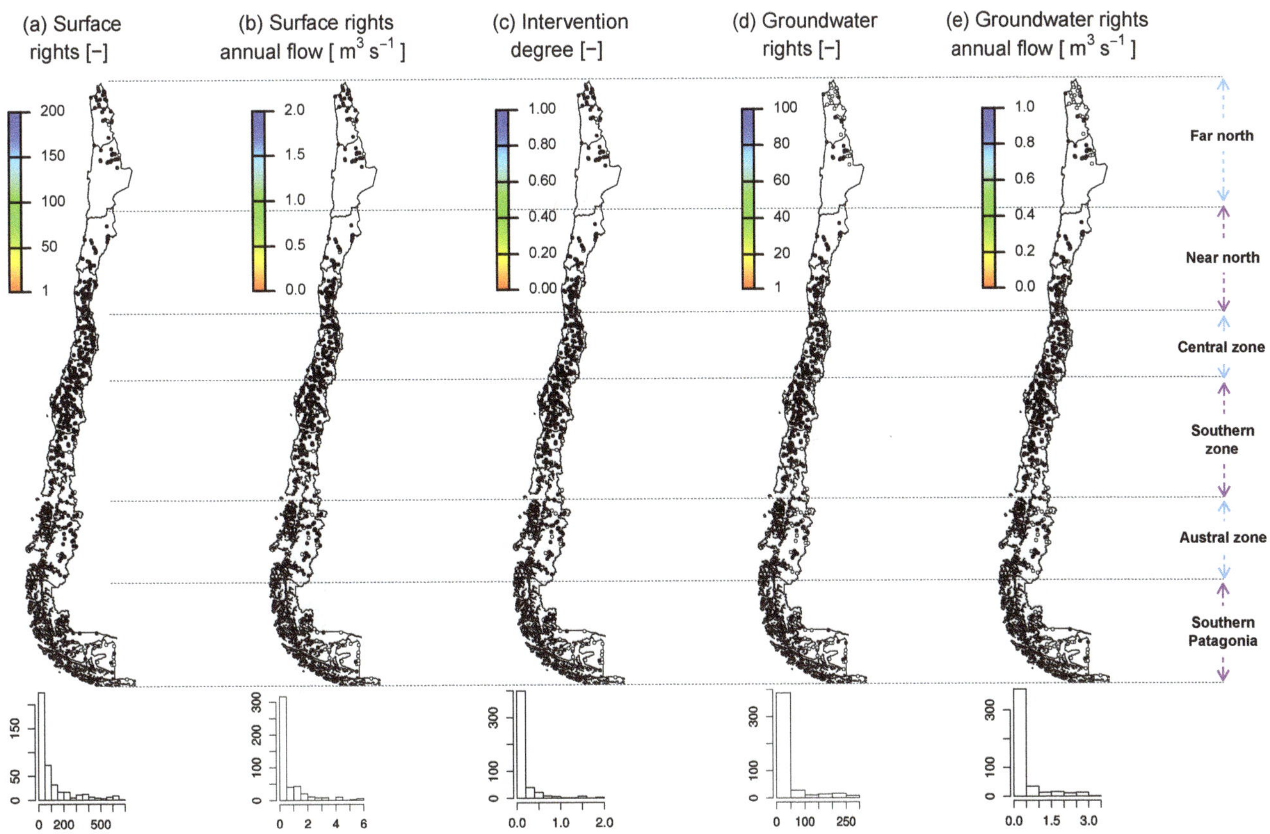

Figure 11. Water rights attributes. For visualisation purposes, the attributes in **(a, b, d, e)** are shown up to their 90th percentile. Attributes below the lower colour bar value are blank, and above the upper colour bar value they are blue. The histograms indicate the number of catchments (out of 516) in each bin.

cially processed at the catchment scale in Chile. To quantify the intervention degree within a catchment, we calculated the interv_degree attribute (described in Table 3 and illustrated in Fig. 11c) as the ratio between the annual surface flow allocated within a catchment and the catchment mean annual runoff. This attribute indicates how much of the annual runoff generated – on average – within a catchment corresponds to the water volume allocated as consumptive surface rights. Further, we defined a binary attribute to characterise the presence of reservoirs within a catchment (large_dam in Table 3), using 0 if there are no dams and 1 if there is at least one dam. A limitation of these human intervention attributes (interv_degree and large_dam) that should be considered for hydrological applications is that they do not incorporate information about groundwater extractions.

To quantify the urbanised fraction of a catchment – another important factor modulating catchment response – we used the impervious fractional area attribute (imp_frac in Table 3), which usually contains urban areas. However, this land cover type is the worst classified class, since urban areas have mixed pixels of vegetation and paved surfaces (Zhao et al., 2016). The urban fraction of the catchments (assumed to be equal to imp_frac) varied between 0 % and 7 % for most catchments (only one catchment had imp_frac = 25 %).

4 Uncertainty in precipitation and PET

4.1 Precipitation assessment

To assess precipitation uncertainty, we looked at the interproduct differences across the study domain. To this end, we defined a precipitation spread attribute (p_mean_spread, Table 3) as the standard deviation of basin-averaged mean annual precipitation from the four different products, normalised by a multi-product mean. To allow such intercomparison, we used data from the concurrent period 1998–2014 (Table 2) and excluded catchments located south of 50° S (since CHIRPS and TMPA cover up to 50° S). Given the different nature of the assessed precipitation products, the spread attribute can be interpreted as a measure of precipitation uncertainty. The underlying assumption is that similar values from different data sources indicate regions with higher confidence in precipitation estimates.

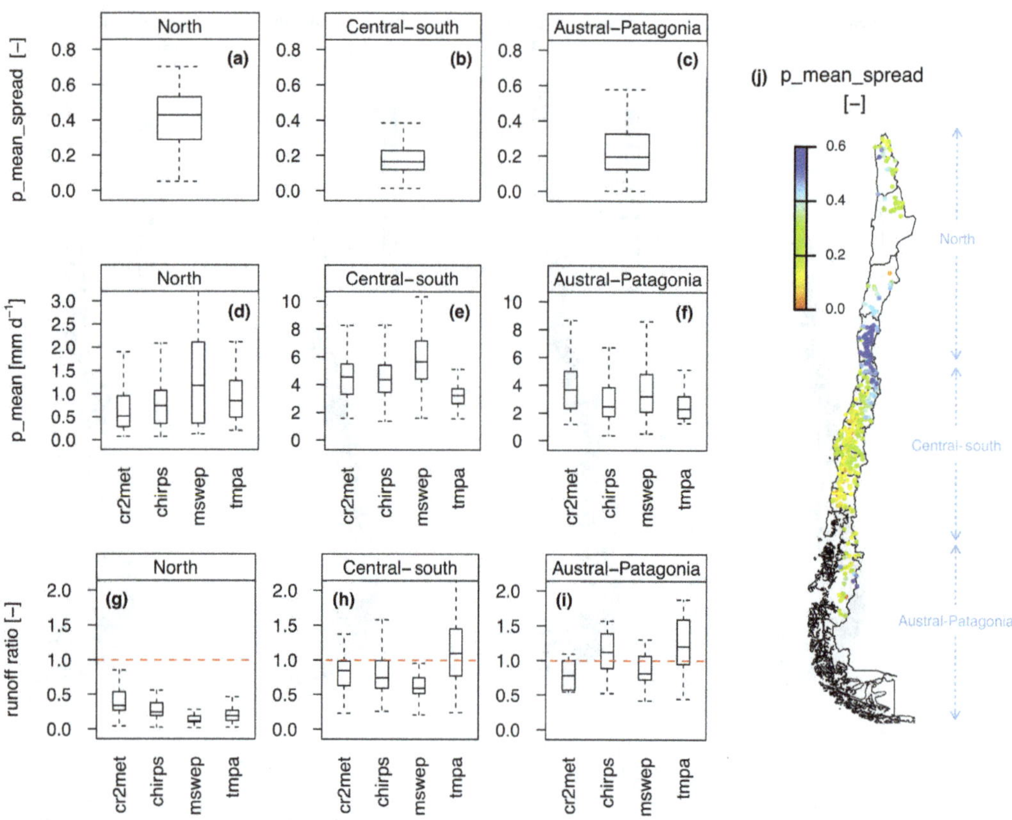

Figure 12. Precipitation spread (p_mean_spread in **a–c**), mean annual precipitation **(d–f)**, and runoff ratio **(g–i)** for the different precipitation products. The domain was divided into three main regions: North (north of 34° S); Central-South; and Austral-Patagonia. **(j)** shows the spatial distribution of p_mean_spread in these sub-regions.

Figure 12 displays catchment-scale mean precipitation and the precipitation spread index for three macro-regions: North (north of 34° S), which includes the Far North and Near North macro-zones; Central-South (between 34 and 43° S); and Austral-Patagonia (south of 43° S). Mean precipitation estimates (p_mean) have a larger spread in the North (Fig. 12a), indicating larger uncertainties in this domain. We attribute these higher relative errors to methodological challenges for detecting events and estimating their intensities in this arid sub-domain, where the occurrence of precipitation events is relatively rare (note the different scale used for p_mean in Fig. 12d). By contrast, considerably larger precipitation amounts (Fig. 12e and f) and lower spread values (Fig. 12b and c) are obtained in Central-South and Austral-Patagonia, which is expected given the relation of p_mean_spread to precipitation mean values. On the other hand, if we look at absolute inter-product differences (Fig. 12d–f), the Central-South region features the largest standard deviation in basin-averaged mean annual precipitation from the four different products (median value of 0.80 mm day^{-1}, compared to 0.29 and 0.51 mm day^{-1} in the North and Austral-Patagonia regions, respectively). This is expected given the larger mean annual values over this humid region.

Although the effects of large precipitation uncertainty on streamflow modelling in the north are not straightforward to determine, some insights can be gained from our analyses. First, surface runoff is not very sensitive to variations in precipitation (i.e. small runoff elasticity values in Fig. 10f), suggesting a weak propagation of precipitation errors by hydrological models. Second, groundwater has the largest contribution to streamflow in this domain (the largest baseflow indices in Fig. 10e and sedimentary rocks as the most common geologic class illustrated in Fig. 7a), especially in the presence of Andean peatlands (represented by the wet_frac land cover attribute). This highlights the need to pursue a realistic representation of groundwater mechanisms in numerical models. Additionally, aquifer boundaries may be quite different from surface catchment boundaries, and therefore accurate delineations are needed to ensure a good representation of surface–groundwater interactions (e.g. Sar et al., 2015; Arkoprovo et al., 2012; Ivkovic et al., 2009).

The ensemble spread of precipitation estimates is a measure of disagreement among the various products rather than a measure of accuracy, which should be quantified using

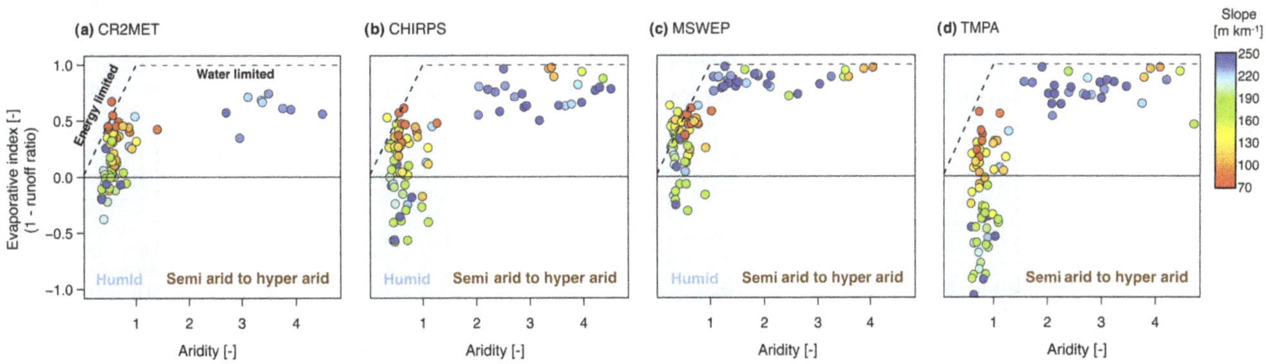

Figure 13. Water balance for 94 near-natural catchments, illustrated in a Budyko scheme for CR2MET **(a)**, CHIRPS **(b)**, MSWEP **(c)**, and TMPA **(d)**. Markers are coloured by the catchment mean slope.

ground observations (e.g. Zambrano-Bigiarini et al., 2017). Such analysis is beyond the scope of this paper, since the assessment of different precipitation products at the basin scale is typically conducted by forcing one or more hydrological models with the different precipitation datasets over the selected study area (e.g. Bisselink et al., 2016; Thiemig et al., 2013; Su et al., 2008).

As an alternative to the model-based approach, we examined the consistency of catchment precipitation estimates based on long-term runoff ratios in 94 near-natural catchments (Fig. 12g–i), selected following the criteria presented in Sect. 3.2.5. Although there are large inter-product differences in runoff ratios in the north (consistent with large p_mean_spread values in Fig. 12a), relatively low runoff ratio values (< 0.4) are obtained, as expected given the arid and semi-arid conditions in this region. By contrast, there are catchments with runoff ratios larger than 1 in central–southern and Austral-Patagonia, indicating that there is more water leaving the catchment than the total amount entering as precipitation. Assuming that streamflow data and catchment area are reliable, and that changes in storage and groundwater contributions are negligible, such cases indicate precipitation underestimation by the various products. In the central–southern (Austral-Patagonia) region, the MSWEP (CR2MET) dataset provides 8 % (20 %) of catchment runoff ratios > 1 – i.e. the smallest among all products. In both domains, the TMPA dataset provides the largest fraction of catchments with runoff ratios > 1 (54 % in central–southern and 70 % in Austral-Patagonia). Such underestimation of TMPA, as well as other satellite precipitation estimates, was also reported by Hobouchian et al. (2017) and Zambrano-Bigiarini et al. (2017).

To further explore differences and systematic biases within the assessed products, we used the Budyko framework (Budyko, 1971) to diagnose the factors affecting the quality of the precipitation datasets. This framework links climate to catchment runoff and evapotranspiration in a simple and easy to interpret visualisation. Figure 13 shows the evaporative index (EI, the ratio of mean annual evapotranspiration to the mean annual precipitation), estimated as 1 minus the runoff ratio (i.e. assuming that changes in storage and groundwater contributions are negligible; Sposito, 2017), as a function of the aridity index for the 94 near-natural catchments over the period 1998–2014. Figure 13 illustrates how the evapotranspiration and runoff rates within this highly diverse set of catchments are governed by the available energy and precipitation; e.g. – for a given amount of precipitation – runoff exceeds evapotranspiration when the available energy and PET are relatively low (points below the energy-limited line in humid regions).

Negative EI values in Fig. 13 represent non-behavioural combinations of precipitation and runoff (Berghuijs et al., 2017). Under the assumption that precipitation estimates represent a relatively larger source of uncertainty compared to runoff, all the points with $\mathrm{EI} < 0$ indicate those catchments where mean annual precipitation is underestimated (i.e. runoff ratios > 1). Therefore, Fig. 13a–d indicate that all precipitation products systematically fail in humid catchments with steep topography (slopes greater than $150\,\mathrm{m\,km^{-1}}$), in agreement with the limitations reported for different satellite precipitation products over the same domain (Hobouchian et al., 2017). The systematic precipitation underestimation can be attributed to the complex topography of headwater catchments and the scarcity of ground stations at high elevations. In fact, 90 % of the 500 rain gauges located south of 34° S are placed below 1000 m a.s.l. These precipitation errors pose challenges for hydrological applications, and the best strategy to address this will depend on the application. It should be noted that 57 catchment attributes (out of 70) and six hydro-meteorological time series (streamflow, min, max and mean temperature, PET, and SWE) do not rely on precipitation. Therefore, some applications may not need to directly address this limitation (e.g. catchment characterisation, water allocation analysis, classification studies not using similarity of climatic indices). Nevertheless, other applications (e.g. hydrological modelling) will have to deal with the uncertainties of this forcing variable. There are strategies for this, including setting up a parameter calibration scheme flexible enough to compensate for errors in meteorological input data (Elsner et al., 2014), or trying to correct precipitation estimates based on streamflow (Henn et al., 2015) and snow observations (Henn et al., 2016).

4.2 PET assessment

To assess the quality of the PET products described in Sect. 3.1.8, we used a different approach than in Sect. 4.1, since a basin-scale PET estimation cannot be evaluated based on observed streamflow. In this case, the evaluation was made with an independent set of PET data, calculated with daily observations for the period 2010–2014 from 75 meteorological stations maintained by the Chilean National Institute of Agricultural Research (INIA, 2017). For each site, we calculated two PET time series: (i) a daily time series obtained through the Hargreaves formulae (Hargreaves and Allen, 2003) fed with INIA temperature observations, called $\mathrm{INIA_{har}}$ hereafter, and (ii) an 8-day accumulated time series based on the FAO Penman–Monteith reference crop evapotranspiration (Allen et al., 1998), called $\mathrm{INIA_{ET_0}}$. $\mathrm{INIA_{har}}$ and $\mathrm{INIA_{ET_0}}$ were used to evaluate the corresponding pixels of $\mathrm{PET_{har}}$ and $\mathrm{PET_{mod}}$, respectively.

The evaluation metrics used in these comparisons, spatially averaged within the macro-zones, are summarised in Table 4. These results indicate good agreement between $\mathrm{PET_{har}}$ and $\mathrm{INIA_{har}}$, with correlation coefficients (r) greater than 0.92 throughout the national territory, except in the Far North macro-zone, where we found a weaker correlation ($r = 0.76$). The ratios between mean $\mathrm{PET_{har}}$ and mean $\mathrm{INIA_{har}}$ indicate that $\mathrm{PET_{har}}$ underestimates (overestimates) by up to 8 % the observation-based PET in the Far and Near North arid regions (the Southern and Austral zones' humid regions).

The comparison between $\mathrm{PET_{mod}}$ and $\mathrm{INIA_{ET_0}}$ led to r values greater than 0.80 within the domain, excepting the Far North, where the correlation was below 0.20 in all available stations. The ratios between $\mathrm{PET_{mod}}$ and $\mathrm{INIA_{ET_0}}$ means indicate that the first one systematically overestimates station-based calculations, which was also found by Westerhoff (2015). Such systematic biases may be explained by the theoretical differences between $\mathrm{INIA_{ET_0}}$ and $\mathrm{PET_{mod}}$ calculated in MOD16. $\mathrm{INIA_{ET_0}}$ represents a potential condition for a regular crop height of 0.12 m and a fixed surface resistance and albedo, which is not the case for $\mathrm{PET_{mod}}$, which includes a more complete parameterisation of those variables based on vegetation characteristics. Further, $\mathrm{INIA_{ET_0}}$ uses local meteorological observations, while $\mathrm{PET_{mod}}$ uses global sources that may not capture meteorological variations at the local scale. If an application requires it (e.g. irrigation or hydrological modelling applications), the reported $\mathrm{PET_{mod}}$ biases can be corrected with conventional statistical methods (e.g. Maraun and Widmann, 2018). Other spatio-temporal analyses (e.g. drought monitoring) may directly apply $\mathrm{PET_{mod}}$ due to its high correlation with ground $\mathrm{ET_0}$ estimates.

Table 4. Evaluation metrics of PET gridded products ($\mathrm{PET_{har}}$ and $\mathrm{PET_{mod}}$). Pearson correlation coefficients (r) and the ratios between gridded PET and observation-based PET ($\mathrm{INIA_{har}}$ and $\mathrm{INIA_{ET_0}}$) were spatially averaged within the macro-zones.

	$\mathrm{PET_{har}}$ compared with $\mathrm{INIA_{har}}$		$\mathrm{PET_{mod}}$ compared with $\mathrm{INIA_{ET_0}}$	
Macro-zone	r	ratio	r	ratio
Far North	0.76	0.96	0.19	1.66
Near North	0.92	0.92	0.91	1.68
Central Zone	0.95	1.00	0.96	1.79
Southern Zone	0.97	1.07	0.96	1.58
Austral Zone	0.97	1.08	0.95	1.14
Southern Patagonia	0.98	0.93	0.96	1.06

Since INIA records were used differently for evaluating $\mathrm{PET_{har}}$ and $\mathrm{PET_{mod}}$, a direct comparison between both assessments is not possible, although they provide valuable information about the quality of PET products across the territory. Furthermore, the formulations behind the two gridded products have different trade-offs. $\mathrm{PET_{mod}}$ is based on the Penman–Monteith equation that solves the surface energy balance, including parameters such as albedo and FPAR/LAI, whereas $\mathrm{PET_{har}}$ is calculated from an empirical approach based only on air temperature. $\mathrm{PET_{har}}$ has a coarser spatial resolution ($5 \times 5\,\mathrm{km}^2$) compared to $\mathrm{PET_{mod}}$ ($\times 1\,\mathrm{km}^2$), which may induce larger errors over complex topography (e.g. mountain catchments) due to the local variations in potential evapotranspiration, with changes in slope and aspect. On the other hand, $\mathrm{PET_{har}}$ covers a longer period (1979–2016, the same as $T_{\min}$ and $T_{\max}$ from Sect. 3.1.7) compared to $\mathrm{PET_{mod}}$ (2000–2014), which is more suitable for characterising climatic trends.

5 Impacts of human intervention on catchment behaviour

Large sample hydrology is a suitable framework to explore anthropic impacts on catchment behaviour through comparative analysis over a broad range of hydroclimatic conditions and catchment characteristics. Such assessment is critical when addressing the question of how climate change will affect global water supply (Vörösmarty et al., 2007). However, it remains unclear how to generalise the results from different studies. For example, Poff et al. (2006) examined the effects of land use on hydrological regimes (e.g. peak and low flows, runoff variability) in 158 basins within the CONUS, finding region-dependent changes in specific metrics. Ochoa-Tocachi et al. (2016) analysed the impacts of land use on the hydrology of 25 Andean catchments, finding that anthro-

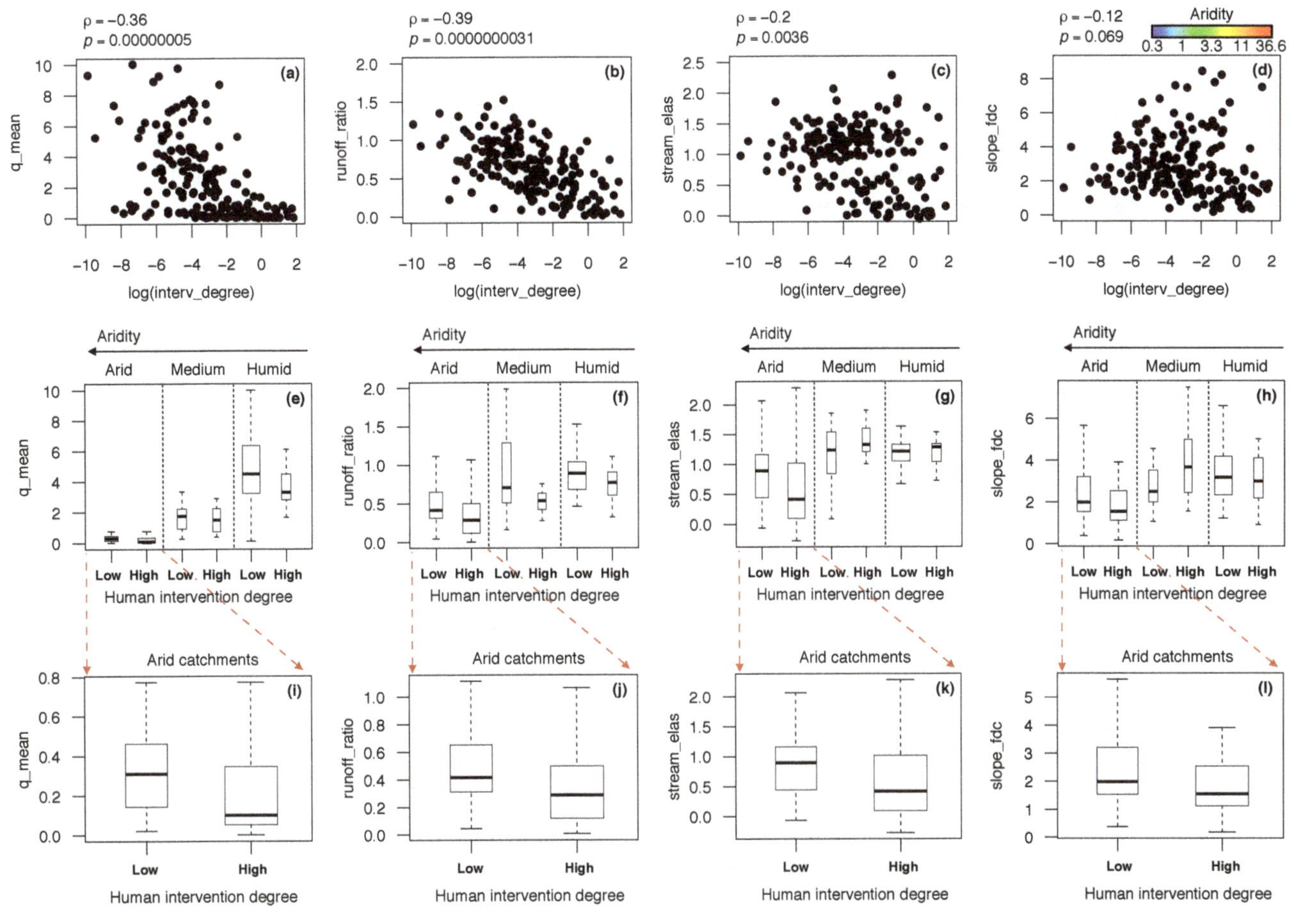

Figure 14. (a)–(d) present the relation between four hydrological signatures and the log-transformed human intervention degree ("interv_degree" from Table 3). The Spearman rank correlation coefficients and their p values at 95 % confidence are shown in each plot. The colour corresponds to the aridity index. (e)–(h) show the boxplots (box widths are proportional to the number of catchments in each box) of the hydrological signatures for the catchments classified by their aridity (humid: aridity below 0.8, medium: aridity between 0.8 and 1.5, and arid: aridity above 1.5) and by their human intervention degree (low: interv_degree below 5 %, and high: interv_degree greater than 5 %). (i)–(l) present a zoomed view of the arid catchments.

pogenic influences propagate towards increased streamflow variability and decreased catchment regulation capacity and water yield. More recently, Tijdeman et al. (2018) examined the effects of human intervention on streamflow drought characteristics across 187 catchments in England and Wales, concluding that most human-influenced catchments did not have drought characteristics different from those expected for near-natural conditions.

In this work, we used hydrological signatures to describe catchment behaviour and the interv_degree attribute (Sect. 3.2.6) to characterise the level of human intervention. Figure 14a–d display scatter plots between four hydrological signatures and the logarithm of the intervention degree index – which accounts for consumptive, continuous surface water rights. Different colours indicate the aridity of each catchment, which is a major driver of hydrological behaviour (as shown in Fig. 13). These plots show that larger human intervention is associated with decreased annual flows and runoff ratios, especially in drier catchments. Interestingly, a larger number of consumptive surface rights (larger interv_degree values) is reflected in decreased elasticity of runoff with respect to precipitation, supported by low p values. Since these scatter plots do not allow one to separate the effects of aridity and human intervention, we binned the data and used boxplots to disentangle such effects. Figure 14e–h show the boxplots with these hydrological signatures for the classified catchments, binned according to their aridity (humid: aridity below 0.8, medium: aridity between 0.8 and 1.5, and arid: aridity above 1.5) and their degree of human intervention. Catchments with low (high) intervention were defined by the interv_degree values lower (greater) than 5 % and the large_dam attribute equal to 0 (1).

The dispersion in hydrological signatures among wet and medium catchments is large, and no significant difference is found between catchments with high and low human intervention. In contrast, dry catchments (zoomed view in

Fig. 14i–l) reveal differences in hydrological signatures for high and low human intervention (i.e. median values in highly disturbed catchments are below the first quartile of low intervention catchments). In agreement with the scatter plots, the annual flows (Fig. 14i) and runoff ratios (Fig. 14j) decrease in catchments with larger numbers of consumptive surface rights, which is expected due to withdrawals of water within water-scarce regions. Further, highly disturbed arid catchments feature lower runoff sensitivities to precipitation compared to less disturbed ones (Fig. 14k), which could be attributed to altered runoff generation mechanisms associated with water withdrawals and reservoirs. Figure 14l shows that there is likely less variation in daily runoff – represented by the mid-segment slope of the flow duration curves – within highly disturbed arid catchments. The results found in arid catchments (i.e. water-scarce regions) provide new evidence of the potential impacts of human intervention on water supply. However, further research is needed to assess the causality of the correlations found here.

6 Concluding remarks

The CAMELS-CL dataset presented here provides novel information at the catchment scale in Chile, within a region that is largely underrepresented in large sample studies. CAMELS-CL includes daily streamflow data and a suite of basin-averaged hydro-meteorological variables, including precipitation, temperature, potential evapotranspiration, and snow water equivalent, for 516 catchments in the country. The dataset also includes shapefiles of drainage area boundaries related to streamflow gauge locations, overcoming the main limitations of the official national hydrographic network (DGA and CIREN, 2014). Further, we synthesised diverse and complementary datasets to compute 70 catchment attributes describing topography, geology, land cover, climate, hydrology, and anthropic intervention.

We described the advantages and main limitations of the datasets used to derive the various catchment attributes, which should be considered when using CAMELS-CL for selecting catchments and interpreting results. The main spatial patterns of catchment attributes and their inter-relationships were analysed across the entire domain (4300 km), which includes high-altitude catchments and five different primary climatic regimes. The main conclusions of this analysis are as follows:

- the Andes position along the country largely explains the elevation and slope gradients from the 516 catchments,
- there is high geological variability between neighbouring catchments and within catchments,
- there is larger heterogeneity in land cover attributes towards southern regions,
- 48 % of the catchments have some degree of glacierisation,
- the climatic attributes show a marked north to south precipitation gradient, combined with an orographic enhancement over the windward slope of the Andes Cordillera,
- hydrological signatures reflect the leading patterns of catchment hydrologic responses, with strong correlations between runoff (daily flows and runoff ratios) and mean annual precipitation and aridity index,
- there are substantial subsurface and groundwater contributions to total runoff in the northern regions (Far North and Near North), and
- hydrological signatures feature, in general, sharper variations compared to patterns in climate indices, which is due to complex process interactions across a range of spatiotemporal scales, enhanced by heterogeneities in topography, soils, vegetation, geology and other landscape properties, and data errors (as also observed in the USA by Addor et al., 2018).

The CAMELS-CL dataset was further used to assess hydrometeorological biases and uncertainties in a large ensemble of watersheds in Chile, based on the comparison of various precipitation products – one national (CR2MET) and three widely used global products (CHIRPS, MSWEP and TMPA). Large discrepancies between products were detected in arid regions, which are explained by the methodological challenges associated with the rare occurrence of precipitation events in this region. Based on a water balance analysis using Budyko curves, we found systematic precipitation underestimation in headwater mountain catchments (high elevations and steep slopes) over humid regions. For these topographic characteristics and climatic conditions, all products failed to provide precipitation estimates that closed the water balance, with the TMPA product featuring the largest errors – in agreement with previous studies over the same domain (Hobouchian et al., 2017; Zambrano-Bigiarini et al., 2017). Such errors were attributed to the complex topography of headwater catchments and the scarcity of ground stations at high elevations (90 % of rain gauges located south of 34° S are placed below 1000 m a.s.l.). These limitations restrict our understanding of hydrological processes, posing challenges for streamflow modelling, water management and allocation. To alleviate these constraints, efforts should be put into improving the surface monitoring network at high elevations (> 1000 m a.s.l.). This would help to obtain more accurate remotely sensed and model-based precipitation estimates in complex terrains. To deal with these precipitation errors in hydrological applications, we suggest exploring suitable strategies based on application requirements and using our findings to better interpret results. Further, we assessed the PET_{har} and PET_{mod} products with an independent observation-based PET. In general, both products showed good correlations with the observation-based PET for the complete domain, except in the Far North arid region. Regarding mean biases, PET_{har} showed slight underestimation (overestimation) in the Far and Near North arid regions (the Southern and Austral zones' humid regions) compared with the observation-based PET. PET_{mod} on the other hand showed a systematic and larger overestimation of the observation-based PET within the complete domain, which was attributed to the theoretical differences between their formulations.

The assessment of precipitation and PET products showed different performances within the domain. Therefore, the choice of these products or similar datasets must be carefully made based on specific application and study requirements.

Finally, we used CAMELS-CL to explore the impact of human activities on catchment behaviour. We showed that larger human intervention is correlated with lower than normal annual flows, runoff ratios, elasticity of runoff with respect to precipitation, and flashiness of runoff, especially in drier catchments. These results not only illustrate how catchment behaviour can change with human intervention, but also reveal the potential of anthropic indices to explain shifts in hydrological systems.

In summary, this paper contributes to hydrological sciences by (i) providing a unique dataset that can be used to advance our understanding of hydrological systems by learning from diversity, (ii) analysing the dominant spatial patterns of physical, climatic and hydrological attributes within the domain, (iii) assessing the quality of one national and three global precipitation datasets based on the observed water balance, (iv) assessing two PET products based on an independent set of PET point values calculated from meteorological records, and (v) examining the interplay between human intervention and changes in observed catchment response.

CAMELS-CL can be used to address research questions related to catchment classification, similarity and regionalisation, model parameter estimation, dominant controls on runoff generation, the impacts of different land cover types on catchment response, characterisation of drought history and projections, and climate change impacts on hydrological processes. CAMELS-CL will be continuously updated to incorporate new records and new datasets, which may include soil characteristics, water quality, seismology records, socio-economic indices and energy generation data. Additionally, new and more detailed information about the Chilean cryosphere will be included, complementing the global inventory processed here with national inventories of Chile and Argentina. Time series of streamflow, meteorological variables, and all catchment attributes described in this paper can be visualised from the CAMELS-CL explorer.

Data availability. ASTER GDEM elevation data were downloaded from the NASA-JPL website (https://asterweb.jpl.nasa.gov/gdem.asp; ASTER GDEM, 2016). Geological data (GLiM) were downloaded from the PANGAEA database (https://doi.pangaea.de/10.1594/PANGAEA.788537; GLiM, 2017). The land cover map was downloaded from the Universidad de Chile website (http://www.gep.uchile.cl/Landcover_CHILE.html; Land cover map, 2017). Glaciers inventory RGI 6.0 was downloaded from the Global Land Ice Measurements from Space initiative (https://www.glims.org/RGI; Glaciers inventory, 2017). The daily streamflow records were obtained from the CR2 website (http://www.cr2.cl/datos-de-caudales; Streamflow, 2018). CR2MET precipitation and temperature products were downloaded from the CR2 website (http://www.cr2.cl/datos-productos-grillados; Precipitation product CR2MET, 2018). CHIRPS, TMPA and MSWEP precipitation products are publicly available and the datasets clipped within the Chilean domain were downloaded from the CR2 website (http://www.cr2.cl/datos-precipitacion-satelital; Satellite precipitation products CHIRPS, MSWEP, TMPA, 2018). MODIS data were downloaded from the USGS website (https://e4ftl01.cr.usgs.gov/MOLT; MODIS, 2017). SWE data were provided by Gonzalo Cortes (Cortés et al., 2016). Water rights were downloaded from the DGA website (http://www.dga.cl/productosyservicios/derechos_historicos; Water rights, 2018). Dam location was downloaded from the Ministerio de Bienes Nacional website (http://www.ide.cl/descarga/capas/item/embalses-2016.html, Dam location, 2017). The time series processed at the basin scale and the catchment attributes introduced in this paper are freely available from https://doi.pangaea.de/10.1594/PANGAEA.894885 and the CAMELS-CL explorer (http://camels.cr2.cl).

Author contributions. This research was conceived by CAG, PAM and JPB. CAG processed the catchment attributes and designed and wrote the manuscript with input from all co-authors. PAM performed the analysis of hydrological signatures. JPB developed and processed the CR2MET products. JPB and RG performed the analysis of climatic indices. NA provided R codes for computing climatic indices and hydrological signatures, and performed the analysis on geological attributes. MG and CP performed the analysis on PET. MZB processed the satellite precipitation products. AL contributed with the analysis of land cover attributes. GC processed the SWE data. JM contributed ideas for analyses. AA processed the glacier data. All the authors have been involved in interpreting the results, discussing the findings, and editing the paper.

Competing interests. The authors declare that they have no conflict of interest.

Acknowledgements. This research emerged from the collaboration with many colleagues at the Center for Climate and Resilience Research (CR2, CONICYT/FONDAP/15110009). Camila Alvarez-Garreton is funded by FONDECYT postdoctoral grant no. 3170428. Pablo Mendoza received additional support from FONDECYT postdoctoral grant no. 3170079. Mauricio Zambrano-Bigiarini thanks FONDECYT 11150861 for financial support. The development of CR2MET was supported by the Chilean Water Directorate (DGA), through National Water Balance Updating Project DGA-2319, and by FONDECYT grant no. 3150492. This study is a contribution to the Large-sample Hydrology working group of the Panta Rhei Research Initiative of the International Association of Hydrological Sciences (IAHS). We thank Guillermo Tapia from the DGA for his advice on stream-gauge locations.

Edited by: Jan Seibert

References

Addor, N., Newman, A. J., Mizukami, N., and Clark, M. P.: The CAMELS data set: Catchment attributes and meteorology for large-sample studies, Hydrol. Earth Syst. Sci., 21, 5293–5313, https://doi.org/10.5194/hess-21-5293-2017, 2017.

Addor, N., Nearing, G., Prieto, C., Newman, A. J., Le Vine, N., and Clark, M. P.: A ranking of hydrological signatures based on their predictability in space, Water Resour. Res., 54, https://doi.org/10.1029/2018WR022606, 2018.

Adler, R. F., Huffman, G. J., Chang, A., Ferraro, R., Xie, P.-P., Janowiak, J., Rudolf, B., Schneider, U., Curtis, S., Bolvin, D., Gruber, A., Susskind, J., Arkin, P., and Nelkin, E.: The Version-2 Global Precipitation Climatology Project (GPCP) Monthly Precipitation Analysis (1979–Present), J. Hydrometeorol., 4, 1147–1167, https://doi.org/10.1175/1525-7541(2003)004<1147:TVGPCP>2.0.CO;2, 2003.

Allen, R. G., Pereira, L. S., Raes, D., and Smith, M.: FAO Penman–Monteith Equation, in: Crop Evapotranspiration: Guidelines for Computing Crop Water Requirements, FAO – Food and Agriculture Organization of the United Nations, Rome, 17–28, 1998.

Allmendinger, R. W., Jordan, T. E., Kay, S. M., and Isacks, B. L.: The evolution of the Altiplano-Puna Plateau of the Central Andes, Annu. Rev. Earth Planet. Sci., 25, 139–174, https://doi.org/10.1146/annurev.earth.25.1.139, 1997.

Andréassian, V., Hall, A., Chahinian, N., and Schaake, J.: Introduction and Synthesis: Why should hydrologists work on a large number of basin data sets?, IAHS-AISH Publ., 307, 1–6, 2006.

Arkoprovo, B., Adarsa, J., and Prakash, S. S.: Delineation of Groundwater Potential Zones using Satellite Remote Sensing and Geographic Information System Techniques: A Case study from Ganjam district, Orissa, India, Res. J. Recent Sci., 1, 59–66, 2012.

Armesto, J. J., Manuschevich, D., Mora, A., Smith-Ramirez, C., Rozzi, R., Abarzúa, A. M., and Marquet, P. A.: From the Holocene to the Anthropocene: A historical framework for land cover change in southwestern South America in the past 15,000 years, Land Use Policy, 27, 148–160, https://doi.org/10.1016/j.landusepol.2009.07.006, 2010.

ASTER GDEM: https://asterweb.jpl.nasa.gov/gdem.asp, last access: December 2016.

Balsamo, G., Albergel, C., Beljaars, A., Boussetta, S., Brun, E., Cloke, H., Dee, D., Dutra, E., Munõz-Sabater, J., Pappenberger, F., De Rosnay, P., Stockdale, T., and Vitart, F.: ERA-Interim/Land: A global land surface reanalysis data set, Hydrol. Earth Syst. Sci., 19, 389–407, https://doi.org/10.5194/hess-19-389-2015, 2015.

Beck, H. E., Van Dijk, A. I. J. M., Levizzani, V., Schellekens, J., Miralles, D. G., Martens, B., and De Roo, A.: MSWEP: 3-hourly 0.25° global gridded precipitation (1979–2015) by merging gauge, satellite, and reanalysis data, Hydrol. Earth Syst. Sci., 21, 589–615, https://doi.org/10.5194/hess-21-589-2017, 2017.

Berghuijs, W. R., Sivapalan, M., Woods, R. A., and Savenije, H. H. G.: Patterns of similarity of seasonal water balances: A window into streamflow variability over a range of time scales, Water Resour. Res., 50, 5638–5661, https://doi.org/10.1002/2014WR015692, 2014.

Berghuijs, W. R., Larsen, J. R., van Emmerik, T. H. M., and Woods, R. A.: A Global Assessment of Runoff Sensitivity to Changes in Precipitation, Potential Evaporation, and Other Factors, Water Resour. Res., 53, 8475–8486, https://doi.org/10.1002/2017WR021593, 2017.

Bisselink, B., Zambrano-Bigiarini, M., Burek, P., and de Roo, A.: Assessing the role of uncertain precipitation estimates on the robustness of hydrological model parameters under highly variable climate conditions, J. Hydrol. Reg. Stud., 8, 112–129, https://doi.org/10.1016/j.ejrh.2016.09.003, 2016.

Blöschl, G., Sivapalan, M., Wagener, T., Viglione, A., and Savenije, H.: Runoff Prediction in Ungauged Basins: Synthesis Across Processes, Places and Scales, Cambridge University Press, Cambridge, 2013.

Budyko, M. I.: Climate and Life, Academic Press, London, 1971.

Carey: General Overview of Water Rights in Chile, available at: http://reformacodigodeaguas.carey.cl/wp-content/uploads/2014/09/general-explanation-of-the-chilean-water (last access: May 2018), 2014.

Cortés, G. and Margulis, S.: Impacts of El Niño and La Niña on interannual snow accumulation in the Andes: Results from a high-resolution 31 year reanalysis, Geophys. Res. Lett., 44, 6859–6867, https://doi.org/10.1002/2017GL073826, 2017.

Cortés, G., Girotto, M., and Margulis, S. A.: Analysis of sub-pixel snow and ice extent over the extratropical Andes using spectral unmixing of historical Landsat imagery, Remote Sens. Environ., 141, 64–78, https://doi.org/10.1016/j.rse.2013.10.023, 2014.

Cortés, G., Girotto, M., and Margulis, S.: Snow process estimation over the extratropical Andes using a data assimilation framework integrating MERRA data and Landsat imagery, Water Resour. Res., 52, 2582–2600, https://doi.org/10.1002/2015WR018376, 2016.

Dam location: http://www.ide.cl/descarga/capas/item/embalses-2016.html, last access: September 2017.

DGA: Mapa Hidrogeológico de Chile, Santiago, Chile, 1986.

DGA: Glaciares de chile, Santiago, Chile, 2014.

DGA: Atlas del Agua – Chile 2016, in: Capítulo 1: Chile en el mundo, Atlas del Agua Chile 2016, Santiago, Chile, 24 pp., 2016a.

DGA: Atlas del Agua – Chile 2016, in Capítulo 4: Gestion del agua, Atlas del Agua Chile 2016, Santiago, Chile, 30 pp., 2016b.

DGA: Actualización del Balance Hídrico Nacional, SIT No. 417, Ministerio de Obras Públicas, Dirección General de Aguas, División de Estudios y Planificación, Santiago, Chile, Realizado por: Universidad de Chile & Pontificia Universidad Católica de Chile, 2017.

DGA and CIREN: Redefinición de la clasificación red hidrográfica a nivel Nacional, Santiago, Chile, 2014.

Di Gregorio, A. and Jansen, L. J.: Land Cover Classification System, Classification concepts and user manual, FAO – Food and Agriculture Organization of the United Nations, Rome, 2005.

Ehret, U., Gupta, H. V., Sivapalan, M., Weijs, S. V., Schymanski, S. J., Blöschl, G., Gelfan, A. N., Harman, C., Kleidon, A., Bogaard, T. A., Wang, D., Wagener, T., Scherer, U., Zehe, E., Bierkens, M. F. P., Di Baldassarre, G., Parajka, J., Van Beek, L. P. H., Van Griensven, A., Westhoff, M. C., and Winsemius, H. C.: Advancing catchment hydrology to deal with predictions under change, Hydrol. Earth Syst. Sci., 18, 649–671, https://doi.org/10.5194/hess-18-649-2014, 2014.

Elsner, M. M., Gangopadhyay, S., Pruitt, T., Brekke, L. D., Mizukami, N., and Clark, M. P.: How Does the Choice of Distributed Meteorological Data Affect Hydrologic Model Calibration and Streamflow Simulations?, J. Hydrometeorol., 15, 1384–1403, https://doi.org/10.1175/JHM-D-13-083.1, 2014.

Figueroa, D. and Moffat, C.: On the influence of topography in the induction of coastal upwelling along the Chilean coast, Geophys. Res. Lett., 27, 3905–3908, https://doi.org/10.1029/1999GL011302, 2000.

Friedl, M., McIver, D., Hodges, J. C., Zhang, X., Muchoney, D., Strahler, A., Woodcock, C., Gopal, S., Schneider, A., Cooper, A., Baccini, A., Gao, F., and Schaaf, C.: Global land cover mapping from MODIS: algorithms and early results, Remote Sens. Environ., 83, 287–302, https://doi.org/10.1016/S0034-4257(02)00078-0, 2002.

Funk, C., Peterson, P., Landsfeld, M., Pedreros, D., Verdin, J., Shukla, S., Husak, G., Rowland, J., Harrison, L., Hoell, A., and Michaelsen, J.: The climate hazards infrared precipitation with stations – A new environmental record for monitoring extremes, Sci. Data, 2, 150066, https://doi.org/10.1038/sdata.2015.66, 2015.

Garreaud, R. D.: The Andes climate and weather, Adv. Geosci., 22, 3–11, https://doi.org/10.5194/adgeo-22-3-2009, 2009.

Garreaud, R. D., Alvarez-Garreton, C., Barichivich, J., Boisier, J. P., Christie, D., Galleguillos, M., LeQuesne, C., McPhee, J., and Zambrano-Bigiarini, M.: The 2010–2015 megadrought in central Chile: impacts on regional hydroclimate and vegetation, Hydrol. Earth Syst. Sci., 21, 6307–6327, https://doi.org/10.5194/hess-21-6307-2017, 2017.

Glaciers inventory: https://www.glims.org/RGI, last access: December 2017.

GLiM: https://doi.pangaea.de/10.1594/PANGAEA.788537, last access: May 2017.

Gong, P., Wang, J., Yu, L., Zhao, Y., Zhao, Y., Liang, L., Niu, Z., Huang, X., Fu, H., Liu, S., Li, C., Li, X., Fu, W., Liu, C., Xu, Y., Wang, X., Cheng, Q., Hu, L., Yao, W., Zhang, H., Zhu, P., Zhao, Z., Zhang, H., Zheng, Y., Ji, L., Zhang, Y., Chen, H., Yan, A., Guo, J., Yu, L., Wang, L., Liu, X., Shi, T., Zhu, M., Chen, Y., Yang, G., Tang, P., Xu, B., Giri, C., Clinton, N., Zhu, Z., Chen, J., and Chen, J.: Finer resolution observation and monitoring of global land cover: First mapping results with Landsat TM and ETM+ data, Int. J. Remote Sens., 34, 2607–2654, https://doi.org/10.1080/01431161.2012.748992, 2013.

Google: Google Earth Pro, Google, available at: https://www.google.com/earth/ (last access: March 2018), 2016.

Gupta, H. V., Perrin, C., Blöschl, G., Montanari, A., Kumar, R., Clark, M., and Andréassian, V.: Large-sample hydrology: A need to balance depth with breadth, Hydrol. Earth Syst. Sci., 18, 463–477, https://doi.org/10.5194/hess-18-463-2014, 2014.

Hargreaves, G. H. and Allen, R. G.: History and evaluation of Hargreaves evapotranspiration equation, J. Irrig. Drain. Eng., 129, 53–63, https://doi.org/10.1061/(ASCE)0733-9437(2003)129:1(53), 2003.

Hargreaves, G. H. and Samani, Z. A.: Reference crop evapotranspiration from temperature, Appl. Eng. Agric., 1, 96–99, https://doi.org/10.13031/2013.26773, 1985.

Hartmann, J. and Moosdorf, N.: The new global lithological map database GLiM: A representation of rock properties at the Earth surface, Geochem. Geophy. Geosy., 13, 1–37, https://doi.org/10.1029/2012GC004370, 2012.

Henn, B., Clark, M. P., Kavetski, D., and Lundquist, J. D.: Estimating mountain basin-mean precipitation fromstreamflow using Bayesian inference, Water Resour. Res., 51, 8012–8033, https://doi.org/0.1002/2014WR016736, 2015.

Henn, B., Clark, M. P., Kavetski, D., McGurk, B., Painter, T. H., and Lundquist, J. D.: Combining snow, streamflow, and precipitation gauge observations to infer basin-mean precipitation, Water Resour. Res., 52, 8700–8723, https://doi.org/10.1002/2015WR018564, 2016.

Hijmans, R. J.: Raster: Geographic Data Analysis and modeling, R Packag. version 2.5-8. https//CRAN.R-project.org/package=raster, 1, r948, 2016.

Hobouchian, M. P., Salio, P., García Skabar, Y., Vila, D., and Garreaud, R.: Assessment of satellite precipitation estimates over the slopes of the subtropical Andes, Atmos. Res., 190, 43–54, https://doi.org/10.1016/j.atmosres.2017.02.006, 2017.

Howell, T. and Evett, S. R.: The Penman–Monteith Method, Bushland, Texas, USDA Agric. Res. Serv., Texas, 1–13, 2001.

Huang, C., Newman, A. J., Clark, M. P., Wood, A. W., and Zheng, X.: Evaluation of snow data assimilation using the ensemble Kalman filter for seasonal streamflow prediction in the western United States, Hydrol. Earth Syst. Sci., 21, 635–650, https://doi.org/10.5194/hess-21-635-2017, 2017.

Huffman, G. J., Bolvin, D. T., Nelkin, E. J., Wolff, D. B., Adler, R. F., Gu, G., Hong, Y., Bowman, K. P., and Stocker, E. F.: The TRMM Multisatellite Precipitation Analysis (TMPA): Quasi-Global, Multiyear, Combined-Sensor Precipitation Estimates at Fine Scales, J. Hydrometeorol., 8, 38–55, https://doi.org/10.1175/JHM560.1, 2007.

Huffman, G. J., Adler, R. F., Bolvin, D. T., and Nelkin, E. J.: The TRMM Multi-satellite Precipitation Analysis (TMPA), in: Satellite Rainfall Applications for Surface Hydrology, Springer, Dordrecht, 3–22, 2010.

Huss, M. and Hock, R.: A new model for global glacier change and sea-level rise, Front. Earth Sci., 3, 54, https://doi.org/10.3389/feart.2015.00054, 2015.

IGM: Hidrografía, in: Geografía de Chile, Santiago, Chile, p. 19, 1984.

INIA: AGROMET, Instituto Nacional de Investigacion Agriola. Ministerio de Agricultura, available at: https://agromet.cl, last access: May 2017.

Ivkovic, K. M., Letcher, R. A., and Croke, B. F. W.: Use of a simple surface-groundwater interaction model to inform water management, Aust. J. Earth Sci., 56, 71–80, 2009.

Jin, Y., Schaaf, C. B., Woodcock, C., Gao, F., Li, X., and Strahler, A. H.: Consistency of MODIS surface bidirectional reflectance distribution function and albedo retrievals: 2. Validation, J. Geophys. Res., 108, 4159, https://doi.org/10.1029/2002JD002804, 2003.

Jones, J., Almeida, A., Cisneros, F., Iroumé, A., Jobbágy, E., Lara, A., de Lima, W. P., Little, C., Llerena, C., Silveira, L., and Villegas, J. C.: Forests and water in South America, Hydrol. Process., 31, 972–980, https://doi.org/10.1002/hyp.11035, 2017.

Kottek, M., Grieser, J., Beck, C., Rudolf, B., and Rubel, F.: World map of the Köppen–Geiger climate classification updated, Meteorol. Z., 15, 259–263, https://doi.org/10.1127/0941-2948/2006/0130, 2006.

Ladson, A., Bronw, R., Neal, B., and Nathan, R.: A standard approach to baseflow separation using the Lyne and Hollick filter, Tech. Pap., Aust. J. Water Resour., 17, 25–34, https://doi.org/10.7158/W12-028.2013.17.1, 2013.

La Moreaux, P. E., Wilson, B. M., and Memon, B. A.: Guide to the hydrology of carbonate rocks, Unesco – United Nations Educational, Scientific and Cultural Organization, Paris, 1984.

Land cover map: http://www.gep.uchile.cl/Landcover_CHILE.html, last access: May 2017.

Lara, A., Little, C., Urrutia, R., McPhee, J., Álvarez-Garretón, C., Oyarzún, C., Soto, D., Donoso, P., Nahuelhual, L., Pino, M., and Arismendi, I.: Assessment of ecosystem services as an opportunity for the conservation and management of native forests in Chile, Forest Ecol. Manage., 258, 415–424, https://doi.org/10.1016/j.foreco.2009.01.004, 2009.

Larraín, S.: El agua en Chile: entre los derechos humanos y las reglas del mercado, http://Polis.Revues.Org (last access: May 2018), 2006.

Le Quesne, C., Acuña, C., Boninsegna, J. A., Rivera, A., and Barichivich, J.: Long-term glacier variations in the Central Andes of Argentina and Chile, inferred from historical records and tree-ring reconstructed precipitation, Palaeogeogr. Palaeocl. Palaeoecol., 281, 334–344, https://doi.org/10.1016/j.palaeo.2008.01.039, 2009.

Liston, G. E.: Representing subgrid snow cover heterogeneities in regional and global models, J. Climate, 17, 1381–1397, https://doi.org/10.1175/1520-0442(2004)017<1381:RSSCHI>2.0.CO;2, 2004.

Lucht, W., Schaaf, C. B., and Strahler, A. H.: An algorithm for the retrieval of albedo from space using semiempirical BRDF models, IEEE T. Geosci. Remote, 38, 977–998, https://doi.org/10.1109/36.841980, 2000.

Maraun, D. and Widmann, M.: Statistical Downscaling and Bias Correction for Climate Research, Cambridge University Press, Cambridge, 170–200, 2018.

Margulis, S. A., Cortés, G., Girotto, M., and Durand, M.: A Landsat-Era Sierra Nevada Snow Reanalysis (1985–2015), J. Hydrometeorol., 17, 1203–1221, https://doi.org/10.1175/JHM-D-15-0177.1, 2016.

Marzeion, B., Jarosch, A. H., and Hofer, M.: Past and future sea-level change from the surface mass balance of glaciers, The Cryosphere, 6, 1295–1322, https://doi.org/10.5194/tc-6-1295-2012, 2012.

McDonnell, J. J. and Woods, R.: On the need for catchment classification, J. Hydrol., 299, 2–3, https://doi.org/10.1016/j.jhydrol.2004.09.003, 2004.

Melsen, L. A., Addor, N., Mizukami, N., Newman, A. J., Torfs, P. J. J. F., Clark, M. P., Uijlenhoet, R., and Teuling, A. J.: Mapping (dis)agreement in hydrologic projections, Hydrol. Earth Syst. Sci., 22, 1775–1791, https://doi.org/10.5194/hess-22-1775-2018, 2018.

Mernild, S. H., Liston, G. E., Hiemstra, C., and Wilson, R.: The Andes Cordillera. Part III: glacier surface mass balance and contribution to sea level rise (1979–2014), Int. J. Climatol., 37, 3154–3174, https://doi.org/10.1002/joc.4907, 2017.

Miller, A.: The climate of Chile, in: World survey of climatology, Elsevier, Amsterdam, 113–145, 1976.

Miranda, A., Altamirano, A., Cayuela, L., Pincheira, F., and Lara, A.: Different times, same story: Native forest loss and landscape homogenization in three physiographical areas of south-central of Chile, Appl. Geogr., 60, 20–28, https://doi.org/10.1016/j.apgeog.2015.02.016, 2015.

Mizukami, N., Clark, M. P., Newman, A. J., Wood, A. W., Gutmann, E. D., Nijssen, B., Rakovec, O., and Samaniego, L.: Towards seamless large-domain parameter estimation for hydrologic models, Water Resour. Res., 53, 8020–8040, https://doi.org/10.1002/2017WR020401, 2017.

MODIS: https://e4ftl01.cr.usgs.gov/MOLT, last access: May 2017.

Mu, Q., Zhao, M., and Running, S.: Brief Introduction to MODIS Evapotranspiration Data Set (MOD16), Water Resour. Res., 45, 0–4, 2005.

Myneni, R. B., Hoffman, S., Knyazikhin, Y., Privette, J. L., Glassy, J., Tian, Y., Wang, Y., Song, X., Zhang, Y., Smith, G. R., Lotsch, A., Friedl, M., Morisette, J. T., Votava, P., Nemani, R. R., and Running, S. W.: Global products of vegetation leaf area and fraction absorbed PAR from year one of MODIS data, Remote Sens. Environ., 83, 214–231, https://doi.org/10.1016/S0034-4257(02)00074-3, 2002.

Neteler, M., Bowman, M. H., Landa, M., and Metz, M.: GRASS GIS: A multi-purpose open source GIS, Environ. Model. Softw., 31, 124–130, https://doi.org/10.1016/j.envsoft.2011.11.014, 2012.

Newman, A. J., Clark, M. P., Sampson, K., Wood, A., Hay, L. E., Bock, A., Viger, R. J., Blodgett, D., Brekke, L., Arnold, J. R., Hopson, T., and Duan, Q.: Development of a large-sample watershed-scale hydrometeorological data set for the contiguous USA: Data set characteristics and assessment of regional variability in hydrologic model performance, Hydrol. Earth Syst. Sci., 19, 209–223, https://doi.org/10.5194/hess-19-209-2015, 2015.

Ochoa-Tocachi, B. F., Buytaert, W., De Bièvre, B., Célleri, R., Crespo, P., Villacís, M., Llerena, C. A., Acosta, L., Villazón, M., Guallpa, M., Gil-Ríos, J., Fuentes, P., Olaya, D., Viñas, P., Rojas, G., and Arias, S.: Impacts of land use on the hydrological response of tropical Andean catchments, Hydrol. Process., 30, 4074–4089, https://doi.org/10.1002/hyp.10980, 2016.

Oudin, L., Andréassian, V., Perrin, C., Michel, C., and Le Moine, N.: Spatial proximity, physical similarity, regression and ungaged catchments: A comparison of regionalization approaches based on 913 French catchments, Water Resour. Res., 44, 1–15, https://doi.org/10.1029/2007WR006240, 2008.

Pellicciotti, F., Ragettli, S., Carenzo, M., and McPhee, J.: Changes of glaciers in the Andes of Chile and priorities for future work, Sci. Total Environ., 493, 1197–1210, https://doi.org/10.1016/j.scitotenv.2013.10.055, 2014.

Poff, N. L. R., Bledsoe, B. P., and Cuhaciyan, C. O.: Hydrologic variation with land use across the contiguous

United States: Geomorphic and ecological consequences for stream ecosystems, Geomorphology, 79, 264–285, https://doi.org/10.1016/j.geomorph.2006.06.032, 2006.

Precipitation product CR2MET: http://www.cr2.cl/datos-productos-grillados, last access: March 2018.

QGIS Development Team: QGIS Geographic Information System, Open Source Geospatial Found. Proj., http://www.qgis.org/ (last access: May 2018), 2015.

RGI Consortium: Randolph Glacier Inventory – A Dataset of Global Glacier Outlines: Version 6.0, available at: https://www.glims.org/RGI, last access: December 2017.

Ropelewski, C. F., Janowiak, J. E., and Halpert, M. S.: The Climate Anomaly Monitoring System (CAMS), in Climate Analysis Center, NSW, NOAA, Washigton, D.C., available from the Climate Prediction Center, Camp Springs, MD, p. 39, 1984.

Sankarasubramanian, A., Vogel, R. M., and Limbrunner, J. F.: Climate elasticity of streamflow in the United States, Water Resour. Res., 37, 1771–1781, https://doi.org/10.1029/2000WR900330, 2001.

Sar, N., Khan, A., Chatterjee, S., and Das, A.: Hydrologic delineation of ground water potential zones using geospatial technique for Keleghai river basin, India, Model, Earth Syst. Environ., 1, 25, https://doi.org/10.1007/s40808-015-0024-3, 2015.

Sarricolea, P., Herrera-Ossandon, M., and Meseguer-Ruiz, Ó.: Climatic regionalisation of continental Chile, J. Maps, 13, 66–73, https://doi.org/10.1080/17445647.2016.1259592, 2017.

Satellite precipitation products CHIRPS, MSWEP, TMPA: http://www.cr2.cl/datos-precipitacion-satelital, last access: March 2018.

Sawicz, K., Wagener, T., Sivapalan, M., Troch, P. A., and Carrillo, G.: Catchment classification: Empirical analysis of hydrologic similarity based on catchment function in the eastern USA, Hydrol. Earth Syst. Sci., 15, 2895–2911, https://doi.org/10.5194/hess-15-2895-2011, 2011.

Sernageomin: Mapa geologico de chile: version digital, Publ. Geol. Digit., 4, 25, 2004.

Sivapalan, M., Takeuchi, K., Franks, S. W., Gupta, V. K., Karambiri, H., Lakshmi, V., Liang, X., McDonnell, J. J., Mendiondo, E. M., O'Conell, P. E., Oki, T., Pomeroy, J. W., Schertzer, D., Uhlenbrook, S., and Zehe, E.: IAHS Decade on Predictions in Ungauged Basins (PUB), 2003–2012: Shaping an exciting future for the hydrological sciences, Hydrolog. Sci. J., 48, 857–880, https://doi.org/10.1623/hysj.48.6.857.51421, 2003.

Sposito, G.: Understanding the budyko equation, Water (Switzerland), 9, 1–14, https://doi.org/10.3390/w9040236, 2017.

Strahler, A. N.: Quantitative analysis of watershed geomorphology, Eos Trans. Am. Geophys. Un., 38, 913–920, https://doi.org/10.1029/TR038i006p00913, 1957.

Streamflow: http://www.cr2.cl/datos-de-caudales, last access: March 2018.

Su, F., Hong, Y., and Lettenmaier, D. P.: Evaluation of TRMM Multisatellite Precipitation Analysis (TMPA) and Its Utility in Hydrologic Prediction in the La Plata Basin, J. Hydrometeorol., 9, 622–640, https://doi.org/10.1175/2007JHM944.1, 2008.

Tachikawa, T., Hato, M., Kaku, M., and Iwasaki, A.: Characteristics of ASTER GDEM version 2, in: International Geoscience and Remote Sensing Symposium (IGARSS), 24–29 July 2011, Vancouver, BC, Canada, 3657–3660, 2011.

Thiemig, V., Rojas, R., Zambrano-Bigiarini, M., and De Roo, A.: Hydrological evaluation of satellite-based rainfall estimates over the Volta and Baro-Akobo Basin, J. Hydrol., 499, 324–338, https://doi.org/10.1016/j.jhydrol.2013.07.012, 2013.

Tian, Y. and Peters-Lidard, C. D.: A global map of uncertainties in satellite-based precipitation measurements, Geophys. Res. Lett., 37, L24407, https://doi.org/10.1029/2010GL046008, 2010.

Tijdeman, E., Hannaford, J., and Stahl, K.: Human influences on streamflow drought characteristics in England and Wales, Hydrol. Earth Syst. Sci., 22, 1051–1064, https://doi.org/10.5194/hess-22-1051-2018, 2018.

Viale, M. and Garreaud, R.: Summer Precipitation Events over the Western Slope of the Subtropical Andes, Mon. Weather Rev., 142, 1074–1092, https://doi.org/10.1175/MWR-D-13-00259.1, 2014.

Vörösmarty, C. J., Vo, C. J., and Green, P.: Global Water Resources: Vulnerability from Climate Change and Population Growth, Science, 80, 284–288, https://doi.org/10.1126/science.289.5477.284, 2007.

Wagener, T., Sivapalan, M., Troch, P., and Woods, R.: Catchment Classification and Hydrologic Similarity, Geogr. Compass, 1, 1–31, https://doi.org/10.1111/j.1749-8198.2007.00039.x, 2007.

Water rights: http://www.dga.cl/productosyservicios/derechos_historicos, last access: May 2018.

Westerberg, I. K. and McMillan, H. K.: Uncertainty in hydrological signatures, Hydrol. Earth Syst. Sci., 19, 3951–3968, https://doi.org/10.5194/hess-19-3951-2015, 2015.

Westerberg, I. K., Wagener, T., Coxon, G., McMillan, H. K., Castellarin, A., Montanari, A., and Freer, J.: Uncertainty in hydrological signatures for gauged and ungauged catchments, Water Resour. Res., 52, 1847–1865, https://doi.org/10.1002/2015WR017635, 2016.

Westerhoff, R. S.: Using uncertainty of Penman and Penman-Monteith methods in combined satellite and ground-based evapotranspiration estimates, Remote Sens. Environ., 169, 102–112, https://doi.org/10.1016/j.rse.2015.07.021, 2015.

Woldemeskel, F. M., Sivakumar, B., and Sharma, A.: Merging gauge and satellite rainfall with specification of associated uncertainty across Australia, J. Hydrol., 499, 167–176, https://doi.org/10.1016/j.jhydrol.2013.06.039, 2013.

Wood, A. W., Hopson, T., Newman, A., Brekke, L., Arnold, J., and Clark, M.: Quantifying Streamflow Forecast Skill Elasticity to Initial Condition and Climate Prediction Skill, J. Hydrometeorol., 17, 651–668, https://doi.org/10.1175/JHM-D-14-0213.1, 2016.

Yang, Z. L., Dickinson, R. E., Robock, A., and Vinnikov, K. Y.: Validation of the snow submodel of the biosphere-atmosphere transfer scheme with Russian snow cover and meteorological observational data, J. Climate, 10, 353–373, https://doi.org/10.1175/1520-0442(1997)010<0353:VOTSSO>2.0.CO;2, 1997.

Zambrano-Bigiarini, M., Nauditt, A., Birkel, C., Verbist, K., and Ribbe, L.: Temporal and spatial evaluation of satellite-based rainfall estimates across the complex topographical and climatic gradients of Chile, Hydrol. Earth Syst. Sci., 21, 1295–1320, https://doi.org/10.5194/hess-21-1295-2017, 2017.

Zhao, Y., Feng, D., Yu, L., Wang, X., Chen, Y., Bai, Y., Hernández, H. J., Galleguillos, M., Estades, C., Biging, G. S., Radke, J. D., and Gong, P.: Detailed dynamic land cover mapping of Chile: Accuracy improvement by integrating multi-temporal data, Remote Sens. Environ., 183, 170–185, https://doi.org/10.1016/j.rse.2016.05.016, 2016.

8

Precipitation downscaling using a probability-matching approach and geostationary infrared data: an evaluation over six climate regions

Ruifang Guo[1,2], **Yuanbo Liu**[1], **Han Zhou**[1,2], **and Yaqiao Zhu**[3]

[1]Key Laboratory of Watershed Geographic Sciences, Nanjing Institute of Geography and Limnology, Chinese Academy of Sciences, No. 73 East Beijing Road, Nanjing 210008, China
[2]University of Chinese Academy of Sciences, No. 19 Yuquan Road, Beijing 100049, China
[3]College of Urban and Environmental Sciences, Hubei Normal University, No. 11 Cihu Road, Huangshi 435002, China

Correspondence: Yuanbo Liu (ybliu@niglas.ac.cn)

Abstract. Precipitation is one of the most important components of the global water cycle. Precipitation data at high spatial and temporal resolutions are crucial for basin-scale hydrological and meteorological studies. In this study, we propose a cumulative distribution of frequency (CDF)-based downscaling method (DCDF) to obtain hourly 0.05° × 0.05° precipitation data. The main hypothesis is that a variable with the same resolution of target data should produce a CDF that is similar to the reference data. The method was demonstrated using the 3-hourly 0.25° × 0.25° Climate Prediction Center morphing method (CMORPH) dataset and the hourly 0.05° × 0.05° FY2-E geostationary (GEO) infrared (IR) temperature brightness (T_b) data. Initially, power function relationships were established between the precipitation rate and T_b for each 1° × 1° region. Then the CMORPH data were downscaled to 0.05° × 0.05°. The downscaled results were validated over diverse rainfall regimes in China. Within each rainfall regime, the fitting functions' coefficients were able to implicitly reflect the characteristics of precipitation. Quantitatively, the downscaled estimates not only improved spatio-temporal resolutions, but also performed better (bias: −7.35–10.35 %; correlation coefficient, CC: 0.48–0.60) than the CMORPH product (bias: 20.82–94.19 %; CC: 0.31–0.59) over convective precipitating regions. The downscaled results performed as well as the CMORPH product over regions dominated with frontal rain systems and performed relatively poorly over mountainous or hilly areas where orographic rain systems dominate. Qualitatively, at the daily scale, DCDF and CMORPH had nearly equivalent performances at the regional scale, and 79 % DCDF may perform better than or nearly equivalently to CMORPH at the point (rain gauge) scale. The downscaled estimates were able to capture more details about rainfall motion and changes under the condition that DCDF performs better than or nearly equivalently to CMORPH.

1 Introduction

Precipitation is a critical component in the global water cycle (Barrett and Martin, 1981; Smith et al., 1998; Tobler, 2004). Precipitation data at spatio-temporal resolutions are favoured mainly for two reasons. First, the poor representativeness and uneven distribution of gauge stations make the data incapable of reflecting the precipitation variation spatially (Hughes, 2006, Collischonn et al., 2008; Javanmard et al., 2010). Second, ground radar systems can provide full coverage spatial data for most regions, but RADAR is very weak in view of the precipitation intensity and is subject to short time series. Moreover, the validation poses a big challenge for hydrological applications (Krajewski and Smith, 2002).

A number of techniques have been developed to estimate or retrieve precipitation (Kidd and Levizzani, 2011). Based on these technologies, precipitation datasets have been produced at various resolutions, including the Global

Precipitation Climatology Project (GPCP) (Huffman et al., 1997, 2001, 2009), the Tropical Rainfall Measuring Mission (TRMM) Multi-Satellite Precipitation (TMPA) (Huffman et al., 2007), the Climate Prediction Center morphing method (CMORPH) (Joyce et al., 2004) and the Global Satellite Mapping of Precipitation (GSMaP) (Ushio et al., 2009), especially over the last 20 years. The typical spatial resolution of these products is $0.25° \times 0.25°$ (Dinku et al., 2007; Ebert et al., 2007; Hirpa et al., 2010; Sohn et al., 2010; Bitew and Gebremichael, 2011; Romilly and Gebremichael, 2011; Thiemig et al., 2012; Hu et al., 2014). This coarse resolution generally impedes the applications of the data for basin-scale hydrological and meteorological studies (Mekonnen et al., 2008). A downscaling procedure would therefore be highly necessary to meet the requirements of small-scale (< 10 km) applications.

Downscaling approaches were first used to interpolate regional-scale atmospheric predictor variables to point-scale meteorological series (Karl et al., 1990; Wigley et al., 1990; Hay et al., 1991, 1992). Currently, downscaling approaches are well developed and can be categorised into regression methods, weather pattern approaches, stochastic weather generators and limited-area climate modelling (Wilby and Wigley, 1997; Cannon, 2008). Most methods are based on meteorological or climate models, and assume that relationships can be established between atmospheric parameters at disparate temporal and/or spatial scales (Giorgi and Mearns, 1999; Willems and Vrac, 2011; Kenabatho et al., 2012). Downscaling approaches can also be categorised into dynamical methods (using regional climate models to translate large-scale weather evolution into physically consistent evolution at a higher resolution) and statistical methods (based on statistical relationships between the regional climate and large-scale predictor variables) (Schmidli et al., 2006). At present, these methods are generally available to downscale data from general circulation models.

Various downscaling techniques have been developed to improve the resolution of satellite precipitation data. Immerzeel et al. (2009) used an exponential relationship between the 1 km Normalized Difference Vegetation Index (NDVI) and precipitation to downscale TRMM 3B43 precipitation data on the Iberian Peninsula. Jia et al. (2011) used a linear regression relationship between a combination of NDVI and a digital elevation model and precipitation to downscale TRMM 3B43 precipitation data in the Qaidam Basin of China. Duan and Bastiaanssen (2013) used a two-degree polynomial regression model between NDVI and precipitation to downscale TRMM 3B43 precipitation data in the Lake Tana basin, Ethiopia, and the Caspian Sea region, Iran. These studies manifest the potential of downscaling methods to obtain fine-resolution precipitation (< 10 km), while mainly focusing on precipitation data with low temporal resolutions (i.e. annual or monthly).

The main objective of this study is to develop a regression-based downscaling method to obtain precipitation estimates with a high spatio-temporal resolution (0.05°, hourly). Barrett et al. (1991) proposed a cumulative histogram method to relate precipitation observations to satellite estimates in an effort to avoid bias problems related to simple regression. In this study, we propose a cumulative distribution of frequency (CDF)-based downscaling method (DCDF) and perform preliminary validation using CMORPH and geostationary (GEO) infrared (IR) temperature brightness (T_b) data. This new method can (1) lead to a better understanding of satellite precipitation data and (2) stimulate scientific interests to engender the development of precipitation data with improved resolutions. The following section introduces study areas and datasets. Section 3 introduces the principles, framework and procedure of the downscaling method. Section 4 presents the major findings followed by discussion in Sect. 5. Finally, Sect. 6 concludes.

2 Study areas and datasets

2.1 Study areas

Existing studies confirmed that the performances of satellite precipitation estimates are highly dependent on the rainfall regime (Arkin et al., 2006; Ebert et al., 2007; Gottschalck et al., 2005), which varies with climate zone, latitude, longitude and elevation. Thus, six $5° \times 5°$ regions were selected for validation (Fig. 1). Their corresponding geographic and climatic characteristics are listed in Table 1. These areas are distributed from south to north and from east to west, and they incorporate most rainfall regimes.

Among the six regions, regions SE (south-east), CE (central-east) and NE (north-east) are located in the eastern monsoon region. It is warm and rainy during the southeast monsoon in June–August, and cold and dry during the northwest monsoon in December–February. These three regions feature low-elevation hills and plains. Regions CW (central-west), NW (north-west) and TP (Tibetan Plateau) are located in the non-monsoon region with a continental climate. CW and NW belong to arid region, with 60–70 % precipitation occurring in June–August. CW has a relatively high elevation, mainly covered by plateaus, mountains and basins. NW is mainly covered by plateaus and basins. TP has a complex climate, mainly covered by plateaus and mountains. The seasonal precipitation distribution has two forms: a unimodal distribution in summer (June–August), and a bimodal distribution in spring (March–May) and autumn (September–November).

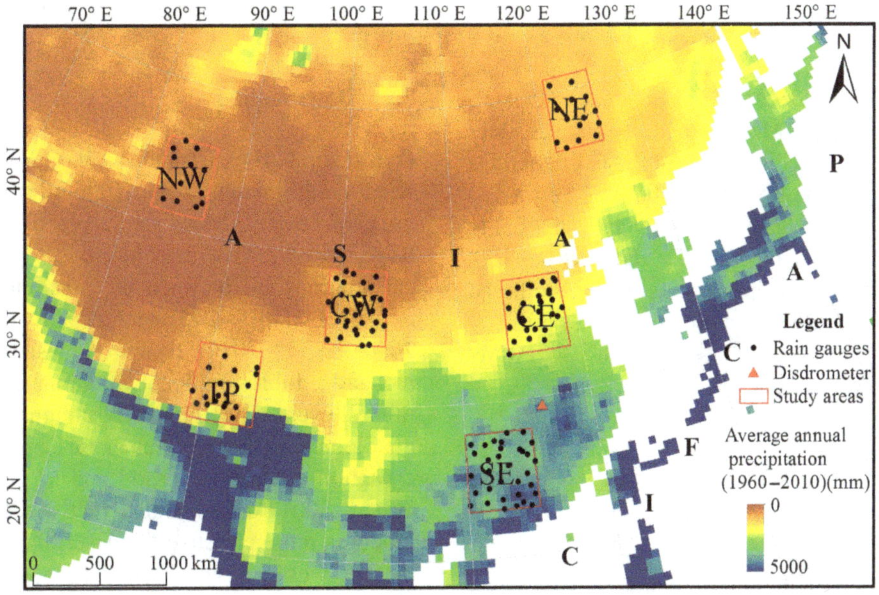

Figure 1. Geographic and climate situations of the six regions. The locations of the rain gauges are superimposed on the map.

Table 1. Geographic and climatic situations of the six regions in China.

Region	Longitude, latitude	Elevation range (m)	Annual precipitation (mm)	Climate zone	
SE	110–115° E, 23–28° N	22–1405	1230	Subtropical humid	Monsoon
CE	114–119° E, 33–38° N	6–1533	670	Warm temperate semi-humid	Monsoon
NE	121–126° E, 46–51° N	147–740	460	Mid-temperate humid	Monsoon
CW	99–104° E, 34–39° N	1368–8500	40	Warm temperate arid	Non-monsoon
NW	82–87° E, 41–46° N	320–2458	70–140	Mid-temperate arid	Non-monsoon
TP	89–94° E, 28–33° N	3552–8260	420	Temperature plateau	Non-monsoon

2.2 Datasets

2.2.1 Meteorological data

Rain gauge data were obtained from the National Meteorological Information Centre of the China Meteorological Administration (CMA) (http://data.cma.cn/, last access: 10 April 2017). The datasets include daily precipitation records at 137 rain gauge stations in 2014 (Fig. 1). Strict quality control has been applied to check extreme values (Ma, 1998). There are 33, 29, 14, 31, 12 and 18 rain gauges in regions SE, CE, NE, CW, NW and TP, respectively. In the case of more than one station located within a pixel, the rain gauge values are averaged to represent the grid value. Statistical analyses were used to evaluate precipitation estimates at the daily scale. In addition, a disdrometer installed at Xingzi station (29.45° N, 116.05° E) in Jiangxi Province (Fig. 1) provided hourly data in 2014, except June and July when the instrument was subject to a transmission error. Disdrometer data is used to evaluate the precipitation estimates.

2.2.2 Satellite data

IR data (10.7 µm) were collected from the Stretched Visible and Infrared Spin Scan Radiometer (S-VISSR) on board the FY2-E satellite. The data are available at the National Satellite Meteorology Center (http://satellite.nsmc.org.cn/portalsite/default.aspx, last access: 12 August 2017). FY2-E provides hourly coverage of eastern Asia from 75° S to 75° N. The IR T_b data were corrected for zenith angle viewing effects.

CMORPH was developed and produced by the Climate Prediction Center (CPC) in the National Oceanic and Atmospheric Administration (NOAA). CMORPH produces 0.25° × 0.25° 3-hourly global precipitation data using PMW and IR data. PMW data are from the Microwave Imager (TMI) on TRMM, the Special Sensor Microwave Imager (SSM/I) on Defense Meteorological Satellite Program (DMSP) satellites 13–15, the Advanced Microwave Scanning Radiometer Earth Observing System (AMSR-E) on Aqua and the Advanced Microwave Sounding Unit-B (AMSU-B) on NOAA satellite 15–18. Precipitation estimates are generated using the algorithms of Ferraro (1997) for SSM/I, Ferraro et al. (2000) for AMSU-B and Kummerow et al. (2001) for TMI. IR data are obtained from the GEO Operational Environmental Satellites (GOES) 8/10, European Meteorological Satellites (Meteosat) 5/7 and Japanese GEO Meteorological Satellites (GMS) 5. CMORPH uses motion vectors derived from GEO satellite IR imagery to propagate the relatively high-quality precipitation estimates derived from PMW data (Joyce et al., 2004). Hence, quantitative precipitation estimates are based solely on PMW data. GEO-IR data are not used to estimate precipitation but rather to interpolate between two PMW-derived precipitation rate fields.

3 Methodology

3.1 CDF matching

The CDF matching is a probability-based process. It assumes a variable (v) should produce a similar CDF to the reference variable (t). The frequencies of t and v are shown in Eqs. (1)–(2), and the cumulative frequencies in Eqs. (3)–(4).

$$P_t = f_1(t) \tag{1}$$

$$P_v = f_2(v) \tag{2}$$

$$C_t(t) = \int_{T_1}^{t} f_1(t)\mathrm{d}t \tag{3}$$

$$C_v(v) = \int_{V_1}^{v} f_2(v)\mathrm{d}v \tag{4}$$

$$C_v^{-1}[C_t(t)] \xrightarrow{f_3} t, \tag{5}$$

where P_t and P_v are the probability of t and v, $f_1(t)$ and $f_2(v)$ are probability density functions of t and v and $C_t(t)$ and $C_v(v)$ are the cumulative density functions of t and v, respectively. $f_3(v)$ represents the relationship between t and v.

The steps for CDF matching are summarised in Fig. 2. First, t and v are shown in histograms (Fig. 2a, b). The frequency of an arbitrary point t_i (or v_i) on the $f_1(t)$ [or $f_2(v)$] curve can be expressed as $P(t = t_i) = f_1(t_i)$ [or $P(v = v_i) = f_2(v_i)$]. Second, these histograms are transformed into cumulative histograms (Fig. 2c, d). The cumulative frequency of an arbitrary point t_i (or v_i) on the $C_t(t)$ [or $C_v(v)$] curve can be expressed as $C(t < t_i) = \int_{T_1}^{t_i} f_1(t)\mathrm{d}t$ [or $C(v < v_i) = \int_{V_1}^{v_i} f_2(v)\mathrm{d}v$]. Third, these cumulative histograms are matched so that v has a cumulative histogram similar to t. The matching process is implemented by a one-to-one mapping CDF of the variable onto that of the reference (Eq. 5). Last, the v–t relationship is established (Eq. 5) (Fig. 2e). Magnusson et al. (2015) demonstrated that CDF matching works better than a histogram-matching method when low values have high frequencies, which is generally the case for precipitation.

3.2 Downscaling

Our method is based on the work of Barrett et al. (1991) and Kidd and Levizzani (2011). Rainfall can be inferred from IR imagery because heavy rainfall tends to be associated with large, tall clouds with cold cloud tops. Therefore, empirical relationships between the precipitation rate and T_b are derived (Arkin and Meisner, 1987; Greene and Morrissey, 2000; Prigent, 2010). However, these relationships are indirect and exhibit significant variations during the lifetime of a rainfall event. They also differ among rain systems and climatological regimes, which causes large uncertainties in precipitation estimations (Kidd and Levizzani, 2011). Ba and Gruce (2001) demonstrated that a two-degree polynomial model is more effective for describing the relationship, and that the coefficients of the model are region-dependent. Overall, the precipitation–T_b relationship is highly variable over time and space.

Microwave (MW) radiation reflects the physical structures of clouds. Emission from rain droplets increases MW radiation, and scattering by precipitating ice particles decreases MW radiation. Although MW techniques are physically more direct than those based on IR radiation, they can both reflect rainfall events. Therefore, we assume that an IR signal produces a similar frequency distribution of precipitation rates to a MW signal over a certain region during a certain period. Barrett et al. (1991) proposed a cumulative-histogram-matching method to relate rainfall observations to satellite precipitation data. Kidd et al. (2003) applied the same method to estimate rainfall using passive microwave (PMW) and IR data over Africa.

The assumptions behind the downscaling method include the following: (1) T_b has a similar cumulative frequency to the precipitation rate at certain spatial and temporal scales, and (2) satellite precipitation products provide relatively accurate estimates with low spatial and temporal resolutions. In contrast, GEO-IR data have a high spatio-temporal resolution, yet low accuracy. Illustrated in Fig. 3, the downscaling method explores the advantages of the satellite precipitation product and GEO-IR data, specifically, (1) to aggregate T_b ($T_{\mathrm{b_h}}$) from a high resolution to a low resolution ($T_{\mathrm{b_l}}$) similar to the precipitation data (Eq. 6), and (2) to apply the CDF matching to the $T_{\mathrm{b_l}}$ and precipitation rate (R_l) to obtain a $T_{\mathrm{b_l}}$–R_l relationship and a rain–no-rain threshold (Eq. 7). The downscaled precipitation rates are estimated based on the $T_{\mathrm{b_l}}$–R_l relationships (Eq. 8).

$$T_{\mathrm{b_l}} = \frac{1}{n}\sum\nolimits_{i=0}^{n} T_{\mathrm{b_h}}(i) \tag{6}$$

$$T_{\mathrm{b_l}} = m \times R_\mathrm{l}^{p} \tag{7}$$

$$R_\mathrm{h} = \left(\frac{T_{\mathrm{b_h}}}{m}\right)^{1/e}, \tag{8}$$

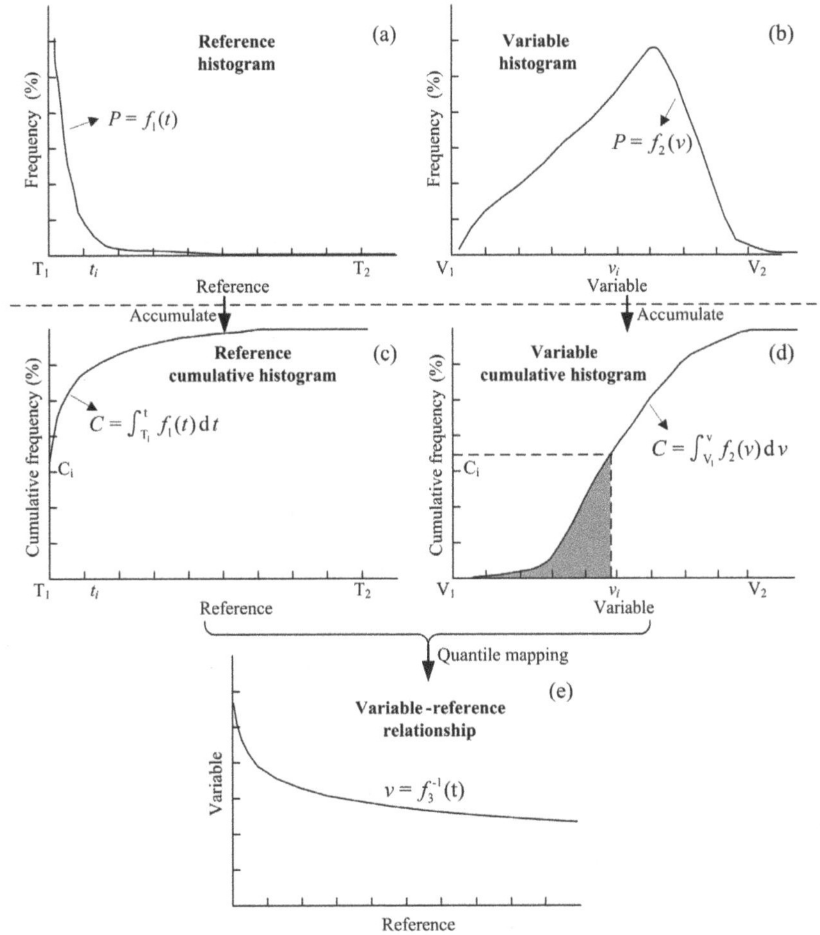

Figure 2. Schematic of the cumulative distribution of frequency (CDF) matching method.

where T_{b_h} denotes high-resolution GEO-IR T_b data, T_{b_l} denotes upscaled T_b data, R_l denotes the low-resolution precipitation product, R_h denotes the derived high-resolution estimates, m and e are coefficients of the T_b–R relationship and n is the number of high-resolution pixels within a low-resolution pixel.

Under the assumption that colder clouds are linked to higher rainfall than warmer clouds, the downscaling method assumes a monotonically increasing precipitation rate with decreasing T_b. Therefore, cumulative histograms of the precipitation rate and T_b are matched, so that the occurrence of the heaviest precipitation is associated with the T_b values linked to the heaviest rainfall. Decreasing T_b values are assigned to increasing precipitation rates so that the final distribution of T_b assigned to the precipitation rates is the same as that determined using precipitation rate data. Specially, all precipitation rate (T_b) are sorted in ascending (descending) order. Then both cumulative probability distributions are obtained. The cumulative probability is defined as critical probability when the precipitation rate equals zero. The rain–no-rain threshold is the T_b with a cumulative probability the same as the critical probability. As shown in Fig. 2c and d (T means precipitation rate; V represents T_b), the rain–no-rain threshold is set at about v_i, where the cumulative probability equals C_i (critical probability).

The specific steps used for downscaling with CMORPH and FY2-E IR data are described as follows.

a. Aggregate IR–T_b data ($T_{b_{0.05}}$) from 0.05 to 0.25° by pixel averaging ($T_{b_{0.25}}$).

   IR–T_b data ($T_{b_{0.05}}$) were aggregated to a 0.25° grid ($T_{b_{0.25}}$) for each 3 h period (00:00–03:00, 03:00–06:00, ..., 21:00–24:00 UTC), in order to match the spatial and temporal resolutions of CMORPH.

b. Generate the histogram database for CDF matching.

   IR–T_b ($T_{b_{0.25}}$) and the CMORPH precipitation rate ($R_{0.25}$) were recorded in a database. The sample area for CDF matching was determined as follows. The horizontal and temporal scales of stratiform precipitation range from 10^1 to 10^3 km and from hours to days (Orlanski, 1975; Trapp, 2013), while those of cumuliform precipitation range from a few kilometres

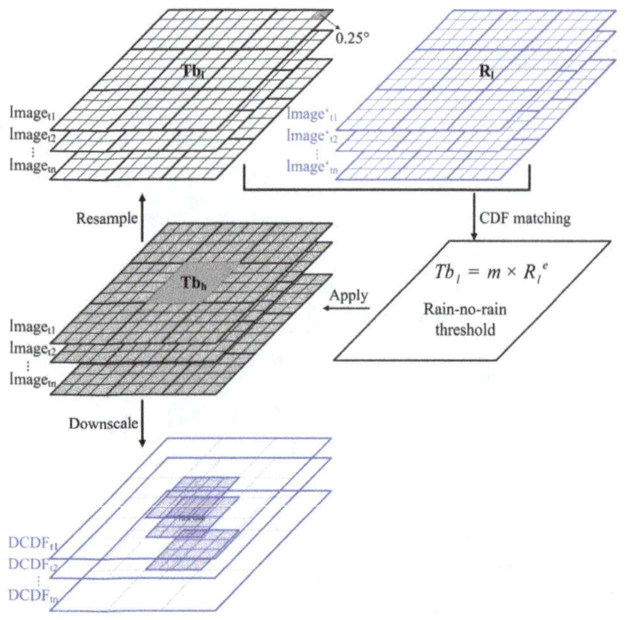

Figure 3. Schematic of the CDF-based downscaling method (DCDF) using CMORPH and FY2-E T_b in this study. R represents the precipitation rate.

to tens of kilometres and from minutes to hours (Orlanski, 1975; Rickenbach, 2008). In combination with previous studies (Kidd et al., 2003; Huffman et al., 2007), the downscaling procedure was conducted at 1° × 1° grids over a 10-day period. To reduce the heterogeneity among grids, a 3 × 3 window was used for smoothing purposes.

c. Build relationships between the precipitation rate and T_b.

The histograms of T_b–precipitation rate were generated and converted to cumulative histograms, and then matched using the CDF matching. (As shown in Fig. 2, the precipitation rate is denoted by T; T_b represents V; v_i is the rain–no-rain threshold.) A power function relationship between the precipitation rate ($R_{0.25}$) and T_b ($T_{b0.25}$) was established for each 1° × 1° area over a 10-day period. Meanwhile, various parameters, including coefficients of the T_b–R relationship, rain–no-rain threshold and R^2, were obtained.

d. Estimate the precipitation rate pixel by pixel at 1 h and 0.05°.

All pixels in the T_b images ($T_{b0.05}$) were divided into two categories, raining ones below the rain–no-rain threshold and non-raining ones above the threshold. T_b–R relationships were applied to these "raining" pixels. Finally, CMORPH data were downscaled to 1 h and 0.05° × 0.05°.

3.3 Variogram

A variogram describes how data correlate with distance. The variogram function $\gamma(h)$ is defined as half of the mean value of the square of the difference between points separated by a distance h (Matheron, 1963). A variogram is generally an increasing function of distance h. The relationship between $\gamma(h)$ and h is commonly described using the nugget effect (C_0), sill ($C_0 + C$) and range (D). C_0 denotes micro-scale variations, equated to of $\gamma(0)$. $C_0 + C$ denotes the limit of the variogram γ $(+\infty)$. D denotes the distance at which the difference of the variogram from the sill becomes negligible. A variogram is used here to describe the spatial structure of satellite precipitation data.

4 Results

4.1 T_b–precipitation rate relationship

Figure 4 shows fitting functions between the precipitation rate and T_b within each 1° × 1° grid. It was observed that T_b had a power function relationship with the precipitation rate. With an increase in the precipitation rate, T_b decreased, and the rate of change also reduced. The model fitting R^2 values were all higher than 0.90. From the region SE to NE, the precipitation rate decreases, mainly subject to latitude. The maximum precipitation rate, rain–no-rain threshold and R^2 all showed decreasing trends. The maximum precipitation rate was 19.9 mm h^{-1} in region SE, 9.8 mm h^{-1} in region CE and 4.3 mm h^{-1} in region NE. The corresponding T_b values were 198, 202 and 210 K, respectively, and the rain–no-rain threshold values were 265, 259 and 249 K. The probability of the precipitation rate was the largest for a given T_b in region SE, followed by region CE and then region NE. Regions CW and NW are arid, while TP is humid. The maximum precipitation rate was 3.5 mm h^{-1} for both regions CW and NW and 11 mm h^{-1} for region TP. The rain–no-rain thresholds for regions CW and NW were approximately 230 K, while it was 254 K for region TP. The probability of the precipitation rate was the largest for a given T_b in region TP because region TP has a complex rain system and high elevation. Generally, the fitting relationships reflected precipitation characteristics well.

4.2 Estimation results

Figure 5 shows a comparison of the spatial distributions of CMORPH and DCDF precipitation estimates for regions SE, NE and TP. The downscaled precipitation showed a similar spatial distribution to CMORPH, yet it reflected more detailed moving and changing processes of rainfall. To demonstrate clouds captured through DCDF and CMORPH, region SE was exemplified (14:00 to 16:00, 21 June 2014). Three cloud centres were observed in the southeastern and mid-eastern parts at 14:00. One hour later, two centres in

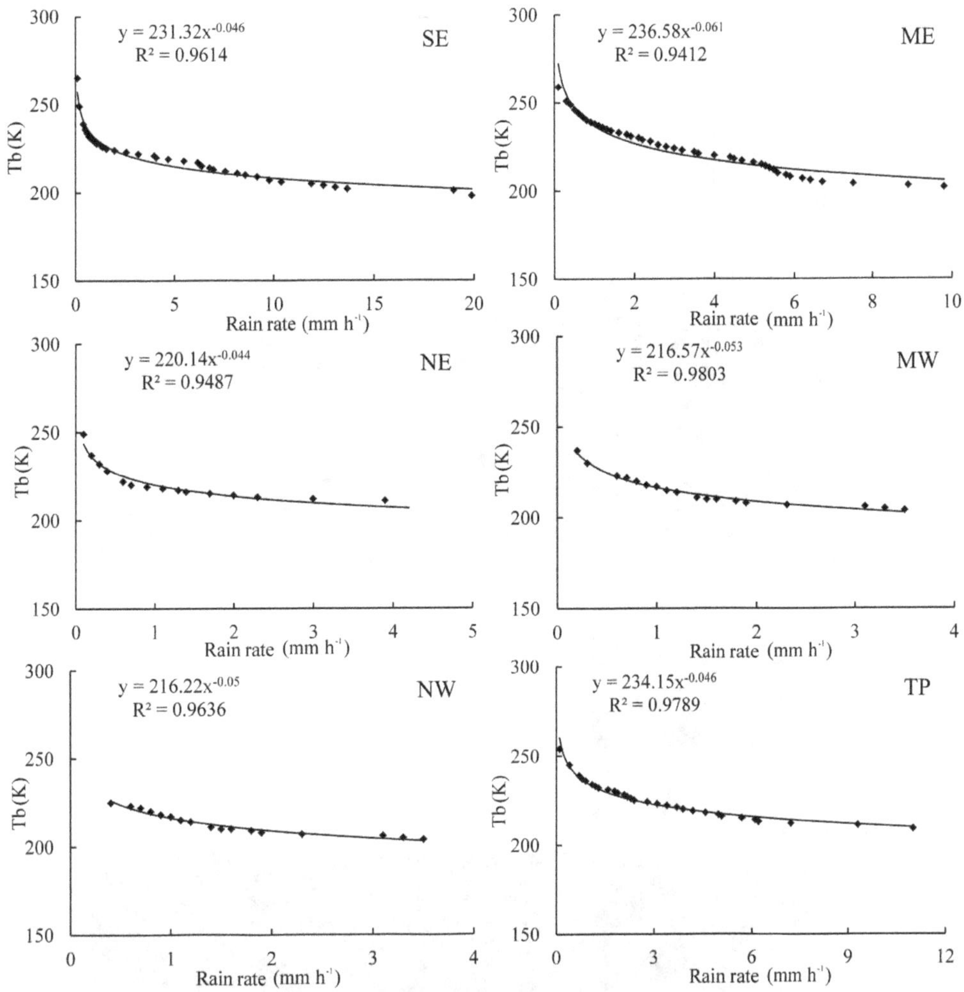

Figure 4. Examples of fitting of the precipitation rate and T_b for each region in China during 9–18 July 2014 for subregion SE (115°35′ E, 27°28′ N), subregion CE (115°39′ E, 36°14′ N), subregion NE (124°20′ E, 51°42′ N), subregion CW (101°38′ E, 37°31′ N), subregion NW (85°43′ E, 46°47′ N) and subregion TP (91°06′ E, 30°29′ N).

the southeast moved eastward and joined together, while another centre moved eastward. Two precipitation centres continued to move eastward at 16:00. In addition, D and sill values of DCDF (2.796 and 1.070) were higher than those of CMORPH (1.614 and 0.489). Large range and sill values indicate a high spatial dependence and high spatial variability. Thus, the spatial dependence and variability for high-resolution data were generally larger than those for low-resolution data.

In region SE, clouds were relatively centralised with a high precipitation rate and were small in size. In region NE, clouds were discrete with a low precipitation rate and were widely distributed. In region TP, both centralised and discrete clouds appeared. Cumuliform cloud is the main type in region SE, while stratiform cloud is dominant in region NE, and both are dominant in region TP. Thus, the cloud distributions obtained through satellite data, especially using the DCDF approach, were consistent with the local characteristics. Sill for cumuliform clouds was larger than that for stratiform clouds. A larger sill value was obtained for region SE (DCDF: 1.070; CMORPH: 0.489) than for region NE (DCDF: 0.007; CMORPH: 0.008). These results indicated that the DCDF method can reflect precipitation characteristics among rain systems and climatological regimes.

4.3 Validation

Figure 6 shows a comparison among the DCDF, CMORPH and disdrometer at the hourly scale. The DCDF and CMORPH were able to capture rainfall events, although they differed in magnitude from the reference data in some cases. The DCDF effectively reflected the peak of each rainfall event, but could not exactly identify same starting and ending times of rainy events, resulting in somewhat delayed or advanced rainfall. The DCDF may detect non-rainy events as rainy events, especially in dry seasons. CMORPH reported low-rain events as non-rainy events. Both of the DCDF and

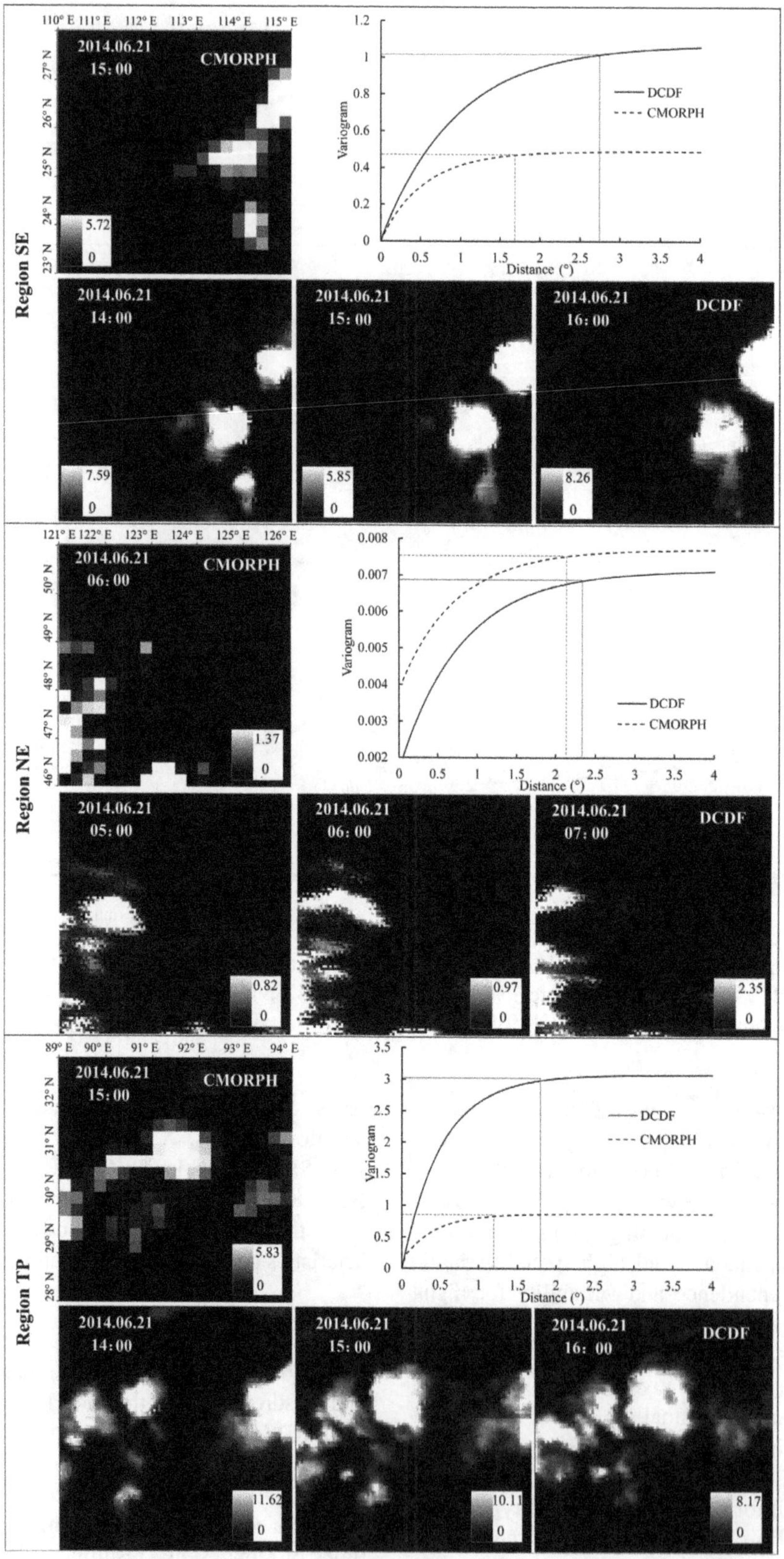

Figure 5. CMORPH precipitation estimates at a nominal resolution of 0.25° and DCDF precipitation maps at a 0.05° resolution for regions SE, NE and TP.

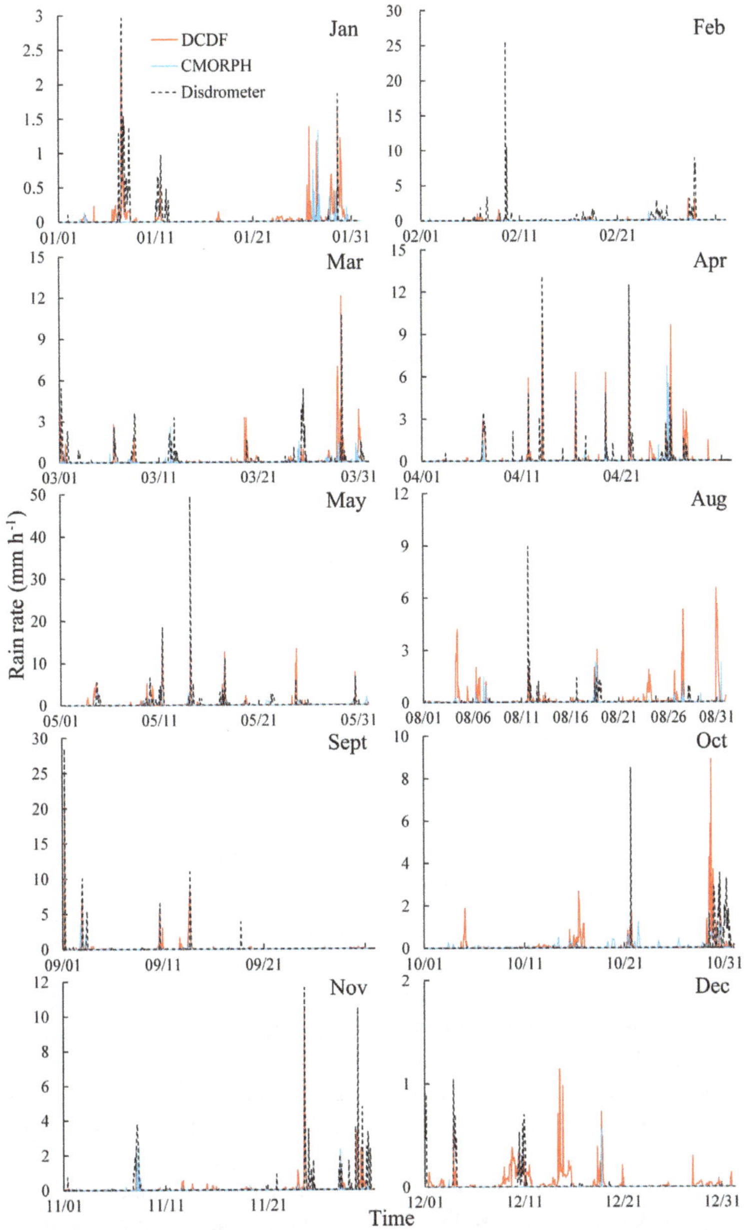

Figure 6. Time series of disdrometer data, original CMORPH and DCDF precipitation at an hourly scale in 2014.

CMORPH estimates coincided with disdrometer data at precipitation rates ranging from 1 to $10\,\mathrm{mm\,h^{-1}}$, such as the events from 10:00 to 14:00 on 9 February and from 21:00 on 13 May to 10:00 on 14 May.

To demonstrate the performance of the DCDF method, a comparison of the DCDF and CMORPH estimates was conducted at the regional scale and at the point (rain gauge) scale. Figure 7 shows the average precipitation of each region derived from the rain gauge, DCDF and CMORPH. The daily average precipitation over each region showed almost identical temporal variations for DCDF and CMORPH. Both DCDF and CMORPH showed similar temporal patterns to the rain gauge observations, but they were probably subject to overestimation for regions CW and NW and underestimation for regions SE and TP. At the point (gauge) scale, the better fit between DCDF and gauge data than that between CMORPH and gauge data is 10 %. The nearly equivalent fit is 69 %. The poorer fit was mainly evident in regions NW, CW and TP. Figure 8 shows that cases of better fit in the time series of DCDF were generally more consistent with the rain gauge data than CMORPH, although the DCDF series occasionally deviated from gauge data or misreported non-rainy events as rainy events. These results indicated that both DCDF and CMORPH demonstrated nearly equivalent

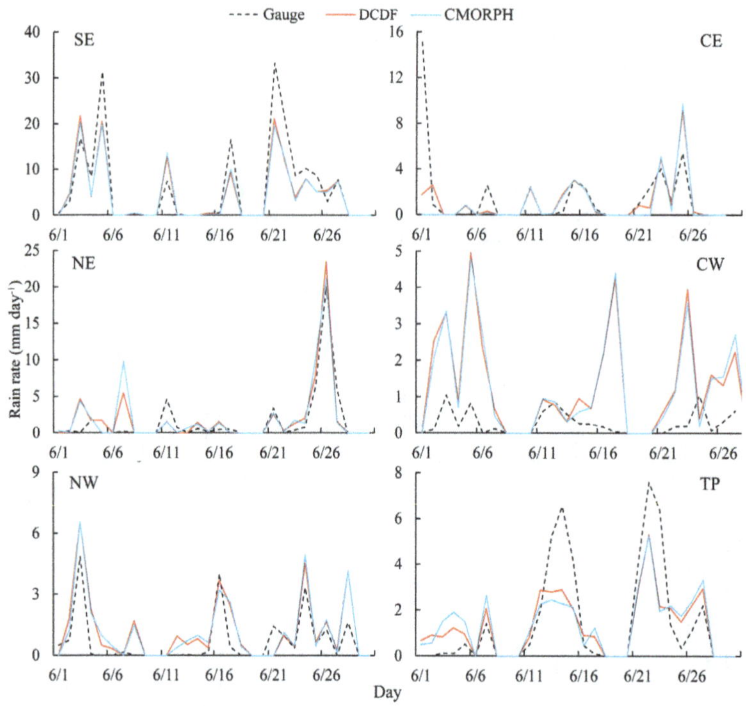

Figure 7. Time series of the average precipitation of each region derived from the gauge, DCDF and CMORPH at the daily scale in June 2014.

performances at the regional scale, and 79 % DCDF may perform better than or nearly equivalent to CMORPH at the point (gauge) scale.

Table 2 lists the seasonal statistics for the six regions at the daily scale. Generally, DCDF performed better than CMORPH in region SE, while it performed equivalently to CMORPH in regions CE and NE. Both of the DCDF and CMORPH showed better performances during the rainy season. The DCDF generally showed the smallest biases between −7.35 and 10.35 % (correlation coefficient, CC: 0.48–0.60) in region SE, and overestimated precipitation by 2.66–33.95 % (CC: 0.05–0.53) in regions CE and NE. CMORPH underestimated precipitation by 20.82–94.19 % (CC: 0.31–0.59) in region SE and showed biases between −93.2 and 6.78 % (CC: 0.00–0.50) in regions CE and NE. DCDF and CMORPH both exhibited poor performances in regions CW, NW and TP, and showed large biases (−73.75–2106 %), low CC values (0.01–0.44) and high false alarm rate (FAR) values (0.33–1.00) during the winter. Further inspection showed that the DCDF overestimation was due to high probability of detection and FAR, which may be caused by a low rain–no-rain threshold. The large biases for regions CW, NW and TP were likely due to the insensitivity of precipitation data to very low precipitation in arid regions and the inability to estimate precipitation over mountainous or hilly areas where orographic rain systems dominate.

5 Discussion

Existing downscaling methods make an assumption that local-scale patterns are driven by large-scale climatic fluctuations (Wilby and Wigley, 1997; Wilby et al., 2002). Most of these methods rely on meteorological or climate models and utilise multiple parameters, such as temperature, humidity, pressure, vorticity and geostrophic airflow. These methods are not used to downscale satellite precipitation products, possibly due to a diversity of parameters and complexity of the meteorological and climate models. In contrast, the DCDF method in this study assumes that the IR retrieval should produce a frequency distribution of precipitation rates similar to that produced by MW retrievals over a certain region during a certain period; that is, IR estimations and MW retrievals from clouds have strong statistical frequency similarities.

Due to high spatial and temporal variability of precipitation, the DCDF method must be conducted over a certain region during a certain period. The area and time period must be large enough for a reasonable sample size, but small enough to represent local characteristics. In the TMPA algorithm, a relationship between IR and the precipitation rate is built within a $1^\circ \times 1^\circ$ area by 3×3 windows over the period of a month (Huffman et al. 2007). Kidd et al. (2003) obtained the relationship within a $1^\circ \times 1^\circ$ area with the use of a $5^\circ \times 5^\circ$ Gaussian filter over a period of 5 days. Based on the horizontal and temporal scales of stratiform and cu-

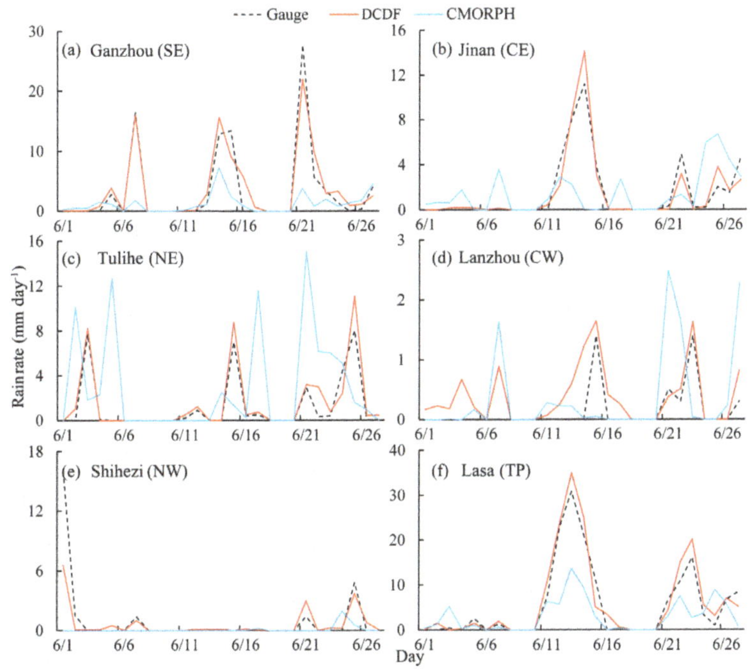

Figure 8. Time series of rain gauge data, original CMORPH and DCDF precipitation for each randomly selected gauge. **(a)** Ganzhou station (SE): 113.1667° E, 25.8667° N. **(b)** Jinan station (CE): 117.05° E, 36.6° N. **(c)** Tulihe station (NE): 121.6833° E, 50.4833° N. **(d)** Lanzhou station (CW): 103.8833° E, 36.05° N. **(e)** Shihezi station (NW): 86.05° E, 44.3167° N. **(f)** Lasa station (TP): 91.1333° E, 29.6667° N.

muliform precipitation (Orlanski, 1975; Rickenbach, 2008, Trapp, 2013) and previous studies (Kidd et al., 2003; Huffman et al. 2007), the DCDF method is applied within a $1° \times 1°$ area by 3×3 windows over a 10-day period. Nevertheless, the same gridded sample area is not the optimal selection. The size of sample area is determined according to local cloud type and varies over space and time. It likely is our future work to improve the precipitation estimates' algorithm.

It seems that IR data are used twice, one for original CMORPH generation and the other for downscaling CMORPH. In fact, IR data serve as an intermediate variable for an interpolation purpose in the first step, while IR data serve as an ancillary variable in the second step for developing a precipitation–T_b relationship. The CMORPH product is essentially derived from MW observations, and therefore the use of IR data is reasonable. We selected CMORPH as reference precipitation data mainly for the following reasons. Products with similar resolutions to GEO-IR data (0.05°) are not used, such as CMORPH at 0.072° and GSMaP at 0.1°. TRMM 3B42 (RT) and the Naval Research Laboratory blended product (NRLB) (Turk, 2005) algorithm combine MW-calibrated IR estimates, which would result in IR reuse.

The DCDF method has two main disadvantages. The physical premise of the DCDF method is that cloud top temperature in the IR imagery is a simple empirical function of cloud top height, and that heavier rainfall tends to be associated with larger, taller clouds with colder cloud tops. Unfortunately, not all cold clouds precipitate, and precipitation does not always fall from cold clouds only (Barrett, 1970). This phenomenon results in misreporting. In addition, the rain–no-rain threshold is very critical for final precipitation estimates. The size of the sample area and the indirect relationship between IR–T_b and the precipitation rate both affect the rain–no-rain threshold. However, both of them have uncertainties among rain systems and climatological regimes, resulting in uncertainties of the rain–no-rain threshold.

Rain gauge measurements represent a space in a very small area, while satellite precipitation products have a spatial resolution of several kilometres or more. Thus, high-resolution data are generally more similar to gauge data than low-resolution data. Furthermore, the characteristic scale is small for convective systems and large for frontal rain systems. Convective precipitation dominates in region SE, while a frontal rain system dominates in regions CE and NE. Thus, a rain gauge measurement can represent a space in a smaller area in region SE than in regions CE and NE. Therefore, discrepancies between rain gauge observations and satellite estimates are lower in region SE than in regions CE and NE. CMORPH performed poorly in regions NW and TP, where orographic rain systems dominate (Hirpa et al., 2010; Romilly and Gebremichael, 2011; Gao and Liu, 2013). Our results are consistent with these findings.

It is expected that the DCDF method also applied to reanalysis precipitation data (e.g. ERA-Interim, 0.75°/6-hourly). First, the assumption that T_b has a similar cumu-

Table 2. Validation results of the daily precipitation for CMORPH and DCDF in 2014 in the six study regions.

Indexes	Time	Type	SE	CE	NE	CW	NW	TP
Bias (%)	1 year	CMORPH	−29.60	−12.82	−7.09	−5.57	120.22	26.41
		DCDF	−3.91	11.54	15.85	32.82.	145.43	52.33
	SP	CMORPH	−20.82	−3.31	−45.50	45.44	159.02	83.32
		DCDF	−7.35	2.94	31.23	50.92	191.79	100.36
	SU	CMORPH	−22.12	3.17	6.78	−43.92	143.11	−9.49
		DCDF	−10.47	2.66	5.94	25.91	217.04	7.53
	FA	CMORPH	−57.75	−33.00	−16.90	10.90	114.44	43.22
		DCDF	5.92	33.95	19.88	25.78	128.51	59.77
	WI	CMORPH	−94.19	−32.83	−96.20	1042	−73.75	1655
		DCDF	10.35	20.54	22.39	1874	54.58	2106
Root mean square error	1 year	CMORPH	12.20	6.69	6.71	3.85	2.32	4.50
		DCDF	7.94	4.38	5.16	4.74	3.96	6.08
	SP	CMORPH	16.23	4.79	3.13	2.81	2.41	2.70
		DCDF	11.81	7.32	2.77	2.80	3.09	3.45
	SU	CMORPH	16.61	10.25	12.39	5.74	3.38	7.43
		DCDF	13.83	10.95	10.64	6.98	5.13	10.27
	FA	CMORPH	6.14	6.90	3.89	3.93	1.94	3.46
		DCDF	0.19	6.44	2.67	4.51	3.72	3.98
	WI	CMORPH	3.80	1.59	0.68	1.61	0.65	2.45
		DCDF	2.86	2.05	0.41	2.47	1.14	3.49
CC	1 year	CMORPH	0.52	0.32	0.32	0.17	0.33	0.28
		DCDF	0.60	0.47	0.42	0.29	0.29	0.33
	SP	CMORPH	0.59	0.34	0.36	0.17	0.07	0.04
		DCDF	0.66	0.40	0.38	0.17	0.05	0.04
	SU	CMORPH	0.36	0.19	0.25	0.17	0.40	0.23
		DCDF	0.48	0.26	0.46	0.44	0.44	0.37
	FA	CMORPH	0.40	0.50	0.36	0.07	0.32	0.11
		DCDF	0.52	0.53	0.46	0.10	0.21	0.08
	WI	CMORPH	0.31	0.02	0.00	0.05	0.03	0.06
		DCDF	0.52	0.17	0.05	0.01	0.02	0.15
Probability of detection	1 year	CMORPH	0.64	0.59	0.51	0.76	0.52	0.80
		DCDF	0.77	0.74	0.62	0.80	0.69	0.87
	SP	CMORPH	0.68	0.52	0.45	0.82	0.51	0.70
		DCDF	0.80	0.66	0.60	0.95	0.63	0.72
	SU	CMORPH	0.86	0.69	0.78	0.82	0.80	0.91
		DCDF	0.99	0.85	0.91	0.87	0.90	1.00
	FA	CMORPH	0.50	0.67	0.46	0.71	0.80	0.72
		DCDF	0.65	0.75	0.59	0.84	0.92	0.89
	WI	CMORPH	0.22	0.19	0.00	0.28	0.59	0.14
		DCDF	1.00	1.00	1.00	1.00	1.00	1.00
False alarm rate	1 year	CMORPH	0.30	0.63	0.48	0.65	0.76	0.65
		DCDF	0.35	0.59	0.55	0.72	0.81	0.64
	SP	CMORPH	0.17	0.76	0.63	0.71	0.85	0.78
		DCDF	0.21	0.70	0.73	0.81	0.92	0.85
	SU	CMORPH	0.31	0.53	0.36	0.33	0.68	0.30
		DCDF	0.43	0.52	0.41	0.57	0.79	0.38
	FA	CMORPH	0.46	0.52	0.58	0.69	0.68	0.73
		DCDF	0.48	0.58	0.66	0.89	0.66	0.91
	WI	CMORPH	0.54	0.90	1.00	0.96	0.76	0.99
		DCDF	0.61	0.95	1.00	1.00	0.97	1.00
Heidke skill score	1 year	CMORPH	0.43	0.25	0.34	0.08	0.14	0.13
		DCDF	0.39	0.31	0.35	0.14	0.08	0.09
	SP	CMORPH	0.41	0.13	0.23	0.00	0.09	0.01
		DCDF	0.44	0.21	0.29	0.03	0.11	0.07
	SU	CMORPH	0.38	0.29	0.40	0.25	0.17	0.22
		DCDF	0.32	0.35	0.37	0.33	−0.08	0.16
	FA	CMORPH	0.38	0.39	0.27	−0.02	0.17	0.04
		DCDF	0.39	0.48	0.34	0.01	−0.06	0.01
	WI	CMORPH	0.14	0.00	−0.01	−0.06	0.16	−0.06
		DCDF	0.21	0.07	0.03	−0.11	0.29	−0.16

lative frequency to the precipitation rate at certain spatial and temporal scales is also applied to reanalysis data. Second, most average R^2 values between T_b and CMORPH are higher than 0.90, which may infer that the poor performance of the DCDF approach in winter and in mountainous regions is mainly caused by the low accuracy of CMORPH. Therefore, using reanalysis data for downscaling may be better than satellite products.

6 Conclusions

Precipitation data with high spatial and temporal resolutions are highly needed in basin-scale hydrological and meteorological studies. Based on the works by Barrett et al. (1991) and Kidd and Levizzani (2011), this study proposed a DCDF method to obtain precipitation data at the hourly, 0.05° scale. The method was demonstrated using the CMORPH dataset and FY2-E GEO-IR T_b data for 2014. With the establishment of a power function relationship, improved precipitation estimates at hourly and 0.05° resolution were produced. The DCDF precipitation estimates were validated using rain gauge data from six 5° × 5° regions in China with different climate and geographical conditions.

There are three key points of the DCDF method. First, it explores the advantages of satellite precipitation estimates and GEO-IR data. The DCDF method assumes a monotonically decreasing T_b rate with an increase of precipitation rate, and it assumes that T_b data have the same cumulative frequency as that of the precipitation rate for certain spatial and temporal scales. The matching process is implemented by quantile-mapping the CDF of T_b onto that of the precipitation rate. Second, the sample area where the CDF matching was conducted needs to be large enough for a reasonable sample size, but small enough to represent the local characteristics. In this study, the size of the sample area was 1° × 1° grid over a 10-day period, based on the characteristic scale of precipitation clouds. Third, a power function relationship between the precipitation rate and T_b was established for each sample area. Meanwhile, a rain–no-rain threshold was obtained as the T_b value with the same cumulative frequency as that of the precipitation rate defined at the critical point of rain–no-rain. Generally, the threshold was the maximum T_b in the CDF-matching procedure.

The established fitting relationships generally reflected the precipitation characteristics well in the six validation regions. For the distributions of precipitation clouds, the DCDF precipitation estimates showed a similar spatial distribution to that produced by CMORPH, but it reflected more detailed moving and changing processes of rainfall under the condition that DCDF performed better than or nearly equivalent to CMORPH. The DCDF method can effectively reflect the precipitation characteristics among rain systems and climatological regimes. At the hourly scale, both DCDF and CMORPH coincided with the disdrometer data at precipitation rates ranging from 1 to 10 mm h^{-1}. The DCDF effectively reflected the peak of each rainfall event, but could not exactly identify the starting and ending times of rainy events. The DCDF may detect non-rainy events as rainy events especially in dry seasons, while CMORPH reported low-rain events as non-rainy events. At the daily scale, DCDF and CMORPH had nearly equivalent performances at the regional scale, and 79 % DCDF may perform better than or nearly equivalent to CMORPH at the point (rain gauge) scale. Generally, the DCDF performed better (bias: 7.35–10.35 %; CC: 0.48–0.60) than the original CMORPH product (bias: 20.82–94.19 %; CC: 0.31–0.59) over the regions where convective precipitation dominates. It performed as well as the CMORPH product over the regions where frontal rain systems dominate and relatively poorly over mountainous or hilly areas where orographic rain systems dominate.

Author contributions. RG and YL developed the method. HZ and YZ were involved in the data processing. RG prepared the manuscript and all co-authors were asked to review the manuscript.

Competing interests. The authors declare that they have no conflict of interest.

Acknowledgements. This work was partially supported by the State Key Program of the National Natural Science Foundation of China under grant 41430855 and by the National High Technology Research and Development Program under grant 2013AA12A301. The authors would like to thank Chris Kidd for providing a report of SSM/I rainfall algorithms, and Pingping Xie for his guidance at the University of Maryland. The authors would like to thank research associates Bo Zhong and Shanlong Wu for data collection and processing at the Institute of Remote Sensing and Digital Earth (RADI), Chinese Academy of Sciences.

Edited by: Matthias Bernhardt

References

Arkin, P. A. and Meisner, B. N.: The relationship between large-scale convective rainfall and cold cloud over the western hemisphere during 1982–1984, Mon. Weather Rev., 115, https://doi.org/10.1175/1520-0493(1987)115<0051:TRBLSC>2.0.CO;2, 1987.

Arkin, P., Turk, J., Ebert, B., Bauer, P., and Sapiano, M.: Evaluation of high resolution precipitation forecasts and analyses from satellite observations, in: AGU Fall Meeting, American Geophysical Union, 1:4, 2006.

Ba, M. B. and Gruber, A.: GOES multispectral rainfall algorithm (GMSRA), J. Appl. Meteorol., 40, 1500–1514, https://doi.org/10.1175/1520-0450(2001)040<1500:GMRAG>2.0.CO;2, 2001.

Barrett, E. C.: The Estimation of Monthly Rainfall from Satellite Data, Mon. Weather Rev., 98, https://doi.org/10.1175/1520-0493(1970)098<0322:TEOMRF> 2.3.CO;2, 1970.

Barrett, E. C. and Martin, D. W.: The Use of Satellite Data in Rainfall Monitoring, Academic Press, London, 1981.

Barrett, E. C., Beaumont, M. J., Brown, K. A., and Kidd, C.: Development and testing of SSM/I rainfall algorithms for regional and global use: NA86AA-H-RA001, Final Rep. to the U.S. Dept. of Commerce, Washington, DC, 77, 1991.

Bitew, M. M. and Gebremichael, M.: Assessment of satellite rainfall products for streamflow simulation in medium watersheds of the Ethiopian highlands, Hydrol. Earth Syst. Sci., 15, 1147–1155, https://doi.org/10.5194/hess-15-1147-2011, 2011.

Cannon, A. J.: Probabilistic Multisite Precipitation Downscaling by an Expanded Bernoulli–Gamma Density Network, J. Hydrometeor., 9, 1284–1300, https://doi.org/10.1175/2008JHM960.1, 2008.

Chen, S., Hong, Y., Cao, Q., Gourley, J. J., Kirstetter, P. E., Yong, B., Tian, Y., Zhang, Z., Shen, Y., Hu, J., and Hardy, J.: Similarity and difference of the two successive v6 and v7 trmm multisatellite precipitation analysis performance over china, J. Geophys. Res.-Atmos., 118, 13060–13074, https://doi.org/10.1002/2013JD019964, 2013

Collischonn, B., Collischonn, W., Carlos, E., and Morelli, T.: Daily hydrological modeling in the Amazon basin using TRMM rainfall estimates, J. Hydrol., 360, 207–216, https://doi.org/10.1016/j.jhydrol.2008.07.032, 2008.

Dinku, T., Ceccato, P., Lemma, M., Connor, S. J., and Ropelewski, C. F.: Validation of satellite rainfall products over east africa's complex topography, Int. J. Remote Sens, 28, 1503–1526, https://doi.org/10.1080/01431160600954688, 2007.

Duan, Z. and Bastiaanssen, W. G. M.: First results from Version 7 TRMM 3B43 precipitation product in combination with a new downscaling-calibration procedure, Remote Sens. Environ., 131, 1–13, https://doi.org/10.1016/j.rse.2012.12.002, 2013.

Ebert, E. E., Janowiak, J. E., and Kidd, C.: Comparison of near real time precipitation estimates from satellite observations and numerical models, B. Amer. Meteor. Soc., 88, 47–64, https://doi.org/10.1175/BAMS-88-1-47, 2007.

Ferraro, R. R., Weng, F., Grody, N. C., and Zhao, L.: Precipitation characteristics over land from the noaa-15 amsu sensor, Geophys. Res. Lett., 27, 2669–2672, https://doi.org/10.1029/2000GL011665, 2000.

Ferraro, R. R.: Special sensor microwave imager derived global rainfall estimates for climatological applications, J. Geophys. Res., 102, 16715–16736, https://doi.org/10.1029/97JD01210, 1997.

Gao, Y. C. and Liu, M. F.: Evaluation of high-resolution satellite precipitation products using rain gauge observations over the Tibetan Plateau, Hydrol. Earth Syst. Sci., 17, 837–849, https://doi.org/10.5194/hess-17-837-2013, 2013.

Giorgi, F. and Mearns, L. O.: Introduction to special section: regional climate modeling revisited, J. Geophys. Res., 104, 6335–6352, https://doi.org/10.1029/98JD02072, 1999.

Gottschalck, J., Meng, J., Rodell, M., and Houser, P.: Analysis of multiple precipitation products and preliminary assessment of their impact on global land data assimilation system land surface states, J. Hydrometeor., 6, 573–598, https://doi.org/10.1175/JHM437.1, 2005.

Greene, J. S. and Morrissey, M. L.: Validation and uncertainty analysis of satellite rainfall algorithms, Prof. Geogr., 52, 247–258, https://doi.org/10.1111/j.0033-0124.2000.t01-1-.x, 2000.

Hay, L. E., Mccabe, G. J., Wolock, D. M., and Ayers, M. A.: Simulation of precipitation by weather type analysis, Water Resour. Res., 27, 493–501, https://doi.org/10.1029/90WR02650, 1991.

Hirpa, F. A., Gebremichael, M., and Hopson, T.: Evaluation of High-Resolution Satellite Precipitation Products over Very Complex Terrain in Ethiopia, J. Appl. Meteorol. Clim., 49, 1044–4051, https://doi.org/10.1175/2009JAMC2298.1, 2009.

Hu, Q., Yang, D., Li, Z., Mishra, A. K., Wang, Y., and Yang, H.: Multi-scale evaluation of six high-resolution satellite monthly rainfall estimates over a humid region in china with dense rain gauges, Int. J. Remote Sens, 35, 1272–1294, https://doi.org/10.1080/01431161.2013.876118, 2014.

Huffman, G. J., Adler, R. F., Arkin, P. A., Chang, A., Ferraro, R., Gruber, A., Janowiak, J., Mcnab, A., Rudolf, B., and Schneider, U.: The Global Precipitation Climatology Project (GPCP) combined precipitation data set, B. Amer. Meteor. Soc., 78, 5–20, https://doi.org/10.1175/1520-0477(1997)078<0005:TGPCPG> 2.0.CO;2, 1997.

Huffman, G. J., Adler, R. F., Bolvin, D. T., and Gu, G.: Improving the global precipitation record: GPCP version 2.1, Geophys. Res. Lett., 36, L17808, https://doi.org/10.1029/2009GL040000, 2009.

Huffman, G. J., Adler, R. F., Morrissey, M. M., Curtis, S., Joyce, R. J., McGavock, B., and Susskind, J.: Global precipitation at one-degree daily resolution from multi-satellite observations, J. Hydrometeor., 2, 36–50, https://doi.org/10.1175/1525-7541(2001)002<0036:GPAODD>2.0.CO;2, 2001.

Huffman, G. J., Adler, R. F., Bolvin, D. T., Gu, G., Nelkin, E. J., Bowman, K. P., Hong, Y., Stocker, E. F., and Wolff, D. B.: The TRMM multisatellite precipitation analysis (TMPA): Quasi-global, multiyear, combined-sensor precipitation estimates at fine scales, J. Hydrometeor., 8, 38–55, https://doi.org/10.1175/JHM560.1, 2007.

Hughes, D. A.: Comparison of satellite rainfall data with observations from gauging station networks, J. Hydrol., 327, 399–410, https://doi.org/10.1016/j.jhydrol.2005.11.041, 2006.

Immerzeel, W. W., Rutten, M. M., and Droogers, P.: Spatial downscaling of TRMM precipitation using vegetative response on the Iberian Peninsula, Remote Sens. Environ., 113, 362–370, https://doi.org/10.1016/j.rse.2008.10.004, 2009.

Javanmard, S., Yatagai, A., Nodzu, M. I., BodaghJamali, J., and Kawamoto, H.: Comparing high-resolution gridded precipitation data with satellite rainfall estimates of TRMM_3B42 over Iran, Adv. Geosci., 25, 11-9-125, https://doi.org/10.5194/adgeo-25-119-2010, 2010.

Jia, S. F., Zhu,W. B., Lu, A. F., and Yan, T. T.: A statistical spatial downscaling algorithm of TRMM precipitation based on NDVI and DEM in the QaidamBasin of China, Remote Sens. Environ., 115, 3069–3079, https://doi.org/10.1016/j.rse.2011.06.009, 2011.

Joyce, R. J., Janowiak, J. E., Arkin, P. A., and Xie, P.: CMORPH: A method that produces global precipitation estimates from passive microwave and infrared data at high spatial and temporal resolution, J. Hydrometeor., 5, 487–503, https://doi.org/10.1175/1525-7541(2004)005<0487:CAMTPG>2.0.CO;2, 2004.

Karl, T. R., Wang, W. C., Schlesinger, M. E., Knight, R. W., and Portman, D.: A method of relating general circulation model simulated climate to the observed local climate. part I: seasonal statistics, J. Climate, 3, 1053–1079, https://doi.org/10.1175/1520-0442(1990)003<1053:AMORGC>2.0.CO;2, 1990.

Kenabatho, P. K., Parida, B. P., and Moalafhi, D. B.: The value of large-scale climate variables in climate change assessment: The case of Botswana's rainfall, Phys. Chem. Earth Parts, 50–52, https://doi.org/10.1016/j.pce.2012.08.006, 2012.

Kidd, C. and Levizzani, V.: Status of satellite precipitation retrievals, Hydrol. Earth Syst. Sci., 15, 1109–1116, https://doi.org/10.5194/hess-15-1109-2011, 2011.

Kidd, C., Kniveton, D. R., Todd, M. C., and Bellerby, T. J.: Satellite rainfall estimation using combined passive microwave and infrared algorithms, J. Hydrometeor., 4, 1088, https://doi.org/10.1175/1525-7541(2003)004<1088:SREUCP>2.0.CO;2, 2003.

Krajewski, W. F. and Smith, J. A.: Radar hydrology: rainfall estimation, Adv. Water Resour., 25, 1387–1394, https://doi.org/10.1016/S0309-1708(02)00062-3, 2002.

Kummerow, C., Hong, Y., Olson, W. S., Yang, S., Adler, R. F., Mccollum, J., Ferraro, R., Petty, G., Shin, D. B., and Wilheit, T. T.: The evolution of the goddard profiling algorithm (gprof) for rainfall estimation from passive microwave sensors, J. Appl. Meteorol, 40, 1801–1820, https://doi.org/10.1175/1520-0450(2001)040<1801:TEOTGP>2.0.CO;2, 2001.

Ma, Y. Z., Liu, X. N., and Xu, S: The description of Chinese radiation data and their quality control procedures, Meteorol. Sci. 2, 53–56, 1998.

Magnusson, M., Vaskevicius, N., Stoyanov, T., Pathak, K., and Birk, A.: Beyond points: evaluating recent 3d scan-matching algorithms, IEEE Int. Conf. Robot., 3631–3637, https://doi.org/10.1109/ICRA.2015.7139703, 2015.

Mekonnen, G., Witold, F. K., Tomas, M. O., Yukarin, T., Phillip, A., and Katayama, M.: Scaling of tropical rainfall as observed by TRMM precipitation radar, Atmos. Res., 88, 337–354, https://doi.org/10.1016/j.atmosres.2007.11.028, 2008.

Orlanski, I.: A rational division of scales for atmospheric processes, B. Am. Meteor. Soc., 56, 527–530, 1975.

Prigent, C.: Precipitation retrieval from space: an overview, Comptes Rendus Geosciences, 342, 380–389, https://doi.org/10.1016/j.crte.2010.01.004, 2010.

Rickenbach, T. M.: Convection in TOGA COARE: Horizontal Scale, Morphology, and Rainfall Production, J. Atmos. Sci., 55, 2715–2729, https://doi.org/10.1175/1520-0469(1998)055<2715:CITCHS>2.0.CO;2, 1998.

Romilly, T. G. and Gebremichael, M.: Evaluation of satellite rainfall estimates over Ethiopian river basins, Hydrol. Earth Syst. Sci., 15, 1505–1514, https://doi.org/10.5194/hess-15-1505-2011, 2011.

Schmidli, J., Frei, C., and Vidale, P. L.: Downscaling from gcm precipitation: a benchmark for dynamical and statistical downscaling methods, Int. J. Climatol., 26, 679–689, https://doi.org/10.1002/joc.1287, 2006.

Smith, E. A., Lamm, J. E., Adler, R., Alishouse, J., Aonashi, K., Barrett, E. C, Bear, W., Chang, A., Ferraro, R., Ferriday, J., Goodman, S., Grpdy, N., Kidd, C., Kniveton, D., Kummerow, C., Liu, G., Marzano, F., Mugnai, A., Olson, W., Petty, G., Shibata, A., Spencer, R., Wentz, F., Wilheit, T., and Zipser, E.: Results of the WetNet PIP-2 Project, J. Atmos. Sci., 55, 1483–1536, https://doi.org/10.1175/1520-0469(1998)055<1483:ROWPP>2.0.CO;2, 1998.

Sohn, B. J., Han, H. J., and Seo, E. K.: Validation of satellite-based high-resolution rainfall products over the korean peninsula using data from a dense rain gauge network, J. Appl. Meteorol. Clim., 49, 367–370, https://doi.org/10.1175/2009JAMC2266.1, 2010.

Thiemig, V., Rojas, R., Zambranobigiarini, M., Levizzani, V., and De Roo, A.: Validation of satellite-based precipitation products over sparsely gauged african river basins, J. Hydrometeor., 13, 1760–1783, https://doi.org/10.1175/JHM-D-12-032.1, 2012.

Tobler, W.: On the first law of geography: A reply, Ann. Assoc. Amer. Geog., 94, 304–310, https://doi.org/10.1111/j.1467-8306.2004.09402009.x, 2004.

Trapp, R. J.: Mesoscale-convective processes in the atmosphere, Cambridge University Press, New York, USA, 346, 2013.

Turk, F. J. and Miller, S. D.: Toward improved characterization of remotely sensed precipitation regimes with modis/amsr-e blended data techniques, IEEE T. Geosci. Remote, 43, 1059–1069, https://doi.org/10.1109/TGRS.2004.841627, 2005.

Ushio, T., Sasashige, K., Kubota, T., Shige, S., Okamoto, K., Aonashi, K., Inoue, T., Takahashi, N., and Iguchi, T., Kachi, M, Oki, R., Morimoto, T., and Kawasaki, Z. I.: A Kalman filter approach to the Global Satellite Mapping of Precipitation (GSMaP) from combined passive microwave and infrared radiometric data, J. Meteorol. Soc. Japan, 87, 137–151, https://doi.org/10.2151/jmsj.87A.137, 2009.

Wigley, T. M. L., Jones, P. D., Briffa, K. R., and Smith, G.: Obtaining sub-grid-scale information from coarse-resolution general circulation model output, J. Geophys. Res.-Atmos., 95, 1943–1953, https://doi.org/10.1029/JD095iD02p01943, 1990.

Wilby, R. L. and Wigley, T. M. L.: Downscaling general circulation model output: a review of methods and limitations, Prog. Phys. Geog., 21, 530–548, https://doi.org/10.1177/030913339702100403, 1997.

Wilby, R. L., Dawson, C. W., and Barrow, E. M.: Sdsm-a decision support tool for the assessment of regional climate change impacts, Environ. Modell. Softw., 17, 145–157, https://doi.org/10.1016/S1364-8152(01)00060-3, 2002.

Wilks, D. S.: Statistical Methods in the Atmospheric Science, Academic, San Diego, Calif, 465, 1995.

Willems, P. and Vrac, M.: Statistical precipitation downscaling for small-scale hydrological impact investigations of climate change, J. Hydrol., 402, 193–205, https://doi.org/10.1016/j.jhydrol.2011.02.030, 2011.

9

Estimating the effect of rainfall on the surface temperature of a tropical lake

Gabriel Gerard Rooney[1], **Nicole van Lipzig**[2], **and Wim Thiery**[3,4]

[1]Met Office, FitzRoy Road, Exeter, EX1 3PB, UK
[2]KU Leuven, Department of Earth and Environmental Sciences, Celestijnenlaan 200E, 3001 Leuven, Belgium
[3]ETH Zurich, Institute for Atmospheric and Climate Science, Universitaetstrasse 16, 8092 Zurich, Switzerland
[4]Vrije Universiteit Brussel, Department of Hydrology and Hydraulic Engineering, Pleinlaan 2, 1050 Brussels, Belgium

Correspondence: Gabriel Gerard Rooney (gabriel.rooney@metoffice.gov.uk)

Abstract. We make use of a unique high-quality, long-term observational dataset on a tropical lake to assess the effect of rainfall on lake surface temperature. The lake in question is Lake Kivu, one of the African Great Lakes, and was selected for its remarkably uniform climate and availability of multi-year over-lake meteorological observations. Rain may have a cooling effect on the lake surface by lowering the near-surface air temperature, by the direct rain heat flux into the lake, by mixing the lake surface layer through the flux of kinetic energy and by convective mixing of the lake surface layer. The potential importance of the rainfall effect is discussed in terms of both heat flux and kinetic energy flux. To estimate the rainfall effect on the mean diurnal cycle of lake surface temperature, the data are binned into categories of daily rainfall amount. They are further filtered based on comparable values of daily mean net radiation, which reduces the influence of radiative-flux differences. Our results indicate that days with heavy rainfall may experience a reduction in lake surface temperature of approximately 0.3 K by the end of the day compared to days with light to moderate rainfall. Overall this study highlights a new potential control on lake surface temperature and suggests that further efforts are needed to quantify this effect in other regions and to include this process in land surface models used for atmospheric prediction.

1 Introduction

Lakes are important features of the terrestrial environment for physical, ecological, economic and recreational reasons. Physically, lake–atmosphere interactions can influence the local weather and climate. Thus their representation in Earth system modelling has increased in complexity in recent years. Lake water surface temperature (LWST) is of particular relevance to atmospheric modelling due to the contrast in temperature, and hence in boundary-layer fluxes, that often exists between lakes and their surroundings (Mironov et al., 2010).

At high latitudes, correct prediction of freezing temperatures, and thus ice-cover periods, is important to obtain accurate boundary-layer fluxes. In the tropics, varying temperature contrasts between lakes and the surrounding land may be associated with cycles of severe weather. For instance, remote sensing data highlight an important impact of the African Great Lakes on the diurnal precipitation and thunderstorm cycle, especially over Lake Victoria and Lake Tanganyika (Camberlin et al., 2017; Thiery et al., 2017). During the afternoon, the typical tropical convective precipitation falls over land, but hardly any rainfall is observed over the lakes. At night, in contrast, very little precipitation is produced over land, while strong convection develops over the lakes, leading to high precipitation amounts. This phenomenon is caused by the diurnal cycle of the lake–land temperature difference, which leads to land breezes converging over the lake surface during the night. When these air masses, moistened by the lake, lift up into the atmosphere, they generate convective precipitation and often thunderstorms (Docquier et al., 2016; Thiery et al., 2015, 2016). The African Great Lakes are thus important regulators of the East African climate, which continues to present a challenge to modellers (James et al., 2018; Woodhams et al., 2018).

Adequately observing and modelling tropical lake–atmosphere interactions often remains a challenge, even though efforts have been made to quantify these exchanges (Verburg and Hecky, 2003; Verburg and Antenucci, 2010; Thiery et al., 2014a, b; Delandmeter et al., 2018; Weyhenmeyer et al., 2017). Moreover, important uncertainties remain present in several of the reference products, notably regarding precipitation (Dinku et al., 2008; Sylla et al., 2013; Awange et al., 2016; Kimani et al., 2017), hence the need for high-quality in situ meteorological measurements over the African Great Lakes (Anyah and Semazzi, 2004; Anyah et al., 2006; Anyah and Semazzi, 2009). To help address this need, a state-of-the-art automatic weather station was installed in 2012 on Lake Kivu (AWS Kivu).

Lakes interact with the atmosphere via a variety of processes. Physical lake models developed for use in a meteorological context have thus far concentrated on lake–atmosphere interaction through turbulent and radiative fluxes. The effects of rain on LWST, both directly from thermal perturbation and indirectly from changing the lake stratification, are little understood or represented to date. Evidence of the significance of rain effects, particularly in the tropics, is beginning to emerge however (Wei et al., 2014). Here, we combine theoretical considerations with analysis of the unique multi-annual dataset from AWS Kivu to estimate the significance of the rainfall effect on LWST in the tropics. As will be shown, the uniformity of the Lake Kivu climate increases its appeal as a location at which to assess this effect.

To summarise the structure of the following sections, in Sect. 2 the study area is described, Sect. 3 discusses the mechanisms of rain effect on LWST, the Kivu data are presented and analysed in Sect. 4, and finally the results are discussed in the conclusion.

2 Description of Lake Kivu

The African Great Lakes are of utmost importance for regional economies, as well as being essential to the survival of the local population. As the largest reservoir of freshwater in the tropics, they provide numerous ecosystem services to local communities, such as fishing grounds, drinking water and electricity. Lake Victoria alone directly supports 200 000 fishermen operating from its shores and sustains the livelihood of more than 30 million people living on its shores (East African Community, 2011). During the last decades, however, the African Great Lakes experienced fast changes in ecosystem structure and functioning, and their future evolution is a major concern (O'Reilly et al., 2003; Verburg et al., 2003; Verburg and Hecky, 2009; Borges et al., 2015). Moreover, outflow dam management, along with climate variability and change, exert a large influence on the water levels of the largest African Great Lake, Lake Victoria (Vanderkelen et al., 2018a, b).

Lake Kivu (01°35′–02° 30′ S, 028°50′–029°23′ E) is situated along the border of Rwanda and the Democratic Republic of the Congo and is one of the seven African Great Lakes (Fig. 1). The lake has a surface area of 2370 km^2, lies 1463 m above sea level and is up to 485 m deep. The outflow is located at the lake's southern tip and forms the Ruzizi River, which flows southwards into Lake Tanganyika. Although the lake is meromictic, the oxic mixolimnion deepens to 60–70 m during the dry season. Below that, the monimolimnion is found to be rich in nutrients and dissolved gases, in particular carbon dioxide and methane (Degens et al., 1973; Borges et al., 2011; Descy et al., 2012; Morana et al., 2015a, b, 2016). Interestingly, temperature and salinity within the monimolimnion increase with depth due to the input of heat and salts from deep geothermal springs (Degens et al., 1973; Spigel and Coulter, 1996; Schmid et al., 2005). Given its high altitude and location close to the Equator, surface water temperatures over Lake Kivu are relatively constant throughout the year.

3 Physical effects of rainfall

Rain may have an effect on the lake surface in four ways: (i) evaporative cooling of the near-surface air during precipitation, which induces an additional upward sensible heat flux from the lake towards the atmosphere; (ii) direct rain heat flux into the lake; (iii) mixing of the lake surface layer through the flux of kinetic energy; and finally (iv) convective mixing of the lake surface layer. The first of these ought to be parametrised by atmospheric models, as with related atmospheric effects like the reduction of insolation by cloud cover. The others lie mainly in the lake modelling domain. Hereafter, we discuss points (i)–(iv) above in some more detail.

3.1 Cooling of near-surface air

Raindrops falling into unsaturated air will cool through evaporation. Their passage through the air leads to heat transfer from the air, hence cooling the air. As the rainfall continues, the air will tend to saturation, and both rain and air will approach the air's original wet-bulb temperature. Thermodynamically, further quantification and parametrisation of this process requires consideration of various factors such as the

Figure 1. Maps of Lake Kivu geography and situation. The lower left corner of the large map is at 2.6° S, 28.7° E, and the upper right corner is at 1.5° S, 29.5° E. The lake is approximately 90 km long and 50 km wide. The purple line along the lake indicates the boundary between the Democratic Republic of the Congo to the west and Rwanda to the east. The weather station is situated approximately 3 km offshore, near Goma at the northern end of the lake. Its position is marked with a red square. The location of Lake Kivu within Africa is marked on the small map by a blue square. (©OpenStreetMap contributors. Mapping data are available under the Open Database License.)

atmospheric moisture and temperature profile, drop-size distribution, drop concentration, etc. (Schlesinger and Oh, 1988; Feingold, 1993).

At the extreme end of intense convection, atmospheric cooling and momentum transfer from rain may produce cold convective downdraughts, which transport cold air to the surface from higher levels (Kamburova and Ludlam, 1968; Knupp and Cotton, 1985; Srivastava, 1987). These additional dynamic effects are potentially less well represented or resolved in atmospheric models but are of importance in terms of gust hazard, as well as for their relevance to modelling of the convective cycle (Rooney, 2015; Thiery et al., 2017).

3.2 Direct heat flux

The specific heat capacity of water is approximately $4.2 \times 10^3\,\mathrm{J\,kg^{-1}\,K^{-1}}$. A rainfall rate of $1\,\mathrm{mm\,h^{-1}}$ is thus equivalent to a heat flux of approximately $1.2\,\Delta T\,\mathrm{W\,m^{-2}}$, where ΔT is the temperature difference between the rain and the surface which absorbs it. Rain temperature variation relative to air or surface temperature is not well known. According to Byers et al. (1949), rain may be much colder than the ambient air at the start of a thunderstorm but possibly comparable at later stages. This is presumably due to evaporative cooling reducing the air temperature to nearer that of the rain over time and is consistent with the approximation of the rain temperature to the wet-bulb temperature (Wei et al., 2014, see also references therein).

On seasonal to decadal timescales, the sensible heat contribution by rainfall is deemed small (Verburg et al., 2011). Wei et al. (2014) have estimated that this flux is largest in the tropics, with seasonal mean values of the order of $-2\,\mathrm{W\,m^{-2}}$, and they also state that neglect of this contribution may partly explain some air temperature biases in climate reanalyses. In their study, van Beek et al. (2012) have noted its significance for correctly estimating the surface temperature of tropical rivers. On much shorter timescales, surface cooling due to rain may affect weather patterns. On land, this process has been widely studied (Seneviratne et al., 2010; Taylor et al., 2012; Guillod et al., 2015; Lorenz et al., 2016). The rain effect in water bodies is more uncertain, although recently there have been some interesting observational studies, e.g. Reverdin et al. (2012).

3.3 Mechanical and convective mixing

As well as the direct effects of an additional heat flux, rainfall may produce a perturbation of LWST by the mechanical and convective mixing of the near-surface portion of the lake.

An early study by Katsaros and Buettner (1969) indicated that large drops (3 mm diameter) produce mixing to depths of 10 cm and smaller drops (≤ 1.2 mm) mixed to perhaps one-third of this depth. Similarly, Green and Houk (1979) found that the presence of some large drops is very important in producing significant subsurface mixing.

Several subsequent studies of artificial rainfall have examined rainfall-generated turbulence in slightly more detail. Artificial heavy rainfall has been observed to produce turbulent mixing over depths of 10–20 cm in the study of Lange et al. (2000), with a drop size of approximately 3 mm, and that of Zappa et al. (2009) with a distribution of drop sizes in the range 0.3–5.3 mm, which was modelled on the measured natural-rain distributions of Marshall and Palmer (1948). By contrast, Harrison and Veron (2017), with drop sizes of approximately 1.3 mm, found turbulent kinetic energy (TKE) to be independent of artificial rainfall intensity, at even higher intensities, and suggested that a significant fraction of the energy went into producing small-length-scale or capillary motions within 1–2 cm of the surface. Again, these results would seem to indicate that drop size is an important factor. While the effects of drop impact in a deep liquid reservoir are quite complex, even for a single drop (Prosperetti et al., 1989; Rein, 1996), it may be reasonable as a first hypothesis to assume that a significant fraction of the kinetic energy of natural heavy rain goes into subsurface TKE production.

The kinetic energy flux of real rainfall has also been estimated in the context of soil erosion studies (Salles et al.,

2002; van Dijk et al., 2002; Yu et al., 2012). A rain rate of $5\,\mathrm{mm\,h^{-1}}$ may give rise to a kinetic energy flux F_K of around $0.02\,\mathrm{W\,m^{-2}}$, for example. The rate of TKE production in the surface boundary layer of a lake may then be assumed to scale as $\phi F_K/(\rho_w \ell)$, where ℓ is the depth to which the rain penetrates, ρ_w is the water density and ϕF_K is some fraction of F_K where $\phi \leq 1$ (see for example Townsend, 1976, chap. 2 for a discussion of TKE evolution). The significance of this rain-driven production of TKE may be gauged by comparison with the kinetic energy input from wind shear.

Mechanical surface forcing by wind upon lakes is usually modelled through matching of stress, so that the aqueous friction velocity at the lake surface u_{*l} is

$$u_{*l} = \left(\frac{\rho_a}{\rho_w}\right)^{1/2} u_*, \tag{1}$$

where u_* is the friction velocity in the atmospheric surface layer, and ρ_a is the air density. A typical atmospheric friction velocity of the order of $1\,\mathrm{m\,s^{-1}}$ then implies an aqueous value of $u_{*l} \approx 0.03\,\mathrm{m\,s^{-1}}$ (Anctil and Donelan, 1996; Csanady, 2001). The lake surface TKE production from wind-driven shear scales as $u_{*l}^3/(\kappa z)$, where z is the depth and $\kappa = 0.4$ is von Kármán's constant (e.g. Skyllingstad and Denbo, 1995).

Thus, setting $z = \ell$, the turbulent mixing from rain may be compared to that from wind shear by comparing $u_{*l}^3/(\kappa\ell)$ with $\phi F_K/(\rho_w \ell)$, or equivalently by comparing u_{*l} with $(\kappa\phi F_K/\rho_w)^{1/3}$. This last term is evaluated as being of the order of $0.02\,\mathrm{m\,s^{-1}}$ for heavy rain, using the value of F_K given above and ϕ in the range of 0.5–1, and is the same order of magnitude as the friction velocity for a moderate to strong wind. Hence, the turbulent mixing rates due to wind shear and rain may at times be comparable in the top few centimetres of a lake.

For cold rain falling onto a relatively warm lake, convective effects will presumably add to the mixing strength and depth. While Green and Houk (1979) concentrated mainly on the case of warm rain falling onto cold water, their experiments with the opposite temperature contrast showed cooling throughout the depth of their reservoir, i.e. to at least 0.4 m. The LWST perturbation caused by mixing effects will depend on the stratification of the lake near-surface region.

4 Data analysis

4.1 Instrumentation and measurements

AWS Kivu is installed on the research platform of the Rwanda Energy Company, approximately 3 km offshore of the cities of Gisenyi (Rwanda) and Goma (Democratic Republic of the Congo; see Fig. 2). Since 9 October 2012, AWS Kivu has provided continuous high-quality observations of near-surface meteorology and four-component radiation. The continuous time series obtained so far from AWS Kivu is, to our knowledge, unique among its kind in the tropics.

AWS Kivu consists of sensors for air temperature (T_a, °C), relative humidity (RH, %), air pressure (p, Pa), precipitation (P, mm), wind speed (U, $\mathrm{m\,s^{-1}}$), and direction (WD, °) and the four radiation components ($\mathrm{SW_{in}}$, $\mathrm{SW_{out}}$, $\mathrm{LW_{in}}$, $\mathrm{LW_{out}}$, all in $\mathrm{W\,m^{-2}}$). Details of the sensors are given in Table 1. LWST is calculated from the upwelling longwave irradiance using the Stefan–Boltzmann law and assuming an emissivity of 0.99 (Wan, 2008). The station is powered by a solar panel, and all sensors are placed at a height of 4.40 m above the water surface (Fig. 2a), except for wind speed and direction, which are measured at 7.20 m above the lake (Fig. 2b). While the station is mounted on a metal container (Fig. 2c), efforts were made to minimise its effect on the meteorological measurements. Notably, temperature, humidity and radiation sensors were mounted 6 m horizontally away from the container edge, making sure that recorded conditions are representative for the water surface.

Variables are sampled every 15 s, from which 30 min averages are calculated and stored. In the case of precipitation, accumulated values are stored, and for wind speed both mean and maximum values are recorded. Moreover, short periods of high-frequency radiation measurements enable an assessment of the potential effect of platform movements. Through a General Packet Radio Service (GPRS), the KU Leuven Regional Climate Studies group receives the observations directly from the station, allowing for remote problem detection.

The time span of measurements used here was between 13 September 2012 and 14 August 2017; however most of the analysis is based on 4 calendar years of data from 1 January 2013 to 30 December 2016.

4.2 Weather and climate

The data indicate that there is a remarkable uniformity of the lake climate. The annual air temperature range is around 14 K. The daily rainfall totals for 4 calendar years show a generally uniform spread but with a slightly drier period around July (Fig. 3). There seem to be two prevailing wind directions which do not vary greatly with the weather, inasmuch as this is represented by rainfall (see Sect. 4.3). This uniformity is beneficial for the following analysis since there are consequently fewer sources of variation upon which the lake behaviour may potentially depend.

The temporal variations in weather may be examined further using power spectra of rainfall and wind speed (Fig. 4). These show that a large part of the variation is on the diurnal scale. This provides justification for the use of average diurnal cycles to explore the behaviour of the system.

We also note the presence of some sub-daily peaks in the wind speed spectrum, the most dominant at a frequency corresponding to a period of approximately 8 h, and the next two corresponding to periods of approximately 6 and 12 h. (As will be shown later, the sub-daily wind fluctuations giving rise to these peaks are evident on plots of mean daily wind

Figure 2. Automatic weather station on Lake Kivu after its installation, 8 October 2012 (© Wim Thiery). Position (a) indicates the location of the temperature, relative humidity and radiation sensors at 4.40 m above the lake surface. Position (b) shows the location of the wind vane at 7.20 m above the lake surface. Position (c) indicates the container on which the station was mounted.

Table 1. AWS Kivu sensor specifications.

Data	Sensor	Range	Accuracy
Air pressure	Campbell Scientific CS100	600–1100 hPa	0.5 hPa
Air temperature	Campbell Scientific CS215	−40 to +70 °C	0.3 K
Relative humidity	Campbell Scientific CS215	0 %–100 %	2 %–4 %
Wind speed	Young 05103	0–100 m s^{-1}	0.3 m s^{-1}
Wind direction	Young 05103	0–360°	3°
Precipitation	Tipping bucket ARG100	0.2–500 mm h^{-1}	95 – 98 %
SW components	Kipp and Zonen CNR4	305–2800 nm	< 5 %
LW components	Kipp and Zonen CNR4	4500–42 000 nm	< 10 %

speed. These fluctuations are probably due to local circulations caused by lake or land breezes, and the largely bimodal distribution of the wind direction, also shown later, appears to support this interpretation.)

4.3 Partitioning by rainfall

To examine the effect of heavy rain, 4 years of data will be analysed (1 January 2013 to 30 December 2016). This amounts to 1457 days, as three days are omitted due to missing data. These data are referred to as ALL data in the following analysis. Based on daily rainfall totals, they may be divided into DRY, WET and VWET days. DRY days are days with no rainfall. The remaining days are partitioned into WET or VWET depending on whether the rainfall total is respectively less or greater than a threshold of 8 mm (Fig. 3). DRY has 690 days, WET has 585 days and VWET has 182 days.

Regarding the distribution of hourly rainfall over the 4 years, 2.4 % of hours had a rainfall total greater than 1 mm, and 0.6 % of hours had a rainfall total greater than 5 mm. Figure 5 shows the average hourly rainfall on WET and VWET days. Both show a minimum in rainfall around 06:00 UTC (08:00 LT, LT = UTC +2 h) and a peak around the middle

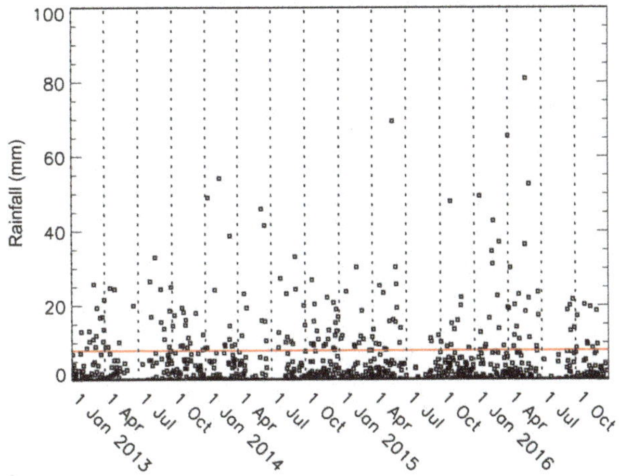

Figure 3. Daily rainfall totals for 4 years of the observational record, beginning on 1 January 2013. The red line marks the 8 mm point, which is used to partition rain days between WET (≤ 8 mm) and VWET(> 8 mm).

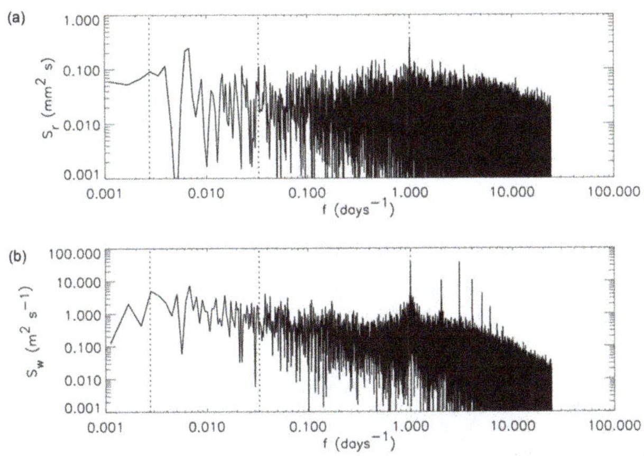

Figure 4. (a) Power spectrum of half-hourly rainfall amount (S_r, mm^2 s) and **(b)** power spectrum of half-hourly mean wind speed (S_w, m^2 s^{-1}), both for the period from 13 September 2012 to 14 August 2017. The vertical dotted lines mark frequencies (f, days^{-1}) corresponding to 1, 30 and 365 days. Both plots show a distinct peak at the daily frequency, with the wind speed also exhibiting several sub-daily peaks.

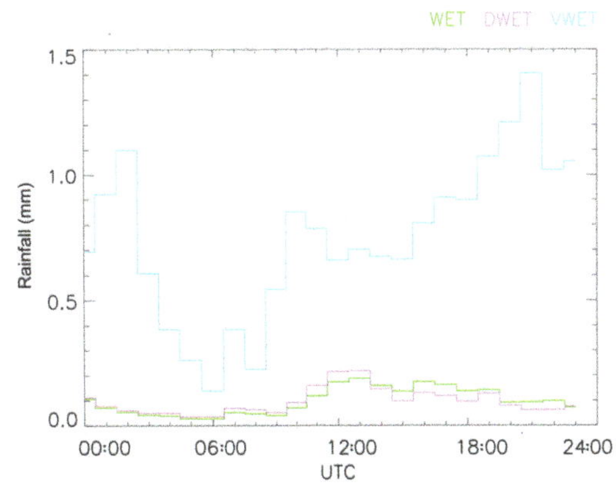

Figure 5. The average rainfall for each hour of the day, for VWET days (cyan), WET days (green) and DWET days (purple). In this and later plots, time is shown as UTC (coordinated universal time), which is 2 h behind LT (local time), i.e. LT = UTC +2 h.

of the day. There is also a later precipitation maximum in VWET, which may indicate the development of nighttime heavy storms.

The effect of daily weather on LWST is summarised in Fig. 6. It can be seen that the average diurnal cycles of air temperature and relative humidity are quite smooth, with a spread related to the rainfall category. Thus DRY days are the warmest and least humid, VWET days are the coldest and most humid, and WET and ALL days lie between these extremes. Both the air temperature and LWST start close together for the WET and VWET categories, but by the end of the day there is a difference in the mean, with VWET being colder than WET. Specifically, the average temperature difference over the last 6 h of the day is 0.42 K.

Atmospheric forcing of LWST is usually characterised in terms of turbulent or radiative fluxes, with turbulent fluxes depending on mean wind speed, lake–air temperature difference and near-surface humidity. For the categories described here, the choice of partitioning threshold between WET and VWET days coincidentally produces extremely similar graphs of mean wind speed. This has the effect of removing an important potential source of variation between these categories. The distributions of wind directions and speeds are also quite uniform; see Figs. 7 and 8.

Figure 9 shows the difference between the WET and VWET cases in terms of mean net radiation, on average over the course of a day. The lake has absorbed approximately 1.7×10^6 J m^{-2} more in the WET case. This is due presumably to WET days having less (or less thick) cloud cover than VWET days on average. This source of difference may be eliminated by adding a further constraint to the total absorbed radiation on WET days. This is described in the next subsection.

4.4 Further partitioning by rainfall and net radiation

WET days have a higher mean net radiation than VWET days (Fig. 9). To separate the effects of rainfall and radiation, the WET days may be further filtered for those with integrated net radiation below a specified amount, so that the mean value is reduced to equal to, or less than, that on VWET days. This subset of "dull" WET days is labelled DWET. With a threshold total radiation of 1.5×10^7 J m^{-2}, the average total radiation of DWET days is approximately 1.03×10^7 J m^{-2}, compared to an average total radiation of 1.05×10^7 J m^{-2} on VWET days. It is worth emphasising that the radiation threshold for DWET days was deliberately cho-

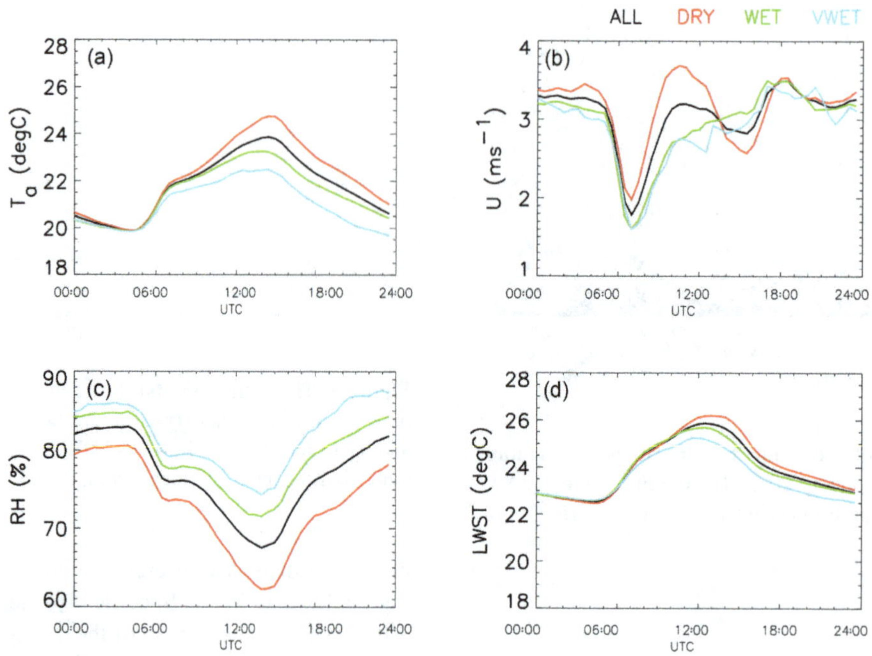

Figure 6. Mean diurnal cycles of **(a)** air temperature (T_a, °C), **(b)** wind speed (U, m s^{-1}), **(c)** relative humidity (RH, %) and **(d)** lake surface temperature (LWST, °C) from the upwelling longwave irradiance, assuming an emissivity of 0.99. The input data are those from the 4 years 2013–2016 of the campaign to represent all seasons equally. The colours correspond to diurnal cycles averaged over ALL days (black), DRY days (red), WET days (green) and VWET days (cyan).

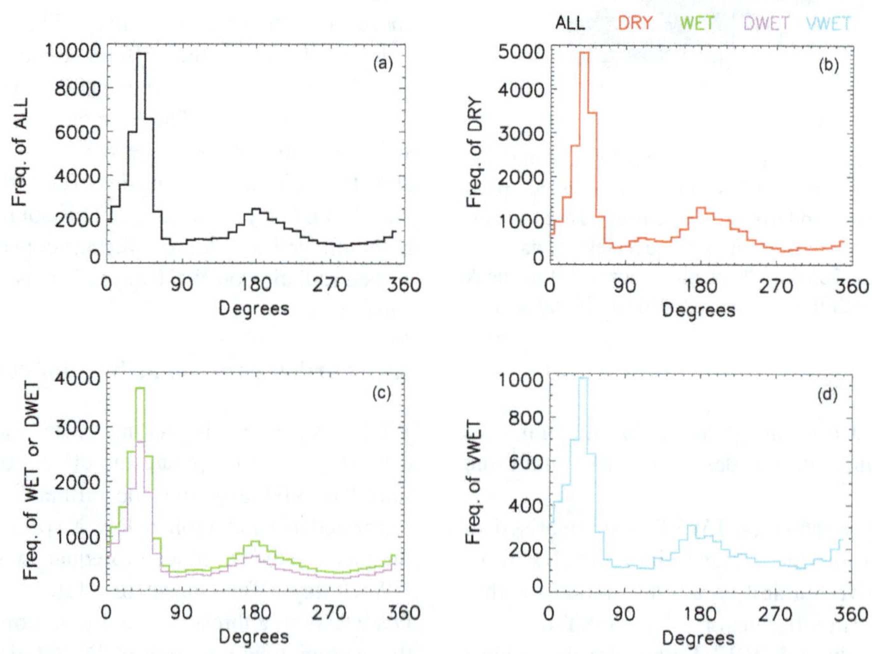

Figure 7. Histograms of half-hourly mean wind direction for 2013–2016. The colours are the same as in Figs. 5 and 6.

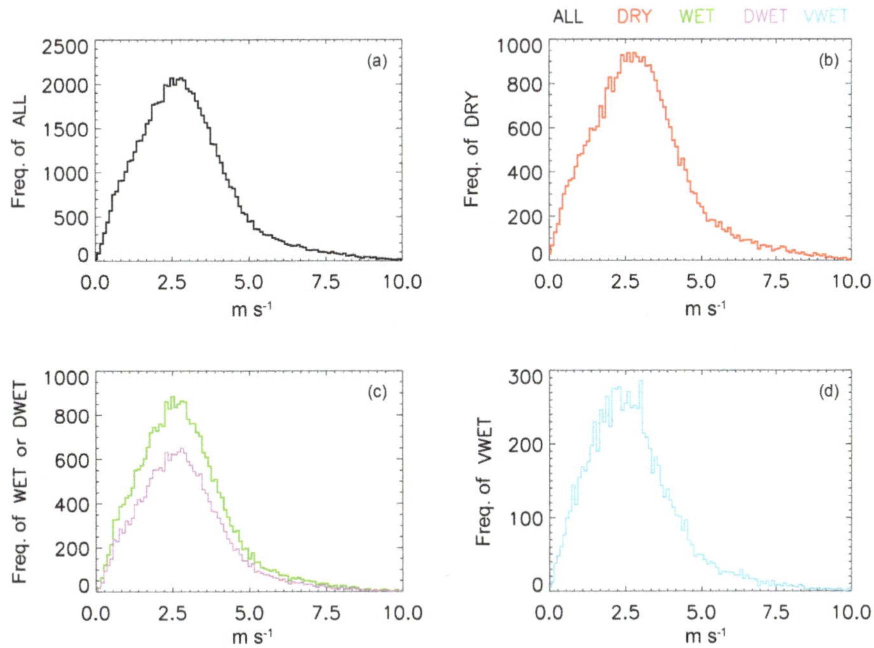

Figure 8. Histograms of half-hourly mean wind speed for 2013–2016. The colours are the same as in Figs. 5 and 6.

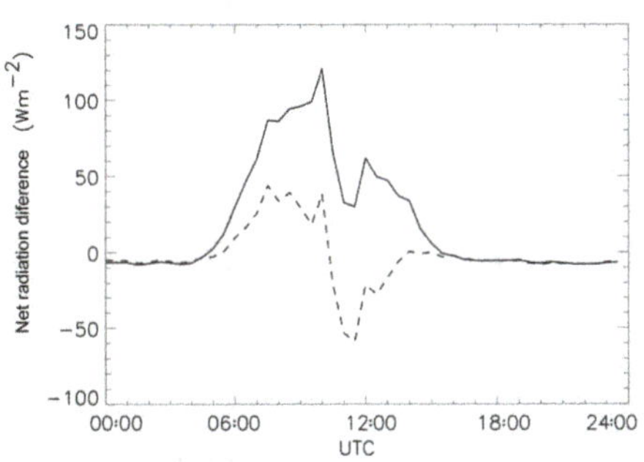

Figure 9. Comparison of the differences in net radiation. WET minus VWET net radiation is the solid line, and DWET minus VWET net radiation is the dashed line.

sen to allow a small margin, so that VWET days absorbed 1 %–2 % more radiation in the mean than DWET days. Thus any extra LWST cooling in VWET days compared to DWET days would be against a background of a slight excess of absorbed radiative energy on VWET days, in the mean.

Using this additional constraint, the number of DWET days in the 4-year period is 425 or 73 % of the WET days. Table 2 summarises the number of days of each type in each year, along with data on which type of day came immediately before and after. The distributions of wind directions and speeds for the DWET category are also plotted in Figs. 7 and 8. The diurnal cycle of the mean net radiation difference between DWET and VWET cases is shown in Fig. 9.

The average daily rainfall on DWET days is 2.31 mm, compared to 2.33 mm on WET days and 17.99 mm on VWET days. Thus, the contrast in rainfall amount is largely preserved by this resampling. The diurnal evolution is also plotted in Fig. 5, again showing that DWET is similar to WET.

The average diurnal evolution for the categories of ALL, DRY, DWET and VWET is shown in Fig. 10. It can be seen that the evolution of diurnal wind speed is reasonably unchanged for DWET days compared to that of WET days (Fig. 6), but the evolution of LWST on DWET days is closer to that of VWET days. However, most of the difference in surface temperature between these categories in the last few hours of the day remains, with an average temperature difference over the last 6 h of 0.29 K.

Considering the reliability of this difference, it may be noted that a difference of 0.3 K against a background at approximately 300 K is equivalent to a difference in upwelling longwave radiation of approximately 1.8 $\mathrm{W\,m^{-2}}$. However, since this is a difference between mean values taken over a minimum of 182 observations at each time of day, it should be compared with the standard error of the mean. That is, it only requires instrumental accuracy of $1.8 \times \sqrt{182} \approx 24\,\mathrm{W\,m^{-2}}$, which is within the instrument specification (Table 1).

In terms of significance, the standard error of the difference between the mean DWET and VWET values is given by

$$\mathrm{SED} = \sqrt{\frac{\sigma^2_{\mathrm{DWET}}}{N_{\mathrm{DWET}}} + \frac{\sigma^2_{\mathrm{VWET}}}{N_{\mathrm{VWET}}}}, \tag{2}$$

Table 2. Percentages of the types of day which came immediately before and after each type, broken down by observation year. N denotes the number of days of each type in each year. OTH denotes days classified as WET but not DWET, so that the OTH and DWET categories combine to make up the WET category. The data from each year are presented separately to give an indication of inter-annual variation. To give an example from the last line of the table, in 2016 there were 45 VWET days, and 27 % of the days in 2016 directly preceding VWET days were DRY, compared to 18 % of the days directly following VWET days.

	% preceding days				N	% following days			
	DRY	OTH	DWET	VWET		DRY	OTH	DWET	VWET
2013									
DRY	66	5	21	7	183	66	7	16	11
OTH	44	22	26	7	27	37	22	22	19
DWET	28	6	46	20	109	36	6	46	12
VWET	43	11	28	17	46	29	4	49	18
2014									
DRY	68	7	20	6	167	68	8	18	6
OTH	31	17	40	12	42	26	17	33	24
DWET	29	13	37	21	106	31	16	37	16
VWET	20	20	34	26	50	20	10	44	26
2015									
DRY	66	6	23	5	158	66	5	20	9
OTH	19	21	48	12	42	24	21	38	17
DWET	26	13	44	18	124	29	16	44	11
VWET	34	17	34	15	41	20	12	54	15
2016									
DRY	68	12	16	4	182	68	9	17	7
OTH	33	14	27	27	49	43	14	22	20
DWET	35	13	35	16	86	35	15	35	15
VWET	27	22	29	22	45	18	29	31	22

Note that calculating percentages to the nearest percent occasionally produces sets that do not sum to 100 exactly.

where σ^2_{DWET} and σ^2_{VWET} are the variances of LWST in the DWET and VWET cases respectively, and the number of observations at any particular time of day are $N_{\text{DWET}} = 425$ and $N_{\text{VWET}} = 182$, as stated earlier. The standard error of the difference (SED) takes values in the range of 0.05–0.07 K during the first and last few hours of the day, climbing to over 0.14 K during daylight hours. The hypothesis that the means are equal may be tested using the difference in mean values divided by SED (e.g. Frank and Althoen, 1994, chap. 10). This is plotted in Fig. 11. It can be seen that, before 16:00 UTC, this statistic takes values in the approximate range of $[-2, 2]$, indicating that the hypothesis of equal means may be accepted at approximately the 5 % level of significance at these times. However, after 17:00 UTC, this statistic climbs to values well above 2, indicating that the hypothesis may reasonably be rejected at later times. It is therefore concluded that the difference in the means during the last few hours of the day is statistically significant.

Finally, the effect of rain on the sensing of LWST should also be considered as a possible cause of observed LWST differences. From an atmospheric modelling viewpoint, the sensed surface temperature is the important quantity in many cases, as has been recently discussed in the context of the introduction of a "skin" temperature into the FLake lake model (Le Moigne et al., 2016). In this particular case, it may be remarked that while the rain rate during VWET days is highest during the final few hours of the day, Fig. 5 shows that the average VWET rain rate is also appreciably higher than that of the DWET category at some earlier times. However, the LWST differences, and their significance, at these earlier times are both much less than during the final few hours of the day, as shown in Figs. 10 and 11. This provides some evidence that a systematic effect of rain on the sensor is not the main cause of LWST differences.

5 Discussion and conclusion

Lake Kivu has a remarkably stable tropical lake climate, and AWS Kivu has yielded a high-quality multi-year over-lake observational record, which is rare and perhaps unique in the

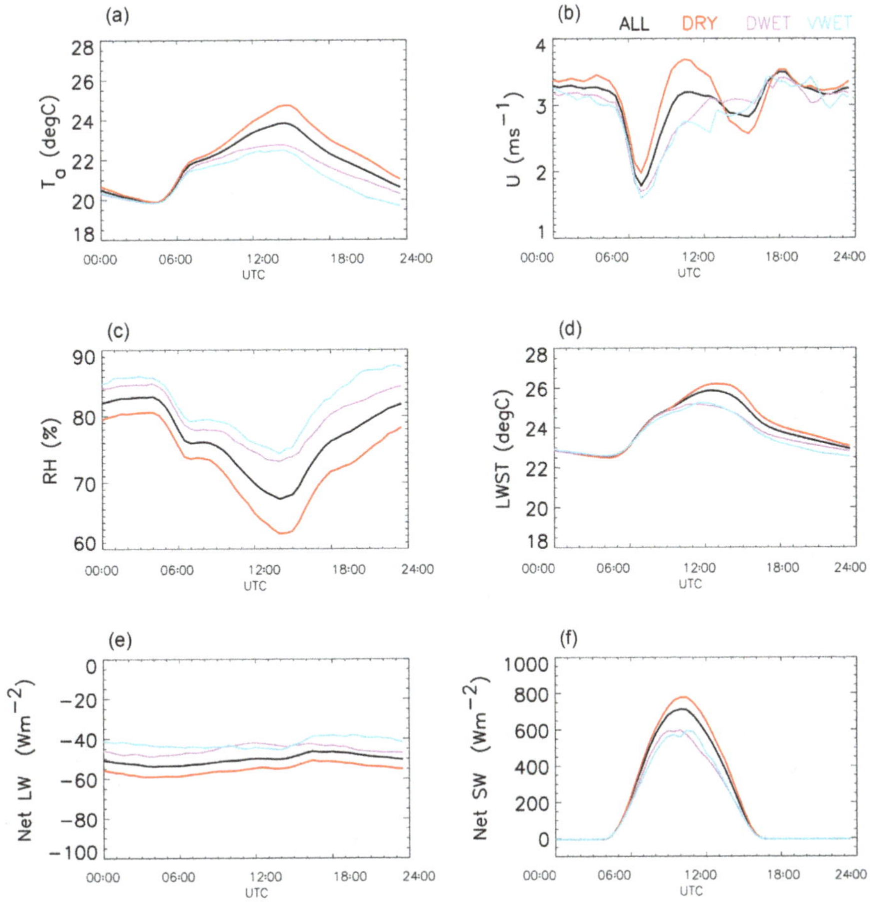

Figure 10. Mean diurnal cycles of **(a)** air temperature (T_a, °C), **(b)** wind speed (U, m s^{-1}), **(c)** relative humidity (RH, %), **(d)** lake surface temperature (LWST, °C) from the upwelling longwave irradiance, assuming an emissivity of 0.99, **(e)** net longwave irradiance (net LW, W m^{-2}) and **(f)** net shortwave irradiance (net SW, W m^{-2}). The input data are those from the 4 years 2013–2016 of the campaign, to represent all seasons equally. The colours correspond to diurnal cycles averaged over ALL days (black), DRY days (red), DWET days (purple) and VWET days (cyan).

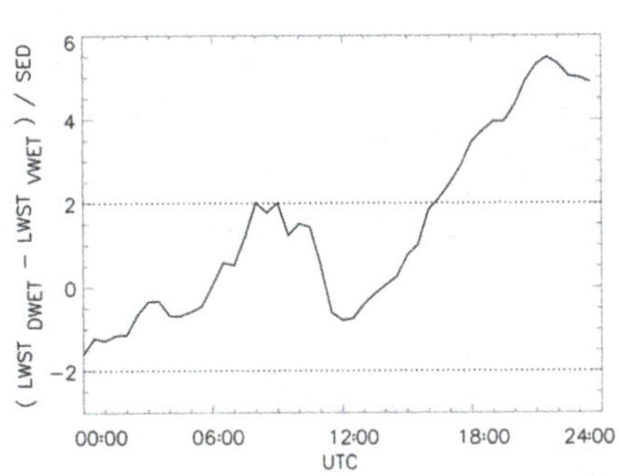

Figure 11. Diurnal behaviour of the difference in mean values of LWST in the DWET and VWET cases, normalised by the standard error of the difference, SED; see Eq. (2). The dotted lines show ±2 standard deviations for this statistic. It may be seen that the value climbs above 2 at later times, indicating the significance of the difference in the means then.

tropics and well suited to the present research question. This study is the first such use of these data.

Data over 4 years from AWS Kivu have been categorised by daily rainfall amount and net radiation to investigate the possible effects of rainfall on lake water surface temperature (LWST), which may be particularly significant in the tropics (Wei et al., 2014). The choice of division between days with heavy or light to moderate rain (respectively greater or less than 8 mm total) has helped minimise the sources of difference among the categories other than that due to rainfall. Spectral analyses have shown that, in this uniform climate, one of the dominant variations is the diurnal cycle, and hence the different categories are compared via their mean diurnal evolution. In the mean data examined here, heavy rain on a tropical lake would seem to have the capability to produce a reduction of a few tenths of 1 K in LWST over the course of several hours at the end of the day, compared to light to moderate rain, and this reduction is statistically significant.

The possible pathways by which this effect may arise are (i) cooling of the air due to contact with evaporatively cooled raindrops, and a subsequent increase in atmospheric sensible heat flux from the lake, (ii) negative heat flux directly to the lake from rain impingement, (iii) mechanical mixing of the lake surface layer by the kinetic energy of rain impact and (iv) convective mixing of the lake surface layer due to the negative heat flux from rain. Of these, the first is the most likely to be a parametrised process in a general circulation model of the atmosphere, although it could be considered the most indirect of the four. The rain heat flux is likely to be proportional to the difference between the air wet-bulb temperature and LWST. We have indicated with scaling arguments that the mechanical mixing due to heavy rain may be comparable to that of a strong wind. The convective mixing will depend on the near-surface temperature structure of the lake, and hence on its recent history.

Unfortunately, the available data do not cover several other process-related quantities that would be useful to have, such as turbulent heat fluxes, rain temperature, fine-scale lake temperature profiles or lake turbulence measurements. Thus, the processes producing this effect are not directly measured. However, through our indirect analysis of the processes, it seems likely that cooling by rain combined with mechanical and convective mixing from droplet impact may have an effect on LWST, in addition to the effect from the more widely studied pathway of evaporative cooling.

Potential avenues of future work would be to examine these processes more closely in a targeted campaign of observations, including the quantities listed above, and to consider how lake models may be modified to include their representation. An intermediate step in the latter might be to re-examine previous modelling studies to explore correlations between lake model errors and rainfall records. Since, as discussed earlier, rainfall may affect not only the surface temperature but potentially also the temperature or depth of any upper mixed layer, some or all of these quantities could be susceptible to rainfall effects. For models that predict vertical fluxes through the water column, comparison of these with any available flux or TKE measurements would be a possible way to estimate rain penetration in real lakes. There is an indication in the data of Reverdin et al. (2012) that rainfall effects may have a sudden onset but a subsequent slower decay, so that some filtering method such as an exponential moving average applied to the rainfall data may be required when considering correlations. The decay timescales of any such filtering could also have a depth dependence.

In large tropical lakes, it is possible that a surface temperature difference of the order of 0.5 K may suppress or enhance the strength of local air circulations, such as lake breezes, and hence have some effect (or even feedback) on the evolution of the local weather (Thiery et al., 2015, 2016). For example, the length of time between severe storms may be partly affected by the recovery timescale of LWST. In the short term it would seem possible to incorporate, perhaps semi-empirically, the effect of rain temperature and rain-induced turbulence into simple lake models as used for weather and climate modelling.

Competing interests. The authors declare that they have no conflict of interest.

Special issue statement. This article is part of the special issue "Modelling lakes in the climate system (GMD/HESS inter-journal SI)". It is a result of the fifth workshop on "Parameterization of Lakes in Numerical Weather Prediction and Climate Modelling", Berlin, Germany, 16–19 October 2017.

Acknowledgements. WT was supported by an ETH Zurich post-doctoral fellowship (Fel-45 15-1). The Uniscientia Foundation and the ETH Zurich Foundation are thanked for their support of this research. The Belgian Science Policy Office (BELSPO) is acknowledged for the support through the research project EAGLES (CD/AR/02A). We thank Stijn Bruggen, who analysed the AWS data in his master's thesis and thereby supported the design of this study and the analysis presented here. GGR thanks John M. Edwards for a helpful discussion of TKE budgets.

Edited by: Miguel Potes

References

Anctil, F. and Donelan, M.: Air–water momentum flux observations over shoaling waves, J. Phys. Oceanogr., 26, 1344–1353, 1996.

Anyah, R. O. and Semazzi, F.: Idealized simulation of hydrodynamic characteristics of Lake Victoria that potentially modulate regional climate, Int. J. Climatol., 29, 971–981, 2009.

Anyah, R. O. and Semazzi, F. H. M.: Simulation of the sensitivity of Lake Victoria basin climate to lake surface temperatures, Theor. Appl. Climatol., 79, 55–69, 2004.

Anyah, R. O., Semazzi, F. H. M., and Xie, L.: Simulated Physical Mechanisms Associated with Climate Variability over Lake Victoria Basin in East Africa, Mon. Weather Rev., 134, 3588–3609, 2006.

Awange, J., Ferreira, V., Forootan, E., Khandu, Andam-Akorful, S., Agutu, N., and He, X.: Uncertainties in remotely sensed precipitation data over Africa, Int. J. Climatol., 36, 303–323, https://doi.org/10.1002/joc.4346, 2016.

Borges, A. V., Abril, G., Delille, B., Descy, J.-P., and Darchambeau, F.: Diffusive methane emissions to the atmosphere from Lake Kivu (Eastern Africa), J. Geophys. Res., 116, G03032, https://doi.org/10.1029/2011JG001673, 2011.

Borges, A. V., Darchambeau, F., Teodoru, C. R., Marwick, T. R., Tamooh, F., Geeraert, N., Omengo, F. O., Guérin, F., Lambert,

T., Morana, C., Okuku, E., and Bouillon, S.: Globally significant greenhouse-gas emissions from African inland waters, Nat. Geosci., 8, 637–642, https://doi.org/10.1038/NGEO2486, 2015.

Byers, H. R., Moses, H., and Harney, P. J.: Measurement of rain temperature, J. Metorol., 6, 51–55, 1949.

Camberlin, P., Gitau, W., Planchon, O., Dubreuil, V., Funatsu, B. M., and Philippon, N.: Major role of water bodies on diurnal precipitation regimes in Eastern Africa, Int. J. Climatol., 38, 613–629, https://doi.org/10.1002/joc.5197, 2017.

Csanady, G. T.: Air-sea interaction: laws and mechanisms, Cambridge University Press, Cambridge, UK, 2001.

Degens, E., Von Herzen, R., Wong, H.-K., Deuser, W., and Jannash, H.: Lake Kivu: Structure, chemistry and biology of an East African rift lake, Geol. Rundsch., 62, 245–277, 1973.

Delandmeter, P., Lambrechts, J., Legat, V., Vallaeys, V., Naithani, J., Thiery, W., Remacle, J.-F., and Deleersnijder, E.: A fully consistent and conservative vertically adaptive coordinate system for SLIM 3D v0.4 with an application to the thermocline oscillations of Lake Tanganyika, Geosci. Model Dev., 11, 1161–1179, https://doi.org/10.5194/gmd-11-1161-2018, 2018.

Descy, J.-P., Darchambeau, F., and Schmid, M.: Lake Kivu: Past and Present, in: Lake Kivu: Limnology and Biogeochemistry of a Tropical Great Lake, edited by: Descy, J.-P., Darchambeau, F., and Schmid, M., Springer, Dordrecht, the Netherlands, 1–11, 2012.

Dinku, T., Chidzambwa, S., Ceccato, P., Connor, S. J., and Ropelewski, C. F.: Validation of high-resolution satellite rainfall products over complex terrain, Int. J. Remote Sens., 29, 4097–4110, 2008.

Docquier, D., Thiery, W., Lhermitte, S., and van Lipzig, N.: Multiyear wind dynamics around Lake Tanganyika, Clim. Dynam., 47, 3191–3202, 2016.

East African Community: Enhancing Safety of Navigation and Efficient Exploitation of Natural Resources over Lake Victoria and its Basin by Strengthening Meteorological Services on the Lake, Tech. Rep. October, Approved by the East African Community Council of Ministers, 2011.

Feingold, G.: Parametrization of the evaporation of rainfall for use in General Circulation Models, J. Atmos. Sci., 50, 3454–3467, 1993.

Frank, H. and Althoen, S. C.: Statistics: concepts and applications, Cambridge University Press, Cambridge, UK, 1994.

Green, T. and Houk, D. F.: The mixing of rain with near-surface water, J. Fluid Mech., 90, 569–588, 1979.

Guillod, B. P., Orlowsky, B., Miralles, D. G., Teuling, A. J., and Seneviratne, S. I.: Reconciling spatial and temporal soil moisture effects on afternoon rainfall, Nat. Commun., 6, 6443, https://doi.org/10.1038/ncomms7443, 2015.

Harrison, E. L. and Veron, F.: Near-surface turbulence and buoyancy induced by heavy rainfall, J. Fluid Mech., 830, 602–630, 2017.

James, R., Washington, R., Abiodun, B., Kay, G., Mutemi, J., Pokam, W., Hart, N., Artan, G., and Senior, C.: Evaluating climate models with an African lens, B. Am. Meteorol. Soc., 99, 313–336, 2018.

Kamburova, P. L. and Ludlam, F. H.: Rainfall evaporation in thunderstorm downdraughts, Q. J. Roy. Meteor. Soc., 94, 510–518, 1968.

Katsaros, K. and Buettner, K. J. K.: Influence of rainfall on temperature and salinity of the ocean surface, J. Appl. Meteorol., 8, 15–18, 1969.

Kimani, M. W., Hoedjes, J. C. B., and Su, Z.: An Assessment of Satellite-Derived Rainfall Products Relative to Ground Observations over East Africa, Remote Sensing, 9, 430, https://doi.org/10.3390/rs9050430, 2017.

Knupp, K. R. and Cotton, W. R.: Convective cloud downdraft structure: an interpretive survey, Rev. Geophys., 23, 183–215, 1985.

Lange, P. A., Van der Graaf, G., and Gade, M.: Rain-induced subsurface turbulence measured using image processing methods, in: Proc. IGARSS, 7, 3175–3177, 2000.

Le Moigne, P., Colin, J., and Decharme, B.: Impact of lake surface temperatures simulated by the FLake scheme in the CNRM-CM5 climate model, Tellus A, 68, 31274, https://doi.org/10.3402/tellusa.v68.31274, 2016.

Lorenz, R., Argüeso, D., Donat, M. G., Pitman, A. J., Hurk, B., Berg, A., Lawrence, D. M., Chéruy, F., Ducharne, A., Hagemann, S., Meier, A., Milly, P. C. D., and Seneviratne, S. I.: Influence of land-atmosphere feedbacks on temperature and precipitation extremes in the GLACE-CMIP5 ensemble, J. Geophys. Res.-Atmos., 121, 607–623, 2016.

Marshall, J. S. and Palmer, W. M.: The distribution of raindrops with size, J. Meteorol., 5, 165–166, 1948.

Mironov, D. V., Heise, E., Kourzeneva, E., Ritter, B., Schneider, N., and Terzhevik, A.: Implementation of the lake parameterisation scheme FLake into the numerical weather prediction model COSMO, Boreal Environ. Res., 15, 218–230, 2010.

Morana, C., Borges, A. V., Roland, F. A. E., Darchambeau, F., Descy, J.-P., and Bouillon, S.: Methanotrophy within the water column of a large meromictic tropical lake (Lake Kivu, East Africa), Biogeosciences, 12, 2077–2088, https://doi.org/10.5194/bg-12-2077-2015, 2015a.

Morana, C., Darchambeau, F., Roland, F. A. E., Borges, A. V., Muvundja, F., Kelemen, Z., Masilya, P., Descy, J.-P., and Bouillon, S.: Biogeochemistry of a large and deep tropical lake (Lake Kivu, East Africa: insights from a stable isotope study covering an annual cycle, Biogeosciences, 12, 4953–4963, https://doi.org/10.5194/bg-12-4953-2015, 2015b.

Morana, C., Roland, F. A. E., Crowe, S. A., Llirós, M., Borges, A. V., Darchambeau, F., and Bouillon, S.: Chemoautotrophy and anoxygenic photosynthesis within the water column of a large meromictic tropical lake (Lake Kivu, East Africa), Limnol. Oceanogr., 61, 1424–1437, 2016.

O'Reilly, C. M., Alin, S. R., Plisnier, P.-D., Cohen, A. S., and McKee, B. A.: Climate change decreases aquatic ecosystem productivity of Lake Tanganyika, Africa., Nature, 424, 766–768, 2003.

Prosperetti, A., Crum, L. A., and Pumphrey, H. C.: The underwater noise of rain, J. Geophys. Res.-Oceans, 94, 3255–3259, 1989.

Rein, M.: The transitional regime between coalescing and splashing drops, J. Fluid Mech., 306, 145–165, 1996.

Reverdin, G., Morisset, S., Boutin, J., and Martin, N.: Rain-induced variability of near sea-surface T and S from drifter data, J. Geophys. Res., 117, C02032, https://doi.org/10.1029/2011JC007549, 2012.

Rooney, G. G.: Descent and spread of negatively buoyant thermals, J. Fluid Mech., 780, 457–479, 2015.

Salles, C., Poesen, J., and Sempere-Torres, D.: Kinetic energy of rain and its functional relationship with intensity, J. Hydrol., 257, 256–270, 2002.

Schlesinger, M. E. and Oh, J.-H.: A parametrization of the evaporation of rainfall, Mon. Weather Rev., 116, 1887–1895, 1988.

Schmid, M., Halbwachs, M., Wehrli, B., and Wüest, A.: Weak mixing in Lake Kivu: New insights indicate increasing risk of uncontrolled gas eruption, Geochem. Geophy. Geosy., 6, Q07009, https://doi.org/10.1029/2004GC000892, 2005.

Seneviratne, S. I., Corti, T., Davin, E. L., Hirschi, M., Jaeger, E. B., Lehner, I., Orlowsky, B., and Teuling, A. J.: Investigating soil moisture-climate interactions in a changing climate: A review, Earth-Sci. Rev., 99, 125–161, 2010.

Skyllingstad, E. D. and Denbo, D. W.: An ocean large-eddy simulation of Langmuir circulations and convection in the surface mixed layer, J. Geophys. Res.-Oceans, 100, 8501–8522, 1995.

Spigel, R. and Coulter, G.: Comparison of Hydorology and physical limnology of the East African Great Lakes: Tanganyika, Malawi, Victoria, Kivu and Turkana (with references to some North American Great Lakes), in: The Limnology, Climatology and Paleoclimatology of the East African lakes, edited by: Johnson, T. and Odada, E., Gordon and Breach Publishers, Amsterdam, the Netherlands, 103–140, 1996.

Srivastava, R. C.: A Model of Intense Downdrafts Driven by the Melting and Evaporation of Precipitation, J. Atmos. Sci., 44, 1752–1773, 1987.

Sylla, M. B., Giorgi, F., Coppola, E., and Mariotti, L.: Uncertainties in daily rainfall over Africa: assessment of gridded observation products and evaluation of a regional climate model simulation, Int. J. Climatol., 33, 1805–1817, 2013.

Taylor, C. M., de Jeu, R. A. M., Guichard, F., Harris, P. P., and Dorigo, W. A.: Afternoon rain more likely over drier soils, Nature, 489, 423–426, 2012.

Thiery, W., Martynov, A., Darchambeau, F., Descy, J.-P., Plisnier, P.-D., Sushama, L., and van Lipzig, N. P. M.: Understanding the performance of the FLake model over two African Great Lakes, Geosci. Model Dev., 7, 317–337, https://doi.org/10.5194/gmd-7-317-2014, 2014a.

Thiery, W., Stepanenko, V. M., Fang, X., Jöhnk, K. D., Li, Z., Martynov, A., Perroud, M., Subin, Z. M., Darchambeau, F., Mironov, D., and van Lipzig, N. P. M.: LakeMIP Kivu: Evaluating the representation of a large, deep tropical lake by a set of 1-dimensional lake models, Tellus A, 66, 21390, https://doi.org/10.3402/tellusa.v66.21390, 2014b.

Thiery, W., Davin, E. L., Panitz, H.-J., Demuzere, M., Lhermitte, S., and van Lipzig, N. P. M.: The Impact of the African Great Lakes on the Regional Climate, J. Climate, 28, 4061–4085, 2015.

Thiery, W., Davin, E. L., Seneviratne, S. I., Bedka, K., Lhermitte, S., and van Lipzig, N. P. M.: Hazardous thunderstorm intensification over Lake Victoria, Nat. Commun., 7, 12786, https://doi.org/10.1038/ncomms12786, 2016.

Thiery, W., Gudmundsson, L., Bedka, K., Semazzi, F. H. M., Lhermitte, S., Willems, P., van Lipzig, N. P. M., and Seneviratne, S. I.: Early warnings of hazardous thunderstorms over Lake Victoria, Environ. Res. Lett., 12, 074012, https://doi.org/10.1088/1748-9326/aa7521, 2017.

Townsend, A. A.: The structure of turbulent shear flow, Cambridge University Press, Cambridge, UK, 1976.

van Beek, L. P. H., Eikelboom, T., van Vliet, M. T. H., and Bierkens, M. F. P.: A physically based model of global freshwater surface temperature, Water Resour. Res., 48, w09530, https://doi.org/10.1029/2012WR011819, 2012.

van Dijk, A., Bruijnzeel, L., and Rosewell, C.: Rainfall intensity-kinetic energy relationships: A critical literature appraisal, J. Hydrol., 261, 1–23, 2002.

Vanderkelen, I., van Lipzig, N. P. M., and Thiery, W.: Modelling the water balance of Lake Victoria (East Africa) – Part 1: Observational analysis, Hydrol. Earth Syst. Sci., 22, 5509–5525, https://doi.org/10.5194/hess-22-5509-2018, 2018a.

Vanderkelen, I., van Lipzig, N. P. M., and Thiery, W.: Modelling the water balance of Lake Victoria (East Africa) – Part 2: Future projections, Hydrol. Earth Syst. Sci., 22, 5527–5549, https://doi.org/10.5194/hess-22-5527-2018, 2018b.

Verburg, P. and Antenucci, J. P.: Persistent unstable atmospheric boundary layer enhances sensible and latent heat loss in a tropical great lake: Lake Tanganyika, J. Geophys. Res.-Atmos., 115, d11109, https://doi.org/10.1029/2009JD012839, 2010.

Verburg, P. and Hecky, R.: Wind patterns, evaporation, and related physical variables in Lake Tanganyika, East Africa, J. Great Lakes Res., 29, 48–61, 2003.

Verburg, P. and Hecky, R.: The physics of the warming of Lake Tanganyika by climate change, Limnol. Oceanogr., 54, 2418–2430, 2009.

Verburg, P., Hecky, R. E., and Kling, H.: Ecological consequences of a century of warming in Lake Tanganyika, Science, 301, 505–507, 2003.

Verburg, P., Antenucci, J. P., and Hecky, R. E.: Differential cooling drives large-scale convective circulation in Lake Tanganyika, Limnol. Oceanogr., 56, 910–926, 2011.

Wan, Z.: New refinements and validation of the MODIS Land-Surface Temperature/Emissivity products, Remote Sens. Environ., 112, 59–74, 2008.

Wei, N., Dai, Y., Zhang, M., Zhou, L., Ji, D., Zhu, S., and Wang, L.: Impact of precipitation-induced sensible heat on the simulation of land-surface air temperature, J. Adv. Model. Earth Sy., 6, 1311–1320, 2014.

Weyhenmeyer, G., Mackay, M., Stockwell, J., Thiery, W., Grossart, H.-P., Augusto-Silva, P., Baulch, H., de Eyto, E., Hejzlar, J., Kangur, K., Kirillin, G., Pierson, D., Rusak, J., Sadro, S., and Woolway, I.: Citizen Science shows systematic changes in the temperature difference between air and inland waters with global warming, Sci. Rep.-UK, 7, 43890, https://doi.org/10.1038/srep43890, 2017.

Woodhams, B. J., Birch, C. E., Marsham, J. H., Bain, C. L., Roberts, N. M., and Boyd, D. F. A.: What Is the Added Value of a Convection-Permitting Model for Forecasting Extreme Rainfall over Tropical East Africa?, Mon. Weather Rev., 146, 2757–2780, 2018.

Yu, N., Boudevillain, B., Delrieu, G., and Uijlenhoet, R.: Estimation of rain kinetic energy from radar reflectivity and/or rain rate based on a scaling formulation of the raindrop size distribution, Water Resour. Res., 48, w04505, https://doi.org/10.1029/2011WR011437, 2012.

Zappa, C. J., Ho, D. T., McGillis, W. R., Banner, M. L., Dacey, J. W. H., Bliven, L. F., Ma, B., and Nystuen, J.: Rain-induced turbulence and air-sea gas transfer, J. Geophys. Res.-Oceans, 114, c07009, https://doi.org/10.1029/2008JC005008, 2009.

10

Impact of capillary rise and recirculation on simulated crop yields

Joop Kroes[1], **Iwan Supit**[1,2], **Jos van Dam**[3], **Paul van Walsum**[1], **and Martin Mulder**[1]

[1]Wageningen University & Research – Wageningen Environmental Research, Wageningen, the Netherlands
[2]Wageningen University & Research – Chair Water Systems and Global Change, Wageningen, the Netherlands
[3]Wageningen University & Research – Chair Soil Physics and Land Management, Wageningen, the Netherlands

Correspondence: Joop G. Kroes (joop.kroes@wur.nl)

Abstract. Upward soil water flow is a vital supply of water to crops. The purpose of this study is to determine if upward flow and recirculated percolation water can be quantified separately, and to determine the contribution of capillary rise and recirculated water to crop yield and groundwater recharge. Therefore, we performed impact analyses of various soil water flow regimes on grass, maize and potato yields in the Dutch delta. Flow regimes are characterized by soil composition and groundwater depth and derived from a national soil database. The intermittent occurrence of upward flow and its influence on crop growth are simulated with the combined SWAP-WOFOST model using various boundary conditions. Case studies and model experiments are used to illustrate the impact of upward flow on yield and crop growth. This impact is clearly present in situations with relatively shallow groundwater levels (85 % of the Netherlands), where capillary rise is a well-known source of upward flow; but also in free-draining situations the impact of upward flow is considerable. In the latter case recirculated percolation water is the flow source. To make this impact explicit we implemented a synthetic modelling option that stops upward flow from reaching the root zone, without inhibiting percolation. Such a hypothetically moisture-stressed situation compared to a natural one in the presence of shallow groundwater shows mean yield reductions for grassland, maize and potatoes of respectively 26, 3 and 14 % or respectively about 3.7, 0.3 and 1.5 t dry matter per hectare. About half of the withheld water behind these yield effects comes from recirculated percolation water as occurs in free-drainage conditions and the other half comes from increased upward capillary rise. Soil water and crop growth modelling should consider both capillary rise from groundwater and recirculation of percolation water as this improves the accuracy of yield simulations. This also improves the accuracy of the simulated groundwater recharge: neglecting these processes causes overestimates of 17 % for grassland and 46 % for potatoes, or 63 and 34 mm yr^{-1}, respectively.

1 Introduction

Crop growth strongly depends on soil moisture conditions. Climate variables determine these conditions through rain that penetrates directly into the root zone or comes available via lateral flow. The moisture distribution in the soil strongly depends on soil physical properties that determine vertical flow. Upward soil water flow becomes an especially vital supply term of a crop when the soil water potential gradient induced by the root extraction manages to bridge the distance to the capillary fringe, inducing increased capillary rise. In this paper we follow the definition of capillary rise given by the Soil Science Society of America (SSSA, 2008), as the "phenomenon that occurs when small pores which reduce the water potential are in contact with free water". This implies that capillary rise as a source for upward flow to crop roots requires the presence of a groundwater table. In conditions without a groundwater table there may also be a contribution of upward flow to crop roots through the process of recirculation. Recirculation is a known process discussed already by Feodoroff (1969) but has never been quantified. We quantified recirculation separately from capillary rise using model experiments.

The contribution of (intermittent) upward flow to the total water budget can be significant. For example Kowalik (2006) mentions that during the grass growing season, in soils with the groundwater close to the soil surface (Aquepts), the capillary rise induced by root extraction varies between 60 and 150 mm yr− 1. Babajimopoulos et al. (2007) found that under the specific field conditions about 3.6 mm d^{-1} of the water in the root zone originated from the shallow water table, which amounts to about 18 % of the water transpired by a maize crop. Fan et al. (2013) analysed the groundwater depth globally and concluded that shallow groundwater influences 22 to 32 % of global land area, and that 7 to 17 % of this area has a water table within or close to plant rooting depths, suggesting a widespread influence of groundwater on crops. This is especially the case in delta areas where high population densities occur and agriculture is the predominant land use.

Wu et al. (2015) showed that capillary rise plays a main role in supplying the vegetation throughout the season with water, hence a strong dependence of vegetation upon groundwater. Han et al. (2015) applied HYDRUS-1D with a simplified crop growth model for cotton in a north-western part of China and concluded that capillary rise from groundwater contributes to almost 23 % of crop transpiration when the average groundwater depth is 1.84 m. According to Geerts et al. (2008) the contribution from capillary rise to the quinoa [*Chenopodium quinoa* Willd.] production in the Irpani region (Bolivia) ranges from 8 to 25 % of seasonal crop evapotranspiration (ETc) of quinoa, depending mostly on groundwater table depth and amount of rainfall during the rainy season. The contribution from a groundwater table located approximately 1.5 to 2 m deep may represent up to 30 % of the soybean [*Glycine max* (L.) Merr.] water requirements in sandy pampas (Videla Mensegue et al., 2015).

In 85 % of the area in the Netherlands the average groundwater table is less than 2 m below the soil surface (De Vries, 2007), where root extraction can induce capillary rise from groundwater. Wesseling and Feddes (2006) report that in summers with a high evapotranspiration demand, crops partially depend on water supply from soil profile storage and induced capillary rise. Van der Gaast et al. (2009), applying the method of Wesseling (1991), found for the Netherlands a maximum capillary flow of 2 mm d^{-1} to the root zone in loamy soils where the groundwater level is at 2.5 m below the soil surface.

Although the contribution of capillary rise to the total water budget can be significant, it is an often neglected part of the crop water demand in situations of shallow groundwater levels (Awan et al., 2014). The capillary properties of a soil strongly depend on soil type. Rijtema (1971) estimated that loamy soils have an almost 2 times higher capillary rise than sandy soils.

Integrated approaches are needed to relate water availability to crop yield prognosis (Van der Ploeg and Teuling, 2013; Norman, 2013). The importance of capillary rise as supplier of water to crops has been shown by many researchers (e.g. Hooghoudt, 1937; Huo et al., 2012; Talebnejad and Sepaskhah, 2015; Han et al., 2015); however, we found only a few studies that use an integrated modelling approach (Xu et al., 2013; Zipper et al., 2015) to quantify capillary rise for different hydrological conditions (including free drainage) using physically based approaches. In this study we explicitly consider the effect of crop type, soil type, weather year and drainage condition on capillary rise. Zipper et al. (2015) introduced the concept of groundwater yield subsidy as the increase in harvested yield (kg ha^{-1}) in the presence of shallow groundwater compared to free-drainage conditions. Following their line we introduce the concept of soil moisture yield subsidy as the additional yield increase in free-drainage conditions due to recirculation of percolated soil moisture.

The driving force for induced capillary rise and recirculation is the difference in soil water potential, referred to as heads, at different soil depths. There are several models available that solve these head differences numerically. Ahuja et al. (2014) evaluated 11 models commonly applied for agricultural water management. Six of these models use simple "bucket" approaches for water storage and have in some cases been extended with more or less empirical options for capillary rise. Five models have the ability to numerically solve Richards equation for water movement in the soil. Examples are HYDRUS (Šimůnek et al., 2008) and SWAP (Feddes et al., 1988; van Dam et al., 2008).

We applied the integrated model SWAP-WOFOST (acronyms for Soil Water Atmosphere Plant – WOrld FOod STudies) to solve head differences and crop yield simulations. Kroes and Supit (2011) applied the same integrated model to quantify the impact of increased groundwater salinity on drought and oxygen of grassland yields in the Netherlands. They recommended further analyses using different crops and different boundary conditions. We now apply this model with different boundary conditions using 45 years of observed weather and three different crops. For the lower boundary we use different hydrologic conditions that influence the vertical flow. For the soil system itself we use a wide range of soil physical conditions. The importance of the soil system was already stated by several authors like Supit (2000). We build on their suggestions and apply the tools for different crops and boundary conditions. Before we applied the model to different boundary conditions, we validated it at the field scale.

This paper quantifies the effects of (intermittent) upward flow on crop growth under different conditions of soil hydrology, soil type and weather. The effects are separately quantified in terms of flow source, namely capillary rise and recirculated percolation water. Therefore, we introduced a synthetic model option and performed a numerical experiment. We studied forage maize, grassland and potatoes and we hypothesize that neglecting upward flow will result in neglecting a considerable amount of soil moisture that is available for crop growth. We quantify this amount and show the importance of including upward flow for crop growth modelling. Our main research questions are the following. (i) Can upward flow with capillary rise and recirculated percolation water as source be quantified separately? (ii) What is the contribution of capillary rise and recirculated water to crop yield and groundwater recharge?

2 Materials and methods

2.1 Modelling approach

We applied the coupled SWAP and WOFOST modelling system, using a 1-day time step for the crop model. SWAP (van Dam et al., 2008; Kroes et al., 2017) is a one-dimensional physically based transport model for water, heat and solute in the saturated and unsaturated zone, and includes modules for simulating irrigation practices. The first version of SWAP, called SWATRE, was developed by Feddes et al. (1978). This version also included a module for crop production, CROPR, that applied principles of De Wit (1965) and is still applied in several countries.

SWAP simulates the unsaturated and saturated water flow in the upper part of the soil system, using a numerical solution of the Richards equation:

$$\frac{\partial \theta}{\partial t} = \frac{\partial \left[K(h) \left(\frac{\partial h}{\partial z} + 1 \right) \right]}{\partial z} - S_a(h) - S_d(h) - S_m(h), \qquad (1)$$

where θ is volumetric water content ($cm^3\,cm^{-3}$), t is time (d), $K(h)$ is hydraulic conductivity ($cm\,d^{-1}$), h is soil water pressure head (cm), z is the vertical coordinate (cm) taken positively upward, $S_a(h)$ is the soil water extraction rate by plant roots (d^{-1}), $S_d(h)$ is the extraction rate by drain discharge in the saturated zone (d^{-1}) and $S_m(h)$ is the exchange rate with macro pores (d^{-1}).

The numerical solution for this equation uses variable time steps that depend on boundary conditions and an iteration scheme. For example, high fluxes require time steps that are much smaller than 1 day (see Kroes et al., 2017, for a detailed explanation).

Root water extraction and lateral exchange with surface water were accounted for. In this study we did not use the option to exchange water flow with macro pores.

The soil hydraulics were described by the Mualem–van Genuchten relations and the potential evapotranspiration was calculated with the Penman–Monteith equation (Allen et al., 1998). At the bottom boundary water fluxes, supplied by a separate regional hydrological model, were used. Drainage and infiltration through the lateral boundary accounted for the flow to surface water. The surface water system was simulated using a simplified, weir controlled, water balance. Note that the surface water system in its turn interacted with the groundwater system. In previous years, SWAP has been successfully used to study soil-water–atmosphere–plant relationships in many locations with various boundary conditions (e.g. Feddes et al., 1988; Bastiaanssen et al., 2007). See van Dam et al. (2008) for an overview. A recent list is available at http://swap.wur.nl/ (last access: 7 May 2018). Eitzinger et al. (2004), Bonfante et al. (2010), Oster et al. (2012), and Rallo et al. (2012) amongst others tested the model performance.

WOFOST is a crop growth simulation model, its principles are explained by Van Keulen and Wolf (1986). Van Diepen et al. (1989) presented the first WOFOST version. WOFOST is applied in many studies (e.g. Rötter, 1993; Van Ittersum et al., 2003; de Wit and Van Diepen, 2008; Supit et al., 2012; De Wit et al., 2012). Crop assimilation was calculated as a function of solar radiation and temperature, using a 3-point Gaussian integration method accounting for leaf angle distribution and extinction of direct and diffuse light. The assimilation was reduced when water stress occurred. Subsequently, the maintenance respiration was subtracted and the remaining assimilates were partitioned between the plant organs (i.e. leaves, stems, roots and storage organs). For maize and potatoes the partitioning was development-stage dependent. For perennial grass, however, a constant partitioning factor was assumed. By integrating the difference between growth and senescence rates over time, dry weights of various plant organs were established.

In SWAP-WOFOST, crop assimilation depends on the ambient CO_2 concentration as well (see: Kroes and Supit, 2011; Supit et al., 2012). To account for unknown residual stress caused by diseases, pests and/or weeds an additional assimilation reduction factor was introduced. The rooting density decreased exponentially with depth. To withdraw water from deeper soil layers for crop uptake, a form of compensatory root uptake was used in case the upper part of the soil was very dry (Jarvis, 1989, 2011). The increasing atmospheric CO_2 concentrations during relatively long historical simulation periods (> 20 years) was accounted for.

2.2 Case studies for validation

SWAP-WOFOST was validated using results of seven case studies at six locations in the Netherlands (Fig. 1), where grassland, maize and potatoes are grown and observations were available from hydrology, soil and crop. The main characteristics of the seven cases are summarized in Table 1. The soil texture ranged from sand to clay. The observations included parameters such as groundwater levels and yields and in some cases soil moisture contents, soil pressure head and evapotranspiration. The weather data were collected from nearby weather stations or from on-site measurements. Observations for case studies 1 and 2 (DM-Grass and DM-Maize in Table 1) were available for a period of 22 years (1992–2013) from one field where grassland and maize were grown for respectively 7 and 15 years.

We used the model calibrations carried out by Kroes et al. (2015) and Hack-ten-Broeke et al. (2016) and limited our calibration efforts to parameter values for drought and management (Table 1), focussing on validation of results. Planting and harvest dates were given. Oxygen and drought stress reduced transpiration, which subsequently reduced crop as-

Table 1. Main characteristics of the case studies used to verify the setup of the model combination SWAP-WOFOST.

Case study[1]	Crop	Location	Period	Soil type	Observations[2]	Reference	Drought stress[3]	MF[4]	RZ[5]	BBC[6]	Lateral boundary
DM-Grass	grass	De Marke	1995–1996, 2005–2008, 2013	dry sandy soil	Gwl, Yield, Theta20cm	Hack-ten-Broeke and Hegmans (1996), Verloop et al. (2014)	h3h = −200.0 cm h3l = −800.0 cm h4 = −8000.0 cm	0.8	40	Cauchy	no drainage
DM-Maize	silage maize	De Marke	1992–1994, 1997–2003, 2009–2012	dry sandy soil	Gwl, Yield, Theta20cm	Hack-ten-Broeke and Hegmans (1996), Verloop et al. (2014)	h3h = −400.0 cm h3l = −500.0 cm h4 = −10 000.0 cm	0.85	40	Cauchy	no drainage
C-Maize	silage maize	Cranendonck	1974–1982	Cumulic Anthrosol	Gwl, Yield	Schröder (1985)	see DM-Maize	0.9	40	Cauchy	no drainage
D-Maize	silage maize	Dijkgraaf	2007	Umbric Gleysol	Gwl, Yield, ET, Theta20cm	Elbers et al. (2009)	see DM-Maize	0.95	100	Cauchy	no drainage
B-Potato	potato	Borgerswold	1992, 1994	sandy loam	Gwl, Yield	Dijkstra et al. (1995)	h3h = −300.0 cm h3l = −500.0 cm h4 = −10 000.0 cm	0.8	100	observed groundwater	no drainage
R-Potato	potato	Rusthoeve	2013	loamy clay	Gwl, Yield, Qdrain	Van Den Brande (2013)	see B-Potato	0.8	100	Cauchy	drain tubes at −90 cm
V-Potato	potato	Vredepeel	2002	sandy loam	Gwl, Yield	De Vos et al. (2006)	see B-Potato	0.8	50	closed	drain ditch at −100 cm

[1] The name of each case study is a combination of an acronym for the location and the crop type, using the acronyms DM = De Marke, C = Cranendonck, D = Dijkgraaf, B = Borgerswold, R = Rusthoeve and V = Vredepeel. [2] Gwl = groundwater level, Yield = actual yield as dry matter of harvested product, Theta20cm = soil moisture content at a depth of 20 cm below surface, Qdrain = drainage from field to surface water via tube drains, ET = evapotranspiration measured via the eddy correlation method. [3] h3h = *h* below which water uptake reduction starts at high Tpot; h3l = *h* below which water uptake red. starts at low Tpot; h4 = no water extraction at lower pressure heads; drought stress was parameterized using the dry part of the reduction function proposed by Feddes et al. (1978), Drought stress is absent when the soil pressure head *h* exceeds the critical value of h3. Drought stress increases linearly between h3 and at h4 (wilting point). The critical pressure head h3 differs between lower and higher potential transpiration (Tpot; respectively h3l and h3h) rates. [4] MF = management factor to account for imperfect management. [5] RZ = maximum depth of root zone (cm). [6] BBC = bottom boundary condition. The Cauchy bottom boundary condition uses a hydraulic head based on piezometer observations from an open data portal (see text).

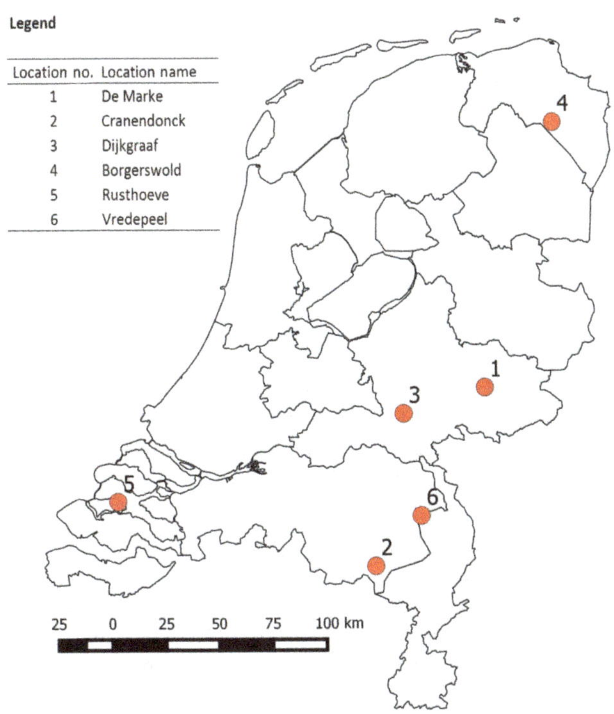

Figure 1. Location of the case studies for grassland, maize and potatoes; location De Marke has a rotation of grassland and maize on the same field.

similation. Oxygen stress was described with the process-based method of Bartholomeus et al. (2008) and parameterized as described by Hack-ten-Broeke et al. (2016). Drought stress was parameterized using the dry part of the reduction function proposed by Feddes et al. (1978). Drought stress was absent when the soil pressure head h exceeded the critical value of h3. Drought stress increased linearly between h3 and at h4 (wilting point). The critical pressure head h3 differed between lower and higher potential transpiration (respectively h3l and h3h) rates. In conditions with drought or oxygen stress, the reduction in stressed parts was partly compensated by extra root water uptake in those parts of the root zone with more favourable soil moisture conditions (Jarvis, 1989, 2011).

For all cases a so-called management factor was used to close the gap between observed and actual yield. The input crop parameters for maize only differed with respect to the management factor which ranges from 0.85 to 0.95. The management factors were relatively high because the case study locations have good management. It is clear that we missed some processes even though our modelling approach is mechanistic, because it is still relatively simple. Some processes like pests and diseases were not included and may have played a role in the field; the calibration was done on experimental farms where the impact from diseases and pests was minimal.

For potatoes the input crop parameters were kept the same for all three cases (Table 1). Maximum rooting depth for grassland, maize and potatoes were respectively 40, 100 and 50 cm.

Soil water conditions were different for all locations and boundary conditions varied, depending on local situation and available data (Table 1). In most cases a Cauchy bottom boundary condition was applied using a hydraulic head based on piezometer observations from the Dutch Geological Survey (https://www.dinoloket.nl/, last access 14 May 2017). Observed groundwater levels were used as a lower boundary condition for Borgerswold (crop: potato). In two cases a lateral boundary condition was applied with drainage to a surface water system (Table 1). The simulation results were analysed using an R package (Bigiarini, 2013) and the statistics are presented in Table 2.

2.3 Soil crop experiment to analyse the role of recirculation and capillary rise

To analyse the impact of soil type on upward soil water flow we modelled soil–crop experiments using 72 soils. Each soil schematization consisted of one or more soil horizons, each with different soil physical properties. The method was described in detail by Wösten et al. (2013a) and the data are available at http://www.wur.nl/nl/show/Bodemfysische-Eenhedenkaart-BOFEK2012.htm (last access: 7 May 2018). The 72 soils were aggregated from 315 soil units of the 1 : 50 000 Dutch Soil Map using soil hydraulic clustering methods and considering the following properties: maximum groundwater depth, saturation deficit between a certain depth and the soil surface, transmissivity for horizontal water flow, resistance for vertical water flow and availability of water in the root zone (Wösten et al., 2013b). The resulting soil hydraulic properties were subsequently used as SWAP-WOFOST input. The bottom of the soil profile was set at 5.5 m below the soil surface. At this depth, the simulated root zone soil water fluxes are not affected anymore by the actual depth of the soil profile bottom. The root zone lower boundary was dynamic, it depended on root growth and consequently varied in time.

For each soil we applied three hydrological conditions (Fig. 2) ranging from relatively dry (condition a; Fig. 2a) to relatively wet (condition c; Fig. 2c). The latter is the natural situation in most of the Netherlands. This hydrological condition had a fluctuating groundwater level derived from a national study (Van Bakel et al., 2008). This national study used simulation units which are unique in land use, crop type and drainage conditions resulting in daily groundwater fluctuations. Lateral infiltration and drainage were accounted for ($q_{\text{infiltration}}$ and q_{drainage} in Fig. 2c). We selected three large simulation units for grassland, maize and potato with long-term average groundwater levels between 40 and 120 cm below the soil surface, covering respectively 1806, 794 and 58 102 ha using data from Van Bakel et al. (2008). See the

Table 2. Results of Case studies: simulated and observed values.

Case study	Name[1]	Unit	Simulated mean	Observed mean	ME[2]	RMSE[3]	NS[4]	d[5]	n[6]
DM-Grass	Yield	$kg\,ha^{-1}\,yr^{-1}$ DM	11 183	11 049	133	1347	0.6	0.9	7
	Gwl	m soil	−1.31	−1.30	−0.01	0.45	0.3	0.9	77
	Theta	$m^3\,m^{-3}$	0.28	0.27	0.01	0.05	0.5	0.9	43
DM-Maize	Yield	$kg\,ha^{-1}\,yr^{-1}$ DM	11 593	11 850	−257	2864	−3.3	0.4	14
C-Maize	Yield	$kg\,ha^{-1}\,yr^{-1}$ DM	14 097	13 788	310	2595	−1.2	0.7	9
	Gwl	m soil	−1.41	−1.36	−0.05	0.25	0.4	0.9	61
D-Maize	Yield	$kg\,ha^{-1}\,yr^{-1}$ DM	15 973	16 306	−333				1
	LAI	$m^2\,m^{-2}$	2.08	2.47	−0.34	0.62	0.7	0.9	10
	ETact	$mm\,yr^{-1}$	1.33	1.93	−0.61	0.89	0.5	0.9	232
	Gwl	m soil	−1.03	−1.07	0.03	0.06	0.9	1.0	112
	Theta	$m^3\,m^{-3}$	0.29	0.27	0.01	0.03	0.5	0.8	219
B-Potato	Yield	$kg\,ha^{-1}\,yr^{-1}$ DM	10 543	9246	1297				2
	Gwl	m soil	−1.10	−1.10	0.00	0.03	1.0	1.0	123
R-Potato	Yield	$kg\,ha^{-1}\,yr^{-1}$ DM	9984	8610	1374				1
	Gwl	m soil	−1.07	−1.10	0.03	0.19	0.6	0.9	887
	qDrain	mm	1.06	0.62	0.44	1.41	0.4	0.8	1084
V-Potato	Yield	$kg\,ha^{-1}\,yr^{-1}$ DM	11 071	11 359	−288				1
	Gwl	m soil	−1.03	−1.07	0.04	0.12	0.8	0.9	353

[1] Gwl = ground water level; Theta = volumic soil moisture content at a depth of 20 cm below the soil surface; LAI = leaf area index; ETact = actual evapotranspiration; qDrain = drainage flux. [2] ME: mean error between simulated (sim) and observed (obs), in the same units of sim and obs, with treatment of missing values. A smaller value indicates better model performance. [3] RMSE: root mean square error between sim and obs, in the same units of sim and obs, with treatment of missing values. RMSE gives the standard deviation of the model prediction error. A smaller value indicates better model performance. [4] NS: Nash–Sutcliffe efficiencies range from $-\infty$ to 1. Essentially, the closer to 1 the more accurate the model is. NS = 1 corresponds to a perfect match of modelled to the observed data. NS = 0 indicates that the model predictions are as accurate as the mean of the observed data. $-\infty <$ NS < 0 indicates that the observed mean is a better predictor than the model. [5] d: the index of agreement (d) developed by as a standardized measure of the degree of model prediction error and varies between 0 and 1. A value of 1 indicates a perfect match, and 0 indicates no agreement at all. The index of agreement can detect additive and proportional differences in the observed and simulated means and variances; however, it is overly sensitive to extreme values due to the squared differences. [6] n: the number of values used with the four statistical criteria to compare simulated and observed results.

Supplement Sect. S2 for more detail and the Supplement of Kroes and Supit (2011) for an additional explanation of the study from Van Bakel et al. (2008).

The other two conditions (a) and (b) were unsaturated and had no groundwater due to a free-draining bottom boundary ($q_{leaching}$, see Fig. 2 conditions a and b). Condition (a) has been included in this study to explicitly demonstrate the role of recirculation as a source of upward flow. A synthetic modelling option has been implemented to stop upward flow from reaching the root zone, without inhibiting percolation. This option was implemented in the numerical solution of the Richards equation and minimized vertical conductivity just below the root zone in situations when the model simulated upward vertical flow. We did use an implicit scheme for the conductivity in such situations. Code adjustment was necessary to carry out the model experiment (no recirculation) and to demonstrate (quantitatively) the added value of simulating more detailed water fluxes in the soil profile in comparison to bucket approaches. When crop models are used for yield forecasting, these detailed processes play an important role; neglecting them may cause large errors.

The upward flux across the bottom of the root zone can either stem from capillary rise or from percolation water that is recirculated (q_{recirc} and $q_{caprise}$, see Fig. 2 conditions b and c). The capillary rise (Fig. 2c) has two sources: (i) groundwater and (ii) recirculated percolation water. In all hydrological conditions, percolation across the root zone and leaching across the lower boundary of the model profile occurs ($q_{percolation}$ and $q_{leaching}$ in Fig. 2). All fluxes were calculated using small variable time steps (< 1 day); however, results were accumulated to daily net fluxes, which implies that small variations within a day cannot be seen from the results. Recirculation depends on crop water demand, soil hydraulic properties and the presence of soil moisture.

The crop parameters were kept the same as for the case studies, with a few exceptions: (i) for grassland an average management factor of 0.9 was used, (ii) timing of grass mowing was done when a dry matter threshold of 4200 $kg\,ha^{-1}$ DM (dry matter) was exceeded and (iii) for maize and potatoes the harvesting dates were respectively set to 25 October and 15 October.

The three crops and three lower boundary conditions resulted in nine combinations. Each combination was simu-

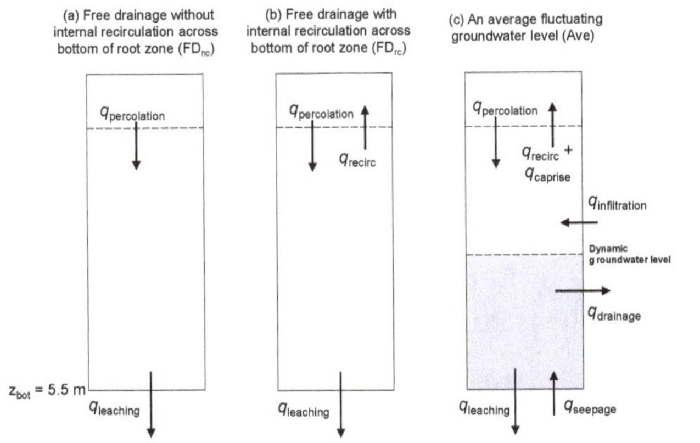

Figure 2. Schematization of three hydrological conditions: **(a)** free drainage without recirculation across the bottom of the root zone (FD_{nc}); **(b)** free drainage with recirculation across bottom of root zone (FDrc); **(c)** average fluctuating groundwater level (Ave). Conditions **(a)** and **(b)** have free-draining bottom boundary conditions without groundwater. Condition **(a)** is artificially created to explicitly demonstrate the role of recirculating percolation resulting in upward flow to the root zone. Condition **(b)** is a common free-drainage situation which includes upward flow due to recirculating percolation water. Condition **(c)** is the natural situation in most of the Netherlands. This hydrological condition has a fluctuating groundwater level derived from a national study (Van Bakel et al., 2008).

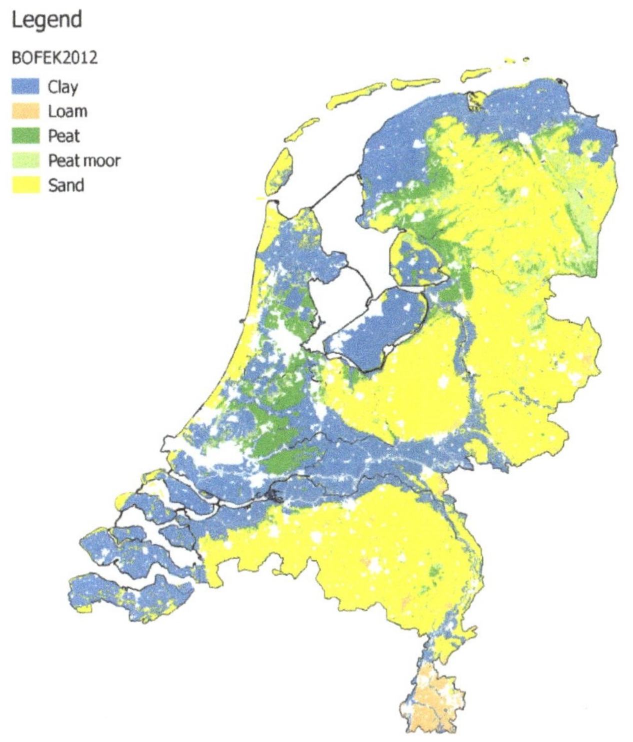

Figure 3. Five grouped soil types, based on 72 soils of the Soil Physical Map of the Netherlands (Wösten et al., 2013).

lated with 72 soils for a period of 45 years (1971–2015) with meteorological data from the station De Bilt (KNMI, 2016). In a subsequent analysis we grouped the results of these 72 soils to five main soil groups: clay, loam, peat, peat moor and sand (Fig. 3) to be able to analyse the impact at grouped soil types.

The implementation of the synthetic modelling option is explained in Sect. S4 with references to the open-source model SWAP version 4.0.1, which was used to carry out all the simulations.

3 Results

3.1 Case studies for validation

The first two case studies are from one location (De Marke), where a grassland–maize rotation was practised. The results show that the hydrological conditions (Fig. 4 and Table 2) were simulated accurately for those years for which observed data were available (1991–1995). From 1995–1997 the groundwater levels dropped as a result of low precipitation (about $700\,\mathrm{mm\,yr^{-1}}$). The autumn of the year 1998 showed rising groundwater levels that corresponded well with very wet conditions at that time. The simulated grassland yields were overestimated by $133\,\mathrm{kg\,ha^{-1}}$ DM and the simulated maize yields were underestimated by $257\,\mathrm{kg\,ha^{-1}}$ DM, and differences were well within acceptable ranges (Fig. 5 and Table 2).

For the other two maize case studies (C-Maize and D-Maize) groundwater levels and soil moisture were well simulated (Table 2). The simulated maize yields (Table 2) were less acceptable for case C-Maize as indicated by a zero or negative Nash–Sutcliffe efficiency (NS), which suggests that the observed mean was a better predictor than the model. One should consider that the NS efficiency is sensitive to sample size and outliers. In 1976, a very dry year, the soil hydrology dynamics and the resulting yield were well captured. The yield of case study D-Maize had a small bias of $333\,\mathrm{kg\,ha^{-1}}$ DM between observed and simulated values.

The simulated hydrological conditions for the three fields of the potato cases – B-Potato, R-Potato and V-Potato – showed a good fit with the observed (Table 2). The simulated yields (Table 2) showed the largest deviation from the observed for the case B-Potato. The more recent experiments of potato case studies, R-Potato and V-Potato, showed differences between simulated and observed yields of respectively 1374 and $-288\,\mathrm{kg\,ha^{-1}}$ DM (Table 2). These case studies unfortunately covered only 1 year. The case R-Potato performed less due to the complex situation in the subsoil with drainage conditions that require more observations to improve the simulations.

However, one has to bear in mind that perfect calibration is not the objective of this study; we used calibration values from earlier studies (Kroes et al., 2015; Hack-ten-Broeke et al., 2016). No detailed assimilation measurements were exe-

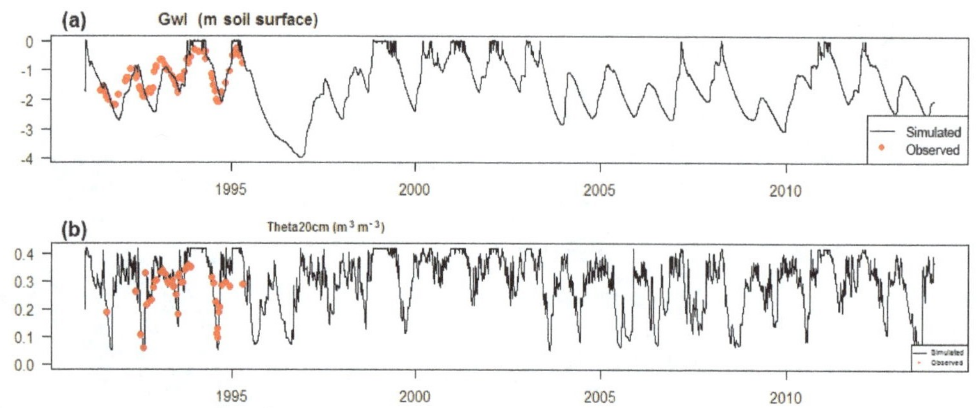

Figure 4. Results of case studies for grassland at location 1 (De Marke): **(a)** for groundwater level (Gwl in metres soil surface); **(b)** for soil moisture content (Theta20cm in $m^3\,m^{-3}$) at 20 cm below the soil surface.

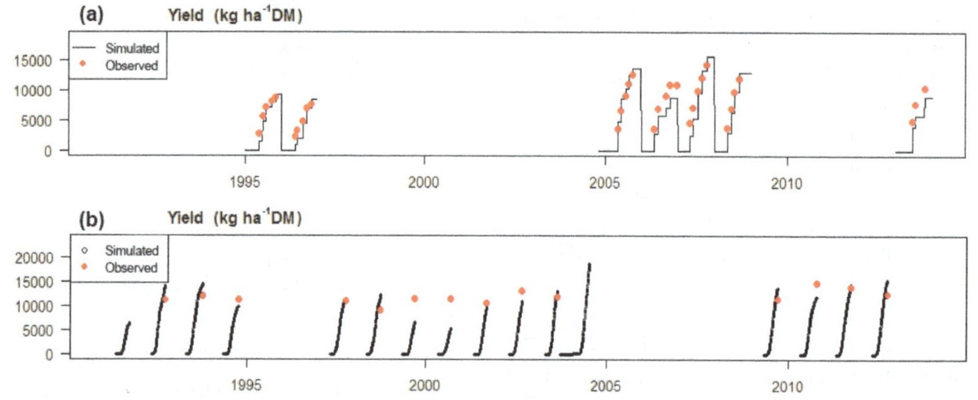

Figure 5. Results of case studies at location 1 (De Marke): observed yields (kg ha^{-1} DM) as red dots and simulated yields as black lines or as black dots: **(a)** for yields (kg ha^{-1} DM) of grassland; **(b)** for yields (kg ha^{-1} DM) of maize.

cuted on the fields and the meteorological data were not measured on site, but taken from meteorological stations sometimes more than 30 km away. Furthermore, no detailed information concerning fertilizer applications and soil carbon was available, therefore we considered it constant in time.

Even though some yields were not accurate enough to satisfy statistical criteria for good model performance, we believe that the dynamics of soil hydrology and crop yield were acceptably captured. With more field information and calibration, a better result could be achieved but we think that current tuning of SWAP-WOFOST for the three crops allowed for an application at a larger scale with various hydrological boundary conditions.

Before the analysis at a larger scale we simulated the impact of upward flow for the case studies. We carried out additional simulations without upward flow towards the root zone, using the specially programmed synthetic model option. Results of these three cases are given in Table 3 for the situation with and without upward flow. This table shows that suppressing upward flow lowered yields by 6, 3 and 20 % respectively for grassland, maize and potato. The groundwater recharge was reduced with respectively 3, 4 and 94 % (Table 3). Detailed results can be found in Sect. S1. In Sect. S4 input data for case 3 (V-Potato) can be found. In a next step, we carried out a larger-scale experiment to quantify this impact for different soil crop and climate conditions.

3.2 Soil crop experiment to analyse the role of capillary rise

The three crops from the case studies were simulated with 72 soils from the national database using three different bottom boundary conditions and 45 years with weather from 1970 to 2015.

Results of simulated upward flow of 45 years of weather, 72 soils and three lower boundary conditions are summarized with mean values in Table 4. The highest values for upward flow to the root zone during crop growth were found for average groundwater conditions (Ave) with long-term mean values for grassland, maize and potatoes of respectively 194, 74 and 112 mm yr^{-1}. Differences among hydrological conditions at the bottom of the root zone were caused by differences in weather, growing season, dynamic position of the

Table 3. Results of the case studies: values and differences in yield, capillary rise and percolation fluxes resulting from simulations with and without capillary rise.

Case study[1]	Model result	Condition A[2]	Condition B[3]	Differences A–B	Unit	Differences (%) 100×(A–B)/A
DM-Grass	Y_{act}	12 928	12 213	715	kg ha^{-1} season^{-1} DM	6
	$q_{caprise}$	30	0	30	mm season^{-1}	100
	$q_{percolation}$	313	305	9	mm season^{-1}	3
DM-Maize	Y_{act}	12 803	12 788	15	kg ha^{-1} season^{-1} DM	0
	$q_{caprise}$	7	0	7	mm season^{-1}	100
	$q_{percolation}$	91	88	3	mm season^{-1}	4
V-Potato	Y_{act}	11 071	8877	2194	kg ha^{-1} season^{-1} DM	20
	$q_{caprise}$	101	0	101	mm season^{-1}	100
	$q_{percolation}$	16	1	15	mm season^{-1}	94

[1] Cases studies DM-Grass and DM-Maize were simulated for limited periods of respectively 2005–2008 and 1991–1994 to have a continuous sequence of years. Case study V-Potato was simulated for 1 year. [2] Condition A has actual bottom boundary conditions according to column BBC in Table 1. [3] Condition B has actual bottom boundary conditions according to column BBC in Table 1, but without capillary rise to root zone.

Table 4. Results of soil crop experiments: mean values of six model results from three different hydrological conditions: FDnc (free drainage with no recirculation), FDrc (free drainage with recirculation) and Ave (average drainage conditions).

Crop	Model Result	FD_{nc}	FD_{rc}	Ave	Unit
Grassland	Y_{act}	10 494	12 147	14 177	kg ha^{-1} season^{-1} DM
	$q_{caprise}$			194	mm season^{-1}
	q_{recirc}	0	78		mm season^{-1}
	$q_{percolation}$	317	338	380	mm season^{-1}
	$q_{seepage}$	0	0	227	mm yr^{-1}
	$q_{leaching}$	301	257	0	mm yr^{-1}
Maize	Y_{act}	12 318	12 378	12 643	kg ha^{-1} season^{-1} DM
	$q_{caprise}$			74	mm season^{-1}
	q_{recirc}	0	17		mm season^{-1}
	$q_{percolation}$	52	57	47	mm season^{-1}
	$q_{seepage}$	0	0	155	mm yr^{-1}
	$q_{leaching}$	396	394	0	mm yr^{-1}
Potato	Y_{act}	8864	9521	10 365	kg ha^{-1} season^{-1} DM
	$q_{caprise}$			112	mm season^{-1}
	q_{recirc}	0	42		mm season^{-1}
	$q_{percolation}$	39	50	73	mm season^{-1}
	$q_{seepage}$	0	0	291	mm yr^{-1}
	$q_{leaching}$	432	416	0	mm yr^{-1}

root zone and demand of root water uptake. Even in free-drainage situations the upward flow to the root zone caused by soil water recirculation was considerable, ranging from 17 to 78 mm long-term average (FD_{rc} in Table 4). In free-draining soils the variation in upward flow to the root zone ranged from about 10 mm in wet and cold to 120 mm in dry and warm years with a high evaporative demand (Fig. 6a). In general, upward flow was highest in loamy soils where soil physical conditions were optimal. Especially in the presence of a groundwater level, differences in upward flow between soils were relatively small compared to differences among years and within one grouped soil type (Fig. 7b).

The upward flow was inversely related to the rooting depth: the larger the rooting depth the smaller the upward flow. Grassland, potatoes and maize had rooting depths of respectively 40, 50 and 100 cm and an upward flow of respec-

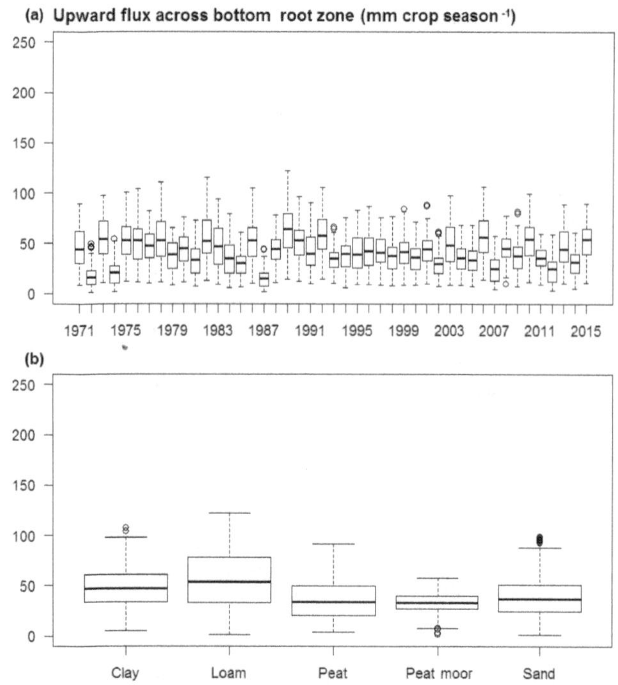

Figure 6. Results of the soil–crop experiment for potato: upward flux across the bottom of the root zone (q_{recirc} in mm crop season^{-1}) for hydrological conditions with free drainage ($\mathrm{FD_{rc}}$): **(a)** for results for all 72 soils for the period 1971–2015; **(b)** for results as box plots for clustered soil types.

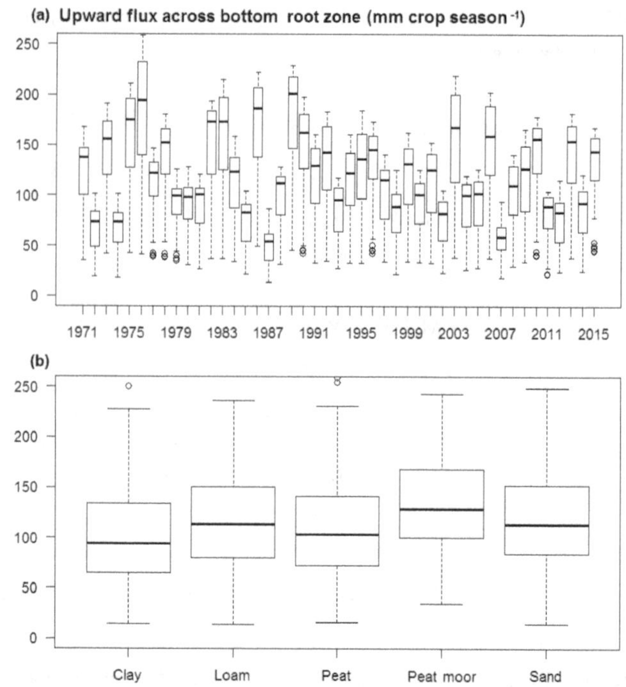

Figure 7. Results of the soil–crop experiment for potato: upward flux across the bottom of the root zone (q_{caprise} in mm crop season^{-1}) for hydrological conditions with average groundwater level (Ave): **(a)** for results for all 72 soils for the period 1971–2015; **(b)** for results as box plots for clustered soil types.

tively 194, 112 and 74 mm per growth season (Table 4). Note that the high value for perennial grassland was also caused by a much longer growing season. The percolation was highest for grassland for the same reasons (Table 4). These high values were largely due to the precipitation excess during winter in the Netherlands.

Upward seepage across the bottom boundary did not occur in the free-drainage conditions (Fig. 2a, b). Leaching was highest (Table 4) in the synthetic free-drainage condition without capillary rise (Fig. 2a). Note that the values in Table 4 for seepage and leaching were given for a calendar year, whereas the other mean values were given for a growing season. Yearly values were used for the bottom boundary because these values give an indication for the yearly deeper groundwater recharge, which may also be influenced by variations in vertical fluxes close to the root zone during the remainder of the year. The leaching flux at 5.5 m depth (Table 4, q_{leaching}) increased when upward flow was suppressed (lower transpiration, more groundwater recharge), with respectively 44, 2 and 16 mm yr^{-1} for grassland, maize and potatoes. The shallow groundwater in Dutch conditions (Fig. 2c) often does not have leaching at greater depth because excess precipitation or upward seepage is discharged via drainage systems. The average condition we used had no leaching but seepage of 227, 155 and 291 mm yr^{-1} for grassland, maize and potatoes (Table 4, q_{seepage}).

As can be expected, the synthetic condition without upward flow and without groundwater (Fig. 2a) had the lowest simulated mean yields for all crops (Table 4). The highest mean yields were simulated when average groundwater situations including capillary rise were considered (Table 4, Ave). The relative mean yield increase was lowest for maize and highest for grassland (Table 5), which was probably caused by the difference in rooting depth.

The simulation results with three different lower boundary conditions (Fig. 2 conditions a, b and c) were also compared by subtraction. The subtraction enables a quantification of the contribution of the two different sources of upward flow: groundwater and recirculating percolation water.

The elimination of recirculating percolation water to the root zone in free-drainage conditions (synthetic condition a compared to b, Fig. 2) reduced grassland, maize and potato yields with respectively 14, 0 and 7 % (Table 5). The higher yields were caused by upward flow using recirculating percolation water as source.

A comparison between situations with free drainage (condition b, Fig. 2) with average groundwater levels (condition c, Fig. 2) showed a similar yield reduction: respectively 14, 2 and 8 %. The higher yields were caused by capillary rise with groundwater and recirculation as source.

When one compares situations with free-drainage conditions without upward flow (synthetic condition a, Fig. 2) with

Table 5. Results of soil–crop experiments: differences (%) between results from three different hydrological conditions: FDnc (free drainage with no recirculation), FDrc (free drainage with recirculation) and Ave (average drainage conditions).

Crop	Model Result	Differences (%)		
		$100\times(FD_{rc}–FD_{nc})$ /FD_{rc}	$100\times(Ave–FD_{rc})$ /Ave	$100\times(Ave–FD_{nc})$ /Ave
grassland	Y_{act}	14	14	26
	$q_{percolation}$	6	11	17
maize	Y_{act}	0	2	3
	$q_{percolation}$	9	−22	−11
potato	Y_{act}	7	8	14
	$q_{percolation}$	22	31	46

average groundwater levels (condition c) yield reductions of grassland, maize and potatoes were respectively 26, 3 and 14 % (Table 5) or respectively about 3.7, 0.3 and 1.5 t ha^{-1} DM (Table 4). These yield differences quantify the contribution to the sum of the two different sources of upward flow: groundwater and recirculating percolation water.

The impact of upward flow on groundwater recharge was highest for potatoes and lowest for maize. For grassland, maize and potatoes differences between downward flux across the bottom of the root zone ($q_{percolation}$ in Fig. 2) of three hydrological conditions were calculated of respectively 17, −11 and 46 % ($q_{percolation}$ in Table 5) or 63, −5 and 34 mm ($q_{percolation}$ in Table 4). Low recharge values for maize were caused by deeper rooting systems which reduced these differences because groundwater levels were closer to the bottom of the root zone. For potatoes this difference in yield did reach values of more than 4 t ha^{-1} DM in stress conditions (Table 6). The results are presented in more detail in the Sect. S3.

4 Discussion

The case studies and soil–crop experiments in this paper demonstrate the combined interaction of recirculation and capillary rise on crop yields. This impact is clearly present in situations where a groundwater level is present (85 % of NL) but also in free-draining situations the impact of upward flow is considerable. According to our simulation results, grassland, maize and potato yields increased with respectively 14, 0 and 7 % in free-drainage conditions when upward flow was included (Table 5). This increase was mainly caused by internal recirculation, i.e. a part of the downward flux past the root zone was redirected upward to the root zone as a result of gradient-driven flow. When upward flow also has groundwater as a source, simulated yields increased by another 14, 2 and 8 % respectively. This increase was supported by a stronger capillary rise due to proximity to the groundwater. Comparing the simple simulations (no upward flow, no groundwater influence) to those with an average groundwater level and capillary rise shows yield increases of 26, 3 and 14 %. About half of these yield increases were caused by internal recirculation as occurs in free-drainage conditions and the other half was caused by an increased upward capillary flow from the groundwater.

Crop models that apply the bucket approach consider the soil system as a reservoir with only percolation and no upward flow (an overview with a model comparison is provided by Ahuja et al., 2014). Such models do not account for soil moisture redistribution within and below the root zone. Similar to Guderle and Hildebrand (2015) our simulation results show that a detailed vertical flow improves predictions of root water uptake. Bucket models generally overestimate drought stress and groundwater recharge and subsequently underestimate crop yield. The irrigation demand may be overestimated as well. The high percolation may also result in overestimation of groundwater recharge (leaching). Groundwater depth is important, because it determines the distance that the capillary flux has to bridge to reach the root zone and should be accounted for in crop modelling.

In the ideal situation one should compare the bucket approach to the approach with full simulation of capillary rise and recirculation using independent data sets. However, the measured data sets were insufficient to calibrate and validate the soil and crop parameters in such detail that they could allow proper statistical evaluation of the two approaches. The calibration of both model approaches had too much freedom with the available data sets, which upset a reliable validation. Therefore, we used the measured data sets to illustrate that with common soil and crop input values, SWAP-WOFOST yielded realistic and plausible results for the crops considered in this study. Further, crop growth and soil water flow were simulated by SWAP-WOFOST with novel concepts. Therefore, we may expect that the model itself can be used to show the effect on crop yield of different boundary conditions with respect to zero flux, recirculation and capillary rise.

Our hypothesis is that the process of recirculation makes crop modelling more accurate.

Table 6. Results for potatoes of soil crop experiments for each clustered soil type: capillary rise, recirculation and yield from three different hydrological conditions: FDnc (Free Drainage with No reCirculation), FDrc (Free Drainage with Recirculation) and Ave (Average Drainage conditions). Results for upward flow of FDnc are zero and therefore not given.

Hydrological condition		Statistic	Values per clustered soil type Clay	Loam	Peat	Moor	Sand	Unit
FD_{rc}	q_{recirc}	min	5	1	4	8	1	mm crop^{-1} season
		lower quartile	34	33	20	27	24	mm crop^{-1} season
		median	47	54	34	33	37	mm crop^{-1} season
		upper quartile	61	78	50	39	51	mm crop^{-1} season
		max	98	122	91	58	88	mm crop^{-1} season
Ave	$q_{caprise}$	min	14	14	15	34	15	mm crop^{-1} season
		lower quartile	65	80	72	100	84	mm crop^{-1} season
		median	94	113	104	129	113	mm crop^{-1} season
		upper quartile	134	151	141	168	152	mm crop^{-1} season
		max	227	236	231	243	249	mm crop^{-1} season
FD_{nc}	Y_{act}	min	3.1	5.4	2.8	2.8	1.2	1000 kg ha^{-1} DM
		lower quartile	7.4	8.7	7.5	6.9	6.9	1000 kg ha^{-1} DM
		median	9.6	10.3	9.8	9.3	9.2	1000 kg ha^{-1} DM
		upper quartile	10.7	10.9	10.7	10.7	10.7	1000 kg ha^{-1} DM
		max	12.2	12.2	12.2	12.2	12.2	1000 kg ha^{-1} DM
FD_{rc}	Y_{act}	min	5.0	7.5	4.8	3.4	3.3	1000 kg ha^{-1} DM
		lower quartile	8.5	9.6	8.4	7.6	7.8	1000 kg ha^{-1} DM
		median	10.2	10.6	10.1	9.8	9.9	1000 kg ha^{-1} DM
		upper quartile	10.9	11.1	10.9	10.8	10.8	1000 kg ha^{-1} DM
		max	12.4	12.4	12.4	12.4	12.4	1000 kg ha^{-1} DM
Ave	Y_{act}	min	7.4	8.0	7.8	7.8	7.7	1000 kg ha^{-1} DM
		lower quartile	9.6	9.8	9.8	9.8	9.7	1000 kg ha^{-1} DM
		median	10.5	10.7	10.7	10.7	10.7	1000 kg ha^{-1} DM
		upper quartile	11.1	11.2	11.1	11.2	11.1	1000 kg ha^{-1} DM
		max	12.6	12.6	12.6	12.6	12.6	1000 kg ha^{-1} DM

To demonstrate and support our hypothesis we added another case study. This is reported in Sect. S5. In this section we demonstrated the difference in soil water pressure head in the upper part of the root zone as caused by drying of the soil due to a lack of recirculating water in the hydrological condition (Fig. 2a). This resulted in a lowering of average yields with 609 kg ha^{-1} (from 7132 to 7741 kg ha^{-1} DM, which is about 9 % yield reduction due to recirculation. This supports the recommendation to use tools that support this process of recirculation in conditions where the vertical water fluxes across the root zone are relatively high. This will clearly be the case in delta regions where you occasionally have a precipitation excess.

A bucket approach generally underestimates water availability in the rooting zone and consequently overestimates drought stress (Boogaard et al., 2013). We suggest to generate additional relations for the contribution of recirculation and capillary rise to upward flow to the root zone. Such an approach has been used in AQUACROP to derive a relation between capillary rise and groundwater (Van Gaelen et al., 2017). Another approach is to calibrate the conceptual parameters of a bucket model with agro-hydrological models like SWAP as done by Romano et al. (2011).

Our analysis shows that soil properties and soil profile layering are important because differences in soil hydraulic properties influence vertical water flow. High upward flow values were found in loamy soils as was expected (Table 6, max row), but if water stress was high and upward flow was low the influence of soil type decreased and low upward flow values were found for loamy soils (Table 6, min row). Comparing the minimum yield values showed that there was a large difference between these soil types in free-drainage conditions with and without upward flow. This means that the storage capacity of loamy soils was larger than the one of sandy soils as could be expected. The yield variation between soil types in water stress conditions was large and illustrated the need for a proper soil schematization, especially in stress full hydrological conditions. As the influence of recirculation increased, the yield variation became less and the influence of soil type decreased. In situations without water stress the soil type was less important. In conditions where groundwater and capillary rise occurred (Ave), yield variation was hardly influenced by soil type.

Therefore, modelling concepts should consider dynamic interactions between soil water and crop growth. Crop models in general should consider recirculation of soil water and, especially in low-lying regions like deltas, groundwater dynamics should be considered as well.

Precipitation, soil texture and water table depth jointly affected the amount of groundwater recharge and time lag between water input and groundwater recharge (Ma et al., 2015). We quantified some of these issues, but several items remain, such as the impact of rooting depth on crop yield and transpiration. Also, soil and water management practises like ploughing and irrigation were not considered. Furthermore, the rooting pattern needed a more detailed analysis; we applied an exponential decrease in root density and compensation of root uptake according to Jarvis (1989, 2011) but the macroscopic root water uptake concept was still simple and may require a more detailed analysis (Dos Santos et al., 2017). Another item we neglected is the preferential flow of water by the occurrence of non-capillary sized macropores (Bouma, 1981; Feddes et al., 1988), which is especially relevant in clay soils. Hysteresis of the water retention function was also not considered. An additional analysis of these issues is recommended, especially the impact of different rooting patterns on capillary rise should be addressed.

The impact of soil type on yield increased when environmental conditions became dryer; situations without groundwater and without recirculation had less yield and higher yield variation than situations where groundwater influenced capillary rise (for detailed information on the results see Sects. S1 and S3).

5 Conclusions

We quantified the impact of upward flow on crop yields of grassland, maize and potatoes in layered soils. We compared situations with average groundwater levels with free-drainage conditions with and without upward flow. The largest impact of upward flow on crop yields was found when one compares situations with average groundwater levels with free-drainage conditions without upward flow. From these differences one may conclude that neglecting upward flow has a large impact on simulated yields and water balance calculations, especially in regions where shallow groundwater occurs. The comparison showed long-term average yield reductions for grassland, maize and potatoes of respectively 26, 3 and 14 % (Table 5) or respectively 3.7, 0.3 and 1.5 t DM per hectare (Table 4). Reduction of the percolation flux can be considerable; for grassland and potatoes the reduction was 17 and 46 % (Table 5) or 63 and 34 mm (Table 4).

About half of the yield increases was caused by internal recirculation as occurs in free-drainage conditions and the other half was caused by an increased upward capillary flow from groundwater. Improved modelling should consider upward flow of soil water which will result in improved estimates of crop yield and percolation.

We believe that the quantification of recirculation for crop yield is a novelty, especially recirculation as part of upward flow and its relation to capillary rise and crop growth. Studies about the relation between soil hydrology and crop growth should quantify this upward flow because neglecting this flow and its impact implies neglecting yield changes, which may have a large economic value in the Dutch delta and in other deltas in general. Another aspect, which could not be found in the referenced studies, is the lack of a quantification of the impact of recirculation on crop yields. Correct quantification of water fluxes contributes to the understanding of crop production and will help the institutions in charge of yield forecasting.

Competing interests. The authors declare that they have no conflict of interest.

Acknowledgements. Part of the case studies has been used before (Hack-ten-Broeke et al., 2016). This project is related to the project WaterVision Agriculture (www.waterwijzer.nl, last access: 7 May 2018) which is financed by a large group of financers: STOWA (Applied Research of the Water Boards), the Ministry of Infrastructure and Environment, ACSG (the Advisory Commission for Damage related to Groundwater), provinces Utrecht and Zuid-Holland, ZON (Zoetwatervoorziening Oost-Nederland), water companies Vitens and Brabant Water, VEWIN, LTO and the Ministry of Economic Affairs (project KB-14-001-046).

We also thank all reviewers for their constructive and valuable comments on earlier versions of this paper.

Edited by: Nunzio Romano

References

Ahuja, L. R., Ma, L., Lascano, R. J., Saseendran, S. A., Fang, Q. X., Nielsen, D. C., Wang, E., and Colaizzi, P. D.: Syntheses of the Current Model Applications for Managing Water and Needs for Experimental Data and Model Improvements to Enhance these Applications, in: Practical Applications of Agricultural System Models to Optimize the Use of Limited Water, Vol. 5, 399–437, 2014.

Allen, R. G., Pereira, L. S., Raes, D., and Smith, M.: FAO Irrigation and Drainage Paper No 56; Crop Evapotranspiration, FAO Irrigation and Drainage Paper, Rome, available at: http://www.fao.org/docrep/X0490E/X0490E00.htm (last access: 7 May 2018), 1998.

Awan, U. K., Tischbein, B., and Martius, C.: A GIS-based approach for up-scaling capillary rise from field to system level under soil-crop-groundwater mix, Irrig. Sci., 32, 449–458, https://doi.org/10.1007/s00271-014-0441-5, 2014.

Awan, U. K., Tischbein, B., and Martius, C.: A GIS-based approach for up-scaling capillary rise from field to system level under soil-crop-groundwater mix, Irrig. Sci., 32, 449–458, https://doi.org/10.1007/s00271-014-0441-5, 2014.

Babajimopoulos, C., Panoras, A., Georgoussis, H., Arampatzis, G., Hatzigiannakis, E., and Papamichail, D.: Contribution to irrigation from shallow water table under field conditions, Agr. Water Manage., 92, 205–210, 2007.

Bartholomeus, R. P., Witte, J. P. M., van Bodegom, P. M., van Dam, J. C., and Aerts, R.: Critical soil conditions for oxygen stress to plant roots: substituting the Feddes-function by a process-based model, J. Hydrol., 360, 147–165, 2008.

Bastiaanssen, W. G. M., Allen, R. G., Droogers, P., Urso, G. D., and Steduto, P.: Twenty-five years modeling irrigated and drained soils?: State of the art, Agr. Water Manage., 92, 111–125, https://doi.org/10.1016/j.agwat.2007.05.013, 2007.

Bigiarini, M. Z.: Package " hydroGOF ", R-package, available at: www.r-project.org/ (last access: 7 May 2018), 2013.

Bonfante, A., Basile, A., Acutis, M., De Mascellis, R., Manna, P., Perego, A., and Terribile, F.: SWAP, CropSyst and MACRO comparison in two contrasting soils cropped with maize in Northern Italy, Agr. Water Manage., 97, 1051–1062, https://doi.org/10.1016/j.agwat.2010.02.010, 2010.

Boogaard, H., Wolf, J., Supit, I., Niemeyer, S., and Ittersum, M.: A regional implementation of WOFOST for calculating yield gaps of autumn-sown wheat across the European Union, Field Crops Res., 1143, 130–142, 2013.

Bouma, J.: Soil morphology and preferential flow along macropores, Agr. Water Manage., 3, 235–250, https://doi.org/10.1016/0378-3774(81)90009-3, 1981.

De Vries, J. J.: Groundwater, in: Geology of the Netherlands, edited by: Wong, T., Batjens, D. A. J., and de Jager, J., KNAW, 295–315, available at: http://www.hydrology.nl/key-publications/231-groundwater-geology-of-the-netherlands.html, 2007.

De Vos, J. A., Sival, F. P., Clevering, O. A., and van Kleef, J.: Stikstof- en fosfaatverliezen naar grond- en oppervlaktewater bij vernatting van landbouwgronden. Veldexperimenten Vredepeel 2003–2005, Alterra, Wageningen & PPO-Akkerbouw, Groene Ruimte en vollegrondsgroente, Alterra-rapport 1392, Lelystad en Wageningen, 2006.

De Wit, C. T.: Photosynthesis of leaf canopies. Versl. Landbouwk. Onderz. (Agric. Res. Rep.) 663, Wageningen, available at: http://edepot.wur.nl/187115 (last access: 7 May 2018), 1965.

De Wit, A. J. W. and van Diepen, C. A.: Crop growth modelling and crop yield forecasting using satellite-derived meteorological inputs, Int. J. Appl. Earth Observ. Geoinform., 10, 414–425, 2008.

De Wit, A. J. W., Duveiller, G., and Defourny, P.: Estimating regional winter wheat yield with WOFOST through the assimilation of green area index retrieved from MODIS observations, Agr. Forest Meteorol., 164, 39–52, 2012.

Dijkstra, J. P., Hack-ten Broeke, M. J. D., and Wijnands, F. G.: Stikstofemissie naar het grondwater van geintegreerde en gangbare bedrijfssystemen in de akkerbouw op de proefboerderijen Borgerswold en Vredepeel: simulatie van de vocht- en nitraathuishouding op de proefboerderij Vredepeel voor de jaren 1990–1993, DLO-Staring Centrum, Rapport 287.3, Wageningen, 1995.

Dos Santos, M. A., de Jong van Lier, Q., van Dam, J. C., and Freire Bezerra, A. H.: Benchmarking test of empirical root water uptake models, Hydrol. Earth Syst. Sci., 21, 473–493, https://doi.org/10.5194/hess-21-473-2017, 2017.

Eitzinger, J., Trnka, M., Hösch, J., Žalud, Z., and Dubrovský, M.: Comparison of CERES , WOFOST and SWAP models in simulating soil water content during growing season under different soil conditions, Ecol. Modell., 171, 223–246, https://doi.org/10.1016/j.ecolmodel.2003.08.012, 2004.

Elbers, J. A., Moors, E. J., and Jacobs, C. M. J.: Gemeten actuele verdamping voor twaalf locaties in Nederland. Alterra-rapport 1920, available at: http://library.wur.nl/WebQuery/wurpubs/445274 (last access: 8 May 2018), 2009.

Fan, Y., Li, H., and Miguez-Macho, G.: Global patterns of groundwater table depth, Science, 339, 940–943, https://doi.org/10.1126/science.1229881, 2013.

Feddes, R. A., Kowalik, P. J., and Zaradny, H.: Simulation of field water use and crop yield, Simul. Monogr., Pudoc, Wageningen, 189 pp., 1978.

Feddes, R. A., Kabat, P., Bakel, P. J. T. Van, Bronswijk, J., and Halbertsma, J.: Modelling soil water dynamics in the unsaturated zone – state of the art, J. Hydrol., 100, 69–111, 1988.

Feodoroff, A.: Redistribution of moisture after infiltration in dry soils, Influence of gravity, 385–389, in: Water in the unsaturated zone; L'eau dans la zone non saturée, edited by: Rijtema, P. E. and Wassink, H., Vol. I, II, 1–525 (I) and 526–995 (II)), Proceedings of the Wageningen Symposium by UNESCO – IASH, available at: http://unesdoc.unesco.org/images/0001/000143/014319mo.pdf (last access: 7 May 2018), 1969.

Geerts, S., Raes, D., Garcia, M., Condori, O., Mamani, J., Miranda, R., Cusicanqui, J., Taboada, C,, Yucra, E., and Vacher, J.: Could deficit irrigation be a sustainable practice for quinoa (Chenopodium quinoa Willd.) in the Southern Bolivian Altiplano?, Agr. Water Manage., 95, 909–917, https://doi.org/10.1016/j.agwat.2008.02.012, 2008.

Guderle, M. and Hildebrandt, A.: Using measured soil water contents to estimate evapotranspiration and root water uptake profiles – a comparative study, Hydrol. Earth Syst. Sci., 19, 409–425, https://doi.org/10.5194/hess-19-409-2015, 2015.

Hack-ten Broeke, M. J. D. and Hegmans, J. H. B. M.: Use of soil physical characteristics from laboratory measurements or standard series for modelling unsaturated water flow, Agr. Water Manage., 29, 201–213, 1996.

Hack-ten-Broeke, M. J. D., Kroes, J. G., Bartholomeus, R. P., Dam, J. C. Van, Wit, A. J. W. De, Supit, I., Walvoort, D. J. J., van Bakel, P. J. T., and Ruijtenberg, R.: Quantification of the impact of hydrology on agricultural production as a result of too dry, too wet or too saline conditions, in: Soil, 2, 391–402, 2016.

Han, M., Zhao, C., Šimůnek, J., and Feng, G.: Evaluating the impact of groundwater on cotton growth and root zone water balance using Hydrus-1D coupled with a crop growth model, Agr. Water Manage., 160, 64–75, https://doi.org/10.1016/j.agwat.2015.06.028, 2015.

Hooghoudt, S. B.: Contributions to the Knowledge of Some Physical Soil Properties, No. 43, 461–676, Determination of the Conductivity of Soils of the Second Kind, in: Dutch: Verslagen van Landbouwkundige Onderzoekingen, no. 43 B, Dep. van Economische Zaken, Directie van den Landbouw, Algemeene Landsdrukkerij, The Hague, 1937.

Huo, Z., Feng, S., Huang, G., Zheng, Y., Wang, Y., and Guo, P.: Effect of groundwater level depth and irrigation amount on water fluxes at the groundwater table and water use of wheat, Irrig. Drain., 61, 348–356, https://doi.org/10.1002/ird.685, 2012.

Jarvis, N. J.: A simple empirical model of root water uptake, J. Hydrol., 107, 57–72, 1989.

Jarvis, N. J.: Simple physics-based models of compensatory plant water uptake: concepts and eco-hydrological consequences, Hydrol. Earth Syst. Sci., 15, 3431–3446, https://doi.org/10.5194/hess-15-3431-2011, 2011.

KNMI: Royal Netherlands Meteorologica l Institute (KNMI), De-Bilt, available at: http://www.knmi.nl/ (last access: 7 May 2018), 2016.

Kowalik, P. J.: Drainage and capillary rise components in water balance of alluvial soils, Agr. Water Manage., 86, 206–211, https://doi.org/10.1016/j.agwat.2006.08.003, 2006.

Kroes, J. G. and Supit, I.: Impact analysis of drought, water excess and salinity on grass production in The Netherlands using historical and future climate data, Agr. Ecosyst. Environ., 144, 370–381, https://doi.org/10.1016/j.agee.2011.09.008, 2011.

Kroes, J., Bartholomeus, R., van Dam, J., Hack-ten Broeke, M., Supit, I., Hendriks, R., de Wit, A., van der Bolt, F., Walvoort, D., Hoving, I., and Van Bakel, J.: Waterwijzer Landbouw, fase 2. Modellering van het bodem-water-plantsysteem met het gekoppelde instrumentarium SWAP-WOFOST (Eng: Watervision Agriculture, phase 2. Modeling of the soil-water-plant system with SWAP-WOFOST), Rapport 2015–16, STOWA, Amersfoort, available at: http://waterwijzer.stowa.nl/Upload/publicatie2014/STOWA%202015%2016%20WEB%20LR.pdf (last access: 14 May 2018), 2015.

Kroes, J. G., Dam, J. C. van, Bartholomeus, R. P., Groenendijk, P., Heinen, M., Hendriks, R. F. A., Mulder, H.M., Supit, I., van, and Walsum, P. E. V.: SWAP version 4, Theory description and user manual, Wageningen Environmental Research, ESG Report 2780, available at: http://library.wur.nl/WebQuery/wurpubs/fulltext/416321 (last access: 7 May 2018), 2017.

Ma, Y., Feng, S., and Song, X.: Evaluation of optimal irrigation scheduling and groundwater recharge at representative sites in the North China Plain with SWAP model and field experiments, Comput. Electro. Agr., 116, 125–136, https://doi.org/10.1016/j.compag.2015.06.015, 2015.

Norman, J. M.: Fifty Years of Study of S-P-A Systems: Past Limitations and a Future Direction, Proc. Environ. Sci., 19, 15–25, https://doi.org/10.1016/j.proenv.2013.06.003, 2013.

Oster, J. D., Letey, J., Vaughan, P., Wu, L., and Qadir, M.: Comparison of transient state models that include salinity and matric stress effects on plant yield, Agr. Water Manage., 103, 167–175, https://doi.org/10.1016/j.agwat.2011.11.011, 2012.

Rallo, G., Agnese, C., Minacapilli, M., and Provenzano, G.: Comparison of SWAP and FAO Agro-Hydrological Models to Schedule Irrigation of Wine Grapes, J. Irrig. Drain. Eng., 138, 581–591, https://doi.org/10.1061/(ASCE)IR.1943-4774.0000435, 2012.

Rijtema, P. E.: Een berekeningsmethode voor de benadering van de landbouwschade ten gevolge van grondwateronttrekking, ICW-Nota 587, Wageningen, available at: http://library.wur.nl/WebQuery/wurpubs/advanced/422164 (last access: 7 May 2018), 1971.

Romano, N., Palladino, M., and Chirico, G. B.: Parameterization of a bucket model for soil-vegetation-atmosphere modeling under seasonal climatic regimes, Hydrol. Earth Syst. Sci., 15, 3877–3893, https://doi.org/10.5194/hess-15-3877-2011, 2011.

Rötter, R.: Simulation of the biophysical limitations to maize production under rainfed conditions in Kenya, Evaluation and application of the Model WOFOST, Materiel zur Ost-Afrika vorschung, Heft 12, 261+, Annexes, 1993.

Šimůnek, J., Sejna, M., Saito, H., Sakai, M., and van Genuchten, M. T.: The HYDRUS-1D Software Package for Simulating the One-Dimensional Movement of Water, Heat, and Multiple Solutes in Variably-Saturated Media, Version 4.0 April 2008, Environmental Sciences, Riverside, California, 2008.

SSSA: Glossary of Soil Science Terms 2008, Soil Science Society of America, Soil Sci., available at: https://www.soils.org/publications/soils-glossary# (last access: 7 May 2018), 2008.

Supit, I.: An exploratory study to improve the predictive capacity of the Crop Growth Monitoring Systems as applied by the European Commission, Doctoraal thesis, Wageningen University, Wageningen, The Netherlands, available at: http://library.wur.nl/WebQuery/wurpubs/fulltext/196579 (last access: 7 May 2018), 2000.

Supit, I., van Diepen, C. A., de Wit, A. J. W., Wolf, J., Kabat, P., Baruth, B., and Ludwig, F.: Assessing climate change effects on European crop yields using the Crop Growth Monitoring System and a weather generator, Agr. Forest Meteorol., 164, 96–111, 2012.

Talebnejad, R. and Sepaskhah, A. R.: Effect of different saline groundwater depths and irrigation water salinities on yield and water use of quinoa in lysimeter, Agr. Water Manage., 148, 177–188, https://doi.org/10.1016/j.agwat.2014.10.005, 2015.

Van Bakel, P. J. T., Massop, H. Th. L., Kroes, J. G., Hoogewoud, J., Pastoors, R., and Kroon, T.: Actualisatie Hydrologie voor STONE 2.3; Aanpassing randvoorwaarden en parameters, koppeling tussen NAGROM en SWAP, en plausibiliteitstoets, (Eng: Updating the hydrology component in STONE 2.3; Adjusting boundary conditions and parameters, linking NAGROM and SWAP, and plausibility test), WOt-rapport 57, Wettelijke Onderzoekstaken Natuur & Milieu (MNP), Alterra, Wageningen, The Netherlands, 2008.

van Dam, J. C., Groenendijk, P., Hendriks, R. F. A., and Kroes, J. G.: Advances of Modeling Water Flow in Variably Saturated Soils with SWAP, Vadose Zone J., 7, 640–653, https://doi.org/10.2136/vzj2007.0060, 2008.

Van den Brande, M.: Remote sensing beelden van NDVI en hydrologisch modelleren, BSc thesis Wageningen UR, Wageningen, 2013.

Van der Gaast, J. W., Massop, H. T. L., and Vroon, H. R. J.: Effecten van klimaatverandering op de watervraag in de Nederlandse groene ruimte, Alterra-rapport 1791, Wageningen, available at: http://edepot.wur.nl/7870 (last access: 7 May 2018), 2009.

Van der Ploeg, M. J. and Teuling, A. J.: Going Back to the Roots: The Need to Link Plant Functional Biology with Vadose Zone Processes, Proc. Environ. Sci., 19, 379–383, https://doi.org/10.1016/j.proenv.2013.06.043, 2013.

Van Diepen, C. A. van, Wolf, J., and van Keulen, H.: WOFOST: a simulation model of crop production, Soil Use Manage., 5, 16–24, 1989.

Van Gaelen, H., Vanuytrecht, E., Willems, P., Diels, J., and Raes, D.: Bridging rigorous assessment of water availability from field to catchment scale with a parsimonious agro-hydrological model, Environ. Modell. Softw., 94, 140–156, https://doi.org/10.1016/j.envsoft.2017.02.014, 2017.

Van Ittersum, M. K., Leffelaar, P. A., van Keulen, H., Kropff, M. J., Bastiaans, L., and Goudriaan, J.: On approaches and applications of the Wageningen crop models, Eur. J. Agro., 18, 201–234, 2003.

Van Keulen, H. and Wolf, J. (Eds): Modelling of agricultural production: weather, soils and crops, Simulation Monographs, Pudoc Wageningen, The Netherlands, 1986.

Verloop, J., Hilhorst, G. J., Oenema, J., Van Keulen, H., Sebek, L. B. J., and Van Ittersum, M. K.: Soil N mineralization in a dairy production system with grass and forage crops, Nut. Cy. Agroecosyst., 98, 267–280, https://doi.org/10.1007/s10705-014-9610-4, 2014.

Videla Mensegue, H., Degioanni, A., and Cisneros, J.: Estimating shallow water table contribution to soybean water use in Argentina, Eur. Sci. J., 11, 23–40, 2015.

Wesseling, J. G.: CAPSEV: steady state moisture flow theory. Program description, user manual. Wageningen, Staring Centre. Report 37, available at: http://edepot.wur.nl/360382 (last access: 7 May 2018), 1991.

Wesseling, J. G. and Feddes, R. A.: Assessing crop water productivity from field to regional scale, Agr. Water Manage., 86, 30–39, https://doi.org/10.1016/j.agwat.2006.06.011, 2006.

Wösten, H., de Vries, F., Hoogland, T., Massop, H., Veldhuizen, A., Vroon, H., Wesseling, J, Heijkers, J., and Bolman, A.: BOFEK2012, the new soil physical schematisation of The Netherlands, (in Dutch: BOFEK2012, de nieuwe, bodemfysische schematisatie van Nederland); Alterra-rapport 2387, Wageningen, 2013a.

Wösten, J. H. M., Verzandvoort, S. J. E., Leenaars, J. G. B., Hoogland, T., and Wesseling, J. G.: Soil hydraulic information for river basin studies in semi-arid regions, Geoderma, 195–196, 79–86, https://doi.org/10.1016/j.geoderma.2012.11.021, 2013b.

Wu, Y., Liu, T., Paredes, P., Duan, L., and Pereira, L. S.: Water use by a groundwater dependent maize in a semi-arid region of Inner Mongolia: Evapotranspiration partitioning and capillary rise, Agr. Water Manage., 152, 222–232, https://doi.org/10.1016/j.agwat.2015.01.016, 2015.

Xu, X., Huang, G., Sun, C., Pereira, L. S., Ramos, T. B., Huang, Q., and Hao, Y.: Assessing the effects of water table depth on water use, soil salinity and wheat yield: Searching for a target depth for irrigated areas in the upper Yellow River basin, Agr. Water Manage., 125, 46–60, https://doi.org/10.1016/j.agwat.2013.04.004, 2013.

Zipper, S. C., Soylu, M. E., Booth, E. G., and Loheide, S. P.: Untangling the Effect of Shallow Groundwater and Soil Texture as Drivers of Subfield-Scale Yield Variability, Water Resour. Res., 51, 1–21, https://doi.org/10.1002/2015WR017522, 2015.

11

Impacts of future climate change on urban flood volumes in Hohhot in northern China: benefits of climate change mitigation and adaptations

Qianqian Zhou[1,2], **Guoyong Leng**[2,a], **and Maoyi Huang**[3]

[1]School of Civil and Transportation Engineering, Guangdong University of Technology, Waihuan Xi Road, Guangzhou 510006, China
[2]Joint Global Change Research Institute, Pacific Northwest National Laboratory, College Park MD 20740, USA
[3]Earth System Analysis and Modeling Group, Pacific Northwest National Laboratory, Richland, WA 99352, USA
[a]now at: Environmental Change Institute, University of Oxford, Oxford OX1 3QY, UK

Correspondence: Guoyong Leng (guoyong.leng@ouce.ox.ac.uk)

Abstract. As China becomes increasingly urbanised, flooding has become a regular occurrence in its major cities. Assessing the effects of future climate change on urban flood volumes is crucial to informing better management of such disasters given the severity of the devastating impacts of flooding (e.g. the 2016 flooding events across China). Although recent studies have investigated the impacts of future climate change on urban flooding, the effects of both climate change mitigation and adaptation have rarely been accounted for together in a consistent framework. In this study, we assess the benefits of mitigating climate change by reducing greenhouse gas (GHG) emissions and locally adapting to climate change by modifying drainage systems to reduce urban flooding under various climate change scenarios through a case study conducted in northern China. The urban drainage model – Storm Water Management Model – was used to simulate urban flood volumes using current and two adapted drainage systems (i.e. pipe enlargement and low-impact development, LID), driven by bias-corrected meteorological forcing from five general circulation models in the Coupled Model Intercomparison Project Phase 5 archive. Results indicate that urban flood volume is projected to increase by 52 % over 2020–2040 compared to the volume in 1971–2000 under the business-as-usual scenario (i.e. Representative Concentration Pathway (RCP) 8.5). The magnitudes of urban flood volumes are found to increase nonlinearly with changes in precipitation intensity. On average, the projected flood volume under RCP 2.6 is 13 % less than that under RCP 8.5, demonstrating the benefits of global-scale climate change mitigation efforts in reducing local urban flood volumes. Comparison of reduced flood volumes between climate change mitigation and local adaptation (by improving drainage systems) scenarios suggests that local adaptation is more effective than climate change mitigation in reducing future flood volumes. This has broad implications for the research community relative to drainage system design and modelling in a changing environment. This study highlights the importance of accounting for local adaptation when coping with future urban floods.

1 Introduction

Floods are one of the most hazardous and frequent disasters in urban areas and can cause enormous impacts on the economy, environment, city infrastructure, and society (Chang et al., 2013; Ashley et al., 2007; Zhou et al., 2017). Urban drainage systems have been constructed to provide carrying and conveyance capacities at a desired frequency to prevent urban flooding. However, the design of drainage systems is often based on historical precipitation statistics for a certain period of time, without considering the potential changes in precipitation extremes for the designed return periods (Yazdanfar and Sharma, 2015; Peng et al., 2015; Zahmatkesh et al., 2015). It is likely that drainage systems would be overwhelmed by additional runoff induced by climate change, which may lead to increased flood frequency and magnitude, disruption of transportation systems, and increased health risks (Chang et al., 2013; Abdellatif et al., 2015). For example, Arnbjerg-Nielsen (2012) reported that the design intensities in Denmark are projected to increase by 10–50 % for return periods ranging from 2 to 100 years. Therefore, it is important to investigate the performance of drainage systems in a changing environment and to assess the potential urban flooding under various scenarios to achieve better adaptations (Mishra, 2015; Karamouz et al., 2013; Yazdanfar and Sharma, 2015; Notaro et al., 2015).

Impacts of climate change on extreme precipitation and urban flooding have been well documented in a number of case studies. For example, Ashley et al. (2005) showed that flooding risks (i.e. occurrence of pluvial floods) in four UK catchments may increase by almost 30 times by the 2080s compared to current conditions around the year 2000, and effective adaptation measures are required to cope with the increasing risks in the UK. Larsen et al. (2009) estimated that future extreme 1 h precipitation will increase by 20–60 % throughout Europe by 2071–2100 relative to 1961–1990. Willems (2013) found that in Belgium the current design storm intensity for the 10-year return period is projected to increase by 50 % by the end of this century. Several studies have also investigated the role of climate change mitigation or adaptation in reducing urban flood damages and risks under climate change scenarios (Alfieri et al., 2016; Arnbjerg-Nielsen et al., 2015; Moore et al., 2016; Poussin et al., 2012). The relationship between changes in precipitation intensity and flood volume has also been well explored, providing additional insights into drainage design strategies (Olsson et al., 2009; Willems et al., 2012; Zahmatkesh et al., 2015). However, previous studies on the effects of climate change mitigation and adaptation are typically conducted separately, and it is unclear which strategy is more effective at reducing urban floods. This study aims to advance our understanding of urban floods within the context of change climate, through investigating the benefits of climate change mitigation (by reducing greenhouse gas, GHG, emissions) and local adaptation (by improving drainage systems) in reducing future urban flood volumes in a consistent framework.

As China increasingly urbanises, flooding has become a regular occurrence in its cities; 62 % of Chinese cities surveyed experienced floods and direct economic losses of up to USD 100 billion between 2011 and 2014 (China Statistical Yearbook, 2015). The 2016 flooding events affected more than 60 million people – across China more than 200 people were killed and USD 22 billion in losses were suffered. Hence, assessing future changes in urban flooding is very important for managing urban floods by designing new urban infrastructure and re-designing existing ones to be resilient in response to the impacts of future climate change. While urban floods are speculated to increase in the future (Yang, 2000; Ding et al., 2006), their magnitudes are hard to assess because of uncertainties associated with future climate change scenarios, as well as the under-representation of plausible climate change mitigation and adaptation strategies in the models.

In this study, we chose a drainage system in a typical city in northern China to illustrate the role of climate change mitigation and local adaptation in coping with future urban flood volumes. Such an investigation of the performance of the present-day drainage system also has important implications for local governments responsible for managing urban flood disasters in the study region. Specifically, we first quantified the effects of future climate change on urban flood volumes as a result of extreme precipitation events for various return periods using the present-day drainage system. We then designed two plausible adaptation strategies for the study region and investigated how much urban flood volume can be reduced with the adapted systems by the 2020s. We also compared the benefits of global-scale climate change mitigation and local adaptation in reducing urban flood volumes to advance our understanding effective measures for coping with future urban floods.

2 Materials and methods

2.1 Study region

The study region (Hohhot) is located in the southern central portion of Inner Mongolia, China. It lies between the Great Blue Mountains to the north and the Hetao Plateau to the south, which has a north-to-south topographic gradient. The region is in a cold semi-arid climate zone, characterised by cold and dry winters and hot and humid summers. The regional annual mean precipitation is approximately 396 mm with large intra-seasonal variations. Most rain falls between June and August, a period that accounts for more than 65 % of the annual precipitation. The drainage area in year 2010 was about 210.72 km^2, and it served a residential population of 1.793 million (Fig. 1a). The land use types in the region can be classified into five categories: agricultural land (8 %), residential areas (38 %), industrial land (13 %), green spaces (7 %), and other facilities (34 %; including municipal squares, commercial districts, institutions). According to local water authorities, the major soil type of the area is a mixture of loam and clay.

The current drainage system can be divided into 3 large sub-basins (Fig. 1c) and 326 sub-catchments with a total pipeline length of 249.36 km. The drainage network has a higher pipeline cover rate in the centre, but a rather low design standard for extreme rainfall events with a return period of less than 1 year. Historical records of stormwater drainage and flood damage indicate that the region has experienced an increase in flood frequency and magnitude within the context of climate change and urbanisation (Zhou et al., 2016). During the major flood event on 11 July 2016, the city, especially

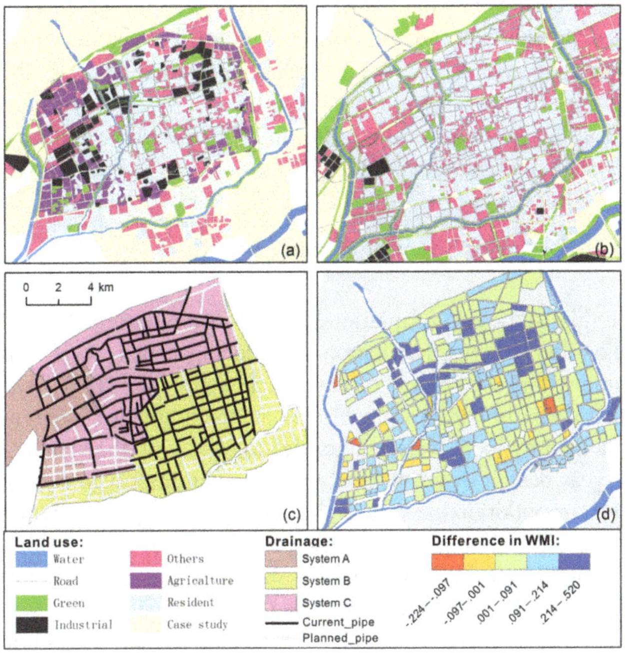

Figure 1. Land use of the study region for the year 2010 **(a)** and 2020 **(b)**. Pipe network description of current and planned drainage systems **(c)**. Difference in weighted mean imperviousness (WMI) between year 2010 and 2020 **(d)**.

the western portion of the watershed, was hit by an extreme rainfall event of more than 100 mm of rain in 3 h. The flood event led to the cancellation of at least 8 flights and 17 trains, and delays of several transportation systems. In particular, in the central area, the flood event caused severe traffic jams on major streets and resulted in a number of flooded residential buildings. A new drainage system is therefore required to cope with increasing urban flood volumes and frequencies in the future. The planned drainage area for 2020s is about 307.83 km^2, which is 50 % larger than the current drainage area (Zhou et al., 2016). The land use categories and distribution are shown in Fig. 1b.

2.2 Climate change scenarios

Climate projections by five general circulation models (GCMs) from the Coupled Model Intercomparison Project Phase 5 (CMIP5) archive were obtained from the Inter-Sectoral Impact Model Intercomparison Project (ISI-MIP; Warszawski et al., 2014). The CMIP5 climate projections were bias-corrected against observed climate for the overlapping period 1950–2000 using a quantile mapping method (Piani et al., 2010; Hempel et al., 2013). Several studies have demonstrated the value of the bias-corrected climate projections in quantifying climate change impacts on global and regional hydrology (e.g. Piontek et al., 2014; Elliott et al., 2014; Haddeland et al., 2014; Leng et al., 2015a, b). In this study, we used the bias-corrected climate from five GCMs (HadGEM2-ES, GFDL-ESM2M, IPSLCM5A-LR, MIROC-ESM-CHEM, and NorESM1-M) under two representative concentration pathways (RCPs; RCP 2.6 and RCP 8.5). The projected urban flood volumes under the business-as-usual scenario RCP 8.5 are compared with those under the climate change mitigation scenario RCP 2.6 to explore the benefits of climate change mitigation in reducing regional urban flood volumes. The possible land–surface–atmosphere interactions that would indirectly affect rainfall and flooding are not considered in this study.

2.3 Urban drainage modelling

The Storm Water Management Model (SWMM 5.1) developed by the U.S. Environmental Protection Administration is a widely used urban stormwater model that can simulate rainfall–runoff routing and pipe dynamics under single or continuous events (Rossman and Huber, 2016). SWMM can be used to evaluate the variation in hydrological and hydraulic processes and the performance of drainage systems under specific mitigation and adaptation scenarios in the context of global warming. The hydrological component requires inputs of precipitation and subcatchment properties including drainage area, subcatchment width, and imperviousness. The pipe network requires inputs from manholes, pipelines, outfalls, and connections to sub-catchments (Zahmatkesh et al., 2015; Chang et al., 2013). Basic flow-routing models include steady flow, kinematic, and dynamic wave methods. Infiltration can be described by the Horton, Green-Ampt, or Soil Conservation Service curve number (SCS-CN) methods. The dynamics of pipe flow are calculated based on the continuity equation and Saint-Venant equations (Rossman and Huber, 2016). Overflow occurs once the surface runoff exceeds the pipe capacity and is expressed as the value of total flood volume (TFV) at each overloaded manhole, i.e. the excess water from manholes after completely filling the pipe system without taking into account the outlet discharges. Other types of model outputs include catchment peak flows, maximum flow rates of pipelines, and flooded hours of manholes. It should be noted that SWMM is not capable of simulating surface inundation dynamics and cannot provide accurate estimation of the inundated zones and depths. The TFV is thus used to approximately reflect the flood condition and drainage system overloading status. Nevertheless, surface inundation models (e.g. Apel et al., 2009; Horritt and Bates, 2002; Vojinovic and Tutulic, 2009) are applicable if more accurate information about overland flow characteristics is available. In this study, the kinematic wave routing and the Horton infiltration model are used for model simulations. The infiltration capacity parameters for the category of "dry loam soils with little or no vegetation" are used in the hydrological model to be consistent with the local soil type (Akan, 1993; Rossman and Huber, 2016; Supplement, Table S1).

Rainfall inputs are calculated based on the regional storm intensity formula (SIF) using historical climatic statistics (Zhang and Guan, 2012; Supplement, Eqs. S1–S4). In this study, we considered 10 return periods: 1-, 2-, 3-, 10-, 20-, 50-, 100-, 200-, 500-, and 1000-year events. A 4 h rainfall time series was generated for each return period at 10 min intervals. The duration of design rainfalls (i.e. 4 h rainfall) is selected based on the time of concentration (ToC) of the watershed in the study region, according to the design principles as outlined in Butler and Davies (2010) and Chow et al. (2013). We assumed that the SIF was constant without considering the non-stationary features in a changing climate. That is, the intensity–duration–frequency (IDF) relationships for estimating design rainfall hydrographs were assumed to remain stable in the future and only changes in the daily mean intensity were considered because of the limited data availability in future sub-hourly climate projections from which these parameters would be derived.

As for future climate, the projected changes (i.e. change factors) in precipitation intensity at various return periods were calculated for each GCM–RCP combination (Table 1). Specifically, for each year, the annual maximum daily precipitation was determined for both historical and future periods. The generalised extreme value (GEV) distribution was then fitted separately to the two sets of daily values (Coles, 2001; Katz et al., 2002). Kolmogorov–Smirnov and Anderson–Darling statistics show that the hypothesis regarding the extreme value distribution is not rejected. That is, the fitted distribution could be well used to describe the extreme precipitation distribution in the study region. The value corresponding to each return period was estimated based on the GEV distribution and the changes between future and historical periods were calculated as the change factors. The derived change factor for each return period was then multiplied by the historical design Chicago Design Storms (CDS) rainfall time series to derive future climate scenarios. We acknowledge that the estimation of changes in extreme precipitation events involves inevitable uncertainties and therefore caution should be exercised when interpreting the relevant results.

2.4 Flood volume assessment

The TFVs of given rainfall events were simulated by the SWMM. A log-linear relationship is assumed to characterise the changes in flood volume with the increase in precipitation intensity as indicated by return periods (Fig. 2a) following Zhou et al. (2012) and Olsen et al. (2015). Generally, more TFV (i.e. system overloading) is expected with an increase in rainfall intensity. In Fig. 2b, the TFVs were linked to their specific occurrence probabilities. The total grey area under the curve denotes the TFVs integrated across various return periods and represents the total expected TFVs per year. The contribution of an individual flood event to total average TFV is dependent not only on the flood volume but also its corresponding probability of occurrence. Intensified precipitation is expected to increase the magnitude of system overflow, resulting in an upward trend in the TFV return period relationship and increased total TFVs.

2.5 Design of adaptation scenarios

In this study, two adaptation scenarios were designed to explore the role of adaptation in reducing urban flood volume within the context of climate change by the 2020s. The first scenario was designed to update the drainage system as planned by local water authorities to cope with the standard 3-year design event. It involved two main improvements of the current drainage system – enhancing the pipeline diameter and expanding the pipe network. The design was implemented in the SWMM model as shown in Fig. 1c. The number of pipelines of the present day and adapted systems was 323 and 488, with a total pipe length of 251.6 and 375.4 km, respectively. In the adapted scenarios, the mean pipeline diameter was about 1.73 m, an increase of 53 % compared to that of the present-day system.

The second adaptation scenario was designed to increase the permeable surfaces (e.g. green spaces) and reduce the regional imperviousness in the study region on the basis of pipe capacity enhancement. This scenario is referred to as the low-impact development (LID) scenario, and it was used to explore the effectiveness of urban green measures, such as the use of permeable pavements, infiltration trenches, and green roofs. Changes in land imperviousness in LID scenario have direct impacts on the performance of drainage systems in managing surface runoff. Due to a lack of detailed information about the permeable soil and coverage rates in the study region, the effects of these specific measures cannot be modelled individually. Here, we used a simplified approach by altering the subcatchment imperviousness to reflect the combined effects of infiltration-related measures. We derived such information by calculating the difference in land use type and imperviousness between the current and planned city maps using GIS. Figure 1d shows the difference in weighted mean imperviousness ($\mathrm{WMI} = \sum_i (\mathrm{IF}_i \times A_i)/\sum_i A_i$, where IF $_i$ and A_i is the impervious factor and area for land use type i, respectively.) for each subcatchment in the current and planned maps, using the commonly applied impervious factors (Pazwash, 2011; Butler and Davies, 2010) for each land use type. The difference in WMI was used to indicate the potential for adaptation based on the city plan. For example, a subcatchment with higher positive changes in the WMI indicates that the area's expected land use will have lower imperviousness and therefore is assumed to be more suitable for LID planning, and vice versa.

Table 1. Projected changes in precipitation intensity under return periods ranging from 1 to 1000 years by five global climate models under two representative concentration pathways (RCPs).

		1	2	3	10	20	50	100	200	500	1000
GFDL-ESM2M	RCP8.5	2.12	1.23	1.34	1.25	1.27	1.21	1.08	1.12	1.24	1.23
	RCP2.6	1.74	1.08	1.03	1.11	1.07	1.15	1.14	1.15	1.19	1.16
HadGEM2-ES	RCP8.5	0.62	1.08	1.09	1.06	1.01	1.03	1.17	1.26	1.23	1.14
	RCP2.6	0.36	1.2	1.19	1.04	1.02	1.11	1.31	1.26	1.37	1.24
IPSL-CM5A-LR	RCP8.5	1.44	1.17	1.28	1.17	1.08	1.09	1.02	1.1	1.12	1.13
	RCP2.6	0.74	1.04	1.18	1.01	1.06	1.03	1.01	0.99	0.95	1
MIROC-ESM-CHEM	RCP8.5	2.13	1.38	1.3	1.51	1.32	1.23	1.17	1.27	1.16	1.31
	RCP2.6	0.71	1.12	1.14	1.18	1.1	1.07	1.01	1.09	1.01	1.09
NorESM1-M	RCP8.5	2.11	0.96	0.8	1.63	1.35	1.15	1.08	1.01	1.04	0.97
	RCP2.6	0.11	1.09	1.05	1.28	1.17	1.08	1.1	1.18	1.09	1.2

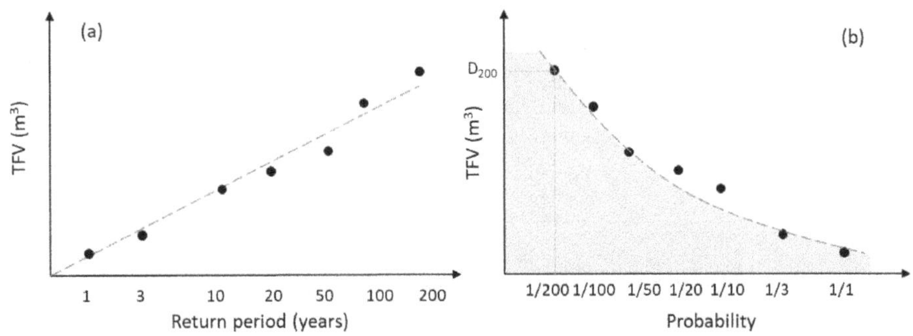

Figure 2. Illustration of total flood volume (TFV) as a function of return periods **(a)** and estimation of average total expected TFV per year (i.e. the grey area in **b)** under a stationary drainage system.

3 Results

3.1 Impacts of future climate change on urban flood volumes

Figure 3 shows the projected climate change (CC) impacts on urban flooding using the present-day drainage system of the near future (i.e. 2020–2040) under the RCP 8.5 scenario. Without climate change mitigation or adaptation, the TFV was projected to increase significantly with the increase of extreme rainfall events for most of investigated return periods (Table 2). Note that the lower bounds for return periods of 1, 3, and 1000 years fall below the current TFV curve due to the decrease in precipitation intensities. Despite the large uncertainty associated with climate projections, in particular with the 1-, 10-, and 1000-year return periods, the poor service performance of the current system in coping with urban flooding was evident. Overall, the urban flood volume was projected to increase by 52 % on average by the multi-model ensemble median by 2020–2040; the largest increase (258 %) was projected for the 1-year event and the smallest increase (12 %) for the 100-year event.

3.2 Benefits of climate change mitigation in reducing urban flood volumes

Figure 4 shows the comparison of TFVs under the RCP 8.5 scenario (i.e. a business-as-usual scenario) and the RCP 2.6 scenario (i.e. a climate change mitigation scenario). Although large uncertainties exist arising from climate models, it is clear that the simulated TFVs are much smaller under the RCP 2.6 scenario than under the RCP 8.5 scenario, demonstrating the benefits of climate mitigation in reducing local urban flood volumes. Such benefits are especially evident for floods for smaller return periods. For example, an increase of 936 m^3 in flood volume is projected with the increase in 1-year extreme rainfall under the business-as-usual climate change scenario (i.e. RCP 8.5), 52 % of which would be reduced if climate change mitigation were in place (i.e. under RCP 2.6). Notably, the peak of the total TFV curve was projected to shift from the 1-year event under the RCP8.5 scenario to the 3-year event under the RCP2.6 scenario (Fig. 4b), indicating a substantial reduction in the TFVs (especially at the 1-year return period; Fig. 4c). The lower total TFVs under RCP2.6 scenario could be attributed to the smaller magnitude of rainfall intensity than in the RCP8.5 scenario (Table 1),

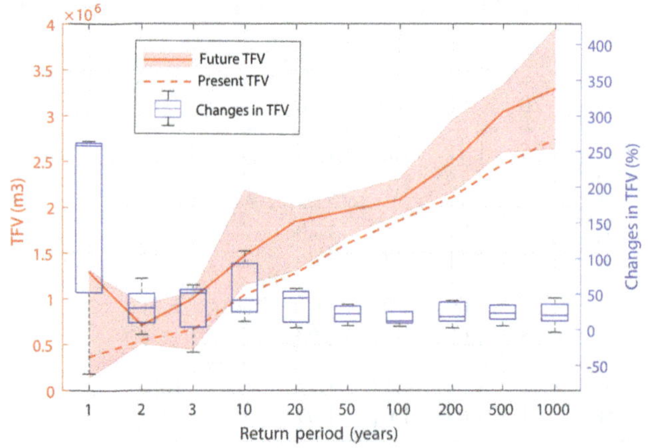

Figure 3. Changes in total flood volume (TFV) as a function of precipitation intensity at various return periods under RCP8.5 scenario without mitigation and adaptation. The red solid line represents the multi-model ensemble median TFV with shaded areas denoting the ensemble range. The red dashed line is the TFV under present condition. Box plots show the relative changes in TFV by 2020–2040 relative to present condition. Box edges illustrate the 25th and 75th percentile, the central mark is the median, and whiskers mark the 5th and 95th percentiles.

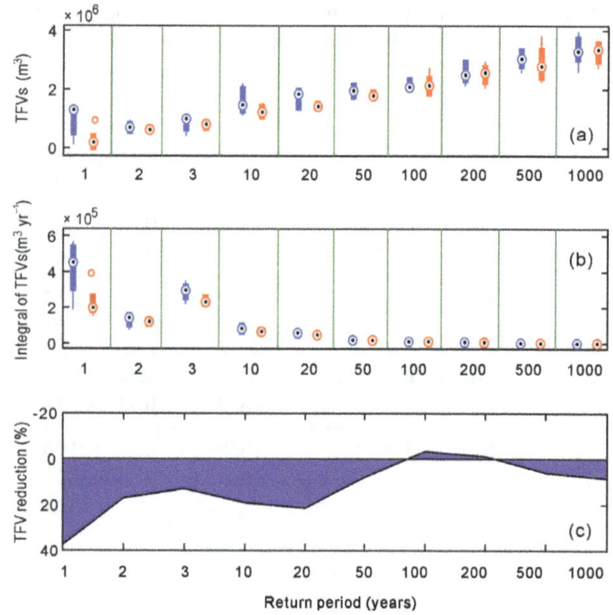

Figure 4. Comparison of **(a)** flood volume and **(b)** total TFVs (i.e. the piece-wise integral of flood volume versus the expected frequency with changes in precipitation intensity of various return periods under RCP8.5, blue, and RCP2.6, red). Panel **(c)** shows the TFV reduction calculated as the percentage difference in TFVs under RCP2.6 compared to RCP8.5 (i.e. benefits of climate mitigation) at various return periods.

demonstrating the important role of climate mitigation in reducing urban flood volumes. Overall, climate change mitigation can reduce future flood volumes by 13 % compared to the scenario without mitigation, as indicated by the multi-model ensemble median.

3.3 Benefits of adaptation in reducing urban flood volumes

Figure 5 shows the overloaded pipelines (red colour) with and without adaptation. The simulated results under the present 3-year event (recommended service level) and 50-year event (one typical extreme event) were selected to illustrate the role of adaptation in coping with floods in the historical period. As shown in Fig. 5a, the simulated locations of overloaded pipelines are in good agreement with historical flood points as recorded by local water authorities. Overall, the percentage of overloaded manholes (POM) and the ratio of flood volume (RFV) to input rainfall volume are up to 37 and 35 % in the current drainage system (Fig. 5a), respectively. When experiencing a 50-year extreme rainfall, the POM and RFV increase to 67 and 38 %, respectively. This indicates that current pipe capacities are insufficient to cope with extreme rainfall events (Fig. 5b). Spatially, the centre of the city is the most affected region due to the low service level in the area. With proposed adaptations, urban floods can be reduced to zero under a 3-year flood event. Such benefits of local adaptations are also evident when experiencing more intense precipitation events (e.g. 50-year events), for which the POM and RFV were reduced from 67 and 50 to 49 and 17 %, respectively.

Figure 6 shows the future changes in urban flood volume (CTFVs; $CTFV = (TFV_c - TFV_{nc})/TFV_{nc}$, where c and nc represent the results with and without climate change, respectively) with changes in extreme rainfall for various return periods. The performance of the current drainage system (no adaptation) was found to be less sensitive to future climate change, as indicated by the flatter slope in Fig. 6. For example, a similar magnitude of changes in flood volume was projected given changes in extreme rainfall for the return periods of 3, 50, and 500 years; the CTFV is 0.62, 0.32, and 0.35 for these periods, respectively. This is because the capacity of the current system is too small to handle extreme rainfall events with return periods larger than 1 year – a condition under which the current drainage system would be flooded completely, not to mention the situations with increased rainfall intensity in the future. Mathematically, the low sensitivity of the current drainage system to changes in extreme rainfall intensity could be attributed to the large value of the denominator in the calculation of CTFV.

With adaptations in place, the flood volume becomes much smaller than that in the current system due to capacity upgrading to hold more water. For example, when experiencing a 10-year extreme rainfall event, the urban flood volumes for the present period (i.e. TFV_{nc}) are 1 041 230, 274 650,

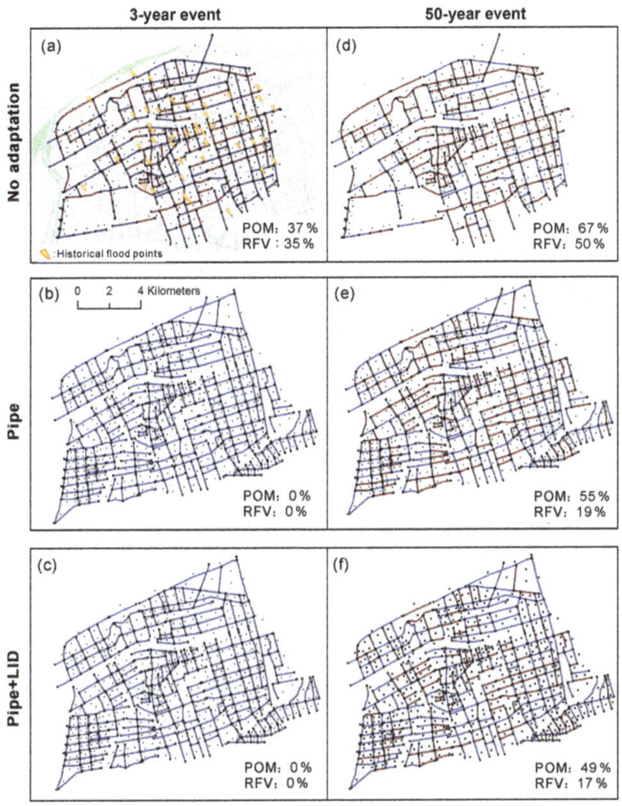

Figure 5. Spatial distribution of overloaded pipelines (red colour) induced by the 3-year **(a–c)** and 50-year extreme events **(d–f)** without and with adaptations. The total percentage of overloaded manholes (POM) and ratio of flood volume (RFV) to input rainfall volume are summarised for each scenario. Historical flood points and local land use, mainly the traffic network and green spaces, are shown in **(a)**.

and 180 610 m^3 in the current and two adapted systems, respectively, while in the future period, the magnitude of flood volume (i.e. TFV_c) is relatively similar among the three drainage systems. Therefore, future CTFVs relative to the historical period are much larger in the adapted systems than in the current system. The larger CTFVs in the adapted systems do not mean worse drainage system performance. Rather, they imply that the capacity (i.e. service level) of adapted drainage systems tends to become lower with climate change, while the current drainage system has already reached its peak capacity in handling extreme rainfall events in the historical period and thus shows a low sensitivity to future increases in rainfall intensity under climate change scenarios. Notably, the considerable increases in the CTFVs for return periods of less than 10 years in the adapted systems imply that the designed adaptations can effectively attenuate extreme rainfall events for small return periods. For more extreme rainfall events of return periods ≥ 50 years, more consistent results were found for both adaptation scenarios. This indicates that although the performance of adapted drainage

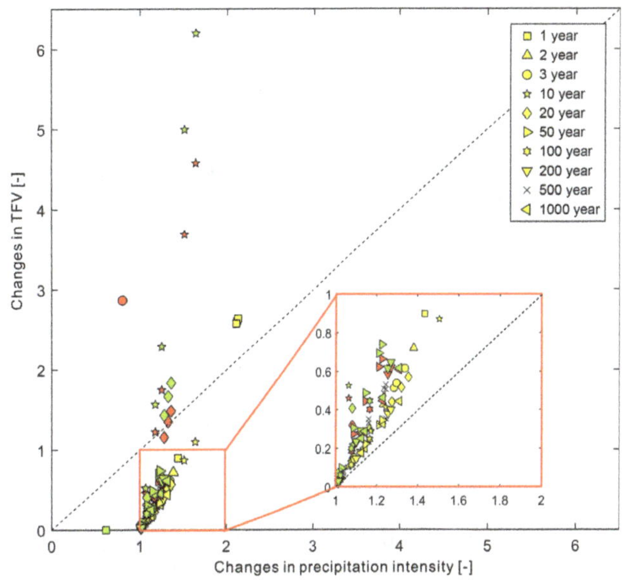

Figure 6. Future changes in flood volumes (CTFVs) relative to historical conditions under the current drainage system (yellow) and two adaptation scenarios (i.e. pipe in red and pipe + LID in green) at various return periods.

systems is significantly improved compared to that of the current system, the flood volume remains large when experiencing extreme rainfall events with return periods larger than 50 years, because flooding in such cases will push the adapted drainage systems to their limits.

3.4 Climate mitigation versus drainage adaptation

Figure 7 shows the reduced TFVs by climate change mitigation and drainage system adaptation as functions of return period. It is evident that both mitigation and adaptation measures are effective at reducing future urban flood volumes. However, such benefits are projected to weaken gradually with the increase in rainfall intensity (i.e. larger return periods). Overall, climate mitigation can lead to a reduction of flood volume by 10–40 % compared to the scenario without mitigation. Notably, there are minor negative values of TFV reductions for the return period of 100 and 200 years under climate mitigation scenario. The negative value is attributed to the slightly higher increase in precipitation intensity under climate mitigation scenario (i.e. RCP26) than the scenario without mitigation (i.e. RCP85) simulated by two of five climate models (i.e. GFDL-ESM2m and NorESM1-M, see Table 1), which translated to slightly larger flood volume under climate mitigation scenario. This climate internal variability is partly cancelled out by the other three climate models, thus leading to very minor negative values by the multi-model ensemble mean. This calls for the use of more climate models to derive more robust projections in the future studies.

Importantly, our results show that the two adaptation systems proposed in this study are found to be more effective

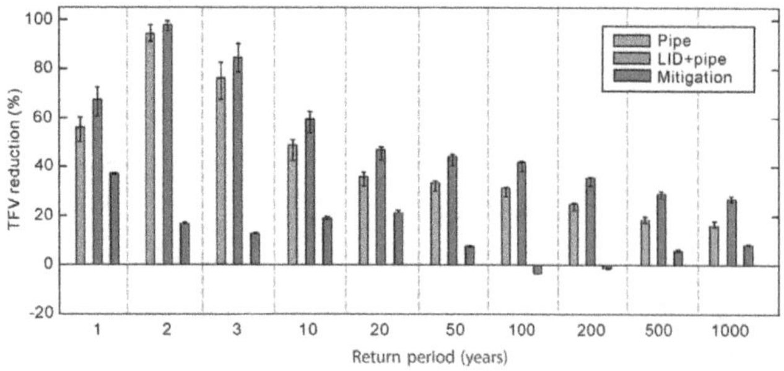

Figure 7. Comparison of benefits of climate mitigation and two adaptation strategies in reducing urban flood volumes with changes in precipitation intensities for various return periods, and with related variations (boundary bars) as a result of uncertainty arising from local soil conditions.

at reducing urban floods than climate change mitigation. In most cases, the benefits of local adaptation are more than double those of mitigation. In extreme cases, the reduction in TFV achieved by adaptation is five times more than that achieved by climate change mitigation (i.e. for the return periods of 2–3 years). Such effectiveness of urban flood reduction through drainage system adaptations has profound implications for local governments charged with managing urban flooding in the future. Notably, the second scenario (LID + pipe) exhibited a higher level of flood volume reduction than the pipe scenario in coping with extreme rainfall events for all investigated return periods. This implies that implementation of LID measures to augment drainage system capacity is more effective through reducing upstream loadings compared to updating the pipe system alone.

It is noted that local soil characteristics could affect the performance of the designed adaptation systems, in particular the LID measures. However, information about soil properties was not available at the subcatchment level in the study region. Here, a set of sensitivity experiments was conducted by adopting different parameters (e.g. infiltration values) associated with possible soil conditions (i.e. dry sand, loam, and clay soils with little or no vegetation in Table S1) for the area. The whiskers in Fig. 7 show the uncertainty range arising from the representation of different soil conditions in the drainage model. The benefits of the designed adaptation measures in reducing urban flood volumes were found to be robust regardless of soil conditions, and such benefits exceeded those of climate change mitigation, confirming our major conclusions found in this study.

4 Uncertainties and limitations

A number of uncertainties and limitations arise from the model structure, parameter inputs, emission scenarios, GCMs, climate downscaling/bias-correction approaches, etc. Specifically, climate projections by GCMs are subject to large uncertainties, in particular regarding precipitation (Covey et al., 2003) at spatial scales which are relevant for urban flood modelling. An alternative approach is to simulate future climate using a regional climate model (RCM) nested within a GCM. Such climate projections by RCMs have added value in terms of higher spatial resolution, which can provide more detailed regional climate information. However, various levels of bias would still remain in RCM simulations and bias corrections of RCM projections would be required; e.g. the European project ENSEMBLES. Running an RCM was not within the scope of this study; instead, we tended to use publicly available climate projections. Here, we obtained the climate projections from the ISI-MIP (Warszawski et al., 2014), which provides spatially downscaled climate data for impact models. The climate projections were also bias-corrected against observations (Hempel et al., 2013) and have been widely used in climate change impact studies on hydrological extremes such as floods and droughts (e.g. Dankers et al., 2014; Prudhomme et al., 2014; Leng et al., 2015a). It should be noted that we used the delta change factor to derive future climate scenarios as inputs into our drainage model instead of using GCM climate directly. This is because the relative climate change signal simulated by GCMs is argued to be more reliable than the simulated absolute values (Ho et al., 2012). Moreover, we used an ensemble of GCM simulations rather than one single climate model in order to characterise the uncertainty range arising from climate projections. However, disadvantages of this method are that transient climate changes cannot be represented and that changes in intra-seasonal or daily climate variability are not taken into account (Leng and Tang, 2014). Such sources of uncertainty can be explored when improved climate models at finer scales become available (Jaramillo and Nazemi, 2017).

In addition, the SIF parameters were assumed to remain stable in the future and only changes in the daily mean intensity were considered, because future sub-hourly climate projections were not readily available. The full climate variability range would also be under-sampled, although we used five climate models to show the possible range. Given the above limitations, we acknowledge that the modelling results represent the first-order potential climate change impacts on urban floods. Future efforts should be devoted to the representation of dynamic rainfall changes at hourly time steps considering non-stationary climate change.

Moreover, several assumptions had to be made due to limitations of the current modelling structure and approach. For example, the conveyance capacities of the drainage system and flood volume would largely depend on the state of drainage systems. Hence, a drainage system obstructed by vegetation, waste, or artefacts (cables, pipes, temporary structures) could make the outcomes of the SWMM calculation significantly different from observations. However, quantifying the impacts of drainage system states on urban flood volumes is not trivial because of the difficulties involved in collecting field data and selecting and using appropriate methods for reasonable assessment of pipe conditions (Ana and Bauwens, 2007; Fenner, 2000), and was not within the scope of this study. With deterioration, such as ageing networks, pipe deterioration, blockage, and construction failures, drainage systems were shown to become more vulnerable to extreme rainfalls as demonstrated in previous studies (Dawson et al., 2008; CIRIA, 1997; Davies et al., 2001). It is very likely that our simulated urban flood volumes would be underestimated without considering the changes in drainage conditions (Pollert et al., 2005).

Further, constrained by the one-dimensional modelling approach using SWMM, the performance of LID measures was mainly evaluated according to their effects at reducing water volume from overloaded manholes (Oraei Zare et al., 2012; Lee et al., 2013). That is, the LID adaptation measure was mainly designed to reduce the amount of water rather than slowing down the water speed, which has been demonstrated to be effective at reducing urban floods (Messner et al., 2006; Ashley et al., 2007; Floodsite, 2009). However, it should be noted that most LID measures can reduce runoff volume and flow speed at the same time, although some of the LID measures are primarily designed to slow down the flow speed, i.e. vegetated swales. To examine whether flood retention of a given event is induced by runoff volume or the internal speed control function in the model is difficult and requires detailed data for model validations. Specifically, the required information about surface roughness, soil conductivity, and seepage rate were unavailable at the subcatchment scale in the study region. Therefore, a simplified modelling approach was used to take advantage of existing data, especially for the design of LID measures. With the aid of more detailed field data and planning documents, the design of LID measures could be significantly improved by implementing more advanced approaches (Elliott and Trowsdale, 2007; Zoppou, 2001).

Evaluation of other potential adaptation strategies, such as flood retention by rain gardens and green roofs, can be explored in the future to gain additional insights into the performance of LID systems. In particular, the cost-effectiveness of the proposed adaptation measures should be accounted for. Indeed, a major limitation of this study is lack of assessment of costs and benefits of adaptation measures from an economic perspective. In fact, besides the effectiveness of proposed adaptation measures in reducing flood volume, assessment of the associated economic costs is essential for flood risk management (Rojas et al., 2013; Veith et al., 2003; Hinkel et al., 2014; Aerts et al., 2014; Ward et al., 2017). For example, Ward et al. (2017) showed that investments in urban flood protections with dykes are not economically attractive everywhere. Higher investment and maintenance costs may prohibit the implementation of adaptation strategies as proposed in this study. Future efforts should therefore be devoted to building a framework for assessing the costs and benefits of urban flood reduction measures and examine whether the reduced losses are higher than the costs of investments and maintenance of these measures.

Nevertheless, given these limitations, this study stands out from previous climate impact assessment studies of urban flood volumes by having proposed two feasible adaptation strategies and compared their benefits to those from global-scale climate change mitigation through GHG reductions within a consistent framework. Depending on the progress at data collection and the demands of local authorities, more advanced methods for pipe assessment (e.g. considering the changing pipe conditions), LID measures (detailed modelling of LID control), and two-dimensional surface flooding for assessment of flood damage and risk are planned in a future study to provide a more comprehensive analysis of the adaptation measures.

5 Summary and conclusions

The potential impacts of future climate change on current urban drainage systems have received increasing attention during recent decades because of the devastating impacts of urban flooding on the economy and society (Chang et al., 2013; Zhou et al., 2012; Abdellatif et al., 2015). However, few studies have explored the role of both climate change mitigation and drainage adaptations in coping with urban flooding in a changing climate. This study investigated the performance of a drainage system in a typical city in northern China in response to various future scenarios. In particular, we assessed the potential changes in urban flood volume and explored the role of both mitigation and adaptation in reducing urban flood volumes in a consistent manner.

Our results show significant increases in urban flood volumes due to increases in precipitation extremes, especially for return periods of less than 10 years. Overall, urban flood volume in the study region is projected to increase by 52 % by the multi-model ensemble median in the period of 2020–2040. Such increases in flood volume can be reduced considerably by climate change mitigation through reduction of GHG emissions. For example, the future TFVs under 1-year extreme rainfall events can be reduced by 50 % when climate change mitigation is in place. Besides global-scale climate change mitigation, regional/local adaptation can be implemented to cope with the adverse impacts of future climate change on urban flood volumes. Here, the adaptation measures as designed in this study were demonstrated to be much more effective at reducing future flood volumes than climate change mitigation measures. In general, the reduced flood volumes achieved by adaptation were more than double those achieved by climate change mitigation.

Through comprehensive investigation of future urban floods, this study provides much-needed insights into urban flood management for similar urban areas in China, most of which are equipped with highly insufficient drainage capacities. By comparing the reduction of flood volume by climate change mitigation (via reduction of GHG emissions) and local adaptation (via improvement of drainage systems), this study highlights the effectiveness of system adaptations in reducing future flood volumes. This has important implications for the research community and decision makers involved in urban flood management. We emphasise the importance of accounting for both global-scale climate change mitigation and local-scale adaptation in assessing future climate impacts on urban flood volumes within a consistent framework.

Competing interests. The authors declare that they have no conflict of interest.

Acknowledgements. This research was supported by the Public Welfare Research and Ability Construction Project of Guangdong Province, China (grant no. 2017A020219003), the Water Conservancy Science and Technology Innovation Project of Guangdong province, China (grant no. 201710), the Natural Science Foundation of Guangdong Province, China (no. 2014A030310121), and the Scientific Research Foundation for the Returned Overseas Chinese Scholars, State Education Ministry. Guoyong Leng and Maoyi Huang were supported by the Integrated Assessment Research program through the Integrated Multi-sector, Multi-scale Modeling (IM^3) Scientific Focus Area (SFA) sponsored by the Biological and Environmental Research Division of Office of Science, U.S. Department of Energy. The Pacific Northwest National Laboratory (PNNL) is operated for the U.S. DOE by Battelle Memorial Institute under contract DE-AC05-76RL01830.

Edited by: Carlo De Michele

References

Abdellatif, M., Atherton, W., Alkhaddar, R., and Osman, Y.: Flood risk assessment for urban water system in a changing climate using artificial neural network, Nat. Hazards, 79, 1059–1077, 2015.

Aerts, J. C., Botzen, W. W., Emanuel, K., Lin, N., de Moel, H., and Michel-Kerjan, E. O.: Evaluating flood resilience strategies for coastal megacities, Science, 344, 473–475, 2014.

Akan, O. A.: Urban stormwater hydrology: a guide to engineering calculations, CRC Press, Basel, Switzerland, 1993.

Alfieri, L., Feyen, L., and Di Baldassarre, G.: Increasing flood risk under climate change: a pan-European assessment of the benefits of four adaptation strategies, Climatic Change, 136, 507–521, https://doi.org/10.1007/s10584-016-1641-1, 2016.

Ana, E. and Bauwens, W.: Sewer network asset management decision-support tools: a review, International Symposium on New Directions in Urban Water Management, 12–14 September 2007, UNESCO Paris, 2007.

Apel, H., Aronica, G. T., Kreibich, H., and Thieken, A. H.: Flood risk analyses-how detailed do we need to be?, Nat. Hazards, 49, 79–98, 2009.

Arnbjerg-Nielsen, K.: Quantification of climate change effects on extreme precipitation used for high resolution hydrologic design, Urban Water J., 9, 57–65, 2012.

Arnbjerg-Nielsen, K., Leonardsen, L., and Madsen, H.: Evaluating adaptation options for urban flooding based on new high-end emission scenario regional climate model simulations, Clim. Res., 64, 73–84, https://doi.org/10.3354/cr01299, 2015.

Ashley, R., Garvin, S., Pasche, E., Vassilopoulos, A., and Zevenbergen, C.: Advances in Urban Flood Management, edited by: Ashley, R., Garvin, S., Pasche, E., Vassilopoulos, A., and Zevenbergen, C., Taylor & Francis/Balkema, London, UK, 2007.

Ashley, R. M., Balmforth, D. J., Saul, A. J., and Blanskby, J. D.: Flooding in the future – predicting climate change, risks and responses in urban areas, Water Sci. Technol., 52, 265–273, 2005.

Butler, D. and Davies, J.: Urban drainage, Third Edn., CRC Press, London, ISBN: 0415455251, 2010.

China Statistical Yearbook: National Bureau of Statistics of China, China Statistics Press, Beijing, 2015.

Chang, H. K., Tan, Y. C., Lai, J. S., Pan, T. Y., Liu, T. M., and Tung, C. P.: Improvement of a drainage system for flood management with assessment of the potential effects of climate change, Hydrolog. Sci. J., 58, 1581–1597, 2013.

Chow, V. T., Maidment, D. R., and Mays, L. W.: Applied Hydrology, 2nd Edn., in, McGraw-Hill, New York, ISBN: 007174391X, 2013.

CIRIA: Risk Management for Real Time Control in Urban Drainage Systems: Scoping Study, Project Report 45, CIRIA, London, 1997.

Coles, S.: An Introduction to Statistical Modeling of Extreme Values, Springer Series in Statistics, Springer, London, 2001.

Covey, C., AchutaRao, K. M., Cubasch, U., Jones, P., Lambert, S. J., Mann, M. E., Phillips, T. J., and Taylor, K. E.: An overview of results from the Coupled Model Intercomparison Project, Global Planet. Change, 37, 103–133, 2003.

Dankers, R., Arnell, N. W., Clark, D. B., Falloon, P. D., Fekete, B. M., Gosling, S. N., Heinke, J., Kim, H., Masaki, Y., Satoh, Y., and Stacke, T.: First look at changes in flood hazard in the Inter-Sectoral Impact Model Intercomparison Project ensemble, P. Natl. Acad. Sci. USA, 111, 3257–3261, 2014.

Davies, J. P., Clarke, B. A., Whiter, J. T., and Cunningham, R. J.: Factors influencing the structural deterioration and collapse of rigid sewer pipes, Urban Water, 3, 73–89, 2001.

Dawson, R. J., Speight, L., Hall, J. W., Djordjevic, S., Savic, D., and Leandro, J.: Attribution of flood risk in urban areas, J. Hydroinform., 10, 275–288, 2008.

Ding, Y., Ren, G., Shi, G., Gong, P., Zheng, X., Zhai, P., Zhang, D., Zhao, Z., Wang, S., Wang, H., Luo, Y., Chen, D., Gao, X., and Dai, X.: National assessment report of climate change (I): climate change in China and its future trend, Adv. Clim. Change Res., 2, 3–8, 2006 (in Chinese).

Elliott, A. H. and Trowsdale, S. A.: A review of models for low impact urban stormwater drainage, Environ. Modell. Softw., 22, 394–405, 2007.

Elliott, J., Deryng, D., Müller, C., Frieler, K., Konzmann, M., Gerten, D., Glotter, M., Flörke, M., Wada, Y., Best, N., and Eisner, S.: Constraints and potentials of future irrigation water availability on agricultural production under climate change, P. Natl. Acad. Sci. USA, 111, 3239–3244, 2014.

Fenner, R. A.: Approaches to sewer maintenance: a review, Urban Water, 2, 343–356, 2000.

Floodsite: Flood risk assessment and flood risk management. An introduction and guidance based on experiences and findings of FLOODsite (an EU-funded Integrated Project), Deltares, Delft Hydraulics, ISBN: 978 90 8 4067 0, 2009.

Haddeland, I., Heinke, J., Biemans, H., Eisner, S., Flörke, M., Hanasaki, N., Konzmann, M., Ludwig, F., Masaki, Y., Schewe, J., and Stacke, T.: Global water resources affected by human interventions and climate change, P. Natl. Acad. Sci. USA, 111, 3251–3256, 2014.

Hempel, S., Frieler, K., Warszawski, L., Schewe, J., and Piontek, F.: A trend-preserving bias correction – the ISI-MIP approach, Earth Syst. Dynam., 4, 219–236, https://doi.org/10.5194/esd-4-219-2013, 2013.

Hinkel, J., Lincke, D., Vafeidis, A. T., Perrette, M., Nicholls, R. J., Tol, R. S., Marzeion, B., Fettweis, X., Ionescu, C., and Levermann, A.: Coastal flood damage and adaptation costs under 21st century sea-level rise, P. Natl. Acad. Sci., 111, 3292–3297, 2014.

Horritt, M. S. and Bates, P. D.: Evaluation of 1D and 2D numerical models for predicting river flood inundation, J. Hydrol., 268, 87–99, 2002.

Ho, C. K., Stephenson, D. B., Collins, M., Ferro, C. A., and Brown, S. J.: Calibration strategies: a source of additional uncertainty in climate change projections, B. Am. Meteorol. Soc., 93, 21–26, 2012.

Jaramillo, P. and Nazemi, A.: Assessing Urban Water Security under Changing Climate: Challenges and Ways Forward, Sustain. Cities Soc., https://doi.org/10.1016/j.scs.2017.04.005, in press, 2017.

Karamouz, M., Nazif, S., and Zahmatkesh, Z.: Self-Organizing Gaussian-Based Downscaling of Climate Data for Simulation of Urban Drainage Systems, J. Irrig. Drainage Eng., 139, 98–112, 2013.

Katz, R. W., Parlange, M. B., and Naveau, P.: Statistics of extremes in hydrology, Adv. Water Resour., 25, 1287–1304, 2002.

Larsen, A. N., Gregersen, I. B., Christensen, O. B., Linde, J. J., and Mikkelsen, P. S.: Potential future increase in extreme one-hour precipitation events over Europe due to climate change, Water Sci. Technol., 60, 2205–2216, 2009.

Lee, J. M., Hyun, K. H., and Choi, J. S.: Analysis of the impact of low impact development on runoff from a new district in Korea, Water Sci. Technol., 68, 1315–1321, 2013.

Leng, G. and Tang, Q.: Modeling the impacts of future climate change on irrigation over China: Sensitivity to adjusted projections, J. Hydrometeorol., 15, 2085–2103, 2014.

Leng, G., Tang, Q., and Rayburg, S.: Climate change impacts on meteorological, agricultural and hydrological droughts in China, Global Planet. Change, 126, 23–34, 2015a.

Leng, G., Huang, M., Tang, Q., and Leung, L. R.: A modeling study of irrigation effects on global surface water and groundwater resources under a changing climate, J. Adv. Model. Earth Sy., 7, 1285–1304, 2015b.

Messner, F., Penning-Rowsell, E., Green, C., Meyer, V., Tunstall, S., and Van der Veen, A.: Guidelines for Socio-economic Flood Damage Evaluation, Report Nr. T9-06-01, FLOOD site, HR Wallingford, UK, 2006.

Mishra, A.: A study on the occurrence of flood events over Jammu and Kashmir during September 2014 using satellite remote sensing, Nat. Hazards, 78, 1463–1467, 2015.

Moore, T. L., Gulliver, J. S., Stack, L., and Simpson, M. H.: Stormwater management and climate change: vulnerability and capacity for adaptation in urban and suburban contexts, Climatic Change, 138, 491–504, https://doi.org/10.1007/s10584-016-1766-2, 2016.

Notaro, V., Liuzzo, L., Freni, G., and La Loggia, G.: Uncertainty Analysis in the Evaluation of Extreme Rainfall Trends and Its Implications on Urban Drainage System Design, Water, 7, 6931–6945, 2015.

Olsen, A., Zhou, Q., Linde, J., and Arnbjerg-Nielsen, K.: Comparing Methods of Calculating Expected Annual Damage in Urban Pluvial Flood Risk Assessments, Water, 7, 255–270, 2015.

Olsson, J., Berggren, K., Olofsson, M., and Viklander, M.: Applying climate model precipitation scenarios for urban hydrological assessment: A case study in Kalmar City, Sweden, Atmos. Res., 92, 364–375, https://doi.org/10.1016/j.atmosres.2009.01.015, 2009.

Oraei Zare, S., Saghafian, B., and Shamsai, A.: Multi-objective optimization for combined quality-quantity urban runoff control, Hydrol. Earth Syst. Sci., 16, 4531–4542, https://doi.org/10.5194/hess-16-4531-2012, 2012.

Pazwash, H.: Urban Storm Water Management, CRC Press, Taylor and Francis, Boca Raton, FL, 2011.

Peng, H. Q., Liu, Y., Wang, H. W., and Ma, L. M.: Assessment of the service performance of drainage system and transformation of pipeline network based on urban combined sewer system model, Environ. Sci. Pollut. R., 22, 15712–15721, 2015.

Piani, C., Weedon, G. P., Best, M., Gomes, S. M., Viterbo, P., Hagemann, S., and Haerter, J. O.: Statistical bias correction of global simulated daily precipitation and temperature for the application of hydrological models, J. Hydrol., 395, 199–215, 2010.

Piontek, F., Müller, C., Pugh, T. A., Clark, D. B., Deryng, D., Elliott, J., González, F. D. J. C., Flörke, M., Folberth, C., Franssen, W., and Frieler, K.: Multisectoral climate impact hotspots in a warming world, P. Natl. Acad. Sci. USA, 111, 3233–3238, 2014.

Pollert, J., Ugarelli, R., Saegrov, S., Schilling, W., and Di Federico, V.: The hydraulic capacity of deteriorating sewer systems, Water Sci. Technol., 52, 207–214, 2005.

Poussin, J. K., Bubeck, P., Aerts, J. C. J. H., and Ward, P. J.: Potential of semi-structural and non-structural adaptation strategies to reduce future flood risk: case study for the Meuse, Nat. Hazards Earth Syst. Sci., 12, 3455–3471, https://doi.org/10.5194/nhess-12-3455-2012, 2012.

Prudhomme, C., Giuntoli, I., Robinson, E. L., Clark, D. B., Arnell, N. W., Dankers, R., Fekete, B. M., Franssen, W., Gerten, D., Gosling, S. N., and Hagemann, S.: Hydrological droughts in the 21st century, hotspots and uncertainties from a global multimodel ensemble experiment, P. Natl. Acad. Sci. USA, 111, 3262–3267, 2014.

Rojas, R., Feyen, L., and Watkiss, P.: Climate change and river floods in the European Union: Socio-economic consequences and the costs and benefits of adaptation, Global Environ. Chang., 23, 1737–1751, https://doi.org/10.1016/j.gloenvcha.2013.08.006, 2013.

Rossman, L. A. and Huber, W. C.: Storm Water Management Model Reference Manual EPA/600/R-15/162A, 2016.

Veith, T. L., Wolfe, M. L., and Heatwole, C. D.: Optimization procedure for cost effective BMP placement at a watershed scale, J. Am. Water Resour. As., 39, 1331–1343, https://doi.org/10.1111/j.1752-1688.2003.tb04421.x, 2003.

Vojinovic, Z. and Tutulic, D.: On the use of 1D and coupled 1D-2D modelling approaches for assessment of flood damage in urban areas, Urban Water J., 6, 183–199, 2009.

Ward, P. J., Jongman, B., Aerts, J. C., Bates, P. D., Botzen, W. J., Loaiza, A. D., Hallegatte, S., Kind, J. M., Kwadijk, J., Scussolini, P., and Winsemius, H. C.: A global framework for future costs and benefits of river-flood protection in urban areas, Nature Climate Change, 7, 642–646, 2017.

Warszawski, L., Frieler, K., Huber, V., Piontek, F., Serdeczny, O., and Schewe, J.: The Inter-Sectoral Impact Model Intercomparison Project [ISI-MIP]: Project framework, P. Natl. Acad. Sci. USA, 111, 3228–3232, 2014.

Willems, P.: Revision of urban drainage design rules after assessment of climate change impacts on precipitation extremes at Uccle, Belgium, J. Hydrol., 496, 166–177, 2013.

Willems, P., Arnbjerg-Nielsen, K., Olsson, J., and Nguyen, V. T. V.: Climate change impact assessment on urban rainfall extremes and urban drainage: Methods and shortcomings, Atmos. Res., 103, 106–118, 2012.

Yang, G.: Historical change and future trends of storm surge disaster in China's coastal area, Journal of Natural Disasters 9, 23–30, (in Chinese) 2000.

Yazdanfar, Z. and Sharma, A.: Urban drainage system planning and design-challenges with climate change and urbanization: a review, Water Sci. Technol., 72, 165–179, 2015.

Zahmatkesh, Z., Karamouz, M., Goharian, E., and Burian, S. J.: Analysis of the Effects of Climate Change on Urban Storm Water Runoff Using Statistically Downscaled Precipitation Data and a Change Factor Approach, J. Hydrol. Eng., 20, 05014022, https://doi.org/10.1061/(ASCE)HE.1943-5584.0001064, 2015.

Zhang, B. and Guan, Y.: Watersupply & Drainage Design Handbook, China Construction Industry Press, Being, China, ISBN: 9787112136803, 2012.

Zhou, Q., Mikkelsen, P. S., Halsnaes, K., and Arnbjerg-Nielsen, K.: Framework for economic pluvial flood risk assessment considering climate change effects and adaptation benefits, J. Hydrol., 414, 539–549, 2012.

Zhou, Q., Ren, Y., Xu, M., Han, N., and Wang, H.: Adaptation to urbanization impacts on drainage in the city of Hohhot, China, Water Sci. Technol., 73, 167–175, https://doi.org/10.2166/wst.2015.478, 2016.

Zhou, Q., Leng, G., and Feng, L.: Predictability of state-level flood damage in the conterminous United States: the role of hazard, exposure and vulnerability, Sci. Rep., 7, 5354, https://doi.org/10.1038/s41598-017-05773-4,, 2017.

Zoppou, C.: Review of urban storm water models, Environ. Modell. Softw., 16, 195–231, 2001.

12

Climate change effects on the hydrology of the headwaters of the Tagus River: implications for the management of the Tagus–Segura transfer

Francisco Pellicer-Martínez[1] **and José Miguel Martínez-Paz**[2]

[1]Department of Civil Engineering, Catholic University of Murcia, Spain
[2]Department of Applied Economics, University of Murcia, Spain

Correspondence: Francisco Pellicer-Martínez (fpellicer@ucam.edu)

Abstract. Currently, climate change is a major concern around the world, especially because of the uncertainty associated with its possible consequences for society. Among them, fluvial alterations can be highlighted in basins whose flows depend on groundwater discharges and snowmelt. This is the case of the headwaters of the Tagus River basin, whose water resources, besides being essential for water uses within this basin, are susceptible to being transferred to the Segura River basin (both basins are in the Iberian Peninsula). This work studies the possible effects that the latest climate change scenarios may have on this transfer, one of the most important ones in southern Europe. In the first place, the possible alterations of the water cycle of the donor basin were estimated. To do this, a hydrological model was calibrated. Then, with this model, three climatic scenarios were simulated, one without climate change and two projections under climate change (Representative Concentration Pathways 4.5 (RCP 4.5) and 8.5 (RCP 8.5)). The results of these three hydrological modelling scenarios were used to determine the possible flows that could be transferred from the Tagus River basin to the Segura River basin, by simulating the water resource exploitation system of the Tagus headwaters. The calibrated hydrological model predicts, for the simulated climate change scenarios, important reductions in the snowfalls and snow covers, the recharge of aquifers, and the available water resources. So, the headwaters of the Tagus River basin would lose part of its natural capacity for regulation. These changes in the water cycle for the climate change scenarios used would imply a reduction of around 70 %–79 % in the possible flows that could be transferred to the Segura basin, with respect to a scenario without climate change. The loss of water resources for the Segura River basin would mean, if no alternative measures were taken, an economic loss of EUR 380–425 million per year, due principally to decreased agricultural production.

1 Introduction

Currently, there are practically no doubts in the scientific community that the Earth is suffering climate change (CC) and that this is due to the anthropic action of greenhouse gas emissions (IPCC, 2014). At the global level, general circulation models predict a warming of the planet of about 2 °C for the year 2050, which will cause a reduction in accumulated ice masses and a rise in the sea level (IPCC, 2014). These changes in the natural environment, which are already causing alterations in the available resources, have a clear socioeconomic repercussion: decreases in fish stocks, increases in energy consumption, changes in the availability of water resources, and land degradation due to erosion. In areas with greater risk and with low capacity for adaptation, the consequences of CC could become critical, the emigration of their population being the only viable solution (Jha et al., 2018).

In the context of CC, water as a resource plays a fundamental role, since – as well as being a basic environmental asset – it is key to human survival and well-being. In general terms, an increase in rainfall in humid areas and a decrease in arid and semi-arid ones are predicted. This situation would be accompanied by an increase in the frequency and intensity of extreme events (droughts and floods), so in areas where already there are shortages of water resources the current situation would be exacerbated (IPCC, 2014). For areas close to polar regions or mountainous areas, the increase in temperature would reduce the precipitation that falls as snow (Szczypta et al., 2015), as well as the volume of ice in the glaciers and the snow cover on the summits (Bajracharya et al., 2018). As a consequence, the fluvial regime for this type of river basin would be modified (Morán-Tejeda et al., 2014), with an increase in the risk of floods and the loss of the part of their natural regulatory capacity provided by the ice and snow covers (Shevnina et al., 2017). Thus, in areas where the water reserves derived from snow covers are used during the summer, new reservoirs would have to be built in order to replace the loss of the natural regulation capacity (Özdoğan, 2011). Similarly, in areas with important aquifers, part of the natural regulatory capacity could also be lost, since changes in precipitation patterns would affect recharge rates (Smerdon, 2017). Indeed, a higher intensity of rainfall would favour surface runoff to the detriment of infiltration (Pulido-Velazquez et al., 2014). In short, the CC predicted for many areas of the planet will suppose an increase in temperature together with a greater availability of surface water – which would increase the water evapotranspiration, accelerating the water cycle and reducing the available water resources that could be used (Wang et al., 2013).

The environmental and social repercussions of these physical effects are being studied from multiple perspectives (Olmstead, 2014; World Bank Group, 2016). For example, there is work related to water quality (Molina-Navarro et al., 2014), the effects on ecosystems and the services they provide (Warziniack et al., 2018), and the impact on the food security (Tumushabe, 2018) and water security (Flörke et al., 2018) of the population, although the majority of the studies deal with the impacts on the economic activities which are more sensitive to the availability of water resources, such as agriculture (Meza et al., 2012), urban supply (Díaz et al., 2017), or the hydroelectric sector (Solaun and Cerdá, 2017).

Water resource transfers between basins (inter-basin water transfer: IBWT) are instruments of water resource allocation that, despite the controversy they sometimes provoke, can play an important role in mitigating the effects of CC in many areas of the world (Shrestha et al., 2017). The IBWTs can be an alternative source of supply to basins affected by a decrease in their available resources and/or an increase in their water use demands (Zhang et al., 2015). In turn, the quantity of available water in the donor basins can change significantly, which makes CC a determining factor that must be analysed when assessing the potential or vulnerability of the IBWT (Zhang et al., 2018). Although there are some works that specifically studied the effects of CC on IBWTs, basically the focus has been on changes in the fluvial regimes in donor basins (Zhang et al., 2012; Li et al., 2015). Moreover, there are scarce examples in the specialized literature of the analysis of the effects of CC within a framework of integrated water resource management (Onagi, 2016). In addition, for an adequate comprehensive study of the effects of CC in an IBWT, and to produce operational indicators for water management plans (Giupponi and Gain, 2017), including the management of water trade (Kahil et al., 2015), it is also necessary to consider the effects in the receiving basin.

The Tagus–Segura Aqueduct (TSA) is one of the most important IBWT projects in southern Europe. This hydraulic infrastructure, in operation since 1979, transfers flows from the Tagus Headwaters River Basin (THRB) to the Segura River basin (SRB). The destination of the volumes transferred, which are variable depending on the available water resources in the donor basin, is basically irrigation, but also urban and tourism uses (Grindlay et al., 2011). The regional models of CC forecast, for both basins, an increase in temperature together with a significant diminution in rainfall, so that a decrease in the available water resources in both is foreseen (CEDEX, 2011a). In addition, the THRB is located in a high, mountainous area where snowfalls are frequent and which extends over important karst aquifers. Then, it is expected that both the precipitation that falls as snow and the aquifer recharge would be reduced. Although there is work in which this problem is explicitly described, with proposals to mitigate the decrease in TSA flows due to CC (Morote et al., 2017), no specific modelling of climate scenarios has been made, neither from a hydrological perspective, to determine the water balance, nor by simulation of the water resource exploitation system.

The overall objective of this work was to determine the hypothetical effects that CC may have on the operation of the TSA. So, the transferable flows were estimated considering explicitly the operating rule of this IBWT – which, basically, is based on the available water storage in the main two reservoirs of the donor basin concerned. For this, first of all, the effects on the water cycle of the donor basin were evaluated by means of hydrological modelling in which the precipitation that falls as snow was included. Then, the historical climate data together with two CC scenarios of the fifth assessment report (AR5) of the Intergovernmental Panel on Climate Change (IPCC, 2013) were recreated in order to analyse the possible alterations in the fluvial regime of the THRB. The results of these three hydrological models were the inputs of the subsequent three simulations of the Tagus Headwaters Water Resources Exploitation System (THWRES), which provided a prediction of the flows that could be transferred to the SRB within a framework of integrated water resource management. Additionally, as another novel contribution of this work, the socioeconomic impacts produced by the climate change on these transferred flows were assessed. Finally, note that the complete methodology was developed by open-source tools and by free software for the scientific community, which facilitates the reproducibility of the work.

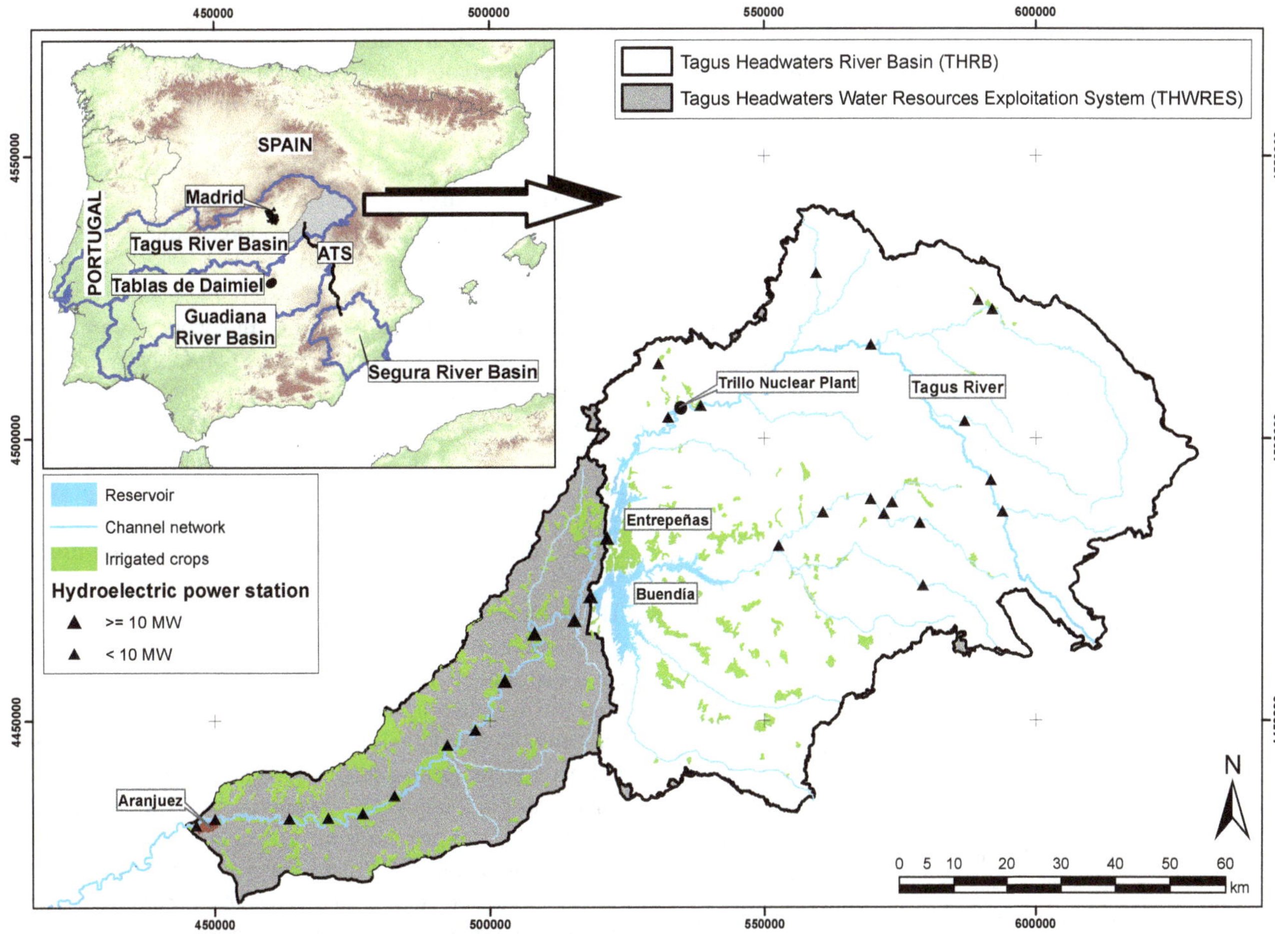

Figure 1. The Tagus Headwaters River Basin (THRB) and Tagus Headwaters Water Resources Exploitation System (THWRES): the main physical characteristics and water uses (including the TSA). Source data: CHT (2018).

2 The water resource exploitation system of the Tagus Headwaters River Basin

The THRB covers an area of $7000\,\text{km}^2$ and is located in the middle of the Iberian Peninsula (Fig. 1). It extends over a high, mountainous area with a continental–Mediterranean (CHJ, 2016) climate with a marked seasonality between the summer (June–September) and winter (December–March) months (Lorenzo-Lacruz et al., 2010; Molina-Navarro et al., 2014). The average annual precipitation is 620 mm and the minimum values occur in summer (June–August). While the average annual temperature is 11 °C, in the coldest months there are values less than zero (November–April), so snowfalls are frequent at the higher altitudes (Lobanova et al., 2016). Moreover, much of the THRB extends over karst aquifers, meaning that groundwater exerts an important influence on the surface flows that circulate along the main river streams (Pellicer-Martínez et al., 2015).

Regarding the water uses within the THRB (urban, industrial, and irrigation), their water necessities represent a very low percentage of the available water resources (around $1000 \times 10^6\,\text{m}^3\,\text{yr}^{-1}$, on average, in the last 70 years): this area has a low population density, is not conducive to agriculture, and its industrial facilities (hydroelectric power stations and an important thermonuclear power station) do not consume much water.

The water resources generated within the THRB are fundamental for the water uses located downstream of the Entrepeñas and Buendía reservoirs (EBR): irrigated agriculture, urban supply (including the city of Madrid), generation of electric power, and maintenance of environmental flows until the city of Aranjuez. Moreover, a large part of these water resources (up to $650 \times 10^6\,\text{m}^3\,\text{yr}^{-1}$) is susceptible to being transferred to the neighbouring Guadiana River basin and, further away, to the SRB. The former can receive up to $50 \times 10^6\,\text{m}^3\,\text{yr}^{-1}$ (BOE, 2014, 2015), of which $20 \times 10^6\,\text{m}^3$ are for the maintenance of the wetland of Tablas de Daimiel and $30 \times 10^6\,\text{m}^3$ are for urban supply to the populations located in the upper Guadiana River basin (CHG, 2016). The SRB can receive up to $600 \times 10^6\,\text{m}^3\,\text{yr}^{-1}$ (gross volume), the maximum monthly flow being $68 \times 10^6\,\text{m}^3$. This IBWT is managed by a complex operating rule that gives priority to the water uses in the Tagus River basin and basically depends on the volume stored in the EBR, which have a total storage capacity of $2494 \times 10^6\,m^3$. The operating rule (Fig. 2) consists of two conditioning factors (BOE, 2014, 2015). The first restricts the maximum volume that can be transferred in each hydrological year (October–September) to $650 \times 10^6\,m^3$. The second establishes the transferable flows for each month according to four levels created from two variables: V_{acu}, the accumulated volume stored in the EBR at the beginning of the month, and A_{acu}, the accumulated volume of the flows that entered into the EBR in the previous 12 months. The four levels are the following.

- Level 4. When V_{acu} is lower than $400 \times 10^6\,m^3$. Transfers are not allowed ($Q_{trans} = 0\,m^3\,month^{-1}$).
- Level 3. When V_{acu} is between $400 \times 10^6\,m^3$ and the values indicated in Fig. 2, which vary between $586 \times 10^6\,m^3$ and $688 \times 10^6\,m^3$, depending on the month. Transfers (Q_{trans}) of $20 \times 10^6\,m^3\,month^{-1}$ are allowed.
- Level 2. When V_{acu} is between the volumes established in Level 3 and $1500 \times 10^6\,m^3$, and in addition A_{acu} is lower than $1000 \times 10^6\,m^3$. Transfers (Q_{trans}) of $38 \times 10^6\,m^3\,month^{-1}$ are allowed.
- Level 1. When V_{acu} is equal to or greater than $1500 \times 10^6\,m^3$, or A_{acu} is equal to or greater than $1000 \times 10^6\,m^3$. Transfers (Q_{trans}) of $68 \times 10^6\,m^3\,month^{-1}$ are allowed.

3 Methodology

The methodology applied to determine the maximum monthly volumes that can be transferred from the THRB to the SRB was structured in two stages (Fig. 3).

The first stage consisted of modelling the hydrology of the THRB until the EBR. For that, a hydrological model was calibrated for the most recent observed flows. Then, with the calibrated model, three scenarios were recreated. In the first one the historical climate series were used (No CC) without climatic correction coefficients. In the others, the data from two characteristic CC scenarios of AR5 (Representative Concentration Pathways 4.5 (RCP 4.5) and 8.5 (RCP 8.5)) were used (IPCC, 2013). In the first (RCP 4.5), CO_2 emissions increase in the future, until in 2050 they stabilize (stabilization scenario), while in the second (RCP 8.5), a continuous and greater increase in emissions of CO_2 is assumed (scenario of very high emissions).

The second stage consisted of simulating the THWRES by means of a decision support system (DSS). This simulation incorporated the future water uses contemplated by the water management board. As water uses downstream of the EBR have priority over possible transferable flows (CHT, 2015), they were included in this simulation.

The methodological framework stages were developed with open-source tools. QGIS (QGIS Development Team, 2016) was used in the data processing of spatial information, R was employed for data analysis and hydrological modelling (R Core Team, 2016), and the DSS SIMGES was used for the simulation of the water resource exploitation system (Pedro-Monzonís et al., 2016a).

Finally, once the series of transferrable flows were calculated, as a complementary goal of this work, an assessment of the socioeconomic consequences that climate change effects have on the main destiny of these flows, the SRB, was made.

3.1 Hydrological modelling

3.1.1 abcd water balance model with snowmelt module

The hydrological modelling was carried out using the abcd water balance model (Thomas, 1981). It was applied in a semi-distributed manner (Pellicer-Martínez and Martínez-Paz, 2014), allowing the use of all the gauging stations in the calibration (and validation) in order to maintain the spatial heterogeneity that defines the parameters and variables. This conceptual model was improved by taking into account the hydrological processes of snow and melting. This water balance model and this structure were selected in order to facilitate the understanding of the developed process, allowing the potential reproducibility of the work.

The hydrological modelling, whose scheme is shown in Fig. 4, began with the snowmelt module proposed by Xu and Singh (1998). This module recreates the hydrological processes of precipitation as snow (S_n), snow accumulation on the summits (S_{np}), and snowmelt (S_m). One equation, with a parameter that depends on the temperature, establishes which part of the precipitation (P) occurs as rain (R_f) and which part occurs as snow (S_n). Snow is accumulated in a storage called snowpack (S_{np}). Then, another equation controlled by one parameter, which also depends on the temperature, establishes the snowmelt (S_m) of the snowpack when the temperature increases. In the modelling, the melting snow ends up forming part of the storage that represents the soil moisture (S). This snowmelt module has been used in previous work, for example, by Li et al. (2011, 2013) in basins of China, and by Pellicer-Martínez and Martínez-Paz (2015) for the Segura headwaters river basin.

The modelling continued with the incorporation of rainfall (R_f) and melting snow (S_m) into the abcd model. This water balance model simplifies the hydrological cycle in two storages, one that simulates the soil moisture balance (S) and another that represents the groundwater storage (G). The model has four parameters – a, b, c, and d (Thomas, 1981) – that give the model its name. The parameters "a" and "b" manage the soil moisture balance (S), establishing evapotranspiration

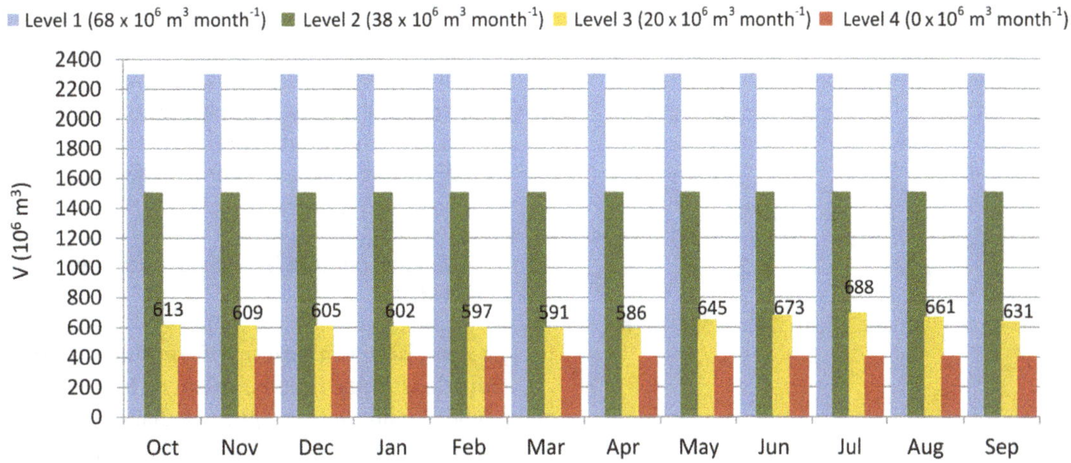

Figure 2. Operating rule of the TSA. Based on (BOE, 2014, 2015).

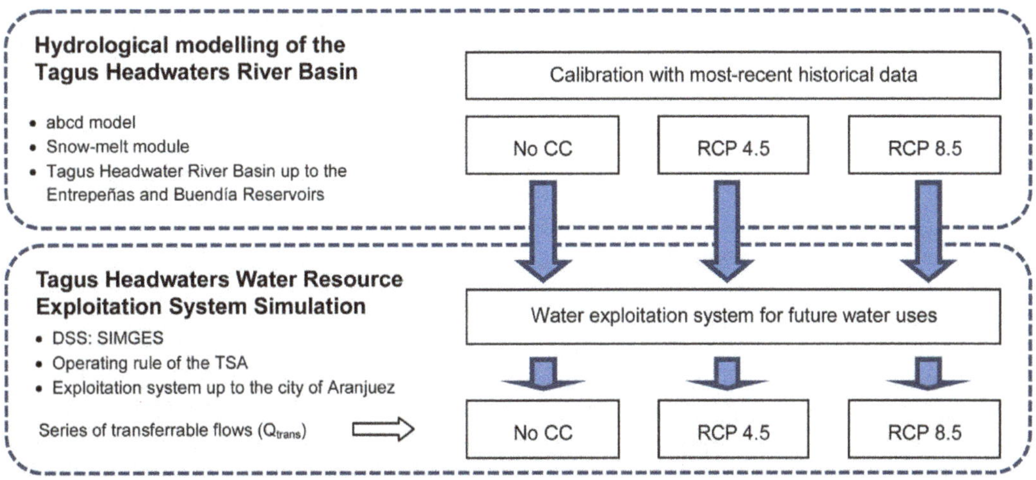

Figure 3. Methodological framework.

(E_T) and the water susceptible to being lost by surface run-off and/or percolation ($Q_s + \Delta G$). The third parameter "c" identifies percolation (ΔG) towards aquifers (G) and surface run-off (Q_s). The fourth parameter "d" determines the discharge from the aquifer (Q_g). The output variables of the model are evapotranspiration (E_T) and surface run-off (Q).

3.1.2 Calibration–validation process

The parameters were calculated by a cascading calibration process (Xue et al., 2016), which consists of determining the parameter values from upstream to downstream. In other words, once the model's parameters are established for upstream catchments, their values become input data in the calibration process of the downstream catchments. The split-sample test, proposed by Klemeš (1986), is carried out by splitting the data series of observed flows into two periods: calibration and validation.

The objective function used in the calibration is the Nash–Sutcliffe efficiency criterion (E_{NS}). This function quantifies the goodness of fit of the model in order to evaluate the performance of the variables selected for study (Nash and Sutcliffe, 1970), which are the flows that go out from each catchment (Q). The E_{NS} varies between $]-\infty, 1]$ and the closer its value is to 1, the better the performance of the model is. The calibration is developed automatically with the Shuffled Complex Evolution Method (SCE-UA) algorithm (Duan et al., 1994; Skøien et al., 2014). Once the semi-distributed model has been calibrated, another three metrics that evaluate the performance of the model, comparing observed with simulated flows, are calculated: the determination coefficient (R^2), the percentage of the bias (P_{BIAS}), and the root mean square error (E_{RMS}) (Gupta and Kling, 2011).

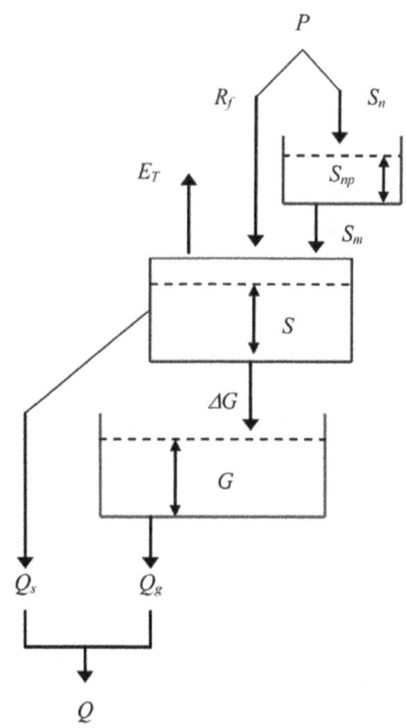

Figure 4. Conceptual scheme of the abcd model.

3.2 Simulation of the water resource exploitation system

The operation of the water exploitation systems for a specific scenario is usually evaluated by a DSS which simulates the part of the water cycle that is anthropically modified. These systems represent the main hydraulic network of an area, generally a river basin, together with its natural and artificial storages (rivers, reservoirs, aquifers, canals, among others), establishing the water uses in them (Pulido-Velazquez et al., 2013). In this hydraulic network, series of flows are introduced, generally in their natural regime. Then, the DSS simulates this kind of system in an integrated manner, fulfilling the priority criteria among the different water uses and the pre-established operating rules (Pedro-Monzonís et al., 2016a). The DSS provides, as results, flow and volume series related to the main fluxes of the hydraulic network (supplies to uses, water consumptions, returns, evaporation in the reservoirs, etc.). Therefore, it is able to estimate the water uses that are not completely met, which are generally those with lower priority and/or those whose spatial location in the hydraulic network does not make it possible to always guarantee their supply (Chavez-Jimenez et al., 2015). For example, water uses located upstream of a reservoir have less guarantee than those located downstream.

There are different DSSs for water exploitation systems (Zare et al., 2017). But, as was advanced in the Methodology section, in this work the model applied is the simulation module of AQUATOOL called SIMGES (Pedro-Monzonís et al., 2016b), which is one of the models applied most in Spanish river basins, as well as in other countries (Chile, Italy, Morocco, etc.). SIMGES simulates the water exploitation system on a monthly basis in a conservative flow network, seeking a compatible solution that accomplishes the defined constraints (Pedro-Monzonís et al., 2016b). This DSS was selected since it is able to reproduce complex operating rules such as the TSA operating rule.

3.3 Socioeconomic impacts at the Segura River basin (SRB)

The precise quantification of the socioeconomic impact of reductions in the volume of water transferred via the TSA would require new integral simulations of the exploitation system of the receiving basin (SRB), which exceed the scope of this work. However, an initial quantification of this impact has been made, based on the work of Martínez-Paz et al. (2016), where supply failures were assigned an economic value in irrigation in the SRB, and of Martínez-Paz and Pellicer-Martínez (2018), who estimated the economic value of the risk associated with droughts in the Region of Murcia. But, given the order of priority of allocation among the water uses, irrigation would suffer the full brunt of any supply deficit. The almost 270 000 ha of irrigated land in the SRB has a net demand of $1363 \times 10^6\ \mathrm{m^3\,yr^{-1}}$ (CHS, 2015), a large part of which is supplied by the TSA. The whole irrigated area in the SRB is divided into seven irrigation zones (IZs); for each of them the water demand curve is estimated from a linear programming model that optimizes the gross value added (GVA) by the optimal cultivation plans according to the water supply. This crop programming includes the irrigation situations of woody crop maintenance, the change from irrigated to rainfed crops and the abandonment of irrigation plots, as well as the impact on employment.

The modelling of the optimal crop plan for each IZ is determined by the following objective function (1) that maximizes the GVA:

$$\max \sum_i (Y_i \cdot P_i - C_i) \cdot L_i, \tag{1}$$

where i denotes crop activities under different management options, Y_i is the yield of each crop i, P_i the price received by the farmer, C_i the direct costs of production per unit area, and L_i the area dedicated to each activity.

The objective function is subject to the following constraints (Eqs. 2–7):

$$\sum_i L_i \leq L_T, \tag{2}$$

$$\sum_i q_i \cdot L_i \leq Q_T, \tag{3}$$

$$L_i^R + L_i^S = L_i^E, \tag{4}$$

$$L_i^R + L_i^M + L_i^P = L_i^E, \tag{5}$$

$$L_i^G \leq L_i^E, \tag{6}$$

$$\sum_i lf_i \cdot L_i \leq LF_T. \tag{7}$$

L_T is the total available surface irrigable in the IZ; q_i is the water requirement of each crop per unit area and Q_T is the availability of water for the entire campaign in the IZ; L_i^R is the surface of irrigated woody crops; L_i^S is the irrigable surface that goes to rainfed; L_i^M is the irrigable surface of woody crops under maintenance irrigation; L_i^P is the surface of the irrigation plots abandoned; L_i^G is the surface of existing irrigated greenhouses; L_i^E is the existing area of each activity in the basin in the reference year; lf_i are the labour requirements for each crop, and LF_T is the availability of agricultural labour in the IZ.

The first constraint (Eq. 2) prevents each unit of demand (IZ) from cultivating more area than the available net irrigable area. The following constraint (Eq. 3) represents the limitation of water availability for each IZ. The set of constraints (Eq. 4), (Eq. 5), and (Eq. 6) allow simulation of specific management options to certain crop groups. Constraint (Eq. 4) fixes the total area of woody crops such as almond, olive, and wine, distributed between irrigated and rainfed depending on the availability of water. Constraint (Eq. 5) represents citrus and fruit trees, whose total area is equal to the area actually irrigated plus, in situations of scarcity of resources, the surface under maintenance irrigation and/or loss of trees because of not being able to perform the minimum maintenance irrigation. Constraint (Eq. 6) sets the maximum available area for greenhouse crops in the reference year. Finally, constraint (Eq. 7) represents the limitation of the available labour for each IZ. The programme used allows estimation of the gross margin generated under different water availability assumptions, as well as derivation of water demand curves and the marginal value of this resource (Griffin, 2006).

The necessary data to characterize the technical coefficients of each IZ have been obtained from the sources indicated in Martínez-Paz et al. (2016) and in Martínez-Paz and Pellicer-Martínez (2018), updating all the economic figures to EUR of 2017. The programme was solved for each IZ and for the three climatic scenarios (No CC, RCP 4.5, and RCP 8.5), so the availability of water in each of them changes (Q_T). The differences in the average volume transferred to the SRB in each CC scenario (RCP 4.5 and RCP 8.5) calculated with SIMGES were distributed proportionally for each IZ, taking as a reference the volume transferred in the scenario without CC. Thereby, the socioeconomic impact due to changes in the availability of water was obtained, *ceteris paribus* the rest of the parameters of the model. Finally, the comparison among the results obtained from GVA and employment for each scenario are presented in the Discussion section.

4 Data source

4.1 Hydrological modelling of the Tagus Headwaters River Basin

The digital elevation model (DEM) employed has a 25 m resolution and is available on the website of the National Geographic Institute of Spain (http://www.cnig.es/, last access: 15 December 2017). This DEM was used as an auxiliary variable in the interpolation models of the climatic variables, and to delimit the main streams and catchments using the D8 algorithm (O'Callaghan and Mark, 1984). The locations of the 12 gauging stations, which have observed flows in the same period, were used to establish the outlet points of the 12 catchments into which the THRB was divided (Fig. 5). The data series of the observed flows are available in the gauging yearbook of the Official Gauging Station Network of Spain (MITECO, 2018) and cover the period from September 1985 to December 2009. These observed flows have been previously naturalized to be used in the calibration–validation process (Wurbs, 2006). For that, the main human alterations located upstream of each gauging station were undone: regulation and evaporation in the reservoirs, as well as the derivations for urban, agricultural, and industrial uses (also, the returns of these uses were considered). As available data for the 12 gauging stations exist for the same period, it is possible to calibrate the model jointly for all the catchments. Since the objective of this work is to obtain a statistically significant calibration and data for its testing, each series of observed flows was divided into two periods (Klemeš, 1986), the first (from September 1985 to July 1995) being used in the validation and the second (from August 1995 to December 2009) being used in the calibration. Thus, the parameters to be used for the CC projections were determined with the most recent data.

The historical climate series used in the calibration–validation comprise the period from September 1980 to December 2009. Thus, the 5 years before the observed flow series were used to warm up the hydrological model. Two information sources were used to obtain the monthly climatic series. The first was the Spain02v5 dataset (Herrera et al., 2016), the historical series used for the calibration–validation of the hydrological model and for its simulation in the scenario without climate change (No CC). For the No CC scenario, these series of data were extended from October 1940 to September 2010 (70 consecutive years), and they were used in the simulation without climatic correction coefficients. The advantages of using Spain02v5 are that the daily series of precipitation and temperature are refined (without outliers and/or in homogeneities), and that they include the spatial variability of the climatic variables in a grid with a 12.5 km resolution (Fig. 5). Since these data are daily, they were aggregated on a monthly basis to apply them in the model. The second source of data was the State Meteorological Agency of Spain (AEMET) and was used for the two CC

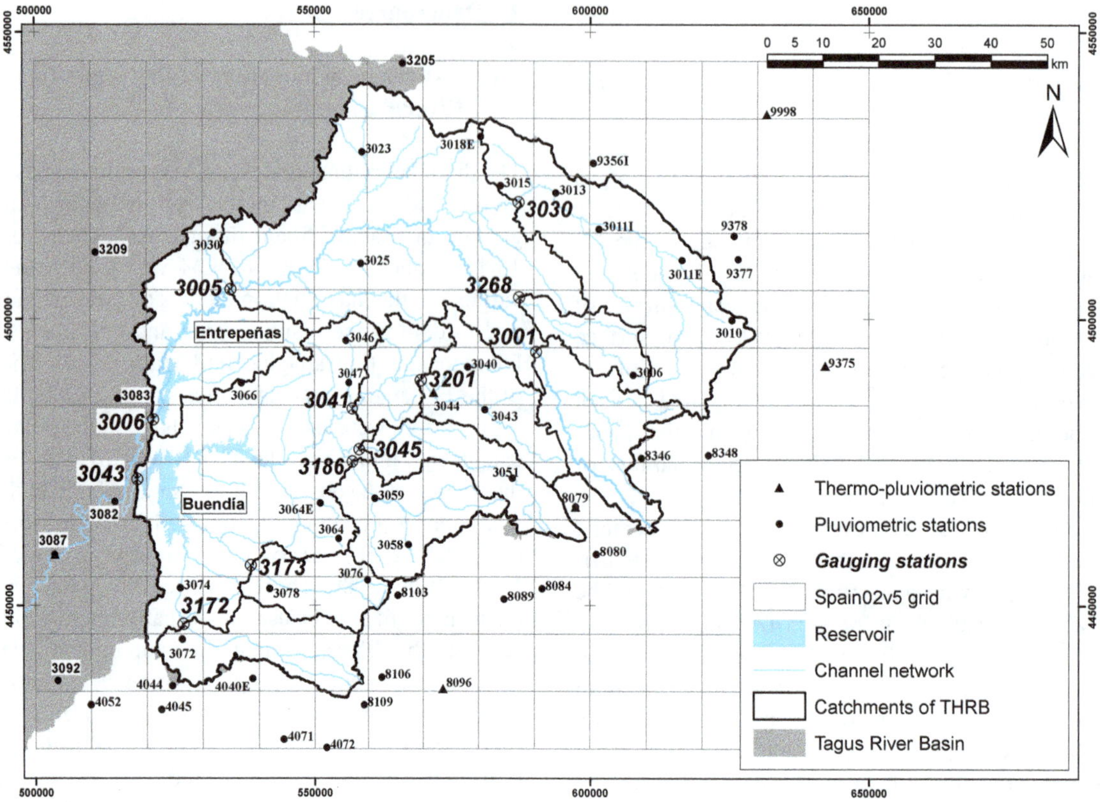

Figure 5. Spatial distribution of climate stations with AR5 information and the grid obtained from Spain02v5.

scenarios of AR5 (IPCC, 2013). This source (AEMET) provides the regionalized projections of 27 models (13 for RCP 4.5 and 14 for RCP 8.5), using the statistical method of analogues (Amblar-Francés et al., 2017). The daily temperature and precipitation series of the 6 thermometric and 48 precipitation stations closest to the THRB were used (Fig. 5). These daily data were also aggregated to monthly series. Next, the reference historical data series of each model were compared with Spain02v5 data, using as a control period the 1971–2005 interval. The 10 models that best fitted for both temperature and precipitation were assembled using the Simple Average Forecast Combination (SA) and Bias-Corrected Eigenvector Forecast Combination (EIG2) (Hsiao and Wan, 2014), available in the GeomComb R-CRAN package (Weiss and Roetzer, 2016). Then, both ensembles were also compared with Spain02v5 data using the same control period (1971–2005). Finally, the series obtained by EIG2 were used, with a lower prediction error compared to the rest of the series. For example, the ENS values obtained with EIG2 for the temperatures were 0.87, while in the separate models they never exceeded 0.77. For the precipitations, the ENS value was 0.30, while in the models it was always lower than 0. In addition, another advantage of the EIG2 method is that it allows one to correct the bias produced by predictive models. The period used in the simulation of CC scenarios (RCPs 4.5 and 8.5) is from October 2020 to September 2090 (also 70 consecutive years).

Based on the average monthly temperature data of the three climatic scenarios, the potential evapotranspiration series were estimated using the Thornthwaite method (Thornthwaite, 1948; Gomariz-Castillo et al., 2018). As this method tends to underestimate the potential evapotranspiration, the series generated were corrected from a linear regression between the estimated series and those used by the SIMPA hydrological model (BOE, 2007). These series used by SIMPA have already regionalized for the Iberian Peninsula based on the Penman–Monteith method (Allen et al., 1998), correcting the underestimation of the Thornthwaite method.

Once the monthly series of precipitation, temperature, and potential evapotranspiration had been calculated, they were spatially interpolated on the cells using the thin-plate splines method (Wahba, 1990), which is based on local interpolation from polynomials. This method turns out to be relatively robust against non-compliance with the statistical assumptions necessary in methods such as kriging, being used with good results for the interpolation of climatic variables such as rainfall (Hutchinson, 1995) and temperatures (McKenney et al., 2006). Finally, each catchment was assigned the average value of the cells over which it extends.

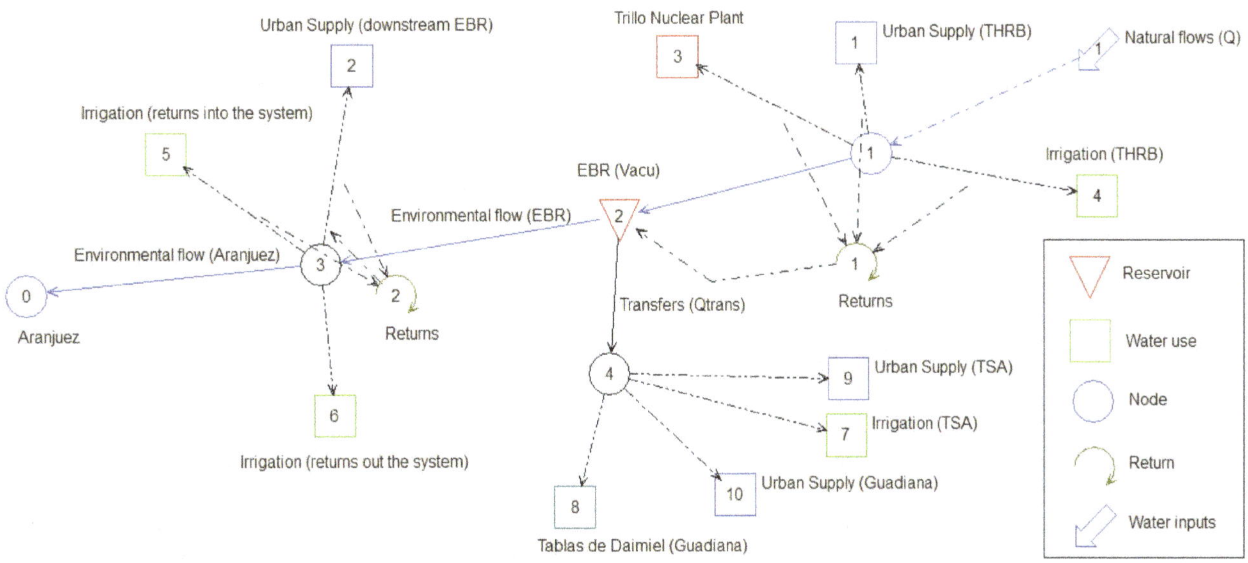

Figure 6. Scheme of the water resource exploitation system of the Tagus Headwaters, as far as Aranjuez.

4.2 Water resource exploitation system of the Tagus Headwaters Basin (THWRES)

The THWRES covers the basin upstream of the EBR and the water uses located in the basin downstream of these reservoirs, up to the city of Aranjuez (Figs. 1 and 6). In its design, the possible future water uses for the years 2016, 2021, and 2033 contemplated by the water management board (CHT, 2015) were analysed. And, as there are no relevant differences between them, the uses predicted for the year 2033 were taken as the reference for the three climatic scenarios (No CC, RCP 4.5, and RCP 8.5) (Table 1). In addition, as the available water resources in the catchments downstream of the EBR are so low with respect to those generated in the EBR drainage basin, they were neglected in the water resource exploitation system simulation.

Regarding water uses, the majority of the urban and irrigation uses are concentrated downstream of the EBR, requiring more than 85 % of the total volume demanded. For urban use, a water return of 80 % was considered with respect to the volume supplied. In 2033 irrigation will have a low water return into the system (< 1 %) due to improvements in irrigation facilities and in the application systems which will be used, so this low return was implemented in the DSS. The main industrial use of water occurs at the Trillo Nuclear Plant. This industrial use requires a constant flow over time that returns 46 % of the water supplied. Finally, an environmental flow of around 11 $m^3\,s^{-1}$ must circulate in the Tagus River from the EBR to the city of Aranjuez. The order of priority in the water resource allocation among the different water uses is urban supplies, Trillo Nuclear Plant, and irrigation. Once the water uses in the THWRES have been supplied, the possibility of authorizing transfers is evaluated (BOE, 2014, 2015). Among the four uses of the transfer there is no stipulated clear criterion in the operating rule that governs it, since sometimes it is done discretionally depending on the needs or level of urgency of the uses. In fact, the transfer to Tablas de Daimiel has occurred only once, to avoid serious damage to this wetland. Therefore, in order to accomplish the general criteria of Spanish legislation, the order of priority followed is urban supply to Segura (TSA), urban supply to the populations located in the upper Guadiana River basin, and irrigation supply to Segura (TSA) and Tablas de Daimiel.

5 Results

5.1 Climate change effects on the hydrology of the Tagus Headwaters River Basin

The values of the criterion coefficients calculated in the hydrological modelling show that the model employed reproduced properly the surface flows in the THRB in the calibration period: high values of E_{NS} and R^2, together with low relative errors (E_{RMS}) and volume errors (P_{BIAS}). However, in the validation period, there are some low values for the goodness-of-fit coefficients calculated, indicating that the results of these catchments have greater uncertainty. These results can be explained by the fact that the validation period was used just after the warming-up process. Thereby, it could cause, in some catchments, the warming to extend to a part of the validation period. This would entail the calibration process using a part of the validation to adjust the initial parameters, obtaining worse adjustments in the validation of the model. However, it is important to highlight that the parameters used in the simulations are adjusted with the more recent

Table 1. Main water uses in the THWRES (unit: 10^6 m^3 yr^{-1}).

Water uses in THWRES	Annual water demanded
Environmental flow in the Tagus River downstream of the EBR	327.0
Environmental flow in the Tagus River in the city of Aranjuez	258.0
Urban water use	35.5
Irrigation water use	159.1
Industrial water use (Trillo Nuclear Plant)	37.8
Transfers Segura (urban and irrigation supplies)	600.0
Transfers Guadiana (Tablas de Daimiel)	20.0
Transfers Guadiana (urban supply)	30.0

data, providing a good performance of the surface flows in the THRB. In addition, the best performance in calibration corresponded to the outlets of the catchments of Entrepeñas and Buendía, which were the flows used as the input in the subsequent simulation of the water resource exploitation system. In fact, both catchments had NSE values around 0.80 and low P_{BIAS} (Table 2).

The simulation of the historical climate series of 1940–2010 with the calibrated model provided an average annual resource of 954.6 × 10^6 m^3 yr^{-1} (Q). This series was temporally moved to the 2020–2090 time period in order to reproduce a future climate scenario without climate change (No CC). The simulations for climate change scenarios RCP 4.5 and RCP 8.5 with the same calibrated model indicated that the THRB could suffer a considerable loss of its natural water resources. The RCP 4.5 scenario forecasted a value of 575.6×10^6 m^3 yr^{-1}, representing a decrease of 39.7 %, while the RCP 8.5 scenario predicted a 46.6 % decline in resources to 508.9 × 10^6 m^3 yr^{-1}, on average (Fig. 7). This is due to a combination of a reduction in precipitation (15 % and 20 % for each scenario, respectively) and an increase in potential evapotranspiration resulting from an increase in temperature of 2.2 and 3.4 °C, respectively, for CC scenarios RCP 4.5 and RCP 8.5.

Regarding the snow-melting modelling, the scenario without CC (historical series) provided an average precipitation as snow (S_n) of 185.4 × 10^6 m^3 month^{-1} (considering the whole year), reaching a maximum value of 406.9 × 10^6 m^3 month^{-1}. The hydrological modelling considered the snow for the months between October and May to be relevant, especially that of December, January, and February. The snow accumulated on the summits (S_{np}) represented an average reserve for each winter of about 112.7 × 10^6 m^3, with a maximum value of 481.5 × 10^6 m^3 (Fig. 8a). For the CC scenarios, the snowfall would significantly decrease; in fact, the models did not detect relevant snow-melting processes for the months of October and May. The snowfall would drop by 68 % for RCP 4.5, the average value being around 59.3 × 10^6 m^3 month^{-1}, while for RCP 8.5 the snowfall would drop by 90 %, giving an average value of about 19.2 × 10^6 m^3 month^{-1}. These values would decrease the snow covers to a similar extent. In the RCP 4.5 scenario the average snow cover of each winter would be 29.5 × 10^6 m^3 (Fig. 8b), which represents a reduction of 74 %, whereas the RCP 8.5 snow cover would suffer a reduction of 80 %, reaching an average value of 22.1 × 10^6 m^3 (Fig. 8c).

The aquifers' recharge estimated by the hydrological modelling (ΔG) had an average value of about 771.0 × 10^6 m^3 yr^{-1} for the historical climate series, with high monthly variability. This value indicates that almost 80 % of the surface flows (Q) have previously passed through aquifers, which underlines the relevance of the groundwater in this river basin. The maximum recharge occurs during the winter and spring months (January–May), the minimum values being found in the summer months (Fig. 9a). For the CC scenarios, the recharge is significantly reduced and concentrated in the months of February and April. In the RCP 4.5 scenario the aquifers' recharge is reduced by 54 % (to 357.7 × 10^6 m^3 yr^{-1}) and represents 62 % of the surface flow (Q) (Fig. 9b). For RCP 8.5 the aquifers' recharge is reduced by 62 % (to 290.0 × 10^6 m^3 yr^{-1}) and would represent only 57 % of the surface flow (Q) (Fig. 9c). The snowfall reduction is one of the reasons for this decrease in aquifer recharge, since the melting snow in the model becomes soil moisture (S), as usually happens in nature. These changes in the aquifers' recharge alter the pattern of the groundwater discharge (Q_{g}), which is part of the surface flows (Q). In fact, the relative differences in the possible groundwater discharges between the summer and spring months would be greater (Fig. 10a). Overall, the CC scenarios predict a significant reduction in the water resources with respect to the historical climate series, together with a loss of the natural capacity of regulation of the river basin itself, manifested as an increase in the relative values of the maximum flows (Fig. 10b).

5.2 Climate change effects on the Tagus Headwaters Water Resources Exploitation System

The simulation of the THWRES was carried out with the flows obtained in the previous hydrological modelling. The

Table 2. Results of the calibration–validation process for the 12 gauging stations in the hydrological modelling.

Gauging stations		Calibration				Validation			
Code	Name	R^2	E_{NS}	P_{BIAS}	E_{RMS}	R^2	E_{NS}	P_{BIAS}	E_{RMS}
3001	Peralejos de las Truchas	0.76	0.74	16.60	6.11	0.51	0.43	−2.90	6.16
3268	Taravillas	0.57	0.57	0.70	1.34	0.43	0.19	−32.20	2.51
3030	Ventosa	0.86	0.86	−2.70	2.90	0.61	0.34	−5.20	4.07
3005	Trillo	0.51	0.50	1.10	0.79	0.47	0.45	−8.10	0.40
3045	Priego Escabas	0.67	0.67	3.50	0.60	0.10	−0.36	19.30	0.49
3172	Huete	0.64	0.64	−1.90	0.90	0.38	0.37	−10.20	1.33
3173	La Peraleja	0.82	0.82	−1.30	2.72	0.31	0.15	34.80	4.95
3186	Priego Trabaque	0.70	0.70	−2.10	1.30	0.69	0.57	−16.20	1.14
3201	Molino de Chincha	0.79	0.78	11.00	15.14	0.61	0.46	−9.50	16.11
3041	Alcantud	0.72	0.71	−8.60	7.69	0.53	0.35	−15.70	6.71
3006	Entrepeñas	0.79	0.78	8.70	16.69	0.65	0.33	1.00	16.93
3043	Buendía	0.87	0.87	0.70	10.44	0.71	0.52	−10.90	11.81

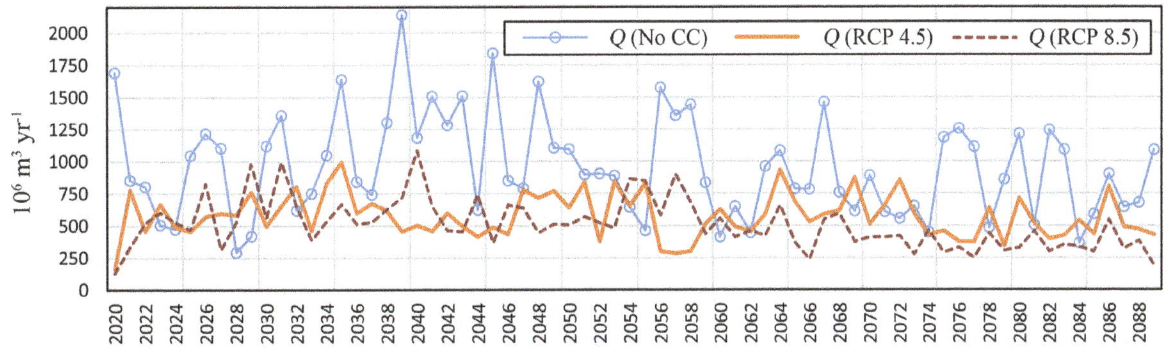

Figure 7. Surface flows that enter into the Entrepeñas and Buendía reservoirs (EBR).

results obtained using the historical climate series (No CC) and the stabilization scenario (RCP 4.5) indicate that the supply to the water uses in the Tagus River basin is guaranteed, in both cases, for the whole of the time period simulated. However, for the scenario of very high emissions (RCP 8.5), the simulation indicates that there would be supply deficits in the THWRES in some years towards the end of the time period considered.

Once the water uses in the Tagus River basin have been supplied, the TSA operating rule comes into effect, providing the monthly volumes that would be possible to transfer in each climatic scenario evaluated. The results are presented in Fig. 11, which shows the volume stored in each month in the EBR (V_{acu}), the transferred volume (Q_{trans}) that leaves the Tagus River basin, and the flow that would go through the TSA to the SRB.

Table 3 summarizes the main statistics of the annual transfer series, specifying which part goes to the SRB. If the future climate were similar to that of past years (No CC), the average transferable volume would be about $450 \times 10^6\,m^3\,yr^{-1}$ (Fig. 11a). This volume would break down, following the priority criterion established, into about $40 \times 10^6\,m^3\,yr^{-1}$ for the Guadiana River basin ($29 \times 10^6\,m^3\,yr^{-1}$ for urban supply and about $10 \times 10^6\,m^3\,yr^{-1}$ for Tablas de Daimiel) and $411 \times 10^6\,m^3\,yr^{-1}$ for the TSA. Thus, given the average losses by infiltration and evaporation from the transfer channel, which are around 10 % (CHS, 2015), the net volume that would reach the SRB would be $370 \times 10^6\,m^3\,yr^{-1}$, a value close to those of the actual series of transfers that have occurred since it came into operation (Morote et al., 2017). So, if there were No CC, despite the fact that the volume reaching the SRB via the TSA would be, on average, much lower than the planned $540 \times 10^6\,m^3\,yr^{-1}$ ($600 \times 10^6\,m^3\,yr^{-1}$ minus 10 % losses), it would be possible to transfer an average of $411 \times 10^6\,m^3\,yr^{-1}$ to the SRB.

In the RCP 4.5 scenario, the transferable volumes drop to $143 \times 10^6\,m^3\,year^{-1}$ on average, $17 \times 10^6\,m^3\,yr^{-1}$ being for the urban supply in the Guadiana River basin, $2.5 \times 10^6\,m^3\,yr^{-1}$ for Tablas de Daimiel, and $123 \times 10^6\,m^3\,yr^{-1}$ for the SRB by means of the TSA infrastructure. So, considering losses of 10 %, this would mean net water resources of $111 \times 10^6\,m^3\,yr^{-1}$ being transferred to the SRB, barely 20 % of the maximum transferable volume. As worrisome as this important decrease in the average value is the existence of

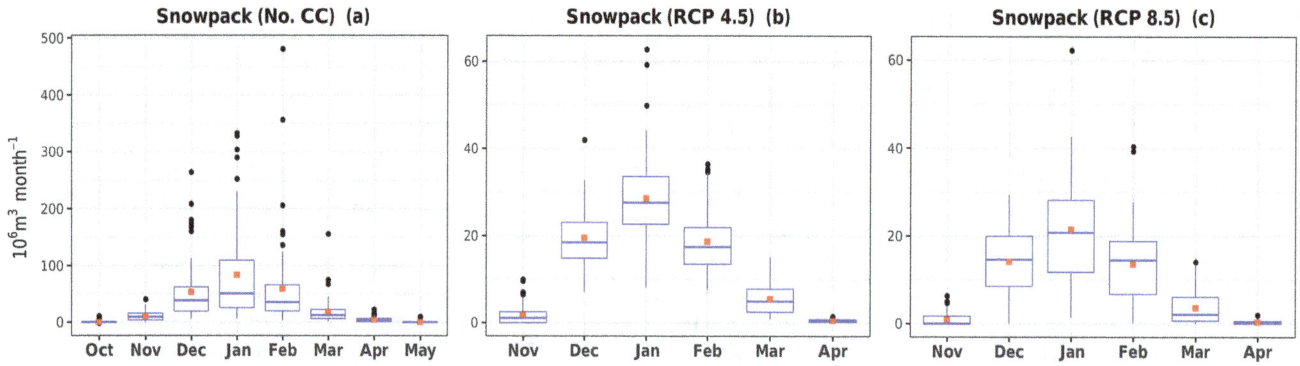

Figure 8. Monthly distribution of the snow cover accumulated on the summits ($S_{\rm np}$) in the Tagus Headwaters River Basin (unit: $10^6\,\mathrm{m}^3\,\mathrm{month}^{-1}$): **(a)** No CC. **(b)** RCP 4.5. **(c)** RCP 8.5.

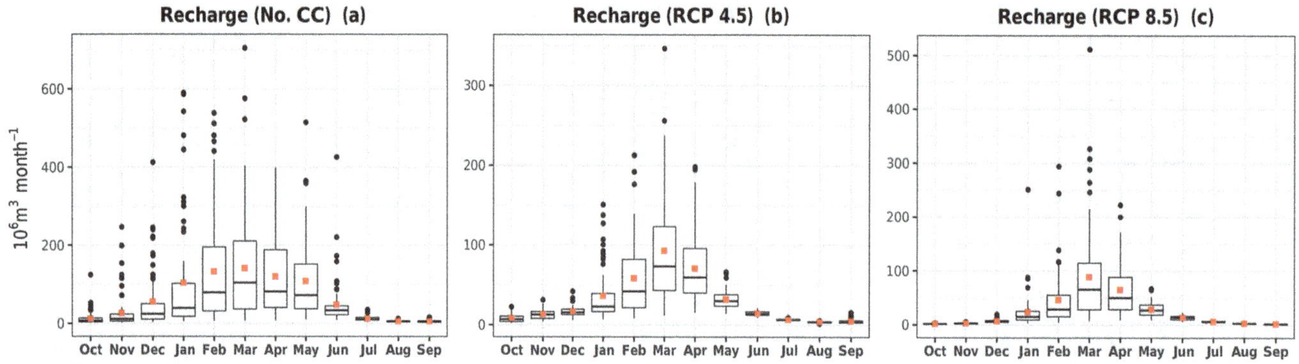

Figure 9. Monthly distribution of the recharge (ΔG) in the Tagus Headwaters River Basin (unit: $10^6\,\mathrm{m}^3\,\mathrm{month}^{-1}$): **(a)** No CC. **(b)** RCP 4.5. **(c)** RCP 8.5.

consecutive periods of 3 and 4 years in which no transfer would occur (Fig. 11b).

This situation would be aggravated for the climatic scenario RCP 8.5 since the transferable volume would be reduced to about $100 \times 10^6\,\mathrm{m}^3\,\mathrm{yr}^{-1}$, on average, distributed as $12 \times 10^6\,\mathrm{m}^3\,\mathrm{yr}^{-1}$ for the urban supply in the Guadiana River basin, $1.5 \times 10^6\,\mathrm{m}^3\,\mathrm{yr}^{-1}$ for the maintenance of the Tablas de Daimiel wetland, and $86 \times 10^6\,\mathrm{m}^3\,\mathrm{yr}^{-1}$ for the TSA. Thus, the SRB would receive approximately throughout the year the volume planned for just 1 month in the operating rule. This scenario worsens the duration and frequency of the no-transfer periods, which intensify over time, reaching a situation of total cessation of the TSA from the year 2067 due to a lack of accumulated volumes ($V_{\rm acu}$) in the EBR (Fig. 11c).

6 Discussion

The hydrological modelling carried out with the AR5 CC scenarios predicts a sharp decrease in the water resources of the THRB. These results are in line with the previous simulations made on the same scale (Lobanova et al., 2018), as well as on a larger scale (Guerreiro et al., 2017; CEDEX, 2011b; Lobanova et al., 2018). The hydrological modelling indicated that the increase in temperature would generate a decline in snowfall, which would lead to a reduction in the snow cover period of 2 months. The relevant snowfalls would have a delay of 1 month and the snowmelt would start a month earlier. The decreases provided by the simulations are similar to those published by CEDEX (2017) for the same mountainous area, in which a decrease of up to 70 % was predicted for the interval 2070–2100 in the RCP 4.5 scenario, and greater than 90 % for the RCP 8.5 scenario (same interval). In addition, CEDEX (2017) also predicted a decrease in the 2-month snow cover period. Regarding the aquifer recharge, the hydrological modelling indicated a decline greater than 50 %. Although these values are slightly higher than those published by CEDEX (2017) for the whole Tajo River basin, they are in line with the provisions in the THRB, since the majority of the projections used for these two scenarios (RCP 4.5 and RCP 8.5) indicated diminishments of around 50 % in aquifer recharge within this area. These alterations of the hydrological processes will probably generate a change in the fluvial regime. In relative terms, the months of autumn (October, November, December) would suffer the major decrease, this change being consistent with the results of Lobanova et al. (2018). At the annual level, these alterations would result in greater variability of the surface flows of the basin due

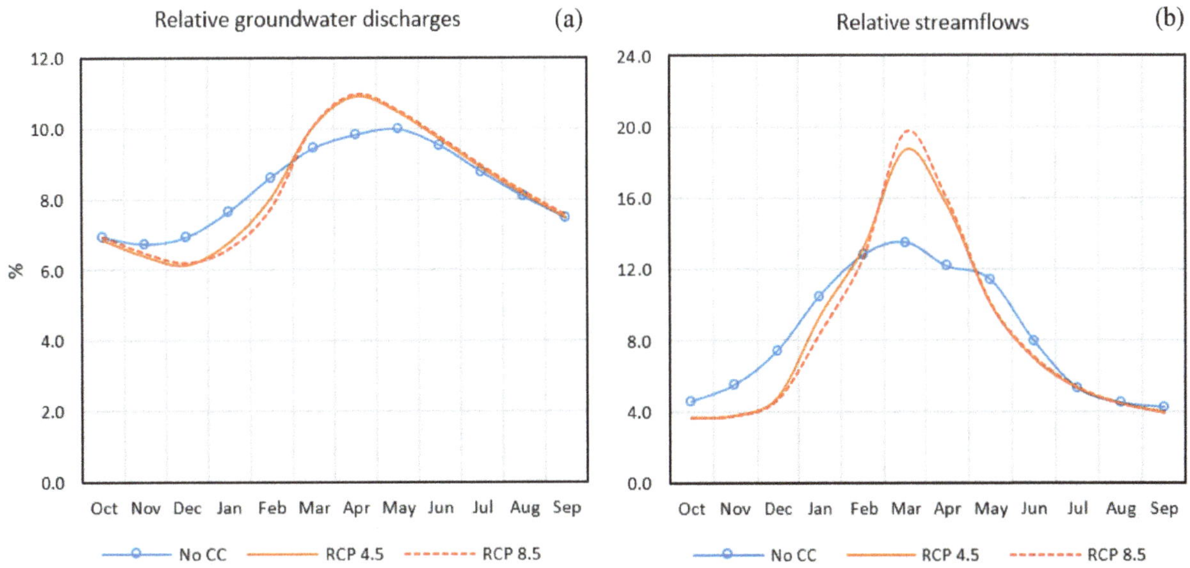

Figure 10. Alteration of the intra-annual pattern of the groundwater discharges from aquifers **(a)** and surface flows **(b)**, in relative terms (%).

Table 3. Water volumes transferred (Q_{trans} and TSA) in each climate scenario (unit: 10^6 m^3 yr^{-1}).

	No CC		RCP 4.5		RCP 8.5	
	Q_{trans}	TSA to SRB	Q_{trans}	TSA to SRB	Q_{trans}	TSA to SRB
Average	449.9	410.7	142.5	123.3	99.6	86.2
Maximum	650.0	601.4	572.6	524.2	456.1	417.3
Minimum	80.0	66.6	0.0	0.0	0.0	0.0
Standard deviation	136.4	130.5	121.5	108.7	115.5	102.5

to increases in the relative difference between the maximum and minimum flows of spring and summer, respectively. In short, these changes are going to diminish the natural capacity of the THRB to regulate its own water resources. However, the results of the THWRES simulations indicate that the EBR will not reach their maximum volume in any situation, and no new infrastructure will be necessary to overcome this loss of regulation.

The effects of these physical changes on the THWRES will translate into average decreases around 70 %–79 % in the volume expected to be available for the TSA in both scenarios evaluated (RCP 4.5 and RCP 8.5). In addition, there would be long periods without transfers for the RCP 4.5 scenario, and this IBWT could even stop operating for the last third of the simulation period under the RCP 8.5 scenario, when there would be deficits regarding the water uses in the Tagus River basin itself.

At this point it is necessary to consider two reflections. The first is the uncertainty of regionalized projections, whose origin is associated with the uncertainties inherited during the different stages of their generation. To this uncertainty must be added the influence of uncontrolled local factors in the regionalization, which have a great relevance in the case of precipitation. In this sense, in Spain, while the regionalized series for temperature are in line with the historical reference series, those of precipitation have strong uncertainties associated with the models (Amblar-Francés et al., 2017; Mitchell and Hulme, 1999). Even so, the use of these data sources is the best way to understand the repercussions of CC for water resources and hence to adapt strategies to CC. The second is that the simulations carried out for the water resource exploitation system represent scenarios in which the uses of the water and the operating rule do not change with time. The flows that enter in the model are the only input data that vary. So, they are not tools to predict future behaviour, since water needs, infrastructure, and water policy are not static. However, they do serve to evaluate the effects that CC could have on the TSA if measures are not taken regarding the management of the demand for water or the incorporation of alternative sources of water resources into this system.

The socioeconomic impact of the decreases in the TSA supply were calculated by the methodology presented in

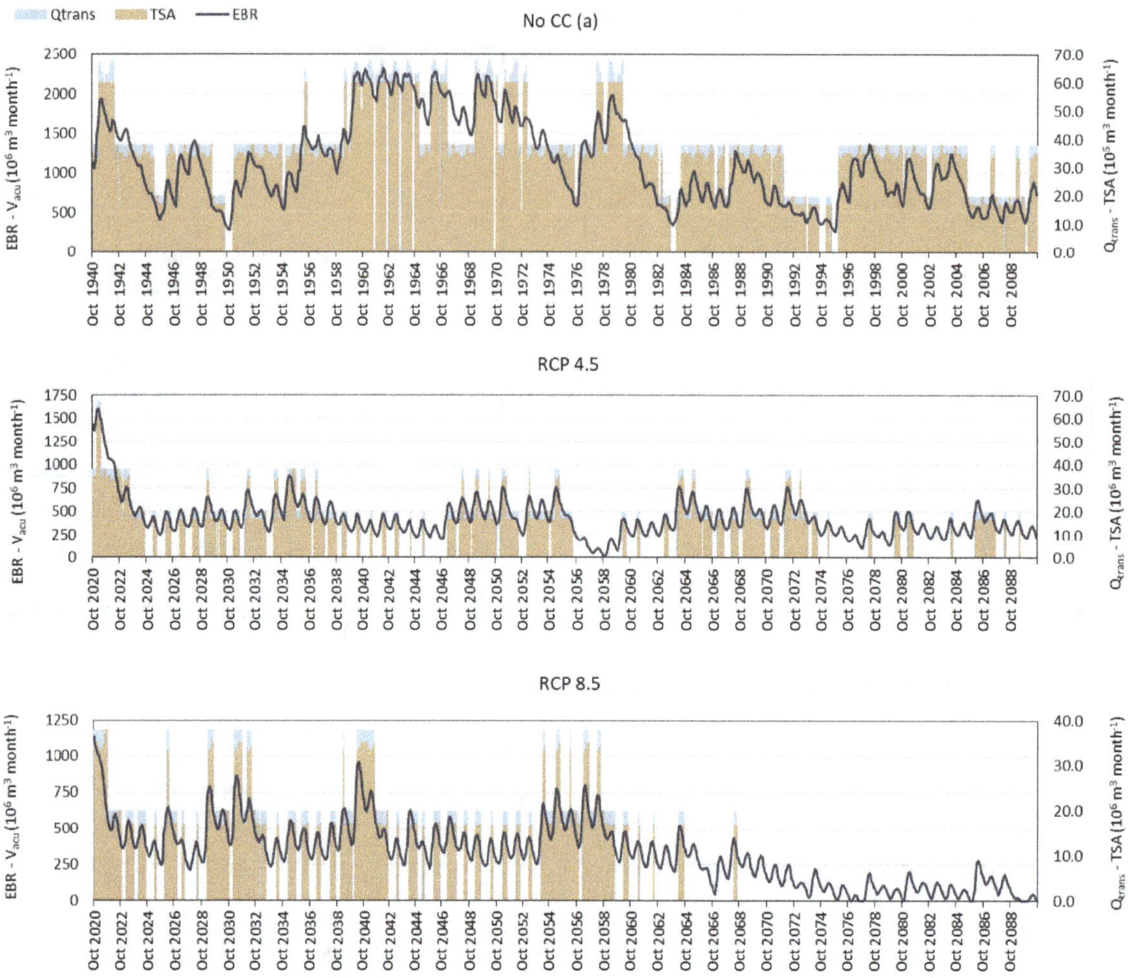

Figure 11. The storage volume in the EBR (V_{acu}), the total transferred flow (Q_{trans}), and the flow to the SRB (TSA) (unit: 10^6 m^3 month^{-1}): **(a)** No CC. **(b)** RCP 4.5. **(c)** RCP 8.5.

Sect. 3.3. First, the global decrease in TSA flows to the SRB was distributed among the seven IZs, according to the results obtained by Martínez-Paz et al. (2016), which were conditioned by the water demand and by the topology of the exploitation system of the SRB. Once these deficits were determined, they were valued using the linear programming presented, calculating GVA and agricultural employment for each IZ according to water allocations available in each scenario. The results are presented in Table 4, which shows the differences in relation to a scenario with no climate change (No CC).

By calculating the ratio between the GVA and TSA the marginal value of water used in irrigation was obtained: 1.29 EUR m^{-3} for the RCP 4.5 scenario and 1.31 EUR m^{-3} for the RCP 8.5 scenario (average values for the whole SRB). The marginal value establishes the upper limit of the marginal productivity of water, and therefore it represents the maximum payment capacity of the sector for water in each scenario. The figures obtained are in line with those presented in other studies for the area (Martínez-Paz et al., 2016; Albaladejo-García et al., 2018).

Thus, the decrease in the TSA supply for the RCP 4.5 scenario would mean a direct loss of 380 million EUR yr^{-1} (at the prices of 2017), while for RCP 8.5 it would amount to 425 million EUR yr^{-1}. Given that the GVA of irrigation in the SRB is around 1870 million EUR yr^{-1} (Martínez-Paz and Pellicer-Martínez, 2018), these figures represent, respectively, direct losses of 20 % and 23 % in terms of GVA. In addition, the decrease in the volumes transferred in the CC scenarios means that the irrigated area is occupied by more labour-extensive crops, causing the loss of around 6690 and 7369 direct full-time agrarian jobs per year in the sector (Table 4). To all these direct effects we should add the drag effects of the primary sector on other economic sectors that depend directly on the use of irrigation in the area (food industry, marketing, transport, inputs, etc.) and make up the

Table 4. Differences of the RCP 4.5 and RCP 8.5 scenarios with respect to No CC (these values represent decreases with respect to the No CC scenario).

IZ	RCP 4.5			RCP 8.5		
	TSA to SRB (10^6 m^3 yr^{-1})	GVA (10^6 EUR yr^{-1})	Labour (full-time jobs yr^{-1})	TSA to SRB (10^6 m^3 yr^{-1})	GVA (10^6 EUR yr^{-1})	Labour (full-time jobs yr^{-1})
1	54.60	79.12	1135	60.14	88.61	1251
2	61.60	58.33	455	67.85	64.26	502
3	22.58	34.04	651	24.87	38.65	717
4	44.99	68.49	1484	49.55	77.26	1635
5	93.12	122.25	2582	102.57	135.06	2845
6	9.41	11.51	378	10.37	13.82	416
7	8.21	5.94	3	9.04	6.79	4
SRB	294.50	379.67	6690	324.40	424.46	7369

well-known agro-industrial cluster of the Region of Murcia (Colino et al., 2014). In this sense, some authors propose a multiplier of no less than 3 to calculate the total economic impact of agricultural production on other related activities. To this economic impact it would be possible to add the environmental one (Martínez-Paz et al., 2018), as it would be, for example, the reduction of the flows circulating in the Segura River, since it is part of the distribution network of the TSA volumes, which would also have effects on the ecosystems associated with this river (Perni and Martínez-Paz, 2017).

7 Conclusions

In this work, the possible effects of CC on the Tagus Headwaters Water Resources Exploitation System have been evaluated. In particular, the work has focused on the hydrological changes that would arise in the Tagus Headwaters River Basin (THRB) if the predicted climatic scenarios arose, and how these would affect the uses of the basin itself and the IBWT (the TSA) that transfers flows to the Segura River basin. The main effects on the water cycle would be the reduction of snowfalls and their accumulation on the summits, and a decrease in aquifer recharge. These changes would generate a significant loss of the natural regulation capacity of the THRB itself, which could not be corrected with new infrastructure since there would also be a significant reduction in the water resources.

The hypothetical flows that can be transferred by the TSA would suffer a significant decrease if the simulated CC scenarios came into being. For RCP 4.5 there would be an average reduction of 70 %, while for RCP 8.5 it would be 79 %, both figures being with respect to the series without CC. These reductions would result in direct economic losses in the irrigation sector of around 20 %. Beyond these average figures, the increase in the zero-transfer periods would have a high impact, since it would further increase the uncertainty associated with this source of supply, preventing adequate planning of water uses in the receiving basin.

Finally, in addition to the undoubted interest that is presented by the case study analysed, this work shows a complete sequential methodological framework which should serve as a guide for the comprehensive evaluation of IBWT within a framework of integrated water resource management. Moreover, the entire methodology has been developed with open-source tools, facilitating its reproducibility in other areas.

Author contributions. Both authors contributed equally to this work.

Competing interests. The authors declare that they have no conflict of interest.

Special issue statement. This article is part of the special issue "Assessing impacts and adaptation to global change in water resource systems depending on natural storage from groundwater and/or snowpacks". It is not associated with a conference.

Acknowledgements. This research work has been supported by project 19342/PI/14 funded by "Fundación Séneca-Agencia de Ciencia y Tecnología de la Región de Murcia" in the framework of PCTIRM 2011–2014. Moreover, the authors thank AEMET and UC for the data provided for this work. We appreciate the valuable comments and suggestions provided by the editor and two anonymous referees.

Edited by: David Pulido-Velazquez

References

Albaladejo-García, J., Martínez-Paz, J., and Colino, J.: Financial evaluation of the feasibility of using desalinated water in the greenhouse agriculture of Campo de Níjar (Almería, Spain), ITEA-Inf. Tec. Econ. Ag., 114, 6, https://doi.org/10.12706/itea.2018.024, 2018 (in Spanish).

Allen, R. G., Pereira, L., Raes, D., and Smith, M.: Crop evapotranspiration: Guidelines for computing crop requirements, FAO Irrigation and drainage paper 56, Rome, 15 pp., available at: https://appgeodb.nancy.inra.fr/biljou/pdf/Allen_FAO1998.pdf (last access: 18 January 2018), 1998.

Amblar-Francés, P., Casado-Calle, M., Saavedra, A., Ramos, P., and Rodríguez, E.: Guía de Escenarios Regionalizados de Cambio Climático sobre España a Partir de los Resultados del IPCC-AR5, Agencia Estatal de Meteorología, Gobierno de España, available at: http://hdl.handle.net/20.500.11765/7956, 17 December 2017.

Bajracharya, A., Bajracharya, S., Shrestha, A., and Maharjan, S.: Climate change impact assessment on the hydrological regime of the Kaligandaki Basin, Nepal, Sci. Total Environ., 625, 837–848, https://doi.org/10.1016/j.scitotenv.2017.12.332, 2018.

BOE: Acuerdo para encomienda de gestión por el Ministerio de Agricultura, Alimentación y Medio Ambiente (Dirección General del Agua) al CEDEX, del Ministerio de Fomento, para la realización de asistencia técnica, investigación y desarrollo tecnológico en materias competencia de la Dirección General del Agua, Boletín Oficial del Estado (BOE), 287, 49436–49458, available at: https://www.boe.es/diario_boe/txt.php?id=BOE-A-2007-20623 (last access: 15 August 2018), 2007 (in Spanish).

BOE: Real Decreto 773/2014, de 12 de septiembre, por el que se aprueban diversas normas reguladoras del trasvase por el acueducto Tajo-Segura, Ministerio de Agricultura, Alimentación y Medio Ambiente, Boletín Oficial del Estado (BOE), 223, 71634–71639, available at: https://www.boe.es/buscar/doc.php?id=BOE-A-2014-9336 (15 August 2018), 2014 (in Spanish).

BOE: Ley 21/2015, de 20 de julio, por la que se modifica la Ley 43/2003, de 21 de noviembre, de Montes, Jefatura del Estado, Boletín Oficial del Estado (BOE), 173, 60234–600272, available at: https://www.boe.es/diario_boe/txt.php?id=BOE-A-2015-8146 (15 August 2018), 2015 (in Spanish).

CEDEX: Estudio de los Impactos del Cambio Climático en los Recursos Hídricos y las Masas de Agua, Informe Final, Centro de Estudios y Experimentación de Obras Públicas, Ministerio de Agricultura, Alimentación y Medio Ambiente, available at: http://www.mapama.gob.es/es/agua/temas/planificacion-hidrologica/planificacion-hidrologica/EGest_CC_RH.aspx (15 August 2018), 2011a.

CEDEX: Estudio de los Impactos del Cambio Climático en los Recursos Hídricos y las Masas de Agua. Ficha 1: Evaluación del Impacto del Cambio Climático en los Recursos Hídricos en Régimen Natural, Centro de Estudios y Experimentación de Obras Públicas, Ministerio de Agricultura, Alimentación y Medio Ambiente, available at: http://www.cedex.es (15 August 2018), 2011b (in Spanish).

CEDEX: Evaluación del Impacto del Cambio Climático en los Recursos Hídricos y Sequías en España, Informe Final, Centro de Estudios Hidrográficos, Centro de Estudios y Experimentación de Obras Públicas, Ministerio de Agricultura, Alimentación y Medio Ambiente, available at: http://www.cedex.es/CEDEX/LANG_CASTELLANO/ORGANISMO/CENTYLAB/CEH/Documentos_Descargas/EvaluacionimpactoCCsequiasEspana2017.htm (15 August 2018), 2017 (in Spanish).

Chavez-Jimenez, A., Granados, A., Garrote, L., and Martín-Carrasco, F.: Adapting water allocation to irrigation demands to constraints in water availability imposed by Climate Change, Water Resour. Manag., 29, 1413–1430, https://doi.org/10.1007/s11269-014-0882-x, 2015.

CHG: Plan Hidrológico de la Parte Española de la Demarcación Hidrográfica del Guadiana, Memoria (Parte I), Confederación Hidrográfica del Guadiana, Ministerio de Agricultura, Alimentación y Medio Ambiente, available at: http://planhidrologico2015.chguadiana.es/?corp=planhidrologico2015&url=61 (15 August 2018), 2016 (in Spanish).

CHJ: Plan Hidrológico de la Demarcación Hidrográfica del Júcar. Memoria, Ciclo de Planificación Hidrológica 2015–2021, Confederación Hidrográfica del Júcar, Ministerio de Agricultura, Alimentación y Medio Ambiente, available at: https://www.chj.es/es-es/medioambiente/planificacionhidrologica/Paginas/PHC-2015-2021-Plan-Hidrologico-cuenca.aspx (15 August 2018), 2016.

CHS: Proyecto de Plan Hidrológico de la Cuenca del Segura, Anejo 02: Recursos Hídricos, Confederación Hidrográfica del Segura, Ministerio de Agricultura, Alimentación y Medio Ambiente, available at: http://www.chsegura.es/chs/planificacionydma/planificacion/, 2015 (in Spanish).

CHT: Plan Hidrológico de la Parte Española de la Demarcación Hidrográfica del Tajo, Memoria, Confederación Hidrográfica del Tajo, Ministerio de Agricultura, Alimentación y Medio Ambiente, available at: http://www.chtajo.es/LaCuenca/Planes/PlanHidrologico/Planif_2015-2021/Paginas/Plan_2015-2021.aspx (last access: 15 August 2018), 2015.

CHT: Centro de descargas de capas en formato shape de la Cuenca Hidrográfica del Tajo, available at: http://www.chtajo.es/LaCuenca/Paginas/DescargaDCapas.aspx (last access: 17 December 2017), 2018 (in Spanish).

Colino, J., Martínez-Carrasco, F., and Martínez-Paz, J. M.: El impacto de la PAC renovada sobre el sector agrario de la Región de Murcia, de la Región de Murcia, available at: https://www.cesmurcia.es/cesmurcia/DocumentServlet?docid=/publicaciones/ficheros/564.pdf (last access: 15 September 2018), 2014.

Díaz, P., Morley, K. M., and Yeh, D. H.: Resilient urban water supply: preparing for the slow-moving consequences of climate change, Water Practice and Technology, 12, 123–138, https://doi.org/10.2166/wpt.2017.016, 2017.

Duan, Q., Sorooshian, S., and Gupta, V. K.: Optimal use of the SCE-UA global optimization method for calibrating watershed models, J. Hydrol., 158, 265–284, https://doi.org/10.1016/0022-1694(94)90057-4, 1994.

Flörke, M., Schneider, C., and McDonald, R. I.: Water competition between cities and agriculture driven by climate change and urban growth, Nature Sustainability, 1, 51–58, https://doi.org/10.1038/s41893-017-0006-8, 2018.

Giupponi, C. and Gain, A. K.: Integrated water resources management (IWRM) for climate change adaptation, Reg. Environ. Change, 17, 1865–1867, https://doi.org/10.1007/s10113-017-1173-x, 2017.

Gomariz-Castillo, F., Alonso-Sarría, F., and Cabezas-Calvo-Rubio, F.: Calibration and spatial modelling of daily ET0 in semiarid areas using Hargreaves equation, Earth Sci. Inform., 11, 325–340, https://doi.org/10.1007/s12145-017-0327-1, 2018.

Griffin, R. C.: Water Resource Economics: The Analysis of Scarcity, Policies and Projects, The MIT Press Cambridge, available at: https://mitpress.mit.edu/books/water-resource-economics (last access: 15 September 2018), 2006.

Grindlay, A., Zamorano, M., Rodríguez, M., Molero, E., and Urrea, M.: Implementation of the European Water Framework Directive: Integration of hydrological and regional planning at the Segura River Basin, southeast Spain, Land Use Policy, 28, 242–256, https://doi.org/10.1016/j.landusepol.2010.06.005, 2011.

Guerreiro, S. B., Birkinshaw, S., Kilsby, C., Fowler, H. J., and Lewis, E.: Dry getting drier – The future of transnational river basins in Iberia, J. Hydrol., 12, 238–252, https://doi.org/10.1016/j.ejrh.2017.05.009, 2017.

Gupta, H. V. and Kling, H.: On typical range, sensitivity, and normalization of Mean Squared Error and Nash-Sutcliffe Efficiency type metrics, Water Resour. Res., 47, w10601, https://doi.org/10.1029/2011WR010962, 2011.

Herrera, S., Fernández, J., and Gutiérrez, J. M.: Update of the Spain02 gridded observational dataset for EURO-CORDEX evaluation: assessing the effect of the interpolation methodology, Int. J. Climatol., 36, 900–908, https://doi.org/10.1002/joc.4391, 2016.

Hsiao, C. and Wan, S. K.: Is there an optimal forecast combination?, J. Econometrics, 178, 294–309, https://doi.org/10.1016/j.jeconom.2013.11.003, 2014.

Hutchinson, M. F.: Interpolating mean rainfall using thin plate smoothing splines, Int. J. Geogr. Inf. Syst., 9, 385–403, https://doi.org/10.1080/02693799508902045, 1995.

IPCC: Climate Change 2013 – The Physical Science Basis: Working Group I Contribution to the Fifth Assessment Report of the Intergovernmental Panel on Climate Change, Cambridge University Press, Cambridge, UK, https://doi.org/10.1017/CBO9781107415324, 2013.

IPCC: Climate Change 2014 – Impacts, Adaptation and Vulnerability: Part B: Regional Aspects: Working Group II Contribution to the IPCC Fifth Assessment Report, vol. 2, Cambridge University Press, Cambridge, UK, https://doi.org/10.1017/CBO9781107415386, 2014.

Jha, C. K., Gupta, V., Chattopadhyay, U., and Sreeraman, B. A.: Migration as adaptation strategy to cope with climate change: A study of farmers' migration in rural India, Int. J Clim. Chang. Str., 10, 121–141, https://doi.org/10.1108/IJCCSM-03-2017-0059, 2018.

Kahil, M. T., Connor, J. D., and Albiac, J.: Efficient water management policies for irrigation adaptation to climate change in Southern Europe, Ecol. Econ., 120, 226–233, https://doi.org/10.1016/j.ecolecon.2015.11.004, 2015.

Klemeš, V.: Operational testing of hydrological simulation models, Hydrolog. Sci. J., 31, 13–24, https://doi.org/10.1080/02626668609491024, 1986.

Li, L., Zhang, L., Xia, J., Gippel, C. J., Wang, R., and Zeng, S.: Implications of modelled climate and land cover changes on runoff in the middle route of the south to north water transfer project in China, Water Resour. Manag., 29, 2563–2579, https://doi.org/10.1007/s11269-015-0957-3, 2015.

Li, Z. L., Xu, Z. X., and Li, Z. J.: Performance of WASMOD and SWAT on hydrological simulation in Yingluoxia watershed in northwest of China, Hydrol. Process., 25, 2001–2008, https://doi.org/10.1002/hyp.7944, 2011.

Li, Z. L., Shao, Q. X., Xu, Z. X., and Xu, C. Y.: Uncertainty issues of a conceptual water balance model for a semi-arid watershed in north-west of China, Hydrol. Process., 27, 304–312, https://doi.org/10.1002/hyp.9258, 2013.

Lobanova, A., Koch, H., Liersch, S., Hattermann, F. F., and Krysanova, V.: Impacts of changing climate on the hydrology and hydropower production of the Tagus River basin, Hydrol. Process., 30, 5039–5052, https://doi.org/10.1002/hyp.10966, 2016.

Lobanova, A., Liersch, S., Nunes, J. P., Didovets, I., Stagl, J., Huang, S., Koch, H., del Rocío Rivas López, M., Maule, C. F., Hattermann, F., and Krysanova, V.: Hydrological impacts of moderate and high-end climate change across European river basins, J. Hydrol., 18, 15–30, https://doi.org/10.1016/j.ejrh.2018.05.003, 2018.

Lorenzo-Lacruz, J., Vicente-Serrano, S., López-Moreno, J., Beguería, S., García-Ruiz, J., and Cuadrat, J.: The impact of droughts and water management on various hydrological systems in the headwaters of the Tagus River (central Spain), J. Hydrol., 386, 13–26, https://doi.org/10.1016/j.jhydrol.2010.01.001, 2010.

Martínez-Paz, J. M. and Pellicer-Martínez, F.: Valoración económica del riesgo asociado a las sequías en la Región de Murcia, in: Riesgos Ambientales en la Región de Murcia, edited by: Conesa, C. and Pérez, P., 269–294, EDITUM, Murcia, 2018 (in Spanish).

Martínez-Paz, J. M., Perni, A., Ruiz-Campuzano, P., and Pellicer-Martínez, F.: Valoración económica de los fallos de suministro en los regadíos de la cuenca del Segura, Revista Española de Estudios Agrosociales y Pesqueros REEAP, 244, 35–67, 2016 (in Spanish).

Martínez-Paz, J. M., Banos-González, I., Martínez-Fernández, J., and Esteve-Selma, M. A.: Assessment of management measures for the conservation of traditional irrigated lands: the case of the Huerta of Murcia (Spain), Land Use Policy, 81, 382–391, https://doi.org/10.1016/j.landusepol.2018.10.050, 2018.

McKenney, D. W., Pedlar, J. H., Papadopol, P., and Hutchinson, M. F.: The development of 1901–2000 historical monthly climate models for Canada and the United States, Agr. Forest Meteorol., 138, 69–81, https://doi.org/10.1016/j.agrformet.2006.03.012, 2006.

Meza, F. J., Wilks, D. S., Gurovich, L., and Bambach, N.: Impacts of Climate Change on irrigated agriculture in the Maipo basin, Chile: Reliability of water rights and changes in the demand for irrigation, J. Water Res. Plan. Man., 138, 421–430, https://doi.org/10.1061/(ASCE)WR.1943-5452.0000216, 2012.

Mitchell, T. D. and Hulme, M.: Predicting regional climate change: living with uncertainty, Prog. Phys. Geog., 23, 57–78, https://doi.org/10.1177/030913339902300103, 1999.

MITECO: Anuario de Aforos de la Red Oficial de Estaciones de Aforos, Ministerio para la Transición Ecológica, available at: https://www.miteco.gob.es/es/agua/temas/evaluacion-de-los-recursos-hidricos/sistema-informacion-anuario-aforos/default.aspx, last access: 15 August 2018.

Molina-Navarro, E., Trolle, D., Martínez-Perez, S., Sastre-Merlin, A., and Jeppesen, E.: Hydrological and water quality impact assessment of a Mediterranean limno-reservoir under climate change and land use management scenarios, J. Hydrol., 509, 354–366, https://doi.org/10.1016/j.jhydrol.2013.11.053, 2014.

Morán-Tejeda, E., Lorenzo-Lacruz, J., Ignacio López-Moreno, J., Rahman, K., and Beniston, M.: Streamflow timing of mountain rivers in Spain: Recent changes and future projections, J. Hydrol., 517, 1114–1127, https://doi.org/10.1016/j.jhydrol.2014.06.053, 2014.

Morote, A. F., Olcina, J., and Rico, A. M.: Challenges and proposals for socio-ecological sustainability of the Tagus-Segura Aqueduct (Spain) under Climate Change, Sustainability, 9, 2058, https://doi.org/10.3390/su9112058, 2017.

Nash, J. and Sutcliffe, J.: River flow forecasting through conceptual models part I – A discussion of principles, J. Hydrol., 10, 282–290, https://doi.org/10.1016/0022-1694(70)90255-6, 1970.

O'Callaghan, J. F. and Mark, D. M.: The extraction of drainage networks from digital elevation data, Comput. Vision Graph., 28, 328–344, 1984.

Olmstead, S. M.: Climate Change adaptation and water resource management: A review of the literature, Energ. Econ., 46, 500–509, https://doi.org/10.1016/j.eneco.2013.09.005, 2014.

Onagi, E.: Climate Change and integrated approach to water resource management in the Murray-Darling basin, in: Sustainable Development and Disaster Risk Reduction, edited by: Uitto, J. I. and Shaw, R., 173–187, Springer Japan, Tokyo, https://doi.org/10.1007/978-4-431-55078-5_11, 2016.

Özdoğan, M.: Climate change impacts on snow water availability in the Euphrates-Tigris basin, Hydrol. Earth Syst. Sci., 15, 2789–2803, https://doi.org/10.5194/hess-15-2789-2011, 2011

Pedro-Monzonís, M., Jiménez-Fernández, P., Solera, A., and Jiménez-Gavilán, P.: The use of AQUATOOL DSS applied to the System of Environmental-Economic Accounting for Water (SEEAW), J. Hydrol., 533, 1–14, https://doi.org/10.1016/j.jhydrol.2015.11.034, 2016a.

Pedro-Monzonís, M., Solera, A., Ferrer, J., Andreu, J., and Estrela, T.: Water accounting for stressed river basins based on water resources management models, Sci. Total Environ., 565, 181–190, https://doi.org/10.1016/j.scitotenv.2016.04.161, 2016b.

Pellicer-Martínez, F. and Martínez-Paz, J. M.: Assessment of interbasin groundwater flows between catchments using a semi-distributed water balance model, J. Hydrol., 519, 1848–1858, https://doi.org/10.1016/j.jhydrol.2014.09.067, 2014.

Pellicer-Martínez, F. and Martínez-Paz, J. M.: Contrast and transferability of parameters of lumped water balance models in the Segura River Basin (Spain), Water Environ. J., 29, 43–50, https://doi.org/10.1111/wej.12091, 2015.

Pellicer-Martínez, F., González-Soto, I., and Martínez-Paz, J. M.: Analysis of incorporating groundwater exchanges in hydrological models, Hydrol. Process., 29, 4361–4366, https://doi.org/10.1002/hyp.10586, 2015.

Perni, A. and Martínez-Paz, J. M.: Measuring conflicts in the management of anthropized ecosystems: Evidence from a choice experiment in a human-created Mediterranean wetland, J. Environ. Manage., 203, 40–50, https://doi.org/10.1016/j.jenvman.2017.07.049, 2017.

Pulido-Velazquez, D., García-Aróstegui, J. L., Molina, J. L., and Pulido-Velazquez, M.: Assessment of future groundwater recharge in semi-arid regions under climate change scenarios (Serral-Salinas aquifer, SE Spain). Could increased rainfall variability increase the recharge rate?, Hydrol. Process., 29, 828–844, https://doi.org/10.1002/hyp.10191, 2014.

Pulido-Velazquez, M., Alvarez-Mendiola, E., and Andreu, J.: Design of efficient water pricing policies integrating basinwide resource opportunity costs, J. Water Res. Plan. Man., 139, 583–592, https://doi.org/10.1061/(ASCE)WR.1943-5452.0000262, 2013.

QGIS Development Team: QGIS Geographic Information System, available at: https://www.osgeo.org/projects/qgis/ (last access: 15 August 2018), 2016.

R Core Team: R: A Language and Environment for Statistical Computing, R Foundation for Statistical Computing, Vienna, Austria, available at: https://www.r-project.org/ (last access: 15 August 2018), 2016.

Shevnina, E., Kourzeneva, E., Kovalenko, V., and Vihma, T.: Assessment of extreme flood events in a changing climate for a long-term planning of socio-economic infrastructure in the Russian Arctic, Hydrol. Earth Syst. Sci., 21, 2559–2578, https://doi.org/10.5194/hess-21-2559-2017, 2017.

Shrestha, S., Shrestha, M., and Babel, M. S.: Assessment of climate change impact on water diversion strategies of Melamchi Water Supply Project in Nepal, Theor. Appl. Climatol., 128, 311–323, https://doi.org/10.1007/s00704-015-1713-6, 2017.

Skøien, J., Blöschl, G., Laaha, G., Pebesma, E., Parajka, J., and Viglione, A.: rtop: An R package for interpolation of data with a variable spatial support, with an example from river networks, Comput. Geosci., 67, 180–190, https://doi.org/10.1016/j.cageo.2014.02.009, 2014.

Smerdon, B. D.: A synopsis of climate change effects on groundwater recharge, J. Hydrol., 555, 125–128, https://doi.org/10.1016/j.jhydrol.2017.09.047, 2017.

Solaun, K. and Cerdá, E.: The impact of Climate Change on the generation of hydroelectric power-A Case study in Southern Spain, Energies, 10, 1343, https://doi.org/10.3390/en10091343, 2017.

Spain02v5: Climatic historical data, available at: http://www.meteo.unican.es/datasets/spain02, last access: 18 January 2018.

Szczypta, C., Gascoin, S., Houet, T., Hagolle, O., Dejoux, J.-F., Vigneau, C., and Fanise, P.: Impact of climate and land cover changes on snow cover in a small Pyrenean catchment, J. Hydrol., 521, 84–99, https://doi.org/10.1016/j.jhydrol.2014.11.060, 2015.

Thomas, H.: Improved Methods for National Water Assessment, Report, contract WR 15249270, U.S. Water Resources Council, available at: https://pubs.er.usgs.gov/publication/70046351 (last access: 15 August 2018), 1981.

Thornthwaite, C. W.: An approach toward a rational classification of climate, Geogr. Rev., 38, 55–94, 1948.

Tumushabe, J. T.: Climate Change, food security and sustainable development in Africa, in: The Palgrave Handbook of African Politics, Governance and Development, edited by: Oloruntoba, S. O. and Falola, T., 853–868, Palgrave Macmillan US, New York, https://doi.org/10.1057/978-1-349-95232-8_53, 2018.

Wahba, G.: Spline Models for Observational Data, CBMS-NSF Regional Conference Series in Applied Mathematics, Society for Industrial and Applied Mathematics, Philadelphia, USA, 1990.

Wang, G. Q., Zhang, J. Y., Xuan, Y. Q., Liu, J. F., Jin, J. L., Bao, Z. X., He, R. M., Liu, C. S., Liu, Y. L., and Yan, X. L.: Simulating the impact of Climate Change on runoff in a typical river catchment of the Loess Plateau, China, J. Hydrometeorol., 14, 1553–1561, https://doi.org/10.1175/JHM-D-12-081.1, 2013.

Warziniack, T., Lawson, M., and Karen Dante-Wood, S.: Effects of Climate Change on ecosystem services in the Northern Rockies, in: Climate Change and Rocky Mountain Ecosystems, edited by: Halofsky, J. E. and Peterson, D. L., 189–208, Springer International Publishing, Cham, https://doi.org/10.1007/978-3-319-56928-4_10, 2018.

Weiss, C. E. and Roetzer, G. R.: GeomComb: (Geometric) Forecast Combination Methods, r package version 1.0, available at:

https://CRAN.R-project.org/package=GeomComb (last access: 15 August 2018), 2016.

World Bank Group: High and Dry: Climate Change, Water, and the Economy, World Bank Group, available at: https://openknowledge.worldbank.org/handle/10986/23665 (last access: 15 September 2018), 2016.

Wurbs, R. A.: Methods for developing naturalized monthly flows at gaged and ungagged sites, J. Hydrol. Eng., 11, 55–64, https://doi.org/10.1061/(ASCE)1084-0699(2006)11:1(55), 2006.

Xu, C.-Y. and Singh, V. P.: A Review on monthly water balance models for water resources investigations, Water Resour. Manag., 12, 20–50, https://doi.org/10.1023/A:1007916816469, 1998.

Xue, X., Zhang, K., Hong, Y., Gourley, J. J., Kellogg, W., McPherson, R. A., Wan, Z., and Austin, B. N.: New multisite cascading calibration approach for hydrological models: Case Study in the Red River Basin using the VIC model, J. Hydrol. Eng., 21, 05015019, https://doi.org/10.1061/(ASCE)HE.1943-5584.0001282, 2016.

Zare, F., Elsawah, S., Iwanaga, T., Jakeman, A. J., and Pierce, S. A.: Integrated water assessment and modelling: A bibliometric analysis of trends in the water resource sector, J. Hydrol., 552, 765–778, https://doi.org/10.1016/j.jhydrol.2017.07.031, 2017.

Zhang, E., Yin, X., Xu, Z., and Yang, Z.: Bottom-up quantification of inter-basin water transfer vulnerability to climate change, Ecol. Indic., 92, 195–206, https://doi.org/10.1016/j.ecolind.2017.04.019, 2018.

Zhang, L., Li, S., Loáiciga, H. A., Zhuang, Y., and Du, Y.: Opportunities and challenges of interbasin water transfers: a literature review with bibliometric analysis, Scientometrics, 105, 279–294, https://doi.org/10.1007/s11192-015-1656-9, 2015.

Zhang, L. P., Qin, L. L., Yang, Z., Xia, J., and Zeng, S.: Climate change impacts on hydrological processes in the water source area of the Middle Route of the South-to-North Water Diversion Project, Water Int., 37, 564–584, https://doi.org/10.1080/02508060.2012.692108, 2012.

13

High-resolution ensemble projections and uncertainty assessment of regional climate change over China in CORDEX East Asia

Huanghe Gu[1,2], Zhongbo Yu[1,2], Chuanguo Yang[1,2], Qin Ju[1,2], Tao Yang[1,2], and Dawei Zhang[3]

[1]State Key Laboratory of Hydrology-Water Resources and Hydraulic Engineering, Hohai University, Nanjing, China
[2]College of Hydrology and Water Resources, Hohai University, Nanjing, China
[3]China Institute of Water Resources and Hydropower Research, Beijing, China

Correspondence: Zhongbo Yu (zyu@hhu.edu.cn)

Abstract. An ensemble simulation of five regional climate models (RCMs) from the coordinated regional downscaling experiment in East Asia is evaluated and used to project future regional climate change in China. The influences of model uncertainty and internal variability on projections are also identified. The RCMs simulate the historical (1980–2005) climate and future (2006–2049) climate projections under the Representative Concentration Pathway (RCP) RCP4.5 scenario. The simulations for five subregions in China, including northeastern China, northern China, southern China, northwestern China, and the Tibetan Plateau, are highlighted in this study. Results show that (1) RCMs can capture the climatology, annual cycle, and interannual variability of temperature and precipitation and that a multi-model ensemble (MME) outperforms that of an individual RCM. The added values for RCMs are confirmed by comparing the performance of RCMs and global climate models (GCMs) in reproducing annual and seasonal mean precipitation and temperature during the historical period. (2) For future (2030–2049) climate, the MME indicates consistent warming trends at around 1 °C in the entire domain and projects pronounced warming in northern and western China. The annual precipitation is likely to increase in most of the simulation region, except for the Tibetan Plateau. (3) Generally, the future projected change in annual and seasonal mean temperature by RCMs is nearly consistent with the results from the driving GCM. However, changes in annual and seasonal mean precipitation exhibit significant inter-RCM differences and possess a larger magnitude and variability than the driving GCM. Even opposite signals for projected changes in average precipitation between the MME and the driving GCM are shown over southern China, northeastern China, and the Tibetan Plateau. (4) The uncertainty in projected mean temperature mainly arises from the internal variability over northern and southern China and the model uncertainty over the other three subregions. For the projected mean precipitation, the dominant uncertainty source is the internal variability over most regions, except for the Tibetan Plateau, where the model uncertainty reaches up to 60 %. Moreover, the model uncertainty increases with prediction lead time across all subregions.

1 Introduction

Globally averaged surface temperature increased by 0.65–1.06 °C from 1880 to 2012 according to several independently produced datasets, and further increases ranging from 0.3 to 4.8 °C are projected for 2081–2100 relative to 1986–2005 using a set of global climate models (GCMs) driven by the Representative Concentration Pathway (RCP) scenarios RCP2.6 to RCP8.5 (IPCC, 2013). Meanwhile, other climate factors, such as precipitation amounts and variability, snow and ice cover patterns, and mean sea level, are also changing (Alfieri et al., 2015; Kerr, 2008; Patz et al., 2005). Reliable projection of regional future climate is critical in evaluating climate change impacts and vulnerability and in developing appropriate mitigation and adaptation measures, especially for developing countries, such as China, which tends to be one of the countries most vulnerable to the adverse effects of climate change (Kreft et al., 2016; Wang et al., 2017b).

The East Asian summer monsoon (EASM) is the most distinctive climate feature in China, and the monsoon area accounts for approximately 60 % of the mainland (Ding and Chan, 2005). EASM system-related precipitation starts around mid-May or even earlier in the Indo-China peninsula, which presents distinct stepwise northward and northeastward advances, with two abrupt northward jumps and three stationary periods, and it begins to withdraw southward in September (Ding, 2004; Hsu, 2005). The rainy seasons of EASM, including the pre-summer rainy season over southern China, mei-yu (in China), normally occurs during the stationary periods, which are imbedded in the northward advance of the summer monsoon. The anomaly of EASM could cause floods and droughts, which are crucial to the livelihoods of more than one billion people (Gu et al., 2015a; Webster et al., 1998; Yu et al., 2018). However, the manner in which climatological rainfall and interannual variation of EASM can be reliably reproduced remains a challenge because of the complex topography and model limitations. The Coupled Model Intercomparison Project Phase 3 (CMIP3) and CMIP5 have problems simulating precipitation in this region. Recent studies have suggested that the new generation of GCMs from the CMIP5 archive exhibits several improvements in reproducing the climatology and interannual variability of the EASM compared with the CMIP3 GCMs, although the simulated biases remained and large inter-model spread existed (Chen and Bordoni, 2014; Gu et al., 2015b; Huang et al., 2013; Yang et al., 2017). For example, the mei-yu rainfall band is missing in GCMs, even though the monsoon circulation is reproduced well.

Considering these deficiencies, high-resolution GCMs have been developed to improve the capabilities in the simulation of monsoon features, including orographic precipitation, low-level jet orientation, and variability, as well as the mei-yu onset and withdrawal (Kitoh et al., 2013; Kusunoki et al., 2006). However, these experiments remain burdensome due to the large computational cost required for multidecadal simulations. Therefore, the regional climate models (RCMs) focusing on a region of interest are commonly used in regional studies of climate projection and climate change impacts (Gao et al., 2006; Giorgi and Mearns, 1999; Gu et al., 2012; Wang et al., 2004; Yira et al., 2017; Yu et al., 2006). The resolution of RCMs is approximately 12–50 km, and they could consider local-scale forcing, e.g., complex terrain features and land cover heterogeneities, in a physically based method. However, RCMs inherit the biases from systematic model errors because of the imperfect conceptualization, discretization, and spatial averaging within grid cells (Dong et al., 2018). Nonetheless, RCM ensembles can be used to understand and characterize uncertainties from different sources, such as future climate scenarios, the driving GCM, and regional model physics, and therefore, reduce the uncertainties and increase credibility in future projections. The ongoing Coordinated Regional Downscaling Experiment (CORDEX) aims to provide high-resolution future regional climate projections for the majority of populated land regions globally by using multi-RCMs and to present an interface for applicants of climate simulations in climate change impact, adaptation, and mitigation studies (Giorgi et al., 2009; Jones et al., 2011). The CORDEX in East Asia (CORDEX-EA) is the East Asian branch of the CORDEX experiment, and it provides ensemble regional climate simulations (http://cordex-ea.climate.go.kr/cordex/download.do, last access: 18 May 2018). A series of studies based on RCMs within CORDEX-EA has been conducted to project extreme and mean precipitation and temperature in East Asia (Jin et al., 2016; Lee et al., 2014; Niu et al., 2015; Park et al., 2016; Tang et al., 2016; Um et al., 2017), but little attention has been paid to quantify the contributions of the uncertainty in future climate projection over China.

Despite large improvements in the simulation of local processes, future climate projections are still accompanied by large uncertainties stemming from different sources, including the forcing GCMs, emission scenarios, downscaling methods (RCMs or statistical downscaling methods), and natural climate internal variability (Déqué et al., 2007; Deser et al., 2012). Numerous studies have demonstrated that GCMs are the main source of uncertainty (Seo et al., 2016). Other uncertainty sources, such as RCMs and internal variability, will become more important than GCMs after exclusion of the outliers from the GCM ensemble (Kay et al., 2009; Wilby and Harris, 2006). In a nonstationary climate, the internal variability of a given GCM–RCM chain can remain high above the trend related to a given emission scenarios forcing (Lafaysse et al., 2014; O'Brien et al., 2011). Little attention has been devoted to quantify the contributions of the uncertainty arising from RCMs and internal variability in future climate projection over China. The objective evaluation of the capability of RCMs and quantification of the uncertainty in future climate projections are necessary.

In this study, we evaluate the performance of five RCMs within CORDEX-EA to reproduce present-day climate and analyze the projected future climate change under the middle emission scenario. More importantly, biases in current climate simulations and uncertainties in future climate projections attributed to RCMs and internal variability are further analyzed. This paper is structured as follows. Data from observations, the model simulation, and the analysis method are described in the succeeding section. Section 3 presents the historical performances of RCMs for temperature and precipitation and future climate change under the RCP4.5 emission scenario in China. The uncertainties in regional future climate projection caused by inter-RCMs and natural climate internal variability are also discussed. The summary and conclusions are presented in Sect. 4.

2 Data and methods

2.1 Observations

The reference temperature data used to evaluate the model results with observation data are developed from the Climate Research Unit Timeseries 3.23 (CRU) of the University of East Anglia, with a spatial resolution of 0.5°, derived from gauge measurements (Harris et al., 2014). Meanwhile, the reference precipitation data, namely the Asian Precipitation-Highly Resolved Observational Data Integration Toward Evaluation (APHRODITE, hereafter APHRO) dataset, with a spatial resolution of 0.25°, were used to evaluate RCMs (Yatagai et al., 2012). To facilitate the comparison, outputs from a host of RCMs were converted to a common grid of 0.5° × 0.5° latitude/longitude as remapped to the CRU and APHRO observations, using bilinear interpolation. The reasons why CRU and APHRO products are used as reference in this study are clarified as below.

Some studies have focused on comparing and evaluating the spatial–temporal similarities and differences of several widely used observed gridded datasets over China (Sun et al., 2014; Wu and Gao, 2013; Yin et al., 2015). Among the widely used gridded datasets, such as the Global Precipitation Climatology Centre (CPCC) product, the University of Delaware (UDEL) product, CRU data, and the National Meteorological Information Center dataset from the China Meteorological Administration, all temperature datasets exhibit similar distribution patterns for the annual average temperature in mainland China. Considering its easier access and wider usage in the evaluation of RCMs used in China and East Asia (Wang et al., 2017), the CRU product is used as the reference temperature data in this study. APHRO's daily gridded precipitation, presently the only long-term, continental-scale, high-resolution daily product, is constructed based on the data collected at 5000–12 000 stations, which represent 2.3–4.5 times the data made available through the stations used for generating global gridded data (i.e., CRU, GPCC, and UDEL) (Yatagai et al., 2012). Thus, the APHRO dataset would give more confidence in the robustness of the results in comparison with other global precipitation datasets and is therefore widely used for evaluating the performance of RCMs in East Asia (Gao et al., 2017; Lau et al., 2017; Um et al., 2017). In addition, the CRU and APHRO products are used instead of station data accessible from the China Meteorological Administration, owing to the study area including in the domain of East Asia, extending beyond the territory of China.

2.2 Models and experiments

In this study, we used five RCMs, namely, HadGEM3-RA (Hadley Centre Global Environmental Model version 3 regional climate model), MM5 (Fifth-Generation Penn State/NCAR Mesoscale Model), WRF (Weather Research and Forecasting model), RegCM4 (Regional Climate Model version 4), and RSM (Regional Spectral Model), for East Asian climate experiments (Table 1). They are derived from the CORDEX East Asia experiment that is able to provide a global holistic framework for regional climate projections so as to understand their uncertainties as well as provide model evaluation. Moreover, the selected five RCMs have been demonstrated to have abilities to reproduce the regional climate over East Asia and have been used for modeling and predicting extreme climate as well as investigating physical processes of East Asia climate (Cha and Lee, 2009; Cha et al., 2011; Hong and Yhang, 2010; Park et al., 2008; Yhang and Hong, 2008). The spatial resolution of the data is 50 km (except HadGEM3-RA is 0.44°), and the whole CORDEX-EA domain includes East Asia, India, the western Pacific Ocean, and the northern part of Australia, as shown in Fig. 1. Model configurations including physical schemes are summarized in Table 1. Please refer to the references Suh et al. (2012) and Park et al. (2016) for more details on RCMs used in this study.

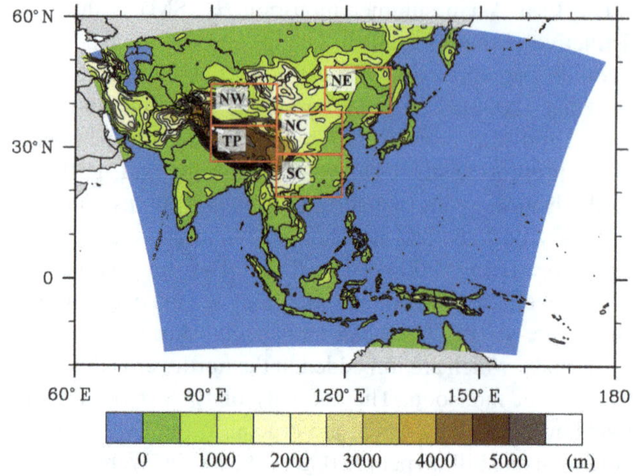

Figure 1. The simulation domain of CORDEX-EA and the topography of the regional climate models (m). The boxes illustrate the five selected subregions over China: northeastern China (NE), northern China (NC), southern China (SC), northwestern China (NW), and the Tibetan Plateau (TP).

In this study, two types of current climate experiments from five RCMs were performed, including the evaluation (hereafter EVAL) experiment from 1989 to 2008 and the historical (HIST) experiment from 1980 to 2005. The EVAL experiment acquires initial and boundary conditions from the National Centers for Environmental Prediction reanalysis, whereas the HIST experiment is forced by the Atmosphere–Ocean coupled Hadley Center Global Environmental Model version 2 (HadGEM2-AO) simulation. HadGEM2-AO (1.875° × 1.25° horizontal resolution) has been used for climate simulations in a CMIP5 set of long-term experiments and has been demonstrated to have a reasonable ability to capture the East Asian climatology (Baek

Table 1. RCMs used in this study* (Park et al., 2016).

	HadGEM3-RA	RegCM4	MM5	WRF	RSM
Resolution	0.44°	50 km	50 km	50 km	50 km
Dynamic process	Nonhydrostatic	Hydrostatic	Nonhydrostatic	Nonhydrostatic	Hydrostatic
Convective scheme	Revised mass flux scheme	MIT–Emanuel	Kain–Fritch II	Kain–Fritch II	Simplified Arakawa–Schubert
Land surface parameterization	MOSES2	CLM3	CLM3	NOAH	NOAH
Planetary boundary layer	MOSES2 nonlocal	Holtslag	YSU	YSU	YSU
Spectral nudging	No	Yes	Yes	Yes	Yes
Center of research	MOHC	ICTP	NCAR	NCAR	YSU
References	Davies et al. (2005)	Giorgi et al. (2012)	Cha and Lee (2009)	Skamarock et al. (2005)	Hong et al. (2013)

* MOSES: Met Office Surface Exchange Scheme, CLM: Community Land Model, NOAH: Noah Land Surface Model, YSU: Yonsei University scheme, MOHC: the Met Office Hadley Centre, ICTP: The International Centre for Theoretical Physics, NCAR: National Center for Atmospheric Research.

et al., 2013; The HadGEM2 Development Team, 2011; Sperber et al., 2013). The future climate simulation is driven by the HadGEM2-AO under the RCP 4.5 scenario, which is an intermediate scenario and a cost-minimizing pathway for which total radiative forcing is stabilized at 4.5 $\mathrm{W\,m^{-2}}$ in the year 2100 (Thomson et al., 2011). The reference period from 1980 to 1999 and the scenario period from 2030 to 2049 are analyzed for climate change research in this study.

The multi-model ensemble (MME) mean, defined as the pointwise arithmetic average over all individual model climatologies, narrows down inter-RCM uncertainties because of their differences in model structures and physics. To further evaluate the model performance on smaller spatial scales, we evaluate the performance of RCMs over five selected subregions (as shown in Fig. 1), namely, northeastern China (40–50° N, 115–130° E), northern China (30–40° N, 105–120° E), southern China (22–30° N, 105–120° E), northwestern China (35–45° N, 80–95° E), and the Tibetan Plateau (28–35° N, 80–95° E).

2.3 Analysis methods

The root-mean-square error (RMSE), bias, and Taylor diagram analysis are selected for statistical measurements of the performance for the individual RCM and the MME. The former two indexes are used for evaluating the ability of models to reproduce annual and seasonal mean of climatology. The Taylor diagram is designed to quantify the degree of correspondence between the modeled and observed behavior by plotting a 2-D graph with three statistics (correlation coefficient, SD, and RMSE). In the Taylor diagram, a small distance between the reference and compared objects indicates close agreement (Baker and Taylor, 2016; Sun et al., 2015). In general, the Taylor diagram enables statistics for different fields (with different units) to be shown in a single plot, facilitating the comparative assessment of different models (Taylor, 2001).

Uncertainty in projected climate change mainly arises from the internal variability of the climate system, the model uncertainty, and the scenario uncertainty (Niu et al., 2015; Woldemeskel et al., 2016). In this study, all RCMs are driven by the same GCM under the same scenario, and thus, the uncertainty of the climate projections is mainly caused by inter-RCM and internal variability. The method developed by Hawkins and Sutton (2009, 2011) is used to separate these two sources of uncertainty. Here we give a brief illustration.

1. Firstly, a smooth fourth-order polynomial is used to fit each individual simulation over the years 1980–2049 by using an ordinary least squares method. Then the raw simulation of each model $X_{m,t}$ for the model m and year t can be expressed by

$$X_{m,t} = z_{m,t} + c_m + \varepsilon_{m,t}, \tag{1}$$

where $z_{m,t}$ represents the simulation from the smooth fit for the model m and year t minus the reference data; the reference data are denoted by c_m, and the residual (internal variability) is denoted by $\varepsilon_{m,t}$. Here the reference data are the mean of the simulation from the smooth fit during the years 1980–1999.

2. The RCMs are weighted by their performance in simulating the current climate from the mean of 1980–1999, up to year 1999. Thus, each model is weighted according to

$$w_m = \frac{1}{z_{\mathrm{obs}} + \left|z_{m,1999} - z_{\mathrm{obs}}\right|}, \tag{2}$$

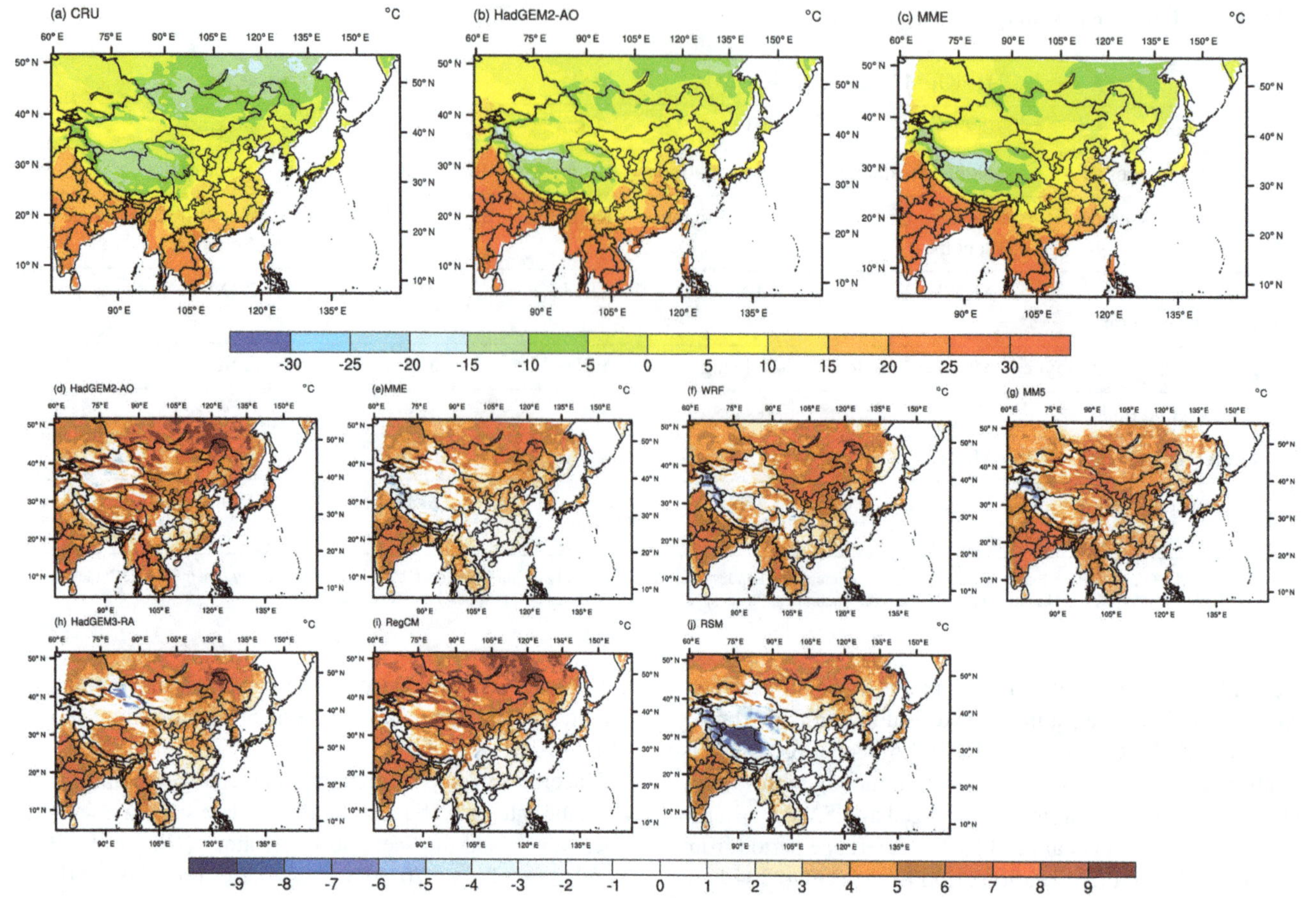

Figure 2. Spatial distributions of annual average temperature (°C) from CRU **(a)**, the driving GCM HadGEM2-AO **(b)**, multi-model ensemble **(c)**, and temperature biases (°C) of the driving GCM HadGEM2-AO **(d)**, multi-RCM ensemble **(e, f)** and five RCMs **(g–k)** during 1980–2005.

where $z_{m,1999}$ is the model climate change in 1999 relative to 1980–1999, and z_{obs} is an observational estimate derived from fitting a similar fourth-order polynomial to observations. The normalized quantities (W_m) of these weightings can be expressed as

$$W_m = \frac{w_m}{\sum_m w_m}. \tag{3}$$

3. Internal variability (V, as shown in Eq. 4) is defined as the multi-model mean of the variance of the residuals from the fit for each model:

$$V = \sum_m W_m \text{var}_t(\varepsilon_{m,t}), \tag{4}$$

$$M(t) = \text{var}_m^{\text{W}}(z_{m,t}). \tag{5}$$

4. Inter-model variability (M, as shown in Eq. 5) is estimated from the weighted variance (var^{W}) in different RCM prediction fits ($z_{m,t}$), where $\text{var}_t(.)$ and $\text{var}_m(.)$ indicate the variance across time and the model, respectively.

5. It was assumed that the two sources of uncertainty can be treated independently (i.e., no interaction exists between them). Thus, the total variability V_{T} is

$$V_{\text{T}}(t) = V + M(t). \tag{6}$$

6. The fraction of variance of internal variability and model uncertainty are defined as $V/V_{\text{T}}(t)$ and $M(t)/V_{\text{T}}(t)$, respectively.

3 Results

3.1 Climatology for the historical climate

3.1.1 Historical annual average climate evaluation

Figure 2 shows the annual average temperature of CRU, the driving GCM HadGEM2-AO, and the multi-model ensemble, as well as the temperature biases of five RCMs driven by HadGEM2-AO from 1980 to 2005. Obviously, both the MME and five RCMs can capture the spatial pattern of annual mean temperature in China, with a decreasing south–

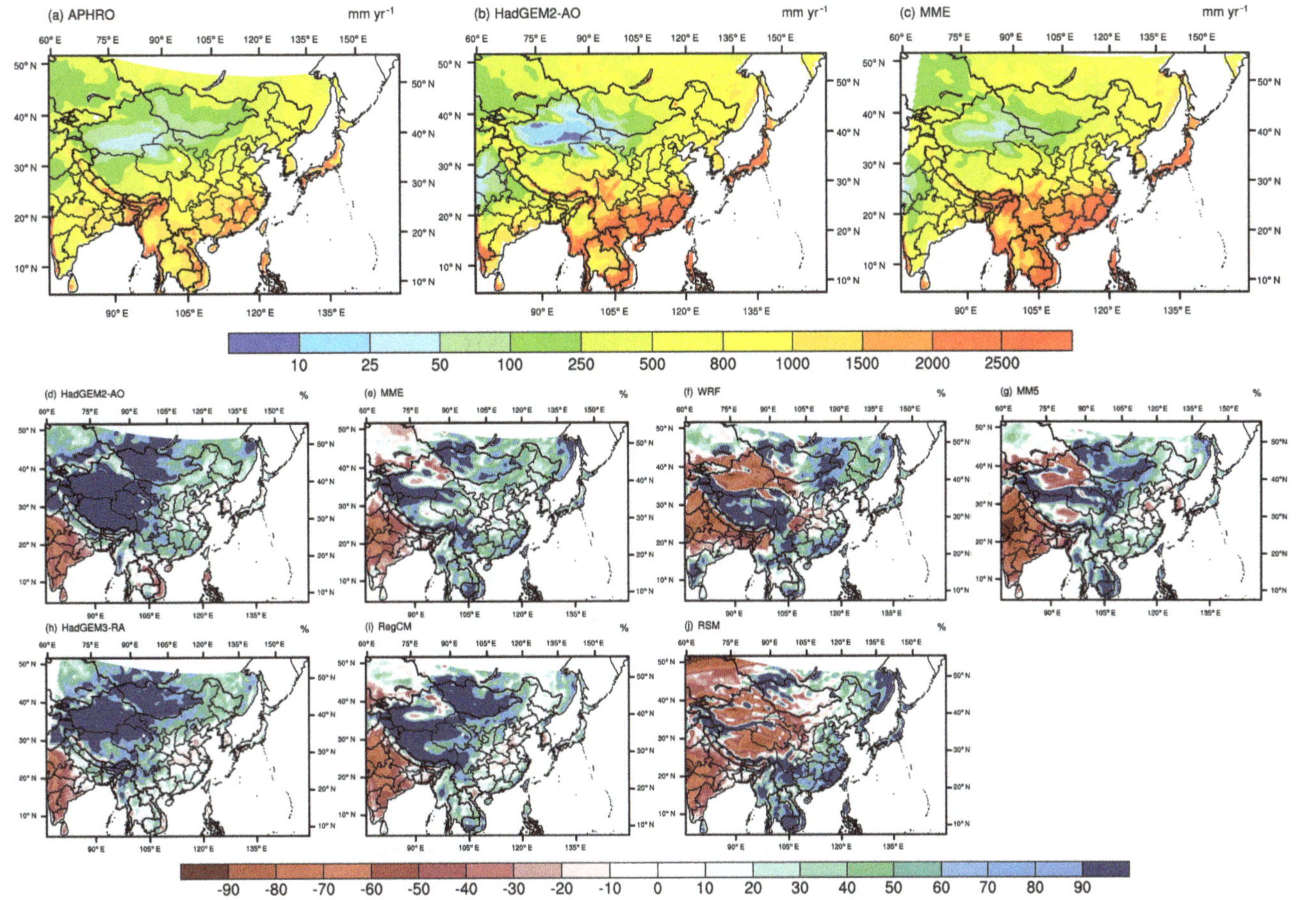

Figure 3. Spatial distributions of annual average precipitation (mm/year) from APHRO **(a)**, the driving GCM HadGEM2-AO **(b)**, and MME **(c)**, and precipitation biases (%) of the driving GCM HadGEM2-AO **(d)**, MME **(e)**, and five RCMs **(f–j)** during 1980–2005.

north gradient and a cold area in the Tibetan Plateau. Moreover, the MME presents the best results overall to reproduce the temperature spatial distribution and provides less than 1 °C temperature biases over most areas of China. However, all RCMs generally overestimated the mean temperature over most of the domain; in particular, warmer mean temperature is simulated by MM5 and HadGEM3-RA. The only exception is that RSM underestimated the mean temperature over the Tibetan Plateau.

The RCMs provide reasonably accurate simulations for mean temperature during the historical period, but they are less successful at reproducing precipitation. Figure 3 shows the annual average precipitation from APHRO, HadGEM2-AO, and MME, as well as the precipitation biases from five RCMs in the current period. It is found that the spatial pattern of annual mean precipitation is characterized by a decreasing southeastern–northwestern gradient over China, which can be successfully simulated by all RCMs. However, quite large precipitation biases are found in different RCMs. For instance, WRF underestimated the annual mean precipitation in northwestern China, where mean precipitation was overestimated by the other RCMs. In comparison with the simulation from each RCM, the MME is better in reproducing annual mean precipitation over most subregions in China.

The spatial variability statistics of the models in reproducing the annual mean temperature and precipitation by the Taylor plot (Taylor, 2001) are exhibited in Fig. 4. The temperature simulations of the five RCMs exhibit a good spatial pattern correlation, ranging from 0.83 to 0.96, whereas the precipitation simulation shows a relatively extensive range of spatial pattern correlations from 0.29 to 0.93. In addition, the MME is superior to most RCMs in capturing the spatial variability of these climate variables, as reflected by the higher spatial correlation coefficient and the lower RMSE. There are several reasons for this phenomenon, as also noted by other scholars in their studies on model intercomparisons (Huttunen et al., 2017; Phillips and Gleckler, 2006; Rozante et al., 2014). On the one hand, the bias of a simulated climate field is symptomatic of random errors to a certain extent, and the MME may reduce or counteract this error from the RCM. On the other hand, the pointwise variations of the climate field are smoothed out by averaging, thereby filtering regional-scale simulations, which are difficult for current climate models to capture.

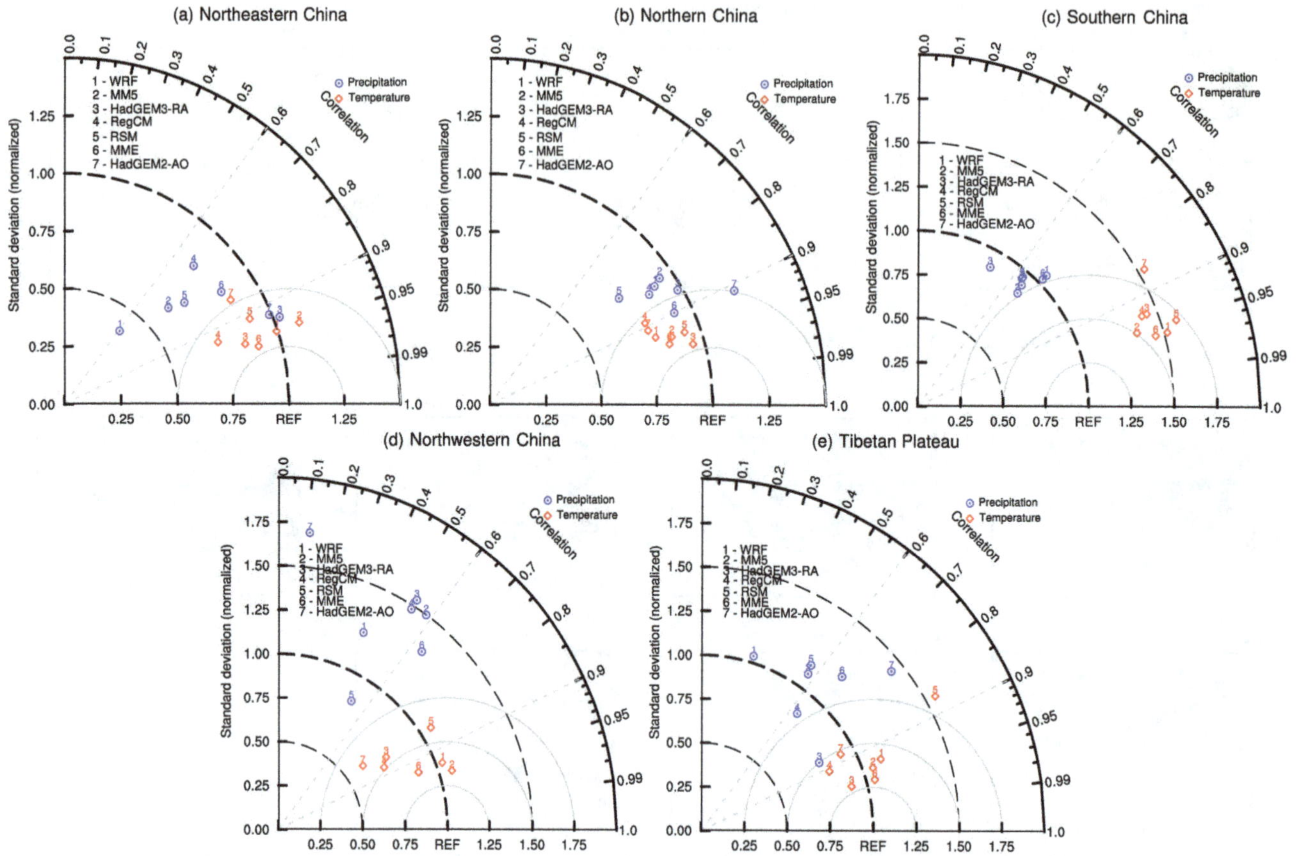

Figure 4. The Taylor diagrams which evaluate the skill of the models in reproducing the annual average temperature and precipitation over the five regions of China, using the CRU (for temperature) and APHRO (for precipitation) data as the reference, are shown. The azimuthal axis shows the pattern spatial correlation. The redial distance from the origin represents the spatial variability, whereas the distance from the reference observation (REF) is the centered RMSE difference between the simulations and observations.

3.1.2 Interannual and seasonal variability

The ability of a climate model to capture realistic interannual variability is a critical measure of its performance. The time series of the annual mean temperature and precipitation from RCMs are compared with CRU and APHRO in Fig. 5. Evidently, the interannual variation of the climatology is generally well reproduced in the MME. In the evaluation experiment for 1989–2008, the correlation coefficient of the annual climatology time series at five subregions between the observation and simulation from the MME ranges from 0.52 to 0.78 for temperature and from 0.50 to 0.87 for precipitation. The correlation coefficient is always lower in western China compared with that in eastern China, especially in the Tibetan Plateau. In the historical experiment from 1980 to 2005, the MME shows a better performance, in comparison with the RCMs, which have difficulty in reproducing the interannual variability for precipitation because of the impact of the driving GCM.

The temporal distributions of precipitation and temperature throughout the year are important for ecosystems and water resource management. To evaluate the RCM's ability to capture the seasonal variability of climatologies, the seasonal cycles of simulated temperature and precipitation averaged over five subregions in China are examined (Fig. 6). It is evident that the seasonal pattern of precipitation features one peak in June over southern China and in July over the rest of the regions, which can be successfully reproduced by all RCMs and MME. However, the inter-model difference in simulated precipitation is large. For instance, monthly precipitation is always underestimated by WRF and overestimated by MM5 and HadGEM3-RA; an especially larger bias is shown in summer. Among the five RCMs, RegCM is the one with the best ability to simulate the seasonal cycles of precipitation. The MME generally provides the most accurate simulation for the temporal distribution of precipitation, in comparison with the RCMs. As for the temperature, the RCMs can capture its temporal pattern over all subregions. Moreover, mean temperatures in different months are always overestimated by most RCMs. However, the MME reduces the bias from the RCMs and therefore generates a more accurate temporal distribution for mean temperature.

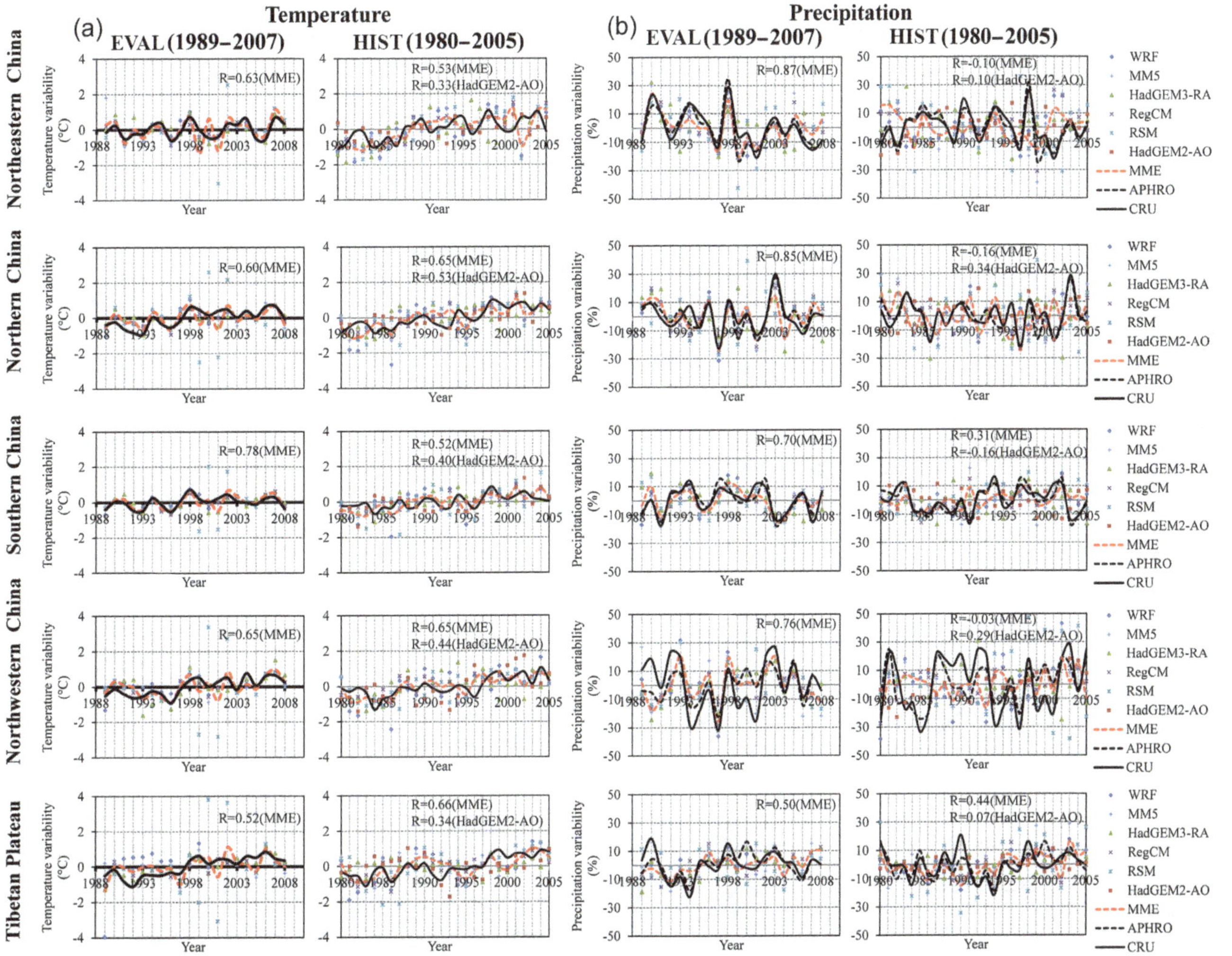

Figure 5. Temporal evolution of the annual mean temperature **(a)** and precipitation **(b)** in RCM simulations and observations over the five subregions during the 1989–2008 (EVAL) and 1980–2005 (HIST) periods. The correlation coefficients between the ensemble of RCMs and the observations are shown at the top right of each panel.

3.1.3 The added values for RCMs

The added values for high-resolution RCMs were confirmed by comparing the performance of RCMs and driving GCM HadGEM2-AO in reproducing annual mean precipitation and temperature during the historical period. According to Figs. 4–6, it is found that the added value for RCMs depends largely on the climate variable and the area of interest. The added value of the RCMs in comparison with the driving GCM was evident in terms of annual mean temperature over all five subregions, with a higher spatial and temporal correlation coefficient and less seasonal bias for all five RCMs. Compared with the driving GCM simulations, the historical precipitation over southern China, northwestern China, and the Tibetan Plateau was improved in most RCMs. The exceptions are over northeastern China and northern China, where a higher performance is shown for the driving global climate model. In reality, the added value in RCM simulations is mainly concerned with a better representation of the spatial variability of surface climate statistics, particularly in areas with small-scale land surface forcing such as orographic and coastal features. Thus, the added value in RCM simulations is commonly significant in regions with fine-scale surface forcing, whereas the performance of RCMs is less improved or even worse than that of the driving GCM over relatively flat regions. For instance, Prömmel and Geyer (Prömmel et al., 2010) also found that some RCMs show larger biases compared to their driving GCM in relatively flat regions surrounding the Alps, especially in summer. In most cases, the five RCMs perform better than the driving GCM HadGEM2-AO. It needs to be emphasized that the better model performance tends to increase confidence in the future climate projections from RCMs.

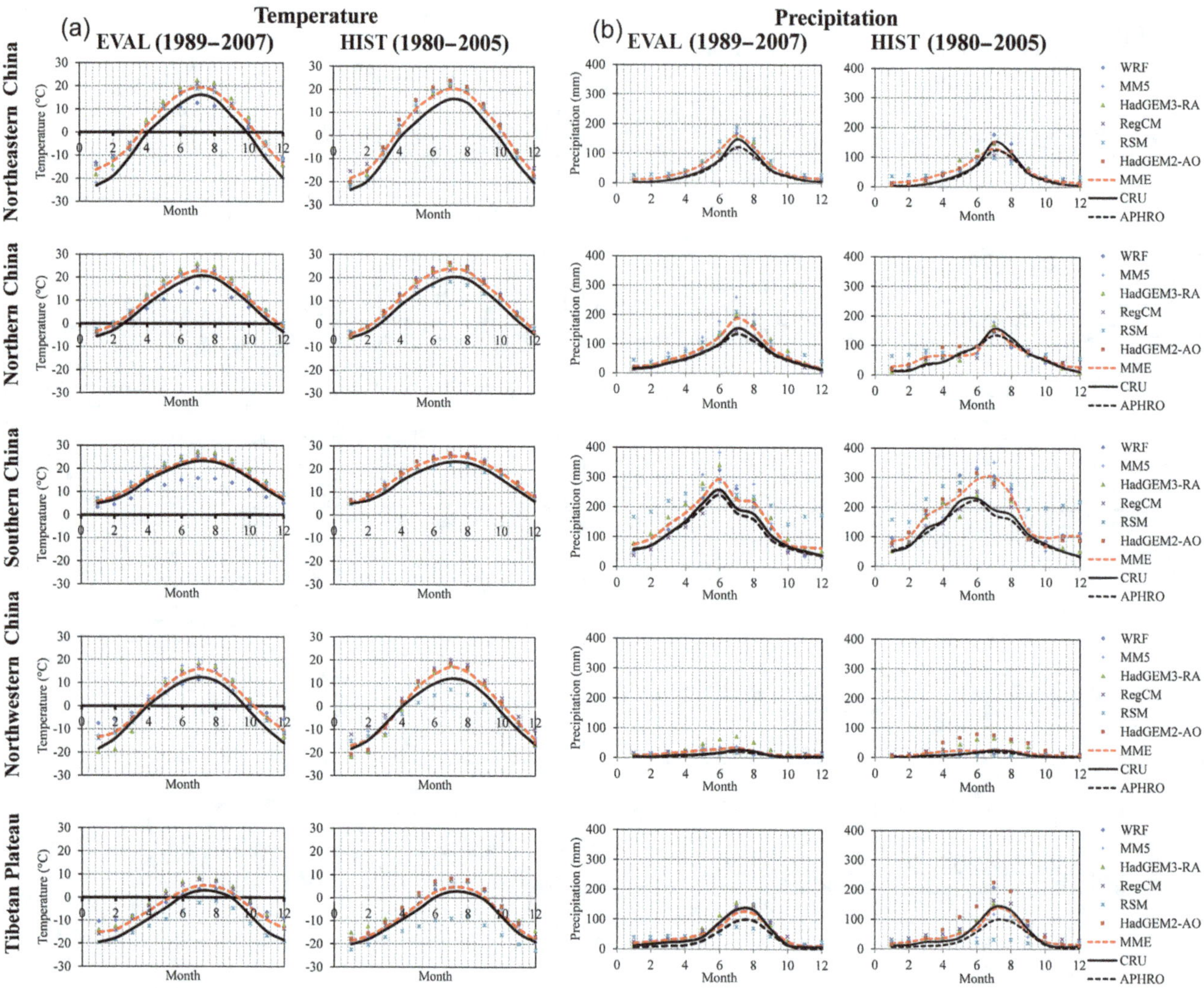

Figure 6. Observed and simulated monthly mean temperature **(a)** and precipitation **(b)** over the five subregions during the 1989–2008 (EVAL) and 1980–2005 (HIST) periods.

3.2 Multi-RCM future climate projection

3.2.1 Future change in climatology

According to Fig. 7, showing the projections for mean temperature from the driving GCM, RCMs, and the MME, similar warming trends are detected over the entire domain from 2030 to 2049 under RCP4.5 scenario. All five models project substantially significant warming while exhibiting different spatial patterns. The increases in annual temperature by the MME are 1.3, 1.0, 0.9, 1.2, and 1.3 °C over northeastern China, northern China, southern China, northwestern China, and the Tibetan Plateau subregions, respectively. The warming in northern and western China is more significant than that in southern China, especially in northeastern China and the Tibetan Plateau, which is similar to the results from previous studies (Sun et al., 2015; You et al., 2014; Zhou and Yu, 2006). Moreover, the magnitude for the increase in annual temperature over a given subregion varies with the RCM. For instance, the projected increase in mean temperature over the Tibetan Plateau ranges from 0.9 to 1.6 °C.

Figure 8 shows the spatial distributions of changes in annual mean precipitation (RCP4.5 – baseline). During the period 2030–2049, increased precipitation is projected by the MME and most RCMs over China. Moreover, the projected spatial pattern from the driving GCM, the MME, and RCMs is nearly consistent, with the most prominent increase in precipitation over northern and northwestern China and slightly increased precipitation over the rest of the regions. The only exception is the results from WRF, which projected a declined mean precipitation over China. In particular, a wider range for the change in projected annual precipitation is shown over the Tibetan Plateau. This is related to the fact that there is a significant difference in the projected precipitation change between WRF and the other RCMs. Therefore,

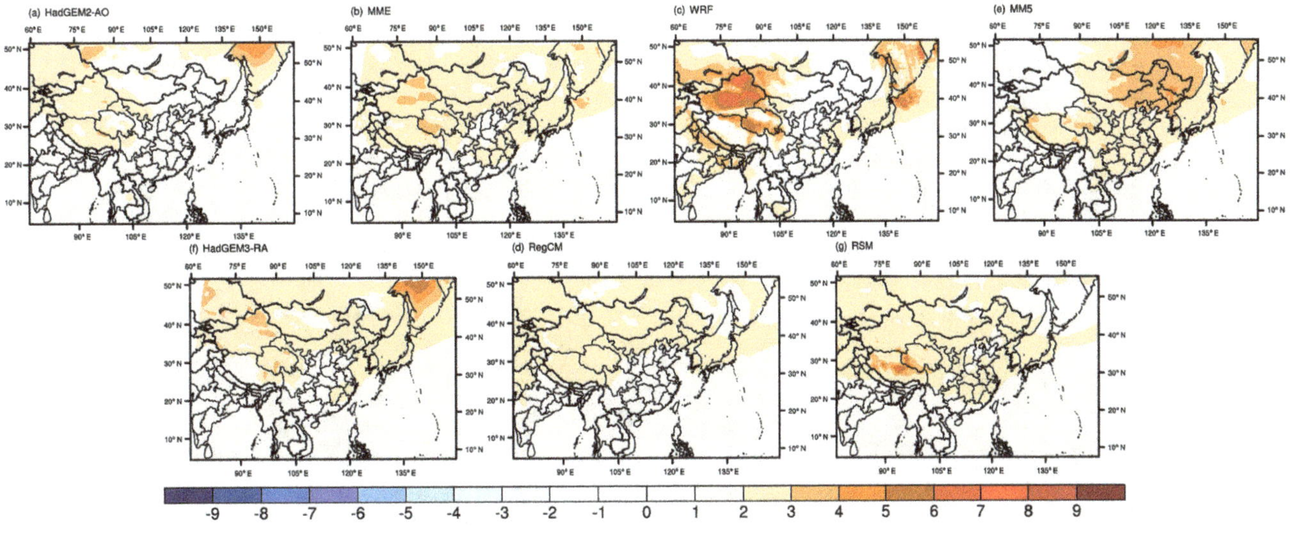

Figure 7. Projected future changes (RCP4.5 – baseline) in surface air temperature by the forcing GCM HadGEM2-AO, the MME, and each of the five RCMs.

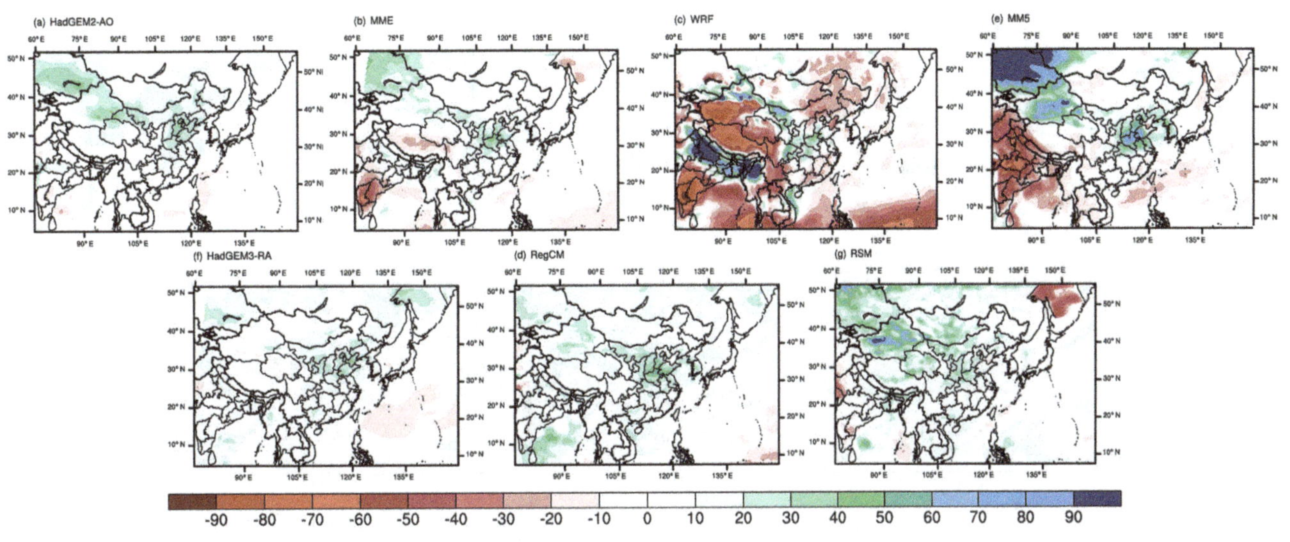

Figure 8. Projected future changes ((RCP4.5 – baseline)/baseline × 100 %) in precipitation by the forcing GCM HadGEM2-AO, the MME, and each of the five RCMs.

the projected change in annual precipitation over the Tibetan Plateau should be treated with caution. Additionally, opposite signals for projected changes in average precipitation between the MME and the driving GCM are detected over southern China, northeastern China, and the Tibetan Plateau (Table 2). Particularly the differences in projection from the two methods above are largest at the Tibetan Plateau, up to about 10 %.

3.2.2 Change in seasonal cycle

The future changes of temperature and precipitation are characteristic of regionality and seasonality. The ensemble projection (Fig. 9) indicates that the monthly temperature change over five subregions in China ranges from 0.3 to 2.2 °C under the RCP4.5 scenario. A more remarkable warming in cold months from November to March is detected by all RCMs. The seasonal cycle of temperature change in MME is also similar to that of the driving GCM HadGEM2-AO. Most RCMs project positive monthly precipitation changes for summer (from June to August) over China, with the exception of the Tibetan Plateau. However, the projected monthly precipitation change by MME has a larger magnitude and variability than the driving GCM. This phenomenon concerns the significance of the model physics and processes for future climate projection. The configuration of each RCM is shown in Table 1. For each RCM, op-

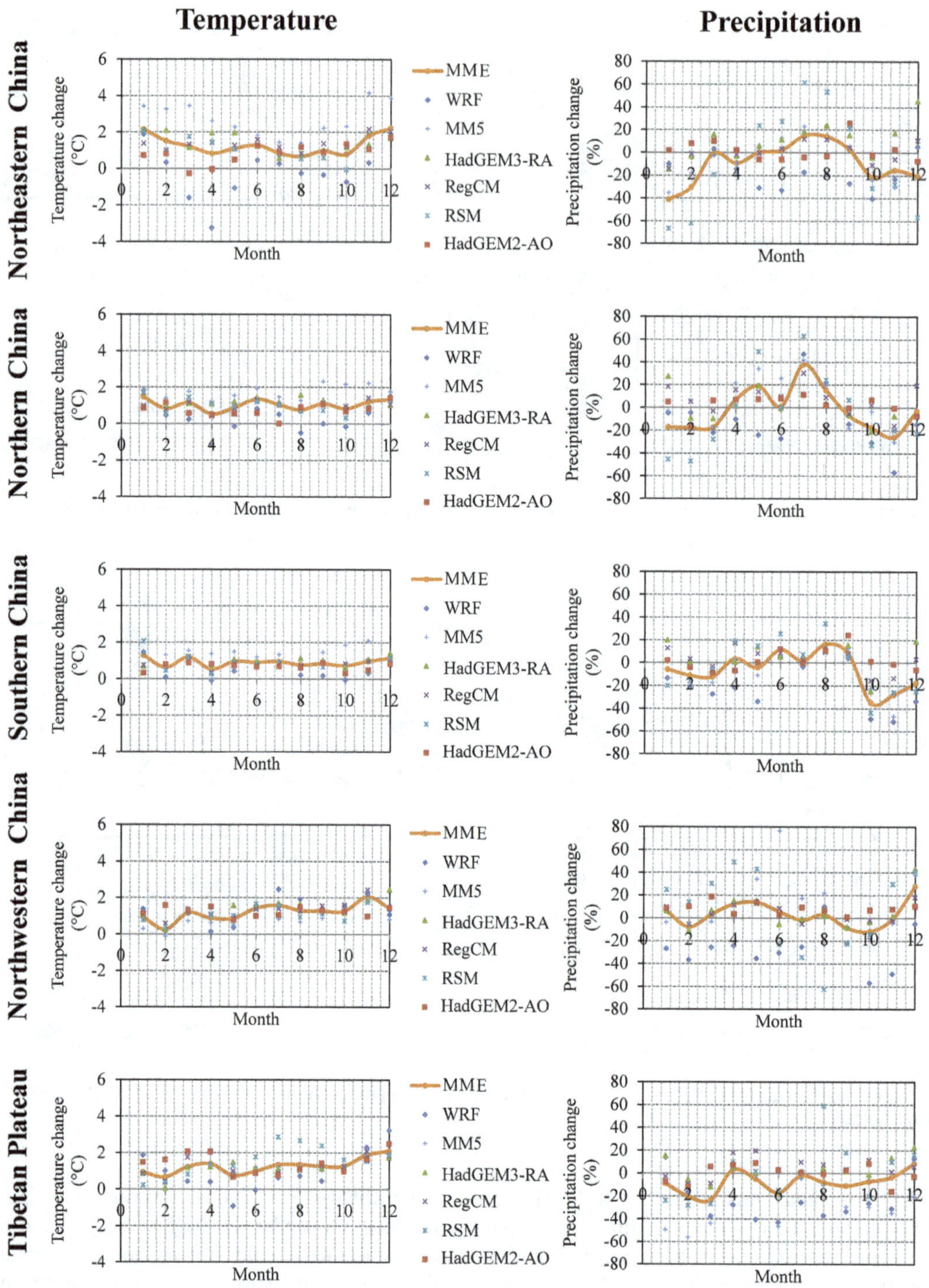

Figure 9. Projected future changes in monthly mean temperature and precipitation by the forcing GCM HadGEM2-AO, the MME, and each of the five RCMs under the RCP4.5 scenario.

timal schemes of the dynamical and physical processes were determined through model sensitivity analysis (Suh et al., 2012). In general, convective parameterization is one of the most important and sensitive processes in a RCM (Huang and Gao, 2017). Land surface parameterizations, as well as parameterizations over the ocean, are also very important because they control the quantity of water vapor flux entering into the atmosphere from the earth's surface (Zhao and Li, 2015). Thus, the phenomenon above could be attributed to the difference in convective parameterization and land sur-

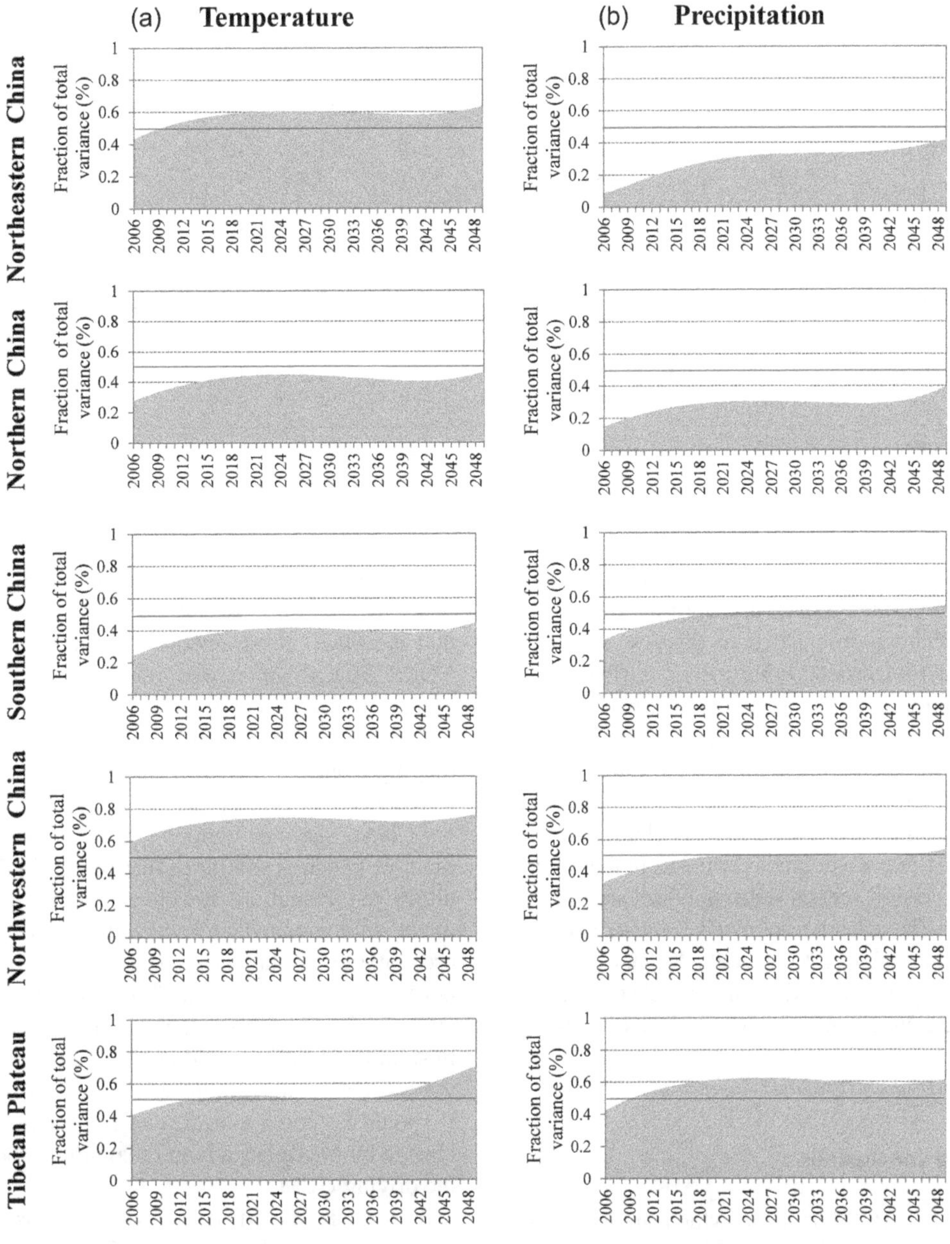

Figure 10. Fraction of total variance in future temperature **(a)** and precipitation **(b)** projections explained by inter-model variability (gray) and internal variability (white) over the five subregions.

face parameterizations, as well as parameterizations over the ocean between GCMs and RCMs. On the other hand, the discrepancies between the RCMs and the driving GCM indicate that the RCM projections are sensitive to local and regional processes and the corresponding methods incorporated in the model (Diallo et al., 2012; Saini et al., 2015).

3.2.3 Inter-RCM variability of multi-RCM projections

The uncertainties of regional climate projection arise from the GCMs, emission scenarios, RCMs, and internal variability for natural climate. In this study, the regional future climate is projected by using five RCMs forced with the same GCM under an intermediate scenario (RCP4.5). Consequently, the contributions of inter-RCM variability and internal variability to total uncertainty in the projections are analyzed in this section.

The contributions of the model uncertainty and natural climate internal variability to the total prediction uncertainty are estimated by the method proposed by Hawkins and Sutton (2009). The results for the five subregions are shown in Fig. 10. The relative importance of model uncertainty increases with prediction lead time over all subregions. For

Table 2. The future changes in average temperature (T; °C) and precipitation (P; %) for the five subregions. The ensemble averages for each statistic are given in the second line. The projections by the forcing GCM are given in the last line.

		WRF	MM5	HadGEM3-RA	RegCM	RSM	Ensemble	HadGEM2-AO
Northeastern China	T (°C)	0.2	2.7	1.4	1.4	1.1	1.3	0.8
	P (%)	−21.7	8.2	13.0	4.4	7.1	1.5	−0.4
Northern China	T (°C)	0.3	1.7	1.1	1.0	1.0	1.0	0.8
	P (%)	−1.5	15.1	3.1	10.2	3.3	6.1	4.9
Southern China	T (°C)	0.5	1.5	1.0	0.8	0.8	0.9	0.7
	P (%)	−14.6	−1.6	4.8	4.9	1.3	−1.5	2.3
Northwestern China	T (°C)	1.3	0.8	1.5	1.3	1.1	1.2	1.2
	P (%)	−27.0	19.4	2.2	4.7	8.9	3.6	7.2
Tibetan Plateau	T (°C)	0.9	1.4	1.2	1.3	1.6	1.3	1.4
	P (%)	−31.6	−17.8	2.4	6.4	7.4	−7.8	2.1

temperature, the model uncertainty is the primary source of uncertainty over northeastern China, northwestern China, and the Tibetan Plateau from 2030 to 2049, reaching up to 70 %. The model uncertainty minimally contributes (approximately 40 %) to the total uncertainty over northern and southern China before the middle of the 21st century. For the uncertainty in projected precipitation, the internal variability is the dominant uncertainty source over most regions, except for the Tibetan Plateau, where the model uncertainties reached up to 60 %. The uncertainties come from the driving GCMs, and the emission scenarios are not discussed in this study, although they have been recognized as important components for total uncertainty (Déqué et al., 2012). Further research on uncertainty quantification on the basis of different GCMs, RCMs, and emission scenarios is needed in the future.

4 Summary and conclusions

In this research, five RCM models, which are simulated within the CORDEX-EA initiative at 50 km resolution, with boundary forcing from a CMIP5 global model, applying the RCP4.5 scenario, are employed to derive the future climate change signal for China and five smaller selected investigation areas. Meanwhile, the contributions of the model uncertainty and natural climate internal variability to the total prediction uncertainty are quantified.

The control runs of CORDEX-EA RCMs revealed an overall reasonable representation of the mean climate properties when compared with the observational gridded dataset. All RCMs generally provide warm biases, whereas the MME demonstrates the overall best performance, with less than a 1 °C annual average temperature bias over most areas in China. Similarly, the MME outperformed the individual RCMs in reproducing the observed spatial pattern of precipitation. Moreover, five RCMs perform better than the driving GCM HadGEM2-AO in reproducing annual and seasonal precipitation over most subregions. Therefore, it is concluded that the MME constructed based on the set of RCMs from CORDEX-EA can be used to provide useful information on climate projections over East Asia.

For 2030 to 2049, MME indicated consistent warming, ranging from 0.9 to 1.6 °C, in the entire domain and more pronounced warming was detected in northern and western China. Seasonal temperature changes drastically in cold months, which is similar to that of the driving GCM. In addition, the annual precipitation is likely to increase in most of the subregions. The projected spatial pattern for annual precipitation is characterized by a prominent increase over northern and northwestern China and slightly increased precipitation over the rest of the regions. Moreover, precipitation in summer months is predicted to consistently increase over the entire domain, with the exception of the Tibetan Plateau. It should be noted that the projected monthly precipitation change by MME has a larger magnitude and variability than the driving GCM.

This study identified the contributions of model uncertainty and internal variability. The uncertainty in projected temperature mainly arises from the internal variability over northern and southern China, whereas the model uncertainty is clearly dominant over the other three subregions, explaining approximately 70 % of the total uncertainty. For precipitation, the internal variability is dominant over most regions except for the Tibetan Plateau, in which model uncertainties reached up to 60 %. Model uncertainty also increases with prediction lead time over all subregions. RCM simulation results are also influenced by the internal physics and boundary conditions from GCMs, as discussed in others' studies (Mariotti et al., 2011; Syed et al., 2012). More reliable future climate information and uncertainty quantification could be provided by coupling large ensembles of GCMs and RCMs under different emission scenarios.

Competing interests. The authors declare that they have no conflict of interest.

Acknowledgements. This work was supported by the National Key R&D Program of China (grant no. 2016YFC0402706, 2016YFC0402710), the National Natural Science Foundation of China (no. 41501015, 51539003, 51421006, 51509263), and the Fundamental Research Funds for the Central Universities (2016B00114). We acknowledge the CORDEX-East Asia databank, which is responsible for the CORDEX dataset, and we thank the National Institute of Meteorological Research (NIMR), three universities in the Republic of Korea (Seoul National Uni., Yonsei Uni., Kongju National Uni.), and other cooperative research institutes in East Asia region for producing and making available their model output. We also thank two anonymous reviewers for their constructive and insightful comments that helped to improve the original manuscript.

Edited by: Thomas Kjeldsen

References

Alfieri, L., Burek, P., Feyen, L., and Forzieri, G.: Global warming increases the frequency of river floods in Europe, Hydrol. Earth Syst. Sci., 19, 2247–2260, https://doi.org/10.5194/hess-19-2247-2015, 2015.

Baek, H. J., Lee, J., Lee, H. S., Hyun, Y. K., Cho, C., Kwon, W. T., Marzin, C., Gan, S. Y., Kim, M. J., Choi, D. H., Lee, J., Lee, J., Boo, K. O., Kang, H. S., and Byun, Y. H: Climate change in the 21st century simulated by HadGEM2-AO under representative concentration pathways, Asia-Pacific J. Atmos. Sci., 49, 603–618, https://doi.org/10.1007/s13143-013-0053-7, 2013.

Baker, N. C. and Taylor, P. C.: A Framework for Evaluating Climate Model Performance Metrics, J. Climate, 29, 1773–1782, https://doi.org/10.1175/JCLI-D-15-0114.1, 2016.

Cha, D. H. and Lee, D. K.: Reduction of systematic errors in regional climate simulations of the summer monsoon over East Asia and the western North Pacific by applying the spectral nudging technique, J. Geophys. Res.-Atmos., 114, D14108, https://doi.org/10.1029/2008JD011176, 2009.

Cha, D. H., Jin, C. S., Lee, D. K., and Kuo, Y. H.: Impact of intermittent spectral nudging on regional climate simulation using Weather Research and Forecasting model, J. Geophys. Res., 116, D10103, https://doi.org/10.1029/2010JD015069, 2011.

Chen, J. and Bordoni, S.: Intermodel spread of East Asian summer monsoon simulations in CMIP5, Geophys. Res. Lett., 41, 1314–1321, https://doi.org/10.1002/2013GL058981, 2014.

Davies, T., Cullen, M. J. P., Malcolm, A. J., Mawson, M. H., Staniforth, A. , White, A. A., and Wood, N.: A new dynamical core for the Met Office's global and regional modelling of the atmosphere, Q. J. Roy. Meteor. Soc., 131, 1759–1782, https://doi.org/10.1256/qj.04.101, 2005.

Déqué, M., Rowell, D. P., Lüthi, D., Giorgi, F., Christensen, J. H., Rockel, B., Jacob, D., Kjellström, E., de Castro, M., and van den Hurk, B.: An intercomparison of regional climate simulations for Europe: assessing uncertainties in model projections, Climatic Change, 81, 53–70, https://doi.org/10.1007/s10584-006-9228-x, 2007.

Déqué, M., Somot, S., Sanchez-Gomez, E., Goodess, C. M., Jacob, D., Lenderink, G., and Christensen, O. B.: The spread amongst ENSEMBLES regional scenarios: regional climate models, driving general circulation models and interannual variability, Clim. Dynam., 38, 951–964, https://doi.org/10.1007/s00382-011-1053-x, 2012.

Deser, C., Phillips, A., and Bourdette, V., and Teng, H.: Uncertainty in climate change projections: the role of internal variability, Clim. Dynam., 38, 527–546, https://doi.org/10.1007/s00382-010-0977-x, 2012.

Diallo, I., Sylla, M. B., Giorgi, F., Gaye, A. T., and Camara, M.: Multimodel GCM-RCM Ensemble-Based Projections of Temperature and Precipitation over West Africa for the Early 21st Century, Int. J. Geophys., 2012, 972896, https://doi.org/10.1155/2012/972896, 2012.

Ding, Y.: Seasonal march of the East-Asian summer monsoon, in: East Asian Monsoon, edited by: Chang, C. P., Mainland Press, Singapore, 3–53, 2004.

Ding, Y. and Chan, J. C. L.: The East Asian summer monsoon: an overview, Meteorol. Atmos. Phys., 89, 117–142, https://doi.org/10.1007/s00703-005-0125-z, 2005.

Dong, N. D., Jayakumar, K. V., and Agilan, V.: Impact of Climate Change on Flood Frequency of the Trian Reservoir in Vietnam Using RCMS, J. Hydrol. Eng., 23, 05017032, https://doi.org/10.1061/(ASCE)HE.1943-5584.0001609, 2018.

Gao, J., Hou, W., Xue, Y., and Wu, S.: Validating the dynamic downscaling ability of WRF for East Asian summer climate, Theor. Appl. Climatol., 128, 241–253, https://doi.org/10.1007/s00704-015-1710-9, 2017.

Gao, X. J., Pal, J. S., and Giorgi, F.: Projected changes in mean and extreme precipitation over the Mediterranean region from a high resolution double nested RCM simulation, Geophys. Res. Lett., 33, L03706, https://doi.org/10.1029/2005GL024954, 2006.

Giorgi, F. and Mearns, L. O.: Introduction to special section: Regional climate modeling revisited, J. Geophys. Res., 104, 6335–6352, https://doi.org/10.1029/98JD02072, 1999.

Giorgi, F., Jones, C., and Asrar, G. R.: Addressing climate information needs at the regional level: the CORDEX framework, WMO Bulletin, 58, 175–183, 2009.

Giorgi, F., Coppola, E., Solmon, F., Mariotti, L., Sylla, M. B., Bi, X., Elguindi, N., Diro, G. T., Nair, V., Giuliani, G., Turuncoglu, U. U., Cozzini, S., Güttler, I., O'Brien, T. A., Tawfik, A. B., Shalaby, A., Zakey, A. S., Steiner, A. L., Stordal, F., Sloan, L. C., and Brankovic, C.: RegCM4: model description and preliminary tests over multiple CORDEX domains, Clim. Res., 52, 7–29, https://doi.org/10.3354/cr01018, 2012.

Gu, H., Wang, G., Yu, Z., and Mei, R.: Assessing future climate changes and extreme indicators in east and south Asia using the RegCM4 regional climate model, Climatic Change, 114, 301–317, https://doi.org/10.1007/s10584-012-0411-y, 2012.

Gu, H., Yu, Z., Wang, G., Wang, J., Ju, Q., Yang, C., and Fan, C.: Impact of climate change on hydrological extremes in the Yangtze River Basin, China, Stoch. Env. Res. Risk A., 29, 693–707, https://doi.org/10.1007/s00477-014-0957-5, 2015a.

Gu, H., Yu, Z., Wang, J., Wang, G., Yang, T., Ju, Q., Yang, C., Xu, F., and Fan, C.: Assessing CMIP5 general circulation model simulations of precipitation and temperature over China, Int. J. Climatol., 35, 2431–2440, https://doi.org/10.1002/joc.4152, 2015b.

Harris, I., Jones, P. D., Osborn, T. J., and Lister, D. H.: Updated high-resolution grids of monthly climatic observations – the CRU TS3.10 Dataset, Int. J. Climatol., 34, 623–642, https://doi.org/10.1002/joc.3711, 2014.

Hawkins, E. and Sutton, R.: The Potential to Narrow Uncertainty in Regional Climate Predictions, B. Am. Meteorol. Soc., 90, 1095–1107, https://doi.org/10.1175/2009BAMS2607.1, 2009.

Hawkins, E. and Sutton, R.: The potential to narrow uncertainty in projections of regional precipitation change, Clim. Dynam., 37, 407–418, https://doi.org/10.1007/s00382-010-0810-6, 2011.

Hong, S. and Yhang, Y.: Implications of a Decadal Climate Shift over East Asia in Winter: A Modeling Study, J. Climate, 23, 4989–5001, https://doi.org/10.1175/2010JCLI3637.1, 2010.

Hong, S. Y., Park, H., Cheong, H. B., Kim, J. E., Koo, M. S., Jang, J., Ham, S., Hwang, S. O., Park, B. K., Chang, E. C., and Li, H.: The Global/Regional Integrated Model system (GRIMs), Asia-Pacific J. Atmos. Sci., 49, 219–243, https://doi.org/10.1007/s13143-013-0023-0, 2013.

Hsu, H.: East Asian monsoon, in: Intraseasonal Variability in the Atmospheric Ocean Climate System, edited by: William, K. M. L. and Waliser, D. E., Springer, Berlin, Heidelberg, 63–94, 2005.

Huang, D. and Gao, S.: Impact of different cumulus convective parameterization schemes on the simulation of precipitation over China, Tellus A, 69, 1406264, https://doi.org/10.1080/16000870.2017.1406264, 2017.

Huang, D. Q, Zhu, J., Zhang, Y. C., and Huang, A. N.: Uncertainties on the simulated summer precipitation over Eastern China from the CMIP5 models, J. Geophys. Res.-Atmos., 118, 9035–9047, https://doi.org/10.1002/jgrd.50695, 2013.

Huttunen, J. M. J., Räisänen, J., Nissinen, A., Lipponen, A., and Kolehmainen, V.: Cross-validation analysis of bias models in Bayesian multi-model projections of climate, Clim. Dynam., 48, 1555–1570, https://doi.org/10.1007/s00382-016-3160-1, 2017.

IPCC: Climate Change 2013: the physical basis, Contribution of Working Group 1 to the Fifth Assessment Report of the IPCC, Cambridge University Press, New York, 2013.

Jin, C. S., Cha, D. H, Lee, D. K, Suh, M. S, Hong, S. Y, Kang, H. S., and Ho, C. H: Evaluation of climatological tropical cyclone activity over the western North Pacific in the CORDEX-East Asia multi-RCM simulations, Clim. Dynam., 47, 765–778, https://doi.org/10.1007/s00382-015-2869-6, 2016.

Jones, C., Giorgi, F., and Asrar, G.: The Coordinated Regional Downscaling Experiment: CORDEX – an international downscaling link to CMIP5, International CLIVAR Project Office Southampton, UK, 2011.

Kay, A. L., Davies, H. N., Bell, V. A., and Jones, R. G.: Comparison of uncertainty sources for climate change impacts: flood frequency in England, Climatic Change, 92, 41–63, https://doi.org/10.1007/s10584-008-9471-4, 2009.

Kerr, R.: Climate change hot spots mapped across the United States, Science, 321, 909, https://doi.org/10.1126/science.321.5891.909, 2008.

Kitoh, A., Endo, H., Krishna Kumar, K., Cavalcanti, I. F. A., Goswami, P., and Zhou, T.: Monsoons in a changing world: A regional perspective in a global context, J. Geophys. Res.-Atmos., 118, 3053–3065, https://doi.org/10.1002/jgrd.50258, 2013.

KMA (Korea Meteorological Administration): Coordinated Regional Climate Downscaling Experiment in East Asia Databank, available at: http://cordex-ea.climate.go.kr/cordex/download.do, last access: 18 May 2018.

Kreft, S., Eckstein, D., and Melchior, I.: Global Climate Risk Index 2017: Who suffers most from extreme weather events? Weather-related loss events in 2015 and 1996 to 2015, Germanwatch e.V., Bonn, Germany, 2016.

Kusunoki, S., Yoshimura, J., Yoshimimura, H., Noda, A., Oouchi, K., and Mizuta, R.: Change of Baiu Rain Band in Global Warming Projection by an Atmospheric General Circulation Model with a 20-km Grid Size, J. Meteorol. Soc. Jpn.. Ser. II, 84, 581–611, https://doi.org/10.2151/jmsj.84.581, 2006.

Lafaysse, M., Hingray, B., Mezghani, A., Gailhard, J., and Terray, L: Internal variability and model uncertainty components in future hydrometeorological projections: The Alpine Durance basin, Water Resour. Res., 50, 3317–3341, https://doi.org/10.1002/2013WR014897, 2014.

Lau, W. K. M., Kim, K., and Leung, L. R.: Changing circulation structure and precipitation characteristics in Asian monsoon regions: greenhouse warming vs. aerosol effects, Geosci. Lett., 4, 28, https://doi.org/10.1186/s40562-017-0094-3, 2017.

Lee, J. W., Hong, S. Y., Chang, E. C, Suh, M. S., and Kang, H. S: Assessment of future climate change over East Asia due to the RCP scenarios downscaled by GRIMs-RMP, Clim. Dynam., 42, 733–747, https://doi.org/10.1007/s00382-013-1841-6, 2014.

Mariotti, L., Coppola, E., Sylla, M. B., Giorgi, F., and Piani, C.: Regional climate model simulation of projected 21st century climate change over an all-Africa domain: Comparison analysis of nested and driving model results, J. Geophys. Res.-Atmos., 116, D15111, https://doi.org/10.1029/2010JD015068, 2011.

Niu, X., Wang, S., Tang, J., Lee, D., Gao, X., Wu, J., Hong, S., Gutowski, W. J., and McGregor, J.: Multimodel ensemble projection of precipitation in eastern China under A1B emission scenario, J. Geophys. Res.-Atmos., 120, 9965–9980, https://doi.org/10.1002/2015JD023853, 2015.

O'Brien, T. A., Sloan, L. C., and Snyder, M. A.: Can ensembles of regional climate model simulations improve results from sensitivity studies?, Clim. Dynam., 37, 1111–1118, https://doi.org/10.1007/s00382-010-0900-5, 2011.

Park, C., Min, S. K., Lee, D., Cha, D. H., Suh, M. S., Kang, H. S., Hong, S. Y, Lee, D. K., Baek, H. J., Boo, K. O., and Kwon, W. T.: Evaluation of multiple regional climate models for summer climate extremes over East Asia, Clim. Dynam., 46, 2469–2486, https://doi.org/10.1007/s00382-015-2713-z, 2016.

Park, E. H., Hong, S. Y., and Kang, H. S.: Characteristics of an East-Asian summer monsoon climatology simulated by the RegCM3, Meteorol. Atmos. Phys., 100, 139–158, https://doi.org/10.1007/s00703-008-0300-0, 2008.

Patz, J. A., Campbell-Lendrum, D., Holloway, T., and Foley, J. A.: Impact of regional climate change on human health, Nature, 438, 310–317, https://doi.org/10.1038/nature04188, 2005.

Phillips, T. J. and Gleckler, P. J.: Evaluation of continental precipitation in 20th century climate simulations: The utility of multimodel statistics, Water Resour. Res., 42, W03202, https://doi.org/10.1029/2005WR004313, 2006.

Prömmel, K., Geyer, B., Jones, J. M., and Widmann, M.: Evaluation of the skill and added value of a reanalysis-driven regional simulation for Alpine temperature, Int. J. Climatol., 30, 760–773, https://doi.org/10.1002/joc.1916, 2010

Rozante, J. R., Moreira, D. S., Godoy, R. C. M., and Fernandes, A. A.: Multi-model ensemble: technique and validation, Geosci.

Model Dev., 7, 2333–2343, https://doi.org/10.5194/gmd-7-2333-2014, 2014.

Saini, R., Wang, G., Yu, M., and Kim, J.: Comparison of RCM and GCM projections of boreal summer precipitation over Africa, J. Geophys. Res.-Atmos., 120, 3679–3699, https://doi.org/10.1002/2014JD022599, 2015.

Seo, S. B., Sinha, T., Mahinthakumar, G., Sankarasubramanian, A., and Kumar, M.: Identification of dominant source of errors in developing streamflow and groundwater projections under near-term climate change, J. Geophys. Res.-Atmos., 121, 7652–7672, https://doi.org/10.1002/2016JD025138, 2016.

Skamarock, W. C., Klemp, J. B., Dudhia, J., Gill, D. O., Barker, D. M., Wang, W., and Powers, J. G.: A Description of the Advanced Research WRF Version 2, National Center for Atmospheric Research, Boulder, 2005.

Sperber, K. R., Annamalai, H., Kang, I. S., Kitoh, A., Moise, A., Turner, A., Wang, B., and Zhou, T.: The Asian summer monsoon: an intercomparison of CMIP5 vs. CMIP3 simulations of the late 20th century, Clim. Dynam., 41, 2711–2744, https://doi.org/10.1007/s00382-012-1607-6, 2013.

Suh, M. S., Oh, S. G., Lee, D. K., Cha, D. H., Choi, S. J., Jin, C. S., and Hong, S. Y.: Development of New Ensemble Methods Based on the Performance Skills of Regional Climate Models over South Korea, J. Climate, 25, 7067–7082, https://doi.org/10.1175/JCLI-D-11-00457.1, 2012.

Sun, Q., Miao, C., Duan, Q., Kong, D., Ye, A., Di, Z., and Gong, W.: Would the "real" observed dataset stand up? A critical examination of eight observed gridded climate datasets for China, Environ. Res. Lett., 9, 015001, https://doi.org/10.1088/1748-9326/9/1/015001, 2014.

Sun, Q., Miao, C., and Duan, Q.: Projected changes in temperature and precipitation in ten river basins over China in 21st century, Int. J. Climatol., 35, 1125–1141, https://doi.org/10.1002/joc.4043, 2015.

Syed, F. S., Yoo, J. H., Körnich, H., and Kucharski, F.: Extratropical influences on the inter-annual variability of South-Asian monsoon, Clim. Dynam., 38, 1661–1674, https://doi.org/10.1007/s00382-011-1059-4, 2012.

Tang, J., Li, Q., Wang, S., Lee, D. K, Hui, P., Niu, X., Gutowski, W. J., Dairaku, K., McGregor, J., Katzfey, J., Gao, X., Wu, J., Hong, S., Wang, Y., and Sasaki, H.: Building Asian climate change scenario by multi-regional climate models ensemble, Part I: surface air temperature, Int. J. Climatol., 36, 4241–4252, https://doi.org/10.1002/joc.4628, 2016.

Taylor, K. E.: Summarizing multiple aspects of model performance in a single diagram, J. Geophys. Res.-Atmos., 106, 7183–7192, https://doi.org/10.1029/2000JD900719, 2001.

The HadGEM2 Development Team: G. M. Martin, Bellouin, N., Collins, W. J., Culverwell, I. D., Halloran, P. R., Hardiman, S. C., Hinton, T. J., Jones, C. D., McDonald, R. E., McLaren, A. J., O'Connor, F. M., Roberts, M. J., Rodriguez, J. M., Woodward, S., Best, M. J., Brooks, M. E., Brown, A. R., Butchart, N., Dearden, C., Derbyshire, S. H., Dharssi, I., Doutriaux-Boucher, M., Edwards, J. M., Falloon, P. D., Gedney, N., Gray, L. J., Hewitt, H. T., Hobson, M., Huddleston, M. R., Hughes, J., Ineson, S., Ingram, W. J., James, P. M., Johns, T. C., Johnson, C. E., Jones, A., Jones, C. P., Joshi, M. M., Keen, A. B., Liddicoat, S., Lock, A. P., Maidens, A. V., Manners, J. C., Milton, S. F., Rae, J. G. L., Ridley, J. K., Sellar, A., Senior, C. A., Totterdell, I. J., Verhoef, A., Vidale, P. L., and Wiltshire, A.: The HadGEM2 family of Met Office Unified Model climate configurations, Geosci. Model Dev., 4, 723–757, https://doi.org/10.5194/gmd-4-723-2011, 2011.

Thomson, A., Calvin, K., Smith, S., Kyle, G. P., Volke, A., Patel, P., Delgado-Arias, S., Bond-Lamberty, B., Wise, M., Clarke, L., and Edmonds, J.: RCP4.5: a pathway for stabilization of radiative forcing by 2100, Climatic Change, 109, 77–94, https://doi.org/10.1007/s10584-011-0151-4, 2011.

Um, M., Kim, Y., and Kim, J.: Evaluating historical drought characteristics simulated in CORDEX East Asia against observations, Int. J. Climatol., 37, 4643–4655, https://doi.org/10.1002/joc.5112, 2017.

Wang, L., Chen, W., Huang, G., and Zeng, G.: Changes of the transitional climate zone in East Asia: past and future, Clim. Dynam., 49, 1463–1477, https://doi.org/10.1007/s00382-016-3400-4, 2017a.

Wang, X., Yang, T., Wortmann, M., Shi, P., Hattermann, F., Lobanova, A., and Aich, V.: Analysis of multi-dimensional hydrological alterations under climate change for four major river basins in different climate zones, Climatic Change, 141, 483–498, https://doi.org/10.1007/s10584-016-1843-6, 2017b.

Wang, Y., Leung, L. R., McGregor, J. L., Lee, D. K., Wang, W. C., Ding, Y. H., and Kimura, F.: Regional climate modeling: progress, challenges, and prospects, J. Meteorol. Soc. Jpn., 82, 1599–1628, https://doi.org/10.2151/jmsj.82.1599, 2004.

Webster, P. J., Magaña, V. O., Palmer, T. N., Shukla, J., Tomas, R. A., Yanai, M., and Yasunari, T.: Monsoons: Processes, predictability, and the prospects for prediction, J. Geophys. Res.-Oceans, 103, 14451–14510, https://doi.org/10.1029/97JC02719, 1998.

Wilby, R. L. and Harris, I.: A framework for assessing uncertainties in climate change impacts: Low-flow scenarios for the River Thames, UK, Water Resour. Res., 42, W02419, https://doi.org/10.1029/2005WR004065, 2006.

Woldemeskel, F. M., Sharma, A., Sivakumar, B., and Mehrotra, R.: Quantification of precipitation and temperature uncertainties simulated by CMIP3 and CMIP5 models, J. Geophys. Res.-Atmos., 121, 3–17, https://doi.org/10.1002/2015JD023719, 2016.

Wu, J. and Gao, X.: A gridded daily observation dataset over China region and comparison with the other datasets, Chinese J. Geophys., 56, 1102–1111, https://doi.org/10.6038/cjg20130406, 2013 (in Chinese).

Yang, T., Cui, T., Xu, C., Ciais, P., and Shi, P.: Development of a new IHA method for impact assessment of climate change on flow regime, Global Planet. Change, 156, 68–79, https://doi.org/10.1016/j.gloplacha.2017.07.006, 2017.

Yatagai, A., Kamiguchi, K., Arakawa, O., Hamada, A., Yasutomi, N., and Kitoh, A.: APHRODITE: Constructing a Long-Term Daily Gridded Precipitation Dataset for Asia Based on a Dense Network of Rain Gauges, B. Am. Meteorol. Soc., 93, 1401–1415, https://doi.org/10.1175/BAMS-D-11-00122.1, 2012.

Yhang, Y. and Hong, S.: Improved Physical Processes in a Regional Climate Model and Their Impact on the Simulated Summer Monsoon Circulations over East Asia, J. Climate, 21, 963–979, https://doi.org/10.1175/2007JCLI1694.1, 2008.

Yin, H., Donat, M. G., Alexander, L. V., and Sun, Y.: Multi-dataset comparison of gridded observed temperature and precipitation extremes over China, Int. J. Climatol., 35, 2809–2827, https://doi.org/10.1002/joc.4174, 2015.

Yira, Y., Diekkrüger, B., Steup, G., and Bossa, A. Y.: Impact of climate change on hydrological conditions in a tropical West African catchment using an ensemble of climate simulations, Hydrol. Earth Syst. Sci., 21, 2143–2161, https://doi.org/10.5194/hess-21-2143-2017, 2017.

You, Q., Min, J., Fraedrich, K., Zhang, W., Kang, S., Zhang, L., and Meng, X.: Projected trends in mean, maximum, and minimum surface temperature in China from simulations, Global Planet. Change, 112, 53–63, https://doi.org/10.1016/j.gloplacha.2013.11.006, 2014.

Yu, Z., Pollard, D., and Cheng, L.: On continental-scale hydrologic simulations with a coupled hydrologic model, J. Hydrol., 331, 110–124, https://doi.org/10.1016/j.jhydrol.2006.05.021, 2006.

Yu, Z., Gu, H., Wang, J., Xia, J., and Lu, B.: Effect of projected climate change on the hydrological regime of the Yangtze River Basin, China, Stoch. Env. Res. Risk A., 32, 1–16, https://doi.org/10.1007/s00477-017-1391-2, 2018.

Zhao, W. and Li, A.: A Review on Land Surface Processes Modelling over Complex Terrain, Adv. Meteorol., 2015, 607181, https://doi.org/10.1155/2015/607181, 2015.

Zhou, T. and Yu, R.: Twentieth-Century Surface Air Temperature over China and the Globe Simulated by Coupled Climate Models, J. Climate, 19, 5843–5858, https://doi.org/10.1175/JCLI3952.1, 2006.

Permissions

All chapters in this book were first published in HESS, by Copernicus Publications; hereby published with permission under the Creative Commons Attribution License or equivalent. Every chapter published in this book has been scrutinized by our experts. Their significance has been extensively debated. The topics covered herein carry significant findings which will fuel the growth of the discipline. They may even be implemented as practical applications or may be referred to as a beginning point for another development.

The contributors of this book come from diverse backgrounds, making this book a truly international effort. This book will bring forth new frontiers with its revolutionizing research information and detailed analysis of the nascent developments around the world.

We would like to thank all the contributing authors for lending their expertise to make the book truly unique. They have played a crucial role in the development of this book. Without their invaluable contributions this book wouldn't have been possible. They have made vital efforts to compile up to date information on the varied aspects of this subject to make this book a valuable addition to the collection of many professionals and students.

This book was conceptualized with the vision of imparting up-to-date information and advanced data in this field. To ensure the same, a matchless editorial board was set up. Every individual on the board went through rigorous rounds of assessment to prove their worth. After which they invested a large part of their time researching and compiling the most relevant data for our readers.

The editorial board has been involved in producing this book since its inception. They have spent rigorous hours researching and exploring the diverse topics which have resulted in the successful publishing of this book. They have passed on their knowledge of decades through this book. To expedite this challenging task, the publisher supported the team at every step. A small team of assistant editors was also appointed to further simplify the editing procedure and attain best results for the readers.

Apart from the editorial board, the designing team has also invested a significant amount of their time in understanding the subject and creating the most relevant covers. They scrutinized every image to scout for the most suitable representation of the subject and create an appropriate cover for the book.

The publishing team has been an ardent support to the editorial, designing and production team. Their endless efforts to recruit the best for this project, has resulted in the accomplishment of this book. They are a veteran in the field of academics and their pool of knowledge is as vast as their experience in printing. Their expertise and guidance has proved useful at every step. Their uncompromising quality standards have made this book an exceptional effort. Their encouragement from time to time has been an inspiration for everyone.

The publisher and the editorial board hope that this book will prove to be a valuable piece of knowledge for researchers, students, practitioners and scholars across the globe.

List of Contributors

D. L. Ficklin
Department of Geography, Indiana University, 701 E. Kirkwood Ave., Bloomington, IN 47405, USA
Center for Geospatial Data Analysis, Indiana Geological Survey, 611 N. Walnut Grove, Bloomington, IN 47405, USA

B. L. Barnhart and G. W. Whittaker
Agricultural Research Service, United States Department of Agriculture, 3450 SW Campus Way, Corvallis, OR 97333, USA

J. H. Knouft
Department of Biology, Saint Louis University, 3507 Laclede Ave., St. Louis, MO 63103, USA
Center for Environmental Sciences, Saint Louis University, 3507 Laclede Ave., St. Louis, MO 63103, USA

I. T. Stewart
Department of Environmental Studies and Sciences, Santa Clara University, 500 El Camino Real, Santa Clara, CA 95053, USA

E. P. Maurer
Civil Engineering Department, Santa Clara University, 500 El Camino Real, Santa Clara, CA 95053, USA

S. L. Letsinger
Center for Geospatial Data Analysis, Indiana Geological Survey, 611 N. Walnut Grove, Bloomington, IN 47405, USA

Zhenhua Li
Global Institute for Water Security, University of Saskatchewan, Saskatoon, Saskatchewan, Canada
Institute of Space and Atmospheric Studies, University of Saskatchewan, Saskatoon, Saskatchewan, Canada

Yanping Li and Lucia Scaff
Global Institute for Water Security, University of Saskatchewan, Saskatoon, Saskatchewan, Canada

Alan H. Manson
Institute of Space and Atmospheric Studies, University of Saskatchewan, Saskatoon, Saskatchewan, Canada

Barrie Bonsal
National Hydrology Research Centre, Environment and Climate Change Canada, Saskatoon, Saskatchewan, Canada

Zhenhua Li
Global Institute for Water Security, University of Saskatchewan, Saskatoon, Saskatchewan, Canada
Institute of Space and Atmospheric Studies, University of Saskatchewan, Saskatoon, Saskatchewan, Canada

Yanping Li and Lucia Scaff
Global Institute for Water Security, University of Saskatchewan, Saskatoon, Saskatchewan, Canada

Barrie Bonsal
National Hydrology Research Centre, Environment and Climate Change Canada, Saskatoon, Saskatchewan, Canada

Alan H. Manson
Institute of Space and Atmospheric Studies, University of Saskatchewan, Saskatoon, Saskatchewan, Canada

Charles W. Downer
Coastal and Hydraulics Laboratory, Vicksburg, MS 39180, USA

Jeffrey D. Niemann
Department of Civil Engineering, Colorado State University, Fort Collins, CO 80523, USA

Julie T. Parno
Cold Regions Research and Engineering Laboratory, Hanover, NH 03755, USA

Michael L. Follum
Coastal and Hydraulics Laboratory, Vicksburg, MS 39180, USA
Department of Civil Engineering, Colorado State University, Fort Collins, CO 80523, USA

Andreas M. Jobst, Daniel G. Kingston and Nicolas J. Cullen
Department of Geography, University of Otago, Dunedin, New Zealand

Josef Schmid
Department of Geography, Ludwig Maximilian University (LMU), Munich, Germany

Hui-Min Wang, Jie Chen and Hua Chen
State Key Laboratory of Water Resources and Hydropower Engineering Science, Wuhan University, Wuhan, 430072, China

Alex J. Cannon
Climate Research Division, Environment and Climate Change Canada, Victoria BC, Canada

Chong-Yu Xu
State Key Laboratory of Water Resources and Hydropower Engineering Science, Wuhan University, Wuhan, 430072, China
Department of Geosciences, University of Oslo, Oslo, Norway

Ulrich Strasser
Department of Geography, University of Innsbruck, Innsbruck, Austria

Florian Hanzer
Department of Geography, University of Innsbruck, Innsbruck, Austria
Wegener Center for Climate and Global Change, University of Graz, Graz, Austria

Kristian Förster
Department of Geography, University of Innsbruck, Innsbruck, Austria
Institute of Hydrology and Water Resources Management, Leibniz Universität Hannover, Hannover, Germany

Johanna Nemec
ENVEO IT GmbH, Innsbruck, Austria

Camila Alvarez-Garreton and Antonio Lara
Center for Climate and Resilience Research (CR2), Santiago, Chile
Instituto de Conservación, Biodiversidad y Territorio, Universidad Austral de Chile, Valdivia, Chile

Pablo A. Mendoza and James McPhee
Department of Civil Engineering, Universidad de Chile, Santiago, Chile
Advanced Mining Technology Center, Universidad de Chile, Santiago, Chile

Juan Pablo Boisier and Rene Garreaud
Center for Climate and Resilience Research (CR2), Santiago, Chile
Department of Geophysics, Universidad de Chile, Santiago, Chile

Nans Addor
Climatic Research Unit, School of Environmental Sciences, University of East Anglia, Norwich, UK

Mauricio Galleguillos and Cristóbal Puelma
Center for Climate and Resilience Research (CR2), Santiago, Chile
Faculty of Agronomic Sciences, Universidad de Chile, Santiago, Chile

Mauricio Zambrano-Bigiarini
Center for Climate and Resilience Research (CR2), Santiago, Chile
Department of Civil Engineering, Faculty of Engineering and Sciences, Universidad de La Frontera, Temuco, Chile

Gonzalo Cortes
Department of Civil and Environmental Engineering, University of California, Los Angeles, California, USA

Alvaro Ayala
Laboratory of Hydraulics, Hydrology and Glaciology (VAW), ETH Zurich, Zurich, Switzerland
Swiss Federal Institute for Forest, Snow and Landscape Research (WSL), Birmensdorf, Switzerland

Ruifang Guo and Han Zhou
Key Laboratory of Watershed Geographic Sciences, Nanjing Institute of Geography and Limnology, Chinese Academy of Sciences, No. 73 East Beijing Road, Nanjing 210008, China
University of Chinese Academy of Sciences, No. 19 Yuquan Road, Beijing 100049, China

Yuanbo Liu
Key Laboratory of Watershed Geographic Sciences, Nanjing Institute of Geography and Limnology, Chinese Academy of Sciences, No. 73 East Beijing Road, Nanjing 210008, China

Yaqiao Zhu
College of Urban and Environmental Sciences, Hubei Normal University, No. 11 Cihu Road, Huangshi 435002, China

Gabriel Gerard Rooney
Met Office, FitzRoy Road, Exeter, EX1 3PB, UK

Nicole van Lipzig
KU Leuven, Department of Earth and Environmental Sciences, Celestijnenlaan 200E, 3001 Leuven, Belgium

Wim Thiery
ETH Zurich, Institute for Atmospheric and Climate Science, Universitaetstrasse 16, 8092 Zurich, Switzerland
Vrije Universiteit Brussel, Department of Hydrology and Hydraulic Engineering, Pleinlaan 2, 1050 Brussels, Belgium

Joop Kroes, Paul van Walsum and Martin Mulder
Wageningen University & Research – Wageningen Environmental Research, Wageningen, the Netherlands

Iwan Supit
Wageningen University & Research - Wageningen Environmental Research, Wageningen, the Netherlands
Wageningen University & Research - Chair Water Systems and Global Change, Wageningen, the Netherlands

Jos van Dam
Wageningen University & Research - Chair Soil Physics and Land Management, Wageningen, the Netherlands

Qianqian Zhou
School of Civil and Transportation Engineering, Guangdong University of Technology, Waihuan Xi Road, Guangzhou 510006, China
Joint Global Change Research Institute, Pacific Northwest National Laboratory, College Park MD 20740, USA

Guoyong Leng
Joint Global Change Research Institute, Pacific Northwest National Laboratory, College Park MD 20740, USA
Environmental Change Institute, University of Oxford, Oxford OX1 3QY, UK

Maoyi Huang
Earth System Analysis and Modeling Group, Pacific Northwest National Laboratory, Richland, WA 99352, USA

Francisco Pellicer-Martínez
Department of Civil Engineering, Catholic University of Murcia, Spain

José Miguel Martínez-Paz
Department of Applied Economics, University of Murcia, Spain

Huanghe Gu, Zhongbo Yu, Chuanguo Yang, Qin Ju and Tao Yang
State Key Laboratory of Hydrology-Water Resources and Hydraulic Engineering, Hohai University, Nanjing, China
College of Hydrology and Water Resources, Hohai University, Nanjing, China

Dawei Zhang
China Institute of Water Resources and Hydropower Research, Beijing, China

Index

www.ingramcontent.com/pod-product-compliance
Lightning Source LLC
LaVergne TN
LVHW060516210926
848485LV00005B/1089
* 9 7 8 1 6 3 9 8 9 0 5 3 8 *